Lecture Notes in Mechanical Engineering

Lecture Notes in Mechanical Engineering (LNME) publishes the latest developments in Mechanical Engineering - quickly, informally and with high quality. Original research reported in proceedings and post-proceedings represents the core of LNME. Volumes published in LNME embrace all aspects, subfields and new challenges of mechanical engineering. Topics in the series include:

- Engineering Design
- Machinery and Machine Elements
- Mechanical Structures and Stress Analysis
- Automotive Engineering
- Engine Technology
- Aerospace Technology and Astronautics
- Nanotechnology and Microengineering
- Control, Robotics, Mechatronics
- MEMS
- Theoretical and Applied Mechanics
- Dynamical Systems, Control
- Fluid Mechanics
- Engineering Thermodynamics, Heat and Mass Transfer
- Manufacturing
- Precision Engineering, Instrumentation, Measurement
- Materials Engineering
- Tribology and Surface Technology

To submit a proposal or request further information, please contact the Springer Editor in your country:

China: Li Shen at li.shen@springer.com
India: Dr. Akash Chakraborty at akash.chakraborty@springernature.com
Rest of Asia, Australia, New Zealand: Swati Meherishi at swati.meherishi@springer.com
All other countries: Dr. Leontina Di Cecco at Leontina.dicecco@springer.com

To submit a proposal for a monograph, please check our Springer Tracts in Mechanical Engineering at http://www.springer.com/series/11693 or contact Leontina.dicecco@springer.com

Indexed by SCOPUS. The books of the series are submitted for indexing to Web of Science.

More information about this series at http://www.springer.com/series/11236

Francisco Cavas-Martínez · Félix Sanz-Adan ·
Paz Morer Camo · Ruben Lostado Lorza ·
Jacinto Santamaría Peña
Editors

Advances in Design Engineering

Proceedings of the XXIX International Congress INGEGRAF, 20–21 June 2019, Logroño, Spain

Springer

Editors
Francisco Cavas-Martínez
Departamento de Estructuras,
Construcción y Expresión Gráfica
Universidad Politécnica de Cartagena
Cartagena, Murcia, Spain

Paz Morer Camo
Departamento de Mecánica
Tecnun - Universidad de Navarra
Donostia-San Sebastián, Guipúzcoa, Spain

Jacinto Santamaría Peña
Departamento de Ingeniería Mecánica
Universidad de la Rioja
Logroño, Spain

Félix Sanz-Adan
Departamento de Ingeniería Mecánica
Universidad de la Rioja
Logroño, La Rioja, Spain

Ruben Lostado Lorza
Departamento de Ingeniería Mecánica
Universidad de la Rioja
Logroño, La Rioja, Spain

ISSN 2195-4356 ISSN 2195-4364 (electronic)
Lecture Notes in Mechanical Engineering
ISBN 978-3-030-41199-2 ISBN 978-3-030-41200-5 (eBook)
https://doi.org/10.1007/978-3-030-41200-5

This Springer imprint is published by the registered company Springer Nature Switzerland AG
The registered company address is: Gewerbestrasse 11, 6330 Cham, Switzerland

Preface and Acknowledgements

The INGEGRAF 2019 Conference originates as the 29th International Conference on GRAPHICS ENGINEERING "The Digital Transformation in the Graphic Engineering".

INGEGRAF 2019 has been organized by the area of graphical expression of the University of La Rioja. Cutting-edge topics in product design and manufacturing, innovative design and computer-aided design were especially encouraged.

The list of topics (and subtopics) covered in the present edition is the following:

- Product design and development: Green engineering and eco-design; User-centred design; Product lifecycle-based design; Robust design, reliability and maintenance; Modelling and simulation-based design; Ergonomics and human factors; Global product development.
- Computer-aided design and interactive design: Virtual approaches for interactive design; Virtual prototyping-based design; CAD, CAE, IFC and BIM; Image processing and analysis; Geometric modelling and analysis; Reverse engineering; Virtual and augmented reality.
- Manufacturing and industrial process design: Integrated/advanced manufacturing; Manufacturing process and production management; Rapid prototyping; Additive manufacturing; Flexible assemblies; Remanufacturing; Industry 4.0.
- Graphical bioengineering: Biomechanics; 3D modelling of biological structures; Computer-aided methods for pathologic diagnosis; Emotional engineering; Biomimicry for product design; Simulation and visualization of biological systems.
- Innovation in design: Creativity and innovation methods; Collaborative engineering; Intellectual and industrial property management; Design and research methods.
- Teaching/learning in graphic engineering: Teaching on graphic expression; Theoretical and applied geometry; Graphic design; New approaches in teaching/learning process; Project-based learning; Interactive 3D modelling.

- Engineering and construction: Sustainable building. Sustainable construction; Building information modelling; Photogrammetry and remote sensing; Geo-information. Data capture; Virtual environments. Augmented reality in AEC; Urban regeneration; Heritage and territory. Industrial heritage conservation.

We would like to thank our main organizers/institutions, High Technical School of Industrial Engineering from the University of La Rioja, and the rest of the sponsoring/collaborating companies and institutions for their support and grants.

We would also like to express our gratitude to the members of the different committees for their support, collaboration and good work. Thanks to all reviewers for their selfless effort reviewing contributions, which positively influenced the quality of the final papers presented at the conference.

Last, but not least, thanks to all the participants of INGEGRAF 2019.

November 2019

Francisco Cavas-Martínez
Félix Sanz-Adan
Paz Morer Camo
Jacinto Santamaría Peña
Ruben Lostado Lorza

Organization Committee

Conference Chair

Francisco Cavas-Martínez	Universidad Politécnica de Cartagena, Spain

Conference Program Chairs

Rubén Lostado Lorza	Universidad de La Rioja, Spain
Jacinto Santamaría Peña	Universidad de La Rioja, Spain

Conference Advisory Chairmen

Paz Morer	Universidad de Navarra—Tecnun, Spain

Scientific Committee

Fernando Aguilar Torres	Universidad de Almería, Spain
Rita Ambu	Università di Cagliari, Italy
Elidia Beatriz Bázquez Parra	Universidad de Málaga, Spain
Alain Bernard	Ecole Centrale Nantes, France
Francesca Campana	Università di Roma "La Sapienza", Italy
Nicola Cappetti	Università degli Studi di Salerno, Italy
Vincent Cheutet	INSA Lyon, France
Marina Corral Bobadilla	Universidad de La Roja, Spain
Alain Daidié	INSA Toulouse, France
Luigi De Napoli	Università di Cagliari, Italy
Francisco Jose Fernández Cañavate	Universidad Politécnica de Cartagena, Spain
Francesco Ferrise	Politecnico di Milano, Italy
Xavier Fisher	ESTIA, France
Daniel G. Fernández-Pacheco	Universidad Politécnica de Cartagena, Spain
Cesar García Hernández	Universidad de Zaragoza, Spain

Valentín Gómez Jáuregui	Universidad de Cantabria, Spain
Tommaso Ingrassia	Università degli Studi di Palermo, Italy
Julien Le Duigou	Université de technologie de Compiègne, France
Francesco Leali	Università degli Studi di Modena e Reggio Emilia, Italy
Carlos León Robles	Universidad de Granada, Spain
Muriel Lombard	Université de Lorraine, France
Cristina Manchado del Val	Universidad de Cantabria, Spain
Cristina Martín Doñate	Univesidad de Jaén, Spain
Marian Martínez-Calvo	Universidad de La Rioja, Spain
Maria Luisa Martínez Muneta	Universidad Politécnica de Madrid, Spain
Massimo Martorelli	Università degli Studi di Napoli "Federico II", Italy
Rikardo Minguez Gabiña	Universidad del País Vasco, Spain
Ramón Miralbes Buil	Universidad de Zaragoza, Spain
Anna Eva Morabito	Università del Salento, Italy
Alessandro Naddeo	Università degli Studi di Salerno, Italy
Manuel Paredes	INSA Toulouse, France
Marcello Pellicciari	Università degli Studi di Modena e Reggio Emilia, Italy
Guillemo Peris	Universidad Politécnica de Valencia, Spain
Eugenio Pezzuti	Università di Roma "Tor Vergata", Italy
Roberto Raffaeli	Università eCampus, Italy
David Ranz	Universidad de Zaragoza, Spain
Roberto Razzoli	Università degli Studi di Genova, Italy
José Ignacio Rojas Sola	Universidad de Jaén, Spain
Bertrand Rose	Université de Strasbourg, France
Félix Sanz-Adan	Universidad de La Rioja, Spain
Irene Sentana Gadea	Universidad de Alicante, Spain
Egoitz Sierra	Universidad del País Vasco, Spain
Eneko Solaberrieta	Universidad del País Vasco, Spain
Domenico Speranza	Università degli Studi di Cassino e del Lazio Meridionale, Italy
Miguel Suffo Pino	Universidad de Cádiz, Spain
Davide Tumino	Università degli Studi di Enna "Kore", Italy

Organizing Committee

President

Félix Sanz-Adan	Universidad de La Rioja, Spain

Secretary

Marian Martínez-Calvo	Universidad de La Rioja, Spain

Members

Contents

Product Design and Development

Use of Composite Materials in the Design of a Ventilation Grille Attending to Eco-Sustainability Factors

F. Xavier Espinach Orús[1], Fernando Julián Pérez[1](✉), Manel Alcalà Vilavella[1], Marc Delgado Aguilar[2], and Quim Tarrés Farrés[2]

[1] Organització Gestió Empresarial i Disseny de Producte, Universitat de Girona, c/ Maria Aurelis Capmany i Farnés 61, 17003 Girona, Spain
fernando.julian@udg.edu
[2] Enginyeria Química, Agrària i Tecnologia Agroalimentària, Universitat de Girona, c/ Maria Aurelis Capmany i Farnés 61, 17003 Girona, Spain

Abstract. This work deals with the design, the engineering of products and the selection of materials destined to the manufacture of a ventilation grille. Additionally, simulates a process of identification and use of agricultural waste for the formulation of composite materials. These materials, once characterized, can be used for the manufacture of new products.

The final product is expected to combine design attributes, technical and legal feasibility, along with the implementation of new biologically based materials that ensure a reduction in the environmental impact.

Taking into account the industrial design, a final concept is proposed and modeled. The product geometry, production technology and legal specifications will be the input data for the product engineering.

The selection of material will be based on the technical and environmental impact requirements. Thermo-mechanical fibers from corn stalks reinforced high-density polyethylene composite was prepared and characterized. These properties are used, together with the boundary conditions, to model and test a virtual model. The final product will be obtained by injection. The new product must be able to deploy under the expected use conditions. The new materials present an economically attractive opportunity to replace other materials, metallic, composite or virgin.

Keywords: Product design · Sustainability eco-design · Composite materials · Use of waste

1 Introduction

Both in the process of designing or redesigning a new product intervene a large number of experts in different fields. The experience of the different actors is devoted to industrial design, product engineering and materials science, among others [1].

F. Cavas-Martínez et al. (Eds.): INGEGRAF 2019, LNME, pp. 3–12, 2020.
https://doi.org/10.1007/978-3-030-41200-5_1

The redesign process includes all the necessary activities and knowledge of the actors previously indicated to adapt an existing design to other requirements, which may be legal, technical or economic. In the case of the study, a ventilation grille is redesigned for its manufacture using another material and another technology. The redesign implies that the resulting product must retain the technical and use [2, 3].

The process of redesigning a ventilation grille is presented taking into account a factor that we consider important, such as sustainability. The original appearance and size have been maintained. The innovation that allows generating value is mainly the change of material.

In order to obtain the reinforcements, an agroforestry waste was used. The fibers were obtained by steam treatment of corn stems. Subsequently, a defibration was carried out and then the fibers were mixed with a high density polyethylene (HDPE). Such composite materials, based on thermo-mechanical fibers of corn and HDPE are recyclable and can save up to 50% or more of the volume of the polymeric matrix, in this case HDPE. At the end of their life cycle, the products manufactured with these materials can be incinerated without the generation of waste. In the construction sector, in the case of the grid, the useful life of the materials is especially long, and it is for this reason that the recycling of its components must be promoted.

In terms of materials selection, the current tendency is to look after natural-based composites in order to take into account the sustainability of the final product [4, 5]. Natural fibers or cellulosic fillers/fibers can be classified under four categories depending on their performance when incorporated in a plastic matrix: (i) wood flour or agroforestry waste flour in general, (ii) wood fibers (iii) natural strands or bundles of strands and (iv) fibers from agroforestry residues [6, 7]. The agricultural residues can be considered as particle fillings that improve the tensile and flexural modulus of the compounds with little effect on the composite resistances [8, 9].

Agroforestry residues, made up of stalks and leaves, use to be chipped and buried, or incinerated in land. The overall exploitation of corn stalks and leaves results in different products suitable for their use as filler or reinforcement of composite materials. Hence, the milling process brings to a corn flour with small aspect ratio (L/d), able for the production of wood plastic composites. However, if corn residues are submitted to a defibering process under aqueous conditions, a fibrous material with higher aspect ratio is obtained, suitable for somewhat better product requirements [10]. Both types of processes show almost 100% of yield, with respect to the initial corn residue. Furthermore, the use of thermo-mechanical processes (vaporization followed to defibering) results in a fibrous matter with intrinsic properties, at a very competitive yield [11]. Finally, the utilization of more aggressive treatments such as organosolv processes or sosa-antraquinone treatments (semi-chemical processes) bring to fibers with better mechanical properties [12, 13].

This work focus on then use of agroforestry reinforced polyolefin. Product design and innovation researchers have worked together with experts in materials science in the development of an ecological ventilation grid. The development of the new product presents innovative objectives in the application of new materials to meet the final requirements of the market sub-sector.

2 Methodology

In order to create value, a process divided by knowledge areas is proposed, converting agricultural waste into product proposals. To get from the starting point to the goal, follow the steps proposed in Fig. 1.

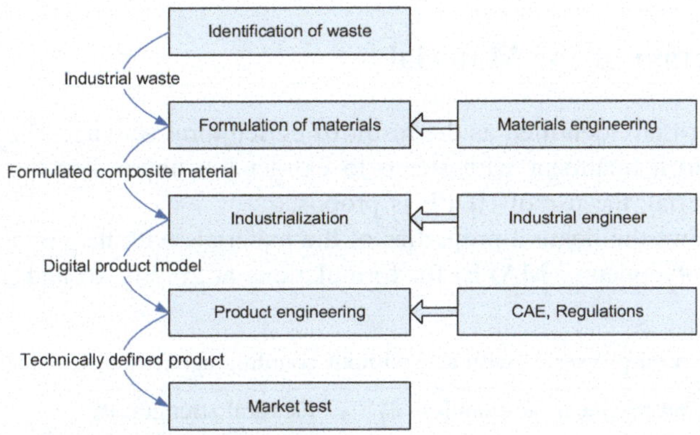

Fig. 1. Research framework for product development.

In the first stage, those materials capable of being recovered are identified. This stage can be the result of an investigation or the proposal of a company that wants to value its waste. It is important the participation of experts in materials science to establish the suitability of the proposed materials.

The second stage comprises the formulation of the composite materials. It is a typically experimental stage. Material science experts have to establish the matrices upon which the composite material will be based. In this way, the uses of the elements to be inertized as loads or reinforcements will be established. All laboratory tests are carried out to establish the physical and chemical characteristics of the formulated materials. At the end of the stage all the graphs and tables characteristic of the composite materials and their variations will be provided to the engineers and designers responsible for the industrialization stage.

The third stage of industrialization combines knowledge in engineering with those of industrial design. Throughout the stage, the product to be designed or industrialized is established, taking into account the production processes. You can start from an existing product or redesign one again. This stage is applied research. At the end of the stage, a digital model of the product is obtained. This model includes all the geometric specifications that must allow adaptation to the use and manufacturability of the product.

The fourth stage has to establish if the design is able to respond positively to the boundary conditions to which it will be subjected. In this way, the loads, pressures, limitations of the degrees of freedom that are either defined in homologation norms or deduced from their use will be applied to the digital model. The stage is typically applied. Computer-aided engineering (CAE) programs are used to reach the results.

During this stage, optimization loops are established with the previous ones since it is typically necessary to make modifications in the geometry or in the specifications of the materials to ensure a correct result. At the end of the stage, a feasible product is obtained from a technical point of view [1].

The next stage includes all market studies and economic analyzes that must ensure economic and market feasibility. This stage is beyond the scope of this investigation.

3 Formulation of the Material

Fibers from corn are identified as the result of agricultural activity. The stems of corn are subjected to a treatment with steam to extract the fibers. For the matrix of the composite material, the use of HDPE is proposed.

Table 1 shows the flexural properties of the materials with the coupling agent (2% de Maleated Polyetylene - MAPE) for formulations at 20, 30, 40 and 50% (w/w).

Table 1. Flexural properties with and without coupling agents for % w/w formulations.

%w/w	Flexural modulus: E_f^c		Flexural strength: σ_f^c	
	2% MAPE	0% MAPE	2% MAPE	0% MAPE
0	0.520 (0.03)	0.52 (0.03)	18.92 (0.35)	18.92 (0.35)
20	1.55 (0.12)	0.51 (0.06)	27.24 (0.56)	28.24 (0.34)
30	1.98 (0.07)	1.92 (0.17)	31.36 (0.4)	34.7 (0.81)
40	2.53 (0.06)	2.61 (0.04)	30.85 (0.34)	39.81 (0.43)
50	2.86 (0.13)	2.78 (0.1)	32.69 (0.66)	42.13 (0.24)

Regarding the σ_f^c, two situations can be verified that refer to its evolution with the reinforcement percentage (Fig. 2). In the case of the formulation without coupling agent, the resistance to bending (σ_f^c) increases linearly up to 30% (w/w) to then inflect. Probably in the first part the increase in σ_f^c is due to the mechanical anchoring of the HDPE on the surface of the fibers. This capacity of diffusion and anchoring of the HDPE on the fibers stabilizes when increasing the percentage of reinforcement, because of the absence of chemical interactions between the fibers and the matrix. On the other hand, it is verified that using only 2% of MAPE the σ_f^c increases linearly with the reinforcement percentage. This is a consequence of the creation of hydrogen bonds between the hydroxyl groups of the surface of the fibers and the maleic anhydride phase of the MAPE. Thus, the interphase of the coupled composites is stronger that uncoupled ones. The highest tensile strength was obtained for the composite with 50% w/w reinforcement content. This composite showed a strength 133% higher than the matrix.

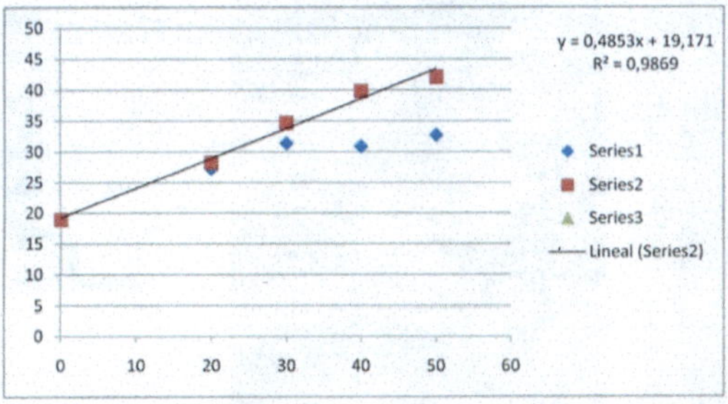

% w/w	0% MAPE	2% MAPE
0	18.92 (0.35)	18.92 (0.35)
20	27.24 (0.56)	28.24 (0.34)
30	31.36 (0.4)	34.7 (0.81)
40	30.85 (0.34)	39.81 (0.43)
50	32.69 (0.66)	42.13 (0.24)

Fig. 2. Flexural strength: σ_f^c.

On the other hand, it was observed that the flexural modulus evolves linearly with the reinforcement percentage and that this evolution is independent of the coupling agent (Fig. 3). Therefore, for this type of application the composite material without coupling agent could be used.

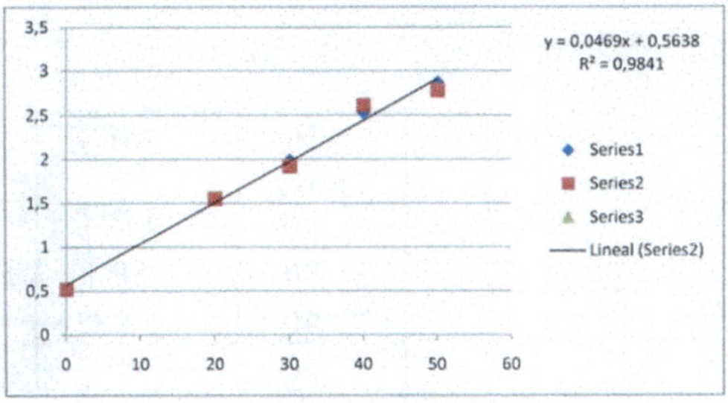

% w/w	2% MAPE	0% MAPE
0	0.520 (0.03)	0.52 (0.03)
20	1.55 (0.12)	0.51 (0.06)
30	1.98 (0.07)	1.92 (0.17)
40	2.53 (0.06)	2.61 (0.04)
50	2.86 (0.13)	2.78 (0.1)

Fig. 3. Module strength: E_f^c.

Figure 4 shows the holes generated by the fibers during the tensile test and can be compared to the interface of a fiber that generates a tensile strength of 20.36 MPa with 2% MAPE (40% w/w)).

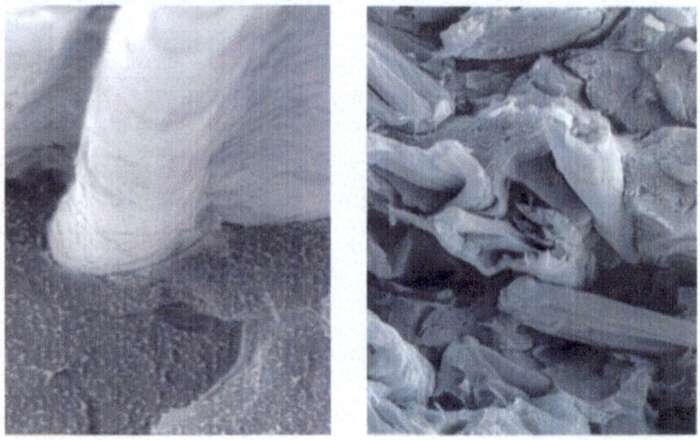

Fig. 4. Gaps that generate fibers, without MAPE and with MAPE.

Regarding the rigidity of the materials, a linear evolution is observed with the reinforcement percentage, typical of this type of materials.

Facing the industrialization of these composites, a very important parameter is the flow index (Melt Flow Index - MFI). In this case, it was evaluated at 210º and 10 kg of weight (Fig. 5). In the case of materials at 50% (w/w) it is observed that at 210 ºC and 10 kg. The pellets do not flow. This would indicate that between 2 and 3% of lubricating agent should be used in this case to drive the MFS to values to 8 ÷ 10 g 10 min.

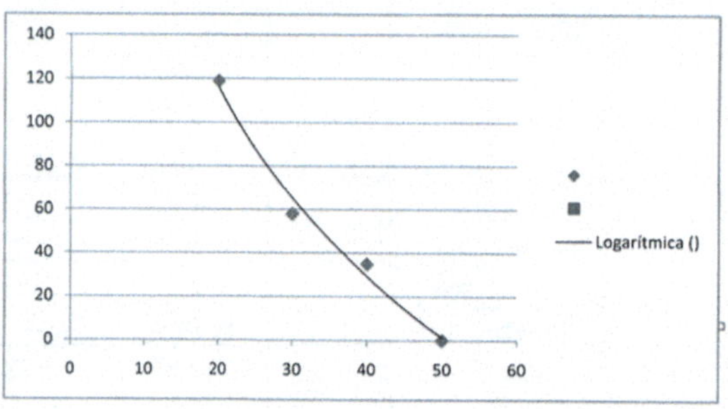

% (w/w)	MFS
20	118.9
30	58.1
40	34.82
50	0

Fig. 5. MFS fluency index.

From the results analyzed and according to economic and sustainability technical criteria, it can be inferred that a good option would be the use of the 40% (w/w) formulation that allows compliance with the required requirements, saving on the consumption of a product like HDPE and inert a significant amount of agricultural waste. From an economic point of view, although the manufacture of the composite material involves an additional transformation operation, the balance is positive.

4 Results and Discussion

The starting points for the choice of the product to be industrialized are, on the one hand, the formulation of the investigated material, the market to which it is addressed, elements and materials for the construction and on the other the manufacturing technology chosen, in this case the injection. These conditions establish that the product to be designed must be of small or medium dimensions, high production, and not subject to great stress or strain. The number of construction elements that respond to these requirements is very high; boxes for meters or electrical connections, mailboxes, answering machines, IP switches, luminaires, cladding…

The choice of the product responds to a search for references and a creativity session with the entire research team. The techniques used were the Synectics and Metaplan to obtain a list of ideas and then an evaluation of them [14–16]. The criteria of choice have been; technical feasibility, market possibilities, and potential improvements in the aspects of industrial design. Under these conditions, the chosen product has been a ventilation grid (Fig. 6).

Fig. 6. Ventilation grid 400 × 400. Reference element and aluminum grid.

The starting product is an aeration grid of 400 × 400 mm. This element is built in concrete. The current product has a mass of 4.5 kg. In the market there are similar products made of aluminum, plastic, steel … but we have not identified any manufacturer that uses composite materials with characteristics similar to the one formulated in this article.

The boundary conditions are established in the Technical Building Code (CTE). The main loads to which it will be subjected are those of the wind. The calculation hypothesis will be an installation in a type building of 25 m in height, located in a C zone and with a degree of roughness V (CTE, Formula 1). For these requirements, the final request is approximately 1000 N/m^2.

$$q_e = q_b \cdot c_e \cdot c_p \tag{1}$$

For the calculations and the redesign we add a coefficient of maximization of loads of 1.5, with which the design solicitation becomes 1500 N/m^2.

Following the steps proposed in the methodology, we modeled the current product, obtaining a 3D model and subjected it to the boundary conditions. The characteristics of the test material are those corresponding to a standard concrete, E: 2.1 GPa. The model is subjected to a load of 1500 N/m^2 perpendicular to the outer face and all its embedded sides are considered (restriction of all degrees of freedom). The most interesting results are an equivalent safety coefficient of 317 and a maximum deformation in the 0.02 mm model (Fig. 7a). The oversized is due to the sections of material used, due to the parameters of manufacturing parts molded with concrete. This type of manufacturing requires large sections to ensure the filling of the molds. The design is acceptable due to the low cost of the material.

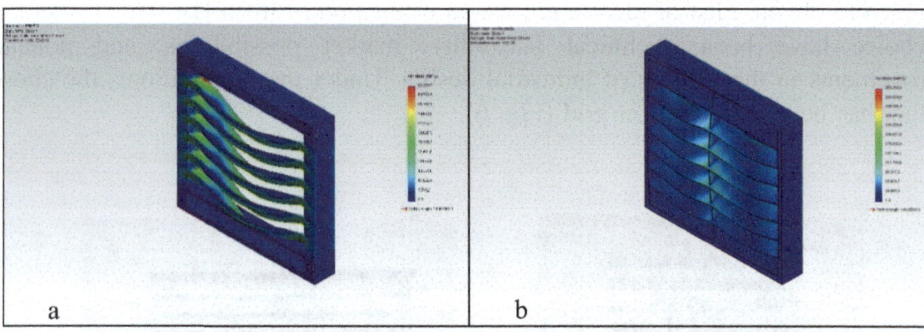

Fig. 7. A: Resulting stresses in the grid subjected to the boundary conditions (def. scale x2242). B: Resulting stresses in the grid subjected to the boundary conditions (def. scale x1151).

To ensure that the product can be manufactured by injection it is important that the thickness of the piece is constant and that there are no barriers to unmold. For the piece to be economical, the mold must be of two pieces, without sliders or inserts. The redesigned product shares the appearance with the starting one, with the inclusion of a central rib that acts as a reinforcement and as a filling channel. The thickness of the wall is 2.5 mm and the final mass of the piece is 1.2 kg. For the realization of the calculations, the small rounding in the edges or the angles of demolding of the piece are not contemplated.

The model is subjected to the same boundary conditions as the original part and the most outstanding results are an equivalent safety coefficient of 41 and a maximum deformation of 0.04 mm (Fig. 7b).

The cost attributable to the material would be, for HDPE virgin 2.04 €, and for HDPE + 40% w/w 1.32 €. In this way, a saving of 35% is established. The final cost of the product would be € 1.62.

With the economic data and the geometric proposal (Fig. 8) it is possible to start the market study stage to verify if the product is totally feasible and can go into production.

Fig. 8. Final proposal

5 Conclusions

A methodology for the design and development of products based on the reuse of waste materials is proposed, which is able to establish a new composite material and a new technically viable product. The application to building materials, regardless of the final product (ventilation grille) opens a wide range of possibilities. The creation of opportunities establishes the need for collaboration between different areas of knowledge. In this case the collaboration between materials science, design and engineering is able to develop a new material and turn it into a product proposal.

The agricultural residues of corn stem are susceptible to be used as reinforcement load of composite materials with an HDPE matrix. With the same and depending on the percentage of reinforcement with and without coupling agent, products capable of satisfying the requirements of a certain gamma of market products can be obtained. This constitutes a clear contribution to sustainability, reducing the consumption of a petroleum product and inerting an agricultural waste that is usually incinerated.

Nonetheless, to stablish clearly the environmental advantages of the proposed composite materials, a life cycle analysis is needed. The use of natural fibers hinders recycling the materials.

References

1. Serrano, A., et al.: Study on the technical feasibility of replacing glass fibers by old newspaper recycled fibers as polypropylene reinforcement. J. Clean. Prod. **65**, 489–496 (2014)
2. Julian, F., et al.: Design and development of fully biodegradable products from starch biopolymer and corn stalk fibres. J. Biobased Mater. Bioenergy **6**(4), 410–417 (2012)
3. Julian, F., et al.: Bio-based composites from stone groundwood applied to new product development. BioResources **7**(4), 5829–5842 (2012)
4. Galan-Marin, C., Rivera-Gomez, C., Garcia-Martinez, A.: Use of natural-fiber bio-composites in construction versus traditional solutions: operational and embodied energy assessment. Materials **9**(6), 465 (2016)

5. Kozlowski, R., Wladyka-Przybylak, M.: Uses of natural fiber reinforced plastics. In: Wallenberger, F.T., Weston, N.E. (eds.) Natural Fibers, Plastics and Composites, pp. 249–274. Springer, Boston (2004)
6. Vallejos, M.E., et al.: Micromechanics of hemp strands in polypropylene composites. Compos. Sci. Technol. **72**(10), 1209–1213 (2012)
7. Espinach, F.X., et al.: Analysis of tensile and flexural modulus in hemp strands/polypropylene composites. Compos. Part B Eng. **47**, 339–343 (2013)
8. Rowell, R.M., et al.: Utilization of natural fibres in plastic composites: problems and opportunities. In: Leao, A.L., Carvalho, F.X., Frollini, E. (eds.) Lignocellulosic-Plastics Composites. Universidade Estadual Paulista, Universidade Estadual Paulista, Sao Paulo (1997)
9. Flandez, J., et al.: Management of corn stalk waste as reinforcement for polypropylene injection moulded composites. BioResources **2**(7), 1836–1849 (2012)
10. Jiménez, A.M., et al.: Tensile strength assessment of injection-molded high yield sugarcane bagasse-reinforced polypropyene. BioResources **11**(3), 6346–6361 (2016)
11. Reixach, R., et al.: Tensile properties of polypropylene composites reinforced with mechanical, thermomechanical, and chemi-thermomechanical pulps from orange pruning. BioResources **10**(3), 4544–4556 (2015)
12. El Mansouri, N.-E., et al.: Research on the suitability of organosolv semi-chemical triticale fibers as reinforcement for recycled HDPE composites. BioResources **7**(4), 5032–5047 (2012)
13. Jimenez, L., et al.: Organosolv ethanolamine pulping of olive wood - Influence of the process variables on the strength properties. Biochem. Eng. J. **39**(2), 230–235 (2008)
14. Ferrer, J.B., et al.: Eco-innovative design method for process engineering. Comput. Chem. Eng. **45**, 137–151 (2012)
15. Howard, T.J., Dekoninck, E.A., Culley, S.J.: The use of creative stimuli at early stages of industrial product innovation. Res. Eng. Des. **21**(4), 263–274 (2010)
16. Julian, F., et al.: Study of the creative tension in students of engineering of two Spanish universities. Interciencia **36**(7), 524–530 (2011)

Improvement in the Design of a Front Hub for the BMX Bicycle

Rafael M. Pinilla-Gutiérrez[1], Laia Miravet-Garret[2(✉)], Juan Franquelo-Soler[2], Fernando Gómez-Hermosa[2], and Francisco J. Ortiz-Zamora[2]

[1] Universidad de Málaga, c/ Doctor Ortiz Ramos s/n, 29071 Málaga, Spain
[2] Dpto. de Expresión Gráfica, Diseño y Proyectos, Escuela de Ingenierías Industriales, Universidad de Málaga, c/ Doctor Ortiz Ramos s/n, 29071 Málaga, Spain
laiamiravet@uma.es

Abstract. Bicycle wheels for BMX (Bicycle Moto Cross) are subjected to considerable stresses during the practice of this sport, particularly in the Freestyle mode where the execution of stunts and large jumps often causes the breakage of certain components.

This study describes the improved design of a front hub for this type of bicycle, which presents two essential advantages in comparison with the existing hubs according to a market survey. Its shell has been optimized to reduce the risk of broken spokes by redesigning the flanges to avoid stress concentration points while increasing their support surface. Moreover, the protection of the spokes is also improved by a new hub cover and the creation of a subdivision of the cone whose assembly provides greater fixation, which allows for maintaining the assembly of all the hub components either with or without the hub cover.

Other aspects of the design have been considered based on several investigations and surveys regarding both aesthetic concepts and other demands of this extreme sport.

Keywords: Bicycle Moto Cross · BMX · Hub · Spokes

1 Introduction

The BMX (Bicycle Moto Cross) emerged in California (USA) in the 60's as teenagers imitated motorcycle racing with their bicycles, which, little by little, increased in popularity until it became a cycling discipline that was recognized by the International Cycling Union in 1993 [1], although there are video references that date back to 1955 in the Netherlands [2, 3].

The "Racing" modality consists of racing at the highest possible speed through bumpy or obstacle-laden courses while the "Freestyle" mode entails the difficulty of performing jumps and acrobatic sequences in the different environments that define its various modes ("park", "Street", "flat", "vertical" and "dirt"), and on different structural elements (benches, stairs, railings, and vertical walls) [1]. Bicycles of a relatively smaller size (frames between 950 and 1025 mm with 20-in. wheels) are used, and these

F. Cavas-Martínez et al. (Eds.): INGEGRAF 2019, LNME, pp. 13–22, 2020.
https://doi.org/10.1007/978-3-030-41200-5_2

are even equipped with 360° handlebars that allow for greater acceleration, precision, and maneuverability [4].

The research presented here proposes an improved design of a BMX Freestyle front wheel hub, particularly indicated for initiation stages, for which the new technical specifications have been defined mainly from personal surveys of BMX bicycle users as well as professionals dedicated to its sale and repair, in an effort to compensate for the lack of studies on BMX. This sport has gained considerable interest in both the medical [5–8] and psychosocial [9–12] fields, whilst the majority of publications on mechanical studies applicable to this specific sport are in the form of patent reports [13–20].

Although its status as a risk activity is reflected by its minority participation [12] and the fact that it is primarily practiced by adolescents [21, 22], in recent years this sport has undergone a notable rise in scale within the sporting arena, culminating in the inclusion of its "Park" specialty in the forthcoming 2020 Olympic Games in Tokyo, preceded by its debut in the Youth Olympic Games held in Buenos Aires (Argentina) in 2018. It is therefore expected that this new projection as an official international sport, will, in addition to increasing the number of followers, lead to an increase in the level of difficulty of the extreme maneuvers [23], which will consequently generate a higher commercial demand for BMX products in general as well as greater demands on the technical components of these bicycles.

2 Market Analysis, Classification, and User Needs

The dimensions of the hub will depend on the fork in which it is installed, specifically the spacing between the dropouts and the thread of the drill located within them that allows for the tightening of the axle of the hub to the fork. For freestyle BMX bicycles, the length of spacing between the inner sides of the dropouts is 100 mm, although regarding the threading these are of two types according to the nominal diameter, 10 mm or 14 mm.

The number of spokes of the front wheels used in this cycling discipline is usually 36, although there are also models that use 48. Following an analysis of the existing BMX front hubs available on the market, it is possible to broadly classify these according to:

1. Type of axle: male or female. The former allows for bearings of a smaller diameter and reduces the size of the assembled set so that it is lighter, although with a greater bearing and equal diameter of the shaft, greater strength is obtained.
2. Type of flanges: these can be vertical, which is the most commonly used model, or horizontal (flangeless), which, although giving greater protection to the shell of the hub when performing stunts where the "pegs" or footrest are used, the spokes suffer more when losing the surface support of the flange.

In the study of patent reports, no previous contributions were found for these types of hubs.

In order to identify the preferences of the users, a survey was carried out (confidence level $p = 0.90$) on a sample of 30 BMX-loving participants in Spain, aged between 15 and 21 years. All the participants were of beginner or intermediate level and were not official or federated competitors. To homogenize the sample specifically in terms of determining the level and frequencies of breakdown of the hubs, in addition to the age and perceived skill level for each athlete, other data were collected, such as their age of initiation into BMX, number of weekly hours dedicated to practicing the sport, as well as a verification of a list of maneuvers that they were able (or not) to perform. The results of the survey on the frequency of breakdowns in BMX front hubs, shown in Fig. 1, where spokes were denoted as the most sensitive elements of the set, were supported by seven professionals with more than five years of experience in the sector, in points of sale and repair of bicycles designed for Freestyle.

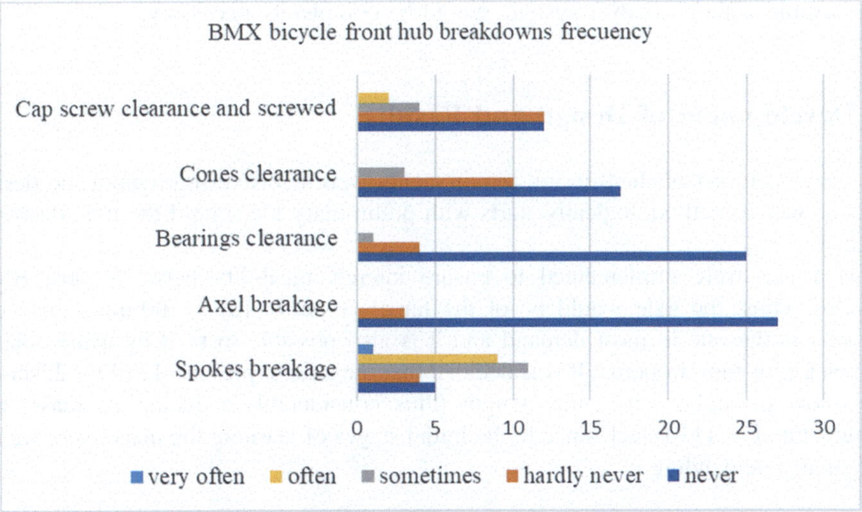

Fig. 1. Results of the survey conducted.

3 Design Conditions

After compiling and analyzing all the scientific documentation, studying the existing products on the market, and having consulted experienced staff in BMX, the initial stages corresponding to the Total Design methodology defined by Stuart Pugh [24, 25] were applied in this work. The main design conditions, derived from both the preferences of the users and the identification of the needs for improvements that the new bicycle hub should offer over those existing in specialized stores, were established as follows:

1. Interchangeability between forks, spokes, rings, and bearings that currently exist in the market.
2. Female hub axle (83.33% preference over the male axle, $p = 0.9$, 71.95–94.71), because when the two front "pegs" are not installed, the ends protrude on the male-type axle from this to the outside, which sometimes causes cyclists to crash and injure themselves.
3. It should be possible to replace the spokes of the wheel in a simple manner, using the least number of tools and preferably without the need to disassemble the wheel from the fork. Damaged and broken spokes is one of the main setbacks for novice users, because when beginning to learn the maneuvers, these breakages occur with sufficient frequency for many users to carry their own tools to the area of practice so that they

themselves can carry out the repair of the wheel "in situ" (76.67% preference over repair in workshop, p = 0.9, 63.75–89.59). This will allow for the advantage of being able to continue practicing without damaging the rim as well as saving both time and costs associated with making arrangements with a specialized workshop.

4. Installation of the hub cover without the need to remove the cone from the assembled set, so that all the components of the hub are always assembled on the wheel (80% preference over incomplete assembly, p = 0.9, 67.78–92.22). In this way, possible loss of the cone is avoided, which, in the case of wanting to re-assemble without a cover system, would be completely necessary.

4 Development of Design and Results

The previously established design requirements were decisive throughout the design process, which methodologically starts with preliminary ideas until the final design is reached.

Measures were implemented to ensure interchangeability between most BMX bicycles. Thus, the axle would be of the female type 9 (10) × 100 mm, since this diameter is the one in most demand and it is also possible to fit it by using adapter sleeves for 14 mm dropouts. It was decided that the axle, represented in Fig. 2, should be hollow throughout its entire length (thus considerably reducing its mass) and manufactured in 4130 steel, since in the initial stages of learning the maneuvers are not particularly demanding.

Fig. 2. Longitudinal section of the axle.

The type of thread for anchoring the hub to the fork was ISO metric screw thread for general use M10 × 0.75 (nominal size and fine pitch) because this meets the most common measurement in both forks and "pegs". In addition, the nuts of this metric are readily available on the market. The decision to use the fine pitch was determined by its better response to vibrations, since they are less likely to become loose in these situations than the thick ones, allowing also for a better adjustment.

The creation of the hub shell began from a design for 36 spokes subject to vertical flanges, since the "flangeless" type involves the insertion of the spokes on the inside and for this the disassembly of the wheel fork is required, whilst the rim tends to decenter more easily with this type of hub because when a lateral force impacts on the wheel cover and transmits this impact to the spokes, only the thickness of the hub shell provides lateral support.

During the design process, we not only considered the needs of the users, but also attempted to anticipate them. In this case, it was decided that the design should be aimed at improving both the protection and the bearing surface of the spokes on the hub in order to reduce the probability of breakage.

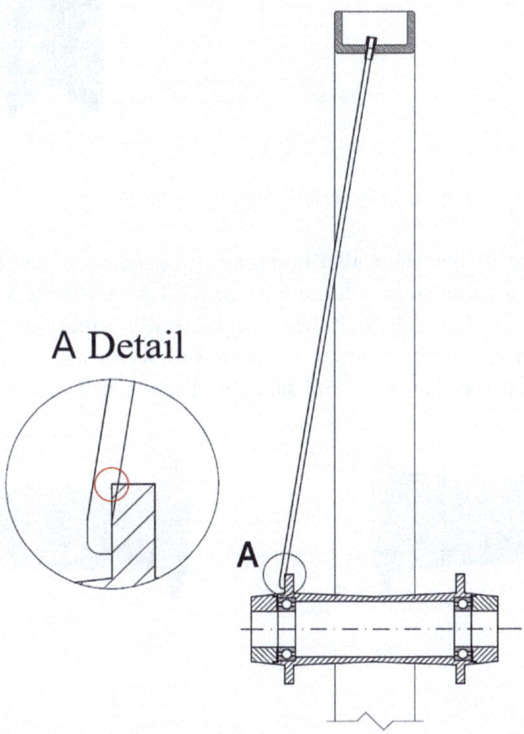

Fig. 3. Dynamic representation of the working position of a spoke mounted on a vertical flange.

Figure 3 provides a dynamic illustration of what happens to a spoke when the flanges of the shell are completely perpendicular to the axis of rotation. In these assemblies, the spokes are in a more forced working position because it is not possible to adopt a greater curvature. The proposed solution is to round both the upper edge of the flange and its outer side thus smoothing the area of tangency between each outer spoke with the flange when assembled. In this way, possible points of high stress concentration that lead to spoke breakage are reduced. Thus, with this new design, the spokes that are pushed in the direction of the hub in response to impact on the "pegs" or the cover, have a larger support surface, being less aggressive and more adaptive in response to possible deformations, and consequently the fatigue in the spokes is reduced.

To unify the design, all edges of the hub shell were rounded off in order to achieve a smooth finish. Figure 4 shows the various phases of the design until the final shape of the flange is achieved.

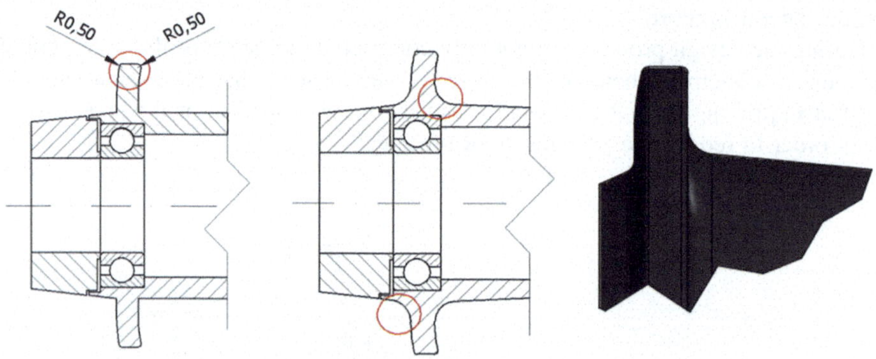

Fig. 4. The design process of the flanges.

The central zone of the shell also underwent a process of evolution in its design (Fig. 5). The use of a cylindrical volume was discarded because it was considered to be coarse and lacking aesthetic appeal, whilst hyperbolic softening was also ruled out because this was more common in the market. Finally, a truncated cone design was chosen that lightened the weight of the piece and made it visually more unique.

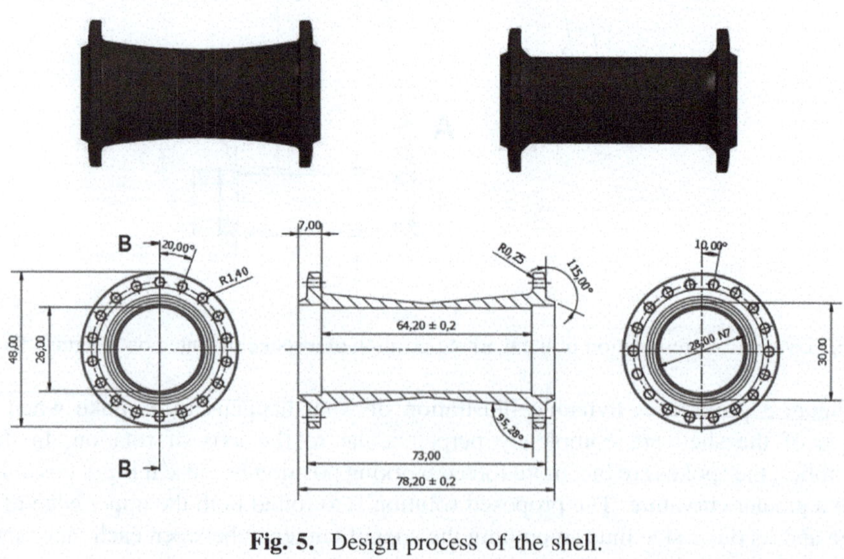

Fig. 5. Design process of the shell.

The choice of bearing was made during the design process of both the shell and the axle, in compliance with ISO 582: 1995 on "Rolling bearings, Chamfer dimensions and Maximum values", opting for a sealed model 15267 - 2RS.

The chosen material for manufacture was 6061 aluminum, which was also used for the other main components.

In order for the hub cones to always be assembled on the wheel, it was necessary to design a hub cover that would allow for this option. Initially the cone, following market trends, was conceived as a single piece. However, in the final design this was divided into sections, turning it into two pieces in order to improve the fixation of the hub cover. Figure 6 shows the mounting position without the hub cover (left) and the arrangement that would be adopted inside the hub cover (right).

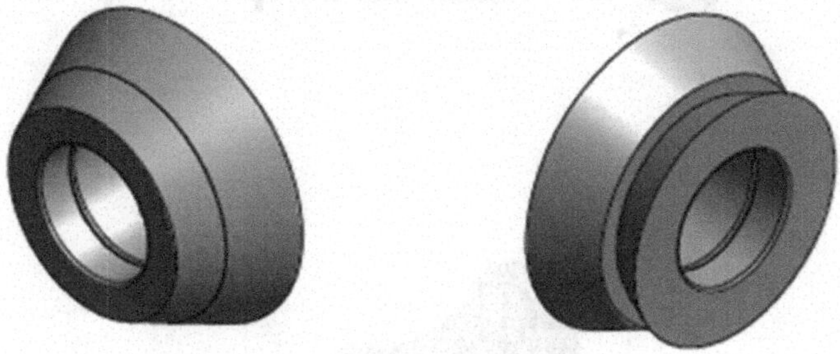

Fig. 6. Assembly positions of the cones.

The hub cover, illustrated in Fig. 7, in addition to providing it with the inner hollow to house the cones, included a recess in the inner side so that this is located closer to the spokes to provide greater protection.

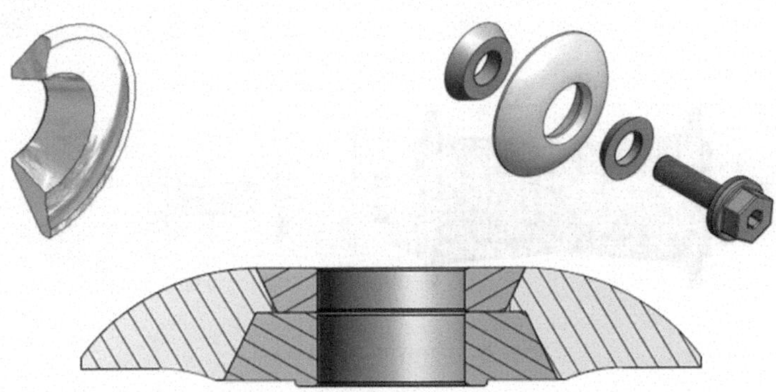

Fig. 7. Design and assembly of the hub covers.

Assembly (Fig. 8) consists of introducing a bearing by means of applying slight pressure inside the shell, while the other bearing is installed once the axle is fitted. Next, the cones are installed according to the operating position, fitting the hub cover on the required side on a pair of cones, and finally the setting is adjusted by screwing the axle to the fork legs. The adjustments were indeterminate in all cases except for the cones that were compatible with both the axle and the hub cover.

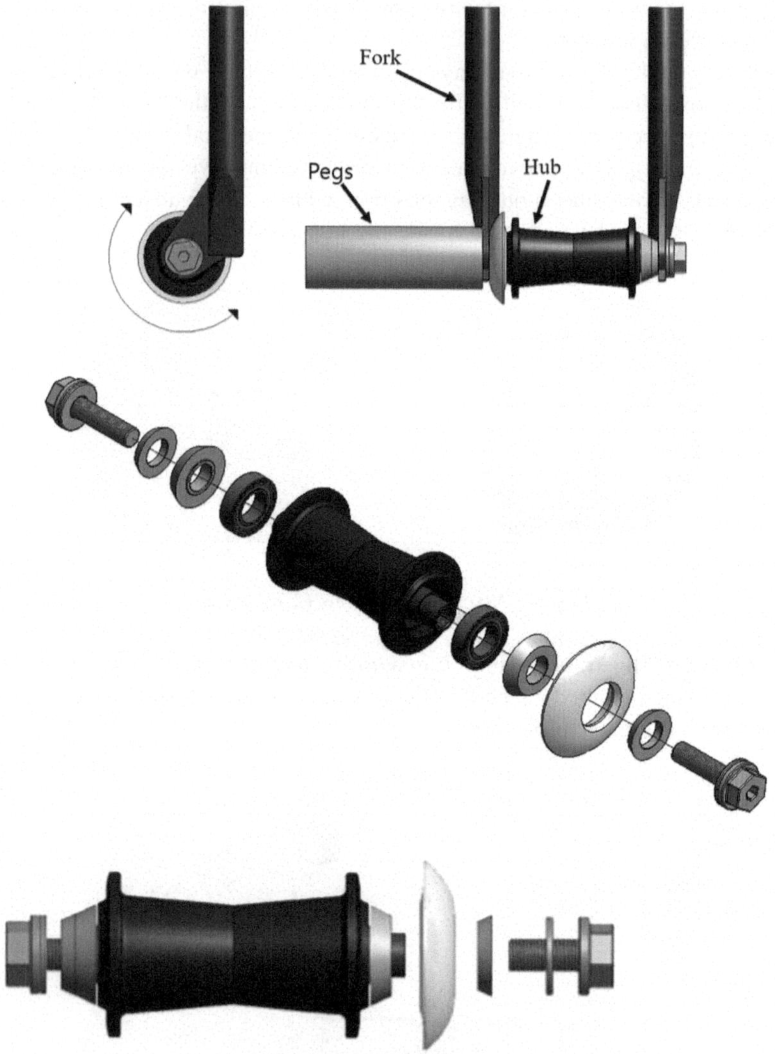

Fig. 8. Drawing of the set.

Finally, the proposed model was validated through an analysis of efforts conducted with the same program used for its conceptual design, i.e. Autodesk Inventor 2014.

Various assumptions were made regarding estimated load values exerted by jumps from a maximum height of 2 meters on the front wheel of a 12 kg bike with an 80 kg cyclist. The minimum safety coefficients obtained in each case were 1.01 for a load of 1420 N (Fig. 9, upper image), 1.44 when applying 1000 N, and 2.05 with 700 N (Fig. 9, lower image).

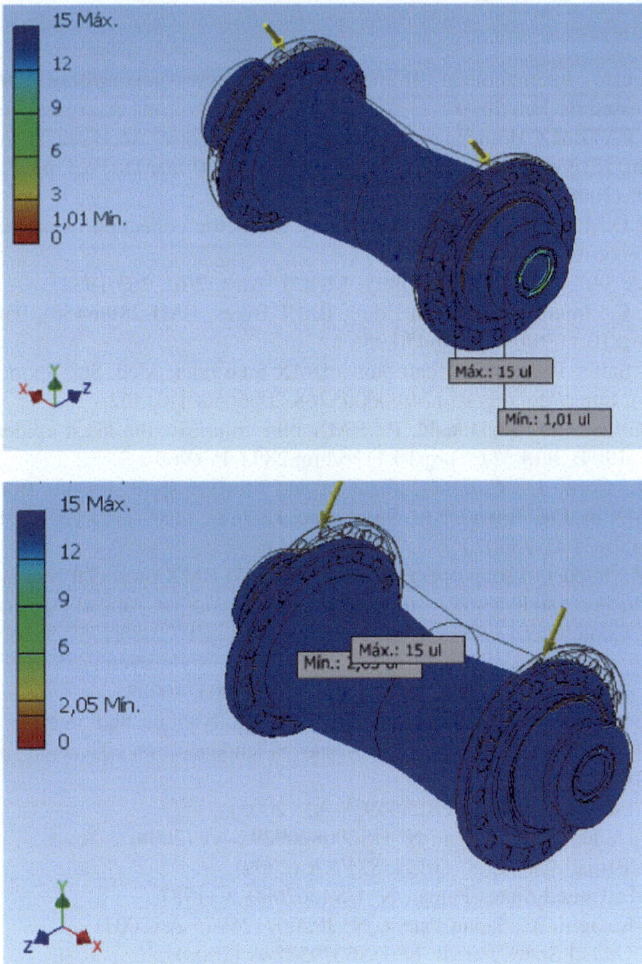

Fig. 9. Safety coefficients obtained in the stress analysis.

5 Conclusions

The present study conducted an analysis of existing products in the market and used the experiences of people dedicated to BMX Freestyle cycling in order to design a front hub with specifications aimed at alleviating one of the main causes of wheel breakage experienced by beginners, i.e. the excessive wear and consequent fatigue of the spokes. In addition, the design of the hub cover has not only been optimized for this function, but also allows the cones to be housed while maintaining the assembly of the complete set. The safety coefficients obtained in the stress analysis [26] ensure an adequate response of the hub in the initial and intermediate learning stages of BMX Freestyle cycling.

References

1. Union Cycliste Internationale Homepage. https://www.uci.org/bmx-racing/about-bmx-racing. Accessed 01 Feb 2019
2. Promotions FATBMX Homepage. https://www.fatbmx.com/. Accessed 15 Feb 2019
3. Roosenboom, H.: Vermaak en sport. Ach Lieve tijd: 2000 Jaar De Peel, pp. 321–344. W. U. Brabant, Ed. (2006)
4. Campillo, P.D.: Pedaling analysis in BMX by telemetric collection of mechanic variables. Braz. J. Biomotricity **1**(2), 15–27 (2007)
5. Burke, A.: A Serious BMX Bike Injury. Med. J. Aust. **2**(6), 263 1982)
6. Illingworth, C.: Injures to children riding BMX Bikes. BMJ **289**(6450), 956–957 (1984). https://doi.org/10.1136/bmj.289.6450.956
7. Schupp, C., Sallis, B.: Posterior calf pain - BMX bike rider. Med. Sci. Sport. Exerc. S150–S151 (2006). https://doi.org/10.1249/00005768-200605001-01562
8. Soysa, S., Grover, M., McDonald, P.: BMX bike injuries - the latest epidemic. Br. Med. J. 960–961 (1984). https://doi.org/10.1136/bmj.289.6450.960-a
9. Edwards, B., Corte, U.: Commercialization and lifestyle sport: lessons from 20 years of freestyle BMX in 'Pro-Town USA'. Sport. Soc. **13**(7–8), 1135–1151 (2010). https://doi.org/10.1080/17430431003780070
10. Mikdar, U.Z.: Impact of soap opera show "Sports" Go BMX on social behavior of children. In: Abdullah, A., Nandiyanto, A., Danuwijaya, A. (eds.) 1st Annual Applied Science and Engineering Conference (AASEC) (2016). https://doi.org/10.1088/1757-899X/180/1/012182
11. Nelson, W.: The historical mediatization of BMX-freestyle cycling. Sport. Soc. **13**(7–8), 1152–1163 (2010). https://doi.org/10.1080/17430431003780088
12. Zubiaur Gonzalez, M., del Riego Casado, M.: Motivations and interests of practicing teenagers of BMX: a pilot study. Retos - Nuevas tendencias en educación física, deporte y recreación 109–113 (2015)
13. Cauwet, J.: French Patent. Nº FR2864938-A1 (2005)
14. Goring, T.: United States Patent. Nº US2008200292-A1 (2008)
15. Goering, T: British Patent. Nº GB2458217-A (2009)
16. Solano, L.M.: United States Patent. Nº US4662090-A (1987)
17. Shioda, R., Katagiri, Y.: Japan Patent. Nº JP2007125975-A (2007)
18. Valdez, R.: United States Patent. Nº US6079522-A (2000)
19. Valdez, R.: United States Patent. Nº US2002020588-A1 (2002)
20. Van Eijk, N.S.: International Patent. Nº WO2009128702-A1 (2009)
21. Mateo, M.S.: Desarrollo de un marco teórico para la iniciación deportiva en la especialidad ciclista de BMX (2008). https://www.efdeportes.com/efd121/iniciacion-deportiva-en-la-especialidad-ciclista-de-bmx.htm. Accessed 10 Feb 2019
22. Michel, G., et al.: Recherche de sensations et conduites à risque chez l'adolescent. Annales Médico-Psychologiques **159**, 708–716 (2001)
23. The International Olympic Committee Homepage. https://www.olympic.org/news/bmx-freestyle-will-bring-a-fresh-look-to-the-olympic-games-tokyo-2020. Accessed 01 Feb 2019
24. Pugh, S., Clausing, D., Andrade, R.: Creating Innovative Products Using Total Design. Addison Wesley Longman, London (1996)
25. Pugh, S.: Total Design: Integrated Methods for Successful Product Engineering. Addison-Wesley, Wokingham (1991)
26. Autodesk Inventor 2014 Help Homepage. https://help.autodesk.com/view/INVNTOR/2014/ESP/?guid=GUID-D541AE00-F9FB-4B7E-A595-402219353010. Accessed 05 Feb 2019

Prosumer and Product Design Through Digital Tools

L. Asión-Suñer(✉) and I. López-Forniés

Universidad de Zaragoza, 50018 Saragossa, Spain
lauraasion@gmail.com

Abstract. Currently, the growing interest of users and consumers in the participation of the creative process has led to the typical "maker culture" practices. Consequently, there is an increasing number of prosumers - users who produce what they consume - who want to be part of the design and transformation process of the products. In order to achieve it, prosumers have begun to use digital tools that greatly facilitate this task. These tools could vary depending on the number of users involved in the process and the freedom of participation that they have on the product. It has been presented a number of qualitative classification of cases involving the end user, individually or collectively, that has influenced as a prosumer in the product design process. The objective is to study the use of digital tools in the creative phase within the design process according to their different levels of participation with respect to the final product. The cases are shown in four tables according to the number of users involved in the process and their level of participation. In these tables, other important aspects related to the study of digital tools such as the type of contribution of the prosumer to the product or the design phase in which he participates will be identified. In conclusion, this work will show if there is a pattern in the use of digital tools according to the number of users involved in the process and the freedom of participation that they have and which are the reasons for their use.

Keywords: Prosumer · Product design · Digital tools · Creative process

1 Introduction

In 1980, Alvin Toffler aimed the term *prosumer* for the first time, referring to mass consumers who, with new communication systems, would assume the role of consumer-producers [1]. Technology has evolved and in the last decades there has been a transition from passive consumers to active prosumers [2]. In this way, we would currently understand as a prosumer all those users that actively participate in the creation of products and services that finally uses [3]. Kotler suggested that this advance towards the postindustrial era would decrease the number of pure consumers, who would be replaced by "prosumers" [4]. However, there exists numerous cases, their claims have not been established in society in a generalized manner.

The consumption model has evolved thanks to new technologies of communication and digital literacy, access to the Internet, the proliferation of social networks and Web 2.0 [5]. Prosumer has emerged to become a central figure in the temporary culture [6],

F. Cavas-Martínez et al. (Eds.): INGEGRAF 2019, LNME, pp. 23–30, 2020.
https://doi.org/10.1007/978-3-030-41200-5_3

empowered and independent comes to self-realization focused on the achievement of their own product that is shared as online content [7]. New technological tools have burst with greater initial impact in the field of design, allowing the democratization of creative processes, centered on the user and allowing their participation in the process, which guarantees their transmission and permanence [8].

On the other hand, the maker movement has been developed [9, 10] as an evolution of the traditional DIY movement, where it has gone from bricolage and assembly to creation and manufacture. The futuristic vision of a user who can become a producer and consumer at the same time, thanks to electronic technology [11], has materialized in some communities that make the prosumer closer to technology and has an easier access to digital tools. In this aspect, it is remarkable the importance of 3D printing, which has been the engine of a new maker current as it is one of the most promising recent technological developments [5]. Currently it is pretended to reach especially the youngest [12] with actions of divulging robotics, 3D design software, agile tools and programming that allow them to develop their own products.

The Toffler model of *prosumer* remains rooted in the era of massive mass media. Roles as consumers and users have long begun to be closely related to those as producer and creator [13]. Specifically, the term *maker* has expanded the motto of DIY (*Do It Yourself*) to DIT (*Do It Together*) or DIWO (*Do It With Others*) and includes disciplines ranging from computer science, electronics and robotics to carpentry and metallurgy [8]. The activities that could be grouped under the category of *maker* range from retail, creation of workspaces and fab labs [14], manufacturing and distribution, exhibition fairs [15] and services [16]. Knott differentiates three types of prosumer according to how they use the tools, materials and indications that companies provide them: the prosumer that follows rules; the one who pursues self-sufficiency; and the one who adapts tools and materials in DIY processes [17].

2 Methodological Description

The main objective of this study focuses on research through a case review in which digital tools has been used throughout the product's creative phases by prosumers. As a result, we hope to find a series of cases that show if there is a pattern in the use of digital tools according to the number of users involved and their different levels of participation with respect to the final product, and which are the reasons for their use.

To achieve this, an initial search of documented cases was carried out but we only found two cases. So, a field search was carried out with the keyword "prosumer" and other related keywords as "collaborative design", "mass customization", "co-creation" and "personalization". After this more precise search, the majority of cases cited in the article were obtained. Finally, a search was carried out through the digital tools identified in the cases found as "design software", "open source" or "tutorials".

A total of 33 cases were identified that, instead of being tested in their entirety, were analyzed and tested superficially to have enough information to make a posterior qualitative classification according to two main aspects. First, the number of participants involved, differentiating between an individual (a single prosumer) or collective (more than one prosumer) participation. And, second, according to the degree of creative freedom, differentiating between free (prosumers could participate freely and establish their own criteria) or restricted (prosumers could opt among the suggestions offered by the company) user participation in the creative phase of the design process. In this way, in order to optimize the classification of cases, these have been divided into four tables: individual and free participation cases (Table 1); individual and restricted participation cases (Table 2); collective and free participation cases (Table 3); and collective and restricted participation cases (Table 4).

In these tables, three aspects were qualitatively defined. First, the type of contribution of the prosumer to the product, differentiating between new design (ND), redesign (RD), copy (C) or selection of features (SF). Second, the design phase, differentiating between concept (C), development (DV), detail (DT) or presentation (P) phase. And third, the digital tool used, such as online platform (OP), design software (DS), tutorials (T), open source (OS), or vote system (VS), being this particular in each case. We conclude with the analysis of the results in form of discussion and conclusions that deal with aspects related to digital tools and prosumers.

3 Results

The qualitative classification of the results is included in Tables 1, 2, 3 and 4. Furthermore, in the cases review we found several websites that offered numerous digital tools for prosumers. Among these portals, it was found a website that offers companies the possibility to customize any of their products through the development of a visual digital tool of selection and filtering [18]; a blog post with a complete list of collaborative resources for 3D modeling and impress [19]; and a page of online resources to do graphic design without being a professional in the field [20].

3.1 Individual and Free Participation Cases

Due to the limitations in the productive systems it is difficult to develop unique products where the prosumer can participate freely. Therefore, some companies give the user the necessary means in order to finish the product by himself [21–23]. There are cases in which the prosumer can follow an example to build his own product and, in addition, add a creative contribution on the design, manufacture or personalization of it [24, 25]. Finally, some of the cases studied also have platforms where they offer free training so that the prosumer learns tasks such as sewing, welding, 3D design and printing, CNC and resins molding, among others [21–23, 25, 26].

Table 1. Characteristics of individual and free participation cases.

Case	Contribution	Phase (Results)	Digital tool
Meccano [21]	ND	C + P (Design)	OP + T + instructions
Burda [22]	RD + C	DV + P (Fabrication)	OP + T + instructions
MODI [23]	RD + C	C + DV (Design)	OP + T + instructions
Spreadshirt [24]	ND + SF	C + P (Design)	OP + DS
Instructables [25]	C + RD	C + DV (Design + Fab.)	OP + catalogue + ideas interchange
Desygner [26]	ND + SF	C + P (Design)	DS + online tools + templates
BeoCafe [27]	ND	C (Conceptual design)	3D Software + design + rendering
Slow photo [28]	ND	C + DV (Design + Fab.)	DS + prototyping tool

3.2 Individual and Restricted Participation Cases

Among the existing tools for a user to participate in the design of a product the most used is mass customization, which allows to configure a product through the choice of suggested options by the company. It is widely used in transportation sector [29, 30], but it is also present in sectors such as sports and textiles [31–33].

Table 2. Characteristics of individual and restricted participation cases.

Case	Contribution	Phase (Results)	Digital tool
Helmets 3D [29]	SF	P (Design)	OP + DS
MINI [30]	SF	P (Design)	OP + DS
NikeID [31]	SF	C + P (Design)	OP + DS
Converse [32]	SF	C + P (Design)	OP + DS
Pro Model Deck [33]	SF	P (Design)	OP + DS
Makercase [34]	SF	D (Detail phase)	Makercase
Ordermade WholeGarment [35]	SF	P (Not design)	Co-design software tool

3.3 Collective and Free Participation Cases

Digital tools such as online platforms and forums generate a common collaborative network where prosumers can co-create and share results [36–38]. Moreover, the phenomenon of 3D printing has overflowed the field of industry and has been set up at other levels [8]. This, together with Open Access, has allowed prosumers to share their creations in forums and file libraries to which any user has access to upload their models and modify others, thus giving rise to models created by various users [19].

Table 3. Characteristics of collective and free participation cases.

Case	Contribution	Phase (Results)	Digital tool
Ikea Hackers [36]	ND + RD	C (Conceptual design)	OP + catalog + ideas interchange
ProsumerFX [37]	ND	C (Conceptual design)	DS for image processing
Braineet [38]	ND	C (Conceptual design)	OP + App
ReViste [39]	RD	C + P (Design)	Not a digital tool (physical)
Github [40]	ND +RD	C (Conceptual design)	OS + CAD + 3D models repository
3D Printeros [41]	ND	C + DV (Design + Fab.)	Cloud + OS + Remote 3D printer
Robotix [42]	ND + RD + C	C + DV (Design)	OP + T + Ideas interchange
Scratch [43]	ND	C (Design)	OP + Software
Maker Shed [44]	ND + RD + C	C + DV (Design)	OP + T + Hardware
Vectary [45]	ND	C + P (Design)	3D software + forum + co-design
OnShape [46]	ND	C + DV (Design)	Learning centre + cloud + library

3.4 Collective and Restricted Participation Cases

In collective cases whose participation on the product is restricted, the user's participation only has relevance as a number, giving types of participations. First, users make design decisions collectively through voting, surveys or user tests, where they value products or services to apply innovations or improvements. In this way, through the choice of features, users can make collective design decisions about the final result of the product [47–49]. Second, users participate through voting systems or crowdfunding to decide if a product already designed by the company or by another user will finally go to the market [50, 51]. And third, there are platforms of products and services that present new contents adjusting to the multiple selection of options made by their viewers [52, 53]. In all three cases, a participation that in principle is individual, ends up becoming a collective result.

Table 4. Characteristics of collective and restricted participation cases.

Case	Contribution	Phase (Results)	Digital tool
Lay's [47]	ND + SF	Market (Fab. + commercialization)	OP + VS + social networks
La Marque du Consommateur [48]	ND + SF	DV (not design)	OP + VS + questionnaire
Manchester City FC [49]	SF (ND + RD)	C + P (Design)	OP + VS + testing
Lego Ideas [50]	ND + SF	Market (Fab. + commercialization)	OP + VS
MADE [51]	ND + SF	Market (Fab. + commercialization)	Cloud + crowdfunding + OP
Netflix [52]	C + SF	P (Not design)	Data processing algorithm
Amazon [53]	C + SF	P (Not design)	Data processing algorithm

4 Discussion

The original idea of Toffler about prosumer [1] has derived in two alternative ways. The first one, known as "mass customization", is to achieve a high degree of personalization by having consumers establishing the design requirements. In this way, the consumer does not participate in the manufacture, but rather limits himself to personalize the product through a filtering of options that affects the final design. The second way is to make the prosumer participate in the chain of transformation of the product either in the creative phase of design, in the production of some of the parts or in the final phases such as assembly or installation. It is necessary to define these two alternatives, since there are many sectors that have used the term prosumer, confusing it with mass customization, personalization at the point of sale or online personalization.

The term self-consumption, which is closely related to prosumer, consists of producing or managing the resources that the consumer himself enjoys. The most well-known examples are associated with the production of food, energy, water or clothing. Consumers could be individuals, groups, companies or public entities. In this way, in the freedom scale we would find self-consumption in the highest rank and mass customization in the lowest. The ideal prosumer and that would fit the original definition of Toffler [1] would be the one between these two extremes. I.e., that prosumer who can act with freedom on the product, but who has certain interdependence of a third party that provides the base on which to act. Two good examples classified that comply these characteristics are MODI [23] and 3D Printer OS [41].

The importance of digital tools in the current era to consolidate the presence of the prosumer tendency is indisputable. These tools allow the user to participate not only in the design phases, but also in the manufacturing and personalization. However, we must also highlight the importance of physical tools such as fab labs [14] or 3D printers [8]. It is these physical tools that complement the work of the digital tools and work in harmony with them to make possible the final materialization of the product.

5 Conclusions

The classification carried out shows a clear dependence of each quadrant with the type of digital tool used. The most notable case is where the prosumer's participation is restricted, since the user's contribution is only made through the selection and filtering of options by means of an online platform to affect the product's presentation phase. That is not the case with free participation, which allows the user to contribute new ideas that affect the conceptual phase of the product through digital tools such as 3D software, tutorials, instructions and specific apps, among other tools.

The difference between hardware product and software product is not only in its tangibility, but also in the way in which the user can participate in its design, transformation or manufacture. Both types of product can share digital tools such as online platforms, design software or open source. However, there are tools that can only be linked to hardware products due to their tangible nature and their need for subsequent manufacturing, such as mass customization, 3D printing or physical tools.

Otherwise, if there is something that characterizes collective cases, it is the use of tools such as online platforms, repositories of files, clouds, online libraries and forums. These open source tools facilitate communication between users to enable the exchange of ideas, the creation of challenges and the realization of collaborative projects at a distance. This is why we can affirm that the presence and development of Web 2.0 has influenced the current prosumer, in which there is also an increasingly tendency closer to remote hardware control, as has been seen in some cases [41].

Despite the large number of tools, there are few documented cases, especially with regard to the ideal prosumer previously defined. This is related to the fact that the prosumer is a very specific small group, that makes fairs [15] and creates eco-systems and working communities [16] only between them. This makes low cultural impact in society despite the large number of available tools, which means that the resulting projects are often not shown to other users, that could even become potential users.

References

1. Toffler, A., Alvin, T.: The Third Wave. Bantam Books, New York (1980)
2. Levine, R., Locke, C., Searls, D.: The Cluetrain Manifesto. Basic Books, New York (2011)
3. Ritzer, G., Dean, P., Jurgenson, N.: The coming of age of the prosumer. Am. Behav. Sci. **56** (4), 379–398 (2012)
4. Kotler, P.: The prosumer movement: a new challenge for marketers. Adv. Consum. Res. **13** (1), 510–513 (1986)
5. Rayna, T., Striukova, L., Darlington, J.: Co-creation and user innovation: the role of online 3D printing platforms. J. Eng. Tech. Manag. **37**, 90–102 (2015)
6. Comor, E.: Contextualizing and critiquing the fantastic prosumer: power, alienation and hegemony. Crit. Sociol. **37**(3), 309–327 (2011)
7. Ritzer, G., Jurgenson, N.: Production, consumption, prosumption: the nature of capitalism in the age of the digital 'prosumer'. J. Consum. Cult. **10**(1), 13–36 (2010)
8. Val Fiel, M.: Prosumer and 3D printing: democratization of the creative process. Revista 180. Arquitectura, arte, diseño (37), 17–22 (2016)
9. Anderson, C.: Makers: la nueva revolución industrial. Ediciones Urano, Barcelona (2013)
10. Hatch, M.: Maker Movement Manifesto: Rules for Innovation in the New World of Crafters. McGraw-Hill Education, New York (2014)
11. McLuhan, M., Nevitt, B.: Take Today: The Executive as Dropout. Harcourt Brace Jovanovich, New York (1972)
12. Herrero-Diz, P., Ramos-Serrano, M., Nó, J.: Minors as creators in the digital age: from prosumer to collaborative creator. Theoretical review 1972–2016. Revista Latina de Comunicacion Social **71**, 1301–1322 (2016)
13. Bruns, A.: From prosumer to produser: understanding user-led content creation. In: Transforming Audiences Conference, London (2009)
14. Mikhak, B., Lyon, C., Gorton, T., Gershenfeld, N., McEnnis, C., Taylor, J.: Fab lab: an alternate model of ICT for development. In: 2nd International Conference on Open Collaborative Design for Sustainable Innovation, pp. 1–7. ThinkCycle, Bangalore (2002)
15. Maker Faire. https://makerfaire.com. Accessed 14 Mar 2019
16. Thingiverse. https://www.thingiverse.com/. Accessed 14 Mar 2019
17. Knott, S.: Design in the age of prosumption: the craft of design after the object. Des. Cult. **5** (1), 45–67 (2013)
18. Smart Customizer. https://www.smartcustomizer.com/. Accessed 14 Mar 2019
19. DIY 3D Printing. http://diy3dprinting.blogspot.com/. Accessed 14 Mar 2019
20. Online Design Tools. https://adespresso.com/blog/online-design-tools/. Accessed 14 Mar 2019
21. Meccano. http://www.meccano.com. Accessed 14 Mar 2019
22. Burda. https://www.burdastyle.es/. Accessed 14 Mar 2019
23. MODI. https://modi.luxrobo.com/. Accessed 14 Mar 2019
24. Spreadshirt. https://www.spreadshirt.es/. Accessed 14 Mar 2019

25. Instructables. https://www.instructables.com/howto/prosumer/. Accessed 14 Mar 2019
26. Desygner. https://desygner.com/es/. Accessed 14 Mar 2019
27. BeoCafe. https://www.behance.net/gallery/34204619/BeoCafe. Accessed 14 Mar 2019
28. Photography the Slow Way. https://www.yankodesign.com/2011/04/22/photography-the-slow-way/. Accessed 14 Mar 2019
29. D Helmets. http://www.3dhelmetsnzi.com/. Accessed 14 Mar 2019
30. MINI Yours Customised. https://yours-customised.mini/. Accessed 14 Mar 2019
31. NikeID. https://www.nike.com/es/es_es/c/nikeid. Accessed 14 Mar 2019
32. Converse Design Your Own. https://www.converse.com/es/es/landing-design-your-own. Accessed 14 Mar 2019
33. Pro Model Deck. http://www.promodeldeck.com/es/. Accessed 14 Mar 2019
34. MakerCase. https://www.makercase.com/. Accessed 14 Mar 2019
35. Peterson, J., Larsson, J., Mujanovic, M., Mattila, H.: Mass customisation of flat knitted fashion products: simulation of the co-design process. Autex Res. J. **11**(1), 6–13 (2011)
36. Ikea Hackers. https://www.ikeahackers.net/. Accessed 14 Mar 2019
37. Dürschmid, T., Söchting, M., Trapp, M.: ProsumerFX: mobile design of image stylization components. In: SA 2017 Symposium on Mobile Graphics and Interactive Applications, Bangkok, Thailand, vol. 1, pp. 1–8 (2015)
38. Braineet. https://www.braineet.com/en/. Accessed 14 Mar 2019
39. ReViste. https://openurbanlab.com/2018/11/09/arranca-el-proyecto-reviste-taller-creativo-de-iniciacion-al-upcycling-denim/. Accessed 14 Mar 2019
40. GitHub. https://github.com/. Accessed 14 Mar 2019
41. D Printer OS. https://www.3dprinteros.com/. Accessed 14 Mar 2019
42. Robotix. https://www.robotix.es/es/. Accessed 14 Mar 2019
43. Scratch. https://scratch.mit.edu/. Accessed 14 Mar 2019
44. Maker Shed. https://www.makershed.com/. Accessed 14 Mar 2019
45. Vectary. https://www.vectary.com/features/. Accessed 14 Mar 2019
46. OnShape. https://www.onshape.com/products/education. Accessed 14 Mar 2019
47. Lay's busca de nuevo sabor del millón de dólares. https://www.hostelvending.com/noticias/noticias.php?n=5647. Accessed 14 Mar 2019
48. La Marque Du Consommateur. https://lamarqueduconsommateur.com/. Accessed 14 Mar 2019
49. Manchester City FC. https://www.mancity.com/news/club-news/club-news/2016/june/new-man-city-website. Accessed 14 Mar 2019
50. Lego Ideas. https://ideas.lego.com. Accessed 14 Mar 2019
51. MADE. https://www.made.com/. Accessed 14 Mar 2019
52. Netflix. https://www.netflix.com/. Accessed 14 Mar 2019
53. Amazon. https://www.amazon.es/. Accessed 14 Mar 2019

Sustainability Assessment in the Implementation Phase of a Retail Space

Natalia Muñoz López(✉), José Luis Santolaya Saénz, Anna Biedermann, and Javier Molina Sánchez-Migallón

Department of Design and Manufacturing Engineering, Universidad de Zaragoza, María Luna 3, 50018 Saragossa, Spain
nataliam@unizar.es

Abstract. Design for sustainability has progressively evolved from the product level, usually referred as product Life Cycle Design or Ecodesign, towards a more complex approach, in which Product-Service Systems (PSS) are considered, and from an analysis only focused on the environmental impact to the simultaneous study of the three dimensions of sustainability: environmental, economic and social. In this work, the sustainability assessment of a PSS is addressed. A small clothing store is projected and all products and activities needed to create the retail space in which commodity is delivered, stored, displaced and tried out by the buyers are analyzed. Sustainability assessment is focused on the implementation phase of the service life cycle. A set of appropriate indicators are used to quantitatively evaluate each sustainability dimension. The greenhouse gas emissions indicator is used to assess the environmental impact, the budget of material execution is selected to evaluate the economic dimension and the working time associated to the category of workers, which is the most affected stakeholder group, is used to value the social dimension.

Keywords: Sustainable design · Product-Service System · Sustainability indicators

1 Introduction

Initiatives to adopt more sustainable approaches in industrial and economic activities have been evolving throughout the last decades. Design methods that do not only focus on the improvement of the environmental impacts associated to products, as Ecodesign [1] are required by the companies. Several authors have proposed providing utility to consumers through the use of services rather than products as a strategy of dematerialization and a possible answer to the sustainability challenge [2]. Since every product offers some kind of service and every service is based on the use of one or several support products, services and products are usually treated as parts of systems that are often referred to as Product Service Systems (PSS) [3]. Tukker and Tischner [4] define PSS as "a mix of tangible products and intangible services designed and combined so that they are jointly capable of fulfilling final customer needs".

In this perspective, it is necessary to develop criteria, methods and tools to support the PSS design process. Although PSS have been recognized as a sustainable solution,

F. Cavas-Martínez et al. (Eds.): INGEGRAF 2019, LNME, pp. 31–39, 2020.
https://doi.org/10.1007/978-3-030-41200-5_4

its effects have not been systematically assessed. In fact, the literature regarding PSS sustainability assessment is still in the early phase and it requires criteria to define whether a PSS is sustainable [5]. In order to ensure PSS sustainability through its life cycle, sustainability assessment is a critical question that needs to be carefully addressed.

On the other hand, a rigorous sustainability assessment requires that three simultaneous dimensions (environmental, economic and social) be considered. This model to address the sustainability is known like Triple Bottom Line (TBL) [6]. The environmental dimension refers to the management of the planet resources, the economic dimension to the control and distribution of benefits and the social dimension to the well-being of living creatures.

The investigations of different authors [7, 8] to evaluate the three sustainability dimensions along the entire life cycle of a product, allowed developing the Life Cycle Sustainability Assessment (LCSA) framework (Fig. 1). It combines three techniques of life cycle assessment: E-LCA, LCC y S-LCA. E-LCA (Environmental Life Cycle Assessment), also called LCA, is focused on environmental aspects. LCC (Life Cycle Costing) assess economic issues and S-LCA (Social Life Cycle Assessment) studies the impacts on organizations and social agents. Thus, three simultaneous life cycles are analyzed, which can be expressed as: LCSA = LCA + LCC + S-LCA. In addition, a set of criteria and indicators are used to evaluate each dimension.

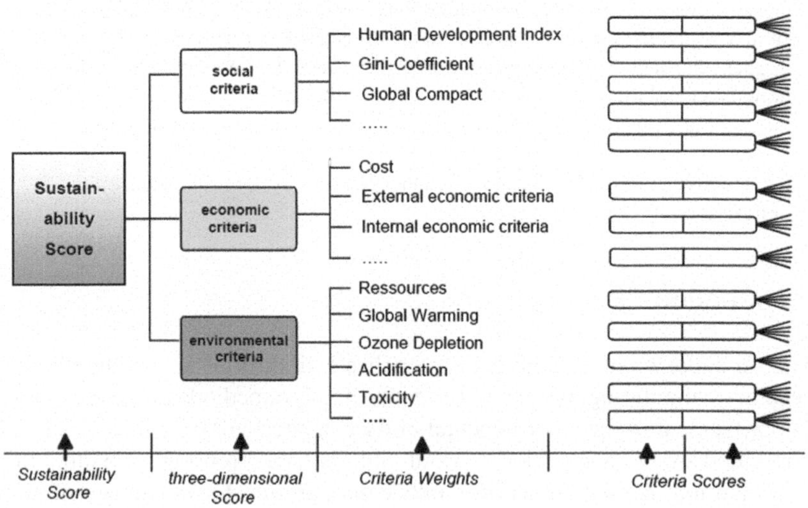

Fig. 1. LCSA diagram addressing three dimensions of sustainability [8].

LCSA methodology is already being applied in product design [9] but it is hardly applied to analyze the sustainability performance in service delivery and Product-Service Systems. In this work, the implementation phase of a clothing retail service is analyzed and its sustainability is assessed using a set of appropriate indicators.

2 Methodology

In this work, LCSA approach has been applied using the same structure of four phases considered in a LCA [10]: goal and scope definition, inventory analysis, impact assessment and interpretation.

First, goal and scope of the study are defined, the functional unit is identified and the system boundaries are established. Next, inventory analysis is carried out through data collection and calculation procedures. Inputs and outputs related with the functional unit, as well as the resources used in the transformation processes are identified and quantified. Environmental, economic and social impacts are obtained in the following phase using the indicators shown in Table 1. The indicators defined are considered particularly useful to develop the suggested methodology in two ways: they can be used to study very different systems and provide quantitative information. The Generator of Construction Prices database of CYPE Engineers software of Spain in 2018 version was used to calculate indicators selected because it enables the evaluation of environmental and socio-economics aspects [11]. Finally, the interpretation phase combines the results of inventory analysis with impact assessment to obtain the required conclusions and recommendations for decision making process. It allows determining the system impact spots and thus, which of them should be improved.

Table 1. Sustainability indicators used to asses PSS sustainability.

Environmental indicators	
GWP_{100} (Kg CO_2-eq)	Global Warming Potential. Total emissions of the greenhouse gases calculating the radiative forcing over a time horizon of 100 years
GE (MJ)	Global Energy. Energy consumption considering electricity as well as net calorific value of resources used
Economic indicators	
BME (€)	Budget of Material Execution. It expresses the total cost to develop a project
EE (€/KgCO_2-eq)	Eco-efficiency. Ratio between BME and GWP_{100}. It allows comparing easily economic and environmental impacts of individual or global activities
Social indicators (workers category)	
Wt (h)	Working time. Average time required by the workers to develop an activity
Lc (€)	Labour cost. Average cost associated to the workers labour

3 Case Study

A clothing retail service is studied. Three stages are considered in its life cycle: implementation or creation, operation or delivery and dismantling or end-of-life. On the other hand, the stages of product, construction, use and end-of-life constitute the life cycle of the clothing retail store, in which the service will be developed. Both, service and product life cycles are shown in Fig. 2.

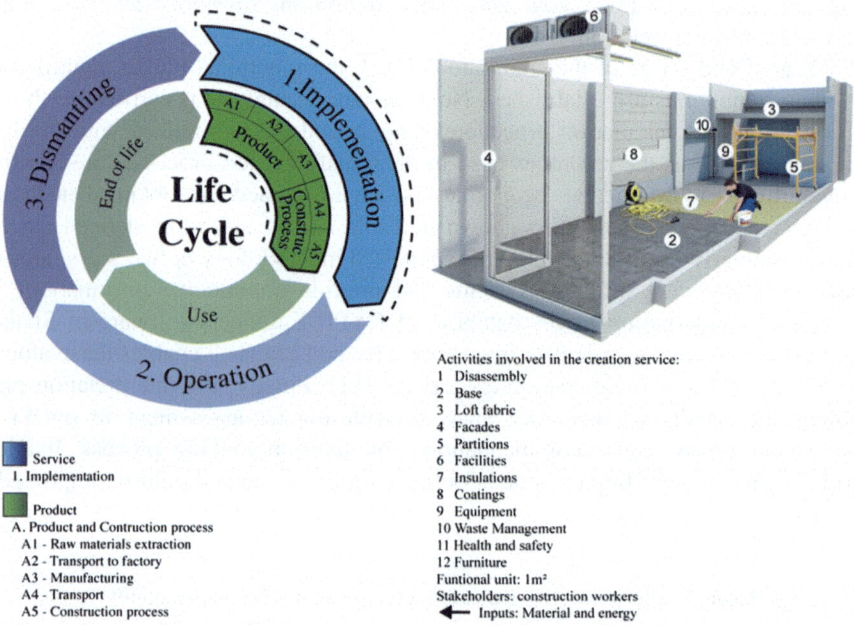

Fig. 2. Life cycle stages of service and product built to develop the service.

3.1 Goal and Scope

The service implementation stage is established as system boundaries. The quantification of environmental, economic and social impacts, taking into account all activities and materials required to create the service, is the objective of this work. As is shown in Fig. 2, two stages in the life cycle of the clothing retail store: product and construction process, are involved in the service creation process. Phases from A1 to A5 can be differentiated within these stages. Obtaining a well-conditioned commercial space of 65 m^2, which will be located in the center of a big city, is the functional unit.

3.2 Inventory

All activities required to create the clothing retail service are shown in Table 2. A total of 12 groups of activities are considered, from 1. Disassembly to 12. Furniture. The

Table 2. Data of the activities required in the creation of the clothing retail service.

Activities	Measure	Units	GE/unit		Workers (A5)			Cost/unit
			A1-2-3 (MJ)	A4-5 (MJ)	N° (units)	Wt/unit (h)	Lc/unit (€)	A1-5 (€)
1. Disassembly								
1. 1. Disassembly of swivelling garage gate	1,00	pc.	0,00	2,12	3,00	1,15	19,48	19,87
2. Base								
2.1. Mass concrete base	65,29	m^2	130,19	7,78	3,00	0,08	1,20	4,74
3. Loft fabric								
3. 1. Reinforced concrete fabric	9,43	m^2	553,63	11,31	6,00	1,49	26,54	50,44
4. Facades								
4.1. Galvanized steel fixed carpentry to window	1,00	pc.	5.581,18	53,48	2,00	0,38	6,60	643,58
4.2. Fixed metal grid to ventilation	1,75	m^2	21,97	0,13	2,00	0,61	10,61	114,45
4.3. Zinc plated steel automatic rolling shutter	1,00	pc.	3.097,24	30,16	6,00	5,46	94,06	1.641,06
4.4. Tempered glass door enclosure	1,00	pc.	15,13	0,28	2,00	2,12	34,38	269,80
4.5. Laminated safety glass	9,23	m^2	589,53	10,35	2,00	0,96	15,57	58,89
5. Partitions								
5.1. Partitioning	33,29	m^2	268,17	21,95	4,00	1,93	33,01	43,31
5.2. Bling folding interior door	1,00	pc.	158,45	4,77	2,00	2,00	34,42	187,31
5.3. Braced self-supporting plasterboard	69,59	m^2	126,51	4,68	2,00	0,48	7,46	22,92
6. Facilities								
6.1. General pipe of distribution SHW	1,10	m	9,45	0,09	2,00	0,24	4,15	15,20
6.2. Air handling unit in false ceilings	1,00	pc.	112,91	1,37	2,00	7,94	137,09	4.084,21
6.3. Plumbing interior installation to toilet	1,00	pc.	157,06	1,82	2,00	5,40	162,29	265,38
6.4. Powder extinguishers	1,00	pc.	754,44	3,45	1,00	0,10	1,62	44,32
6.5. Fixed and movable luminaires	31,00	pc.	127,07	0,35	2,00	0,82	14,15	143,54
7. Insulations								
7.1. Thermoacoustic insulation under the floor	51,74	m^2	33,97	1,38	2,00	0,28	4,36	9,89
8. Coatings								
8.1. Tiling	10,16	m^2	208,06	8,40	2,00	0,70	11,89	23,64
8.2. Plaster in vertical surfaces	33,85	m^2	61,06	0,86	2,00	0,51	8,73	10,78
8.3. Pavement base to ramp	1,20	m^2	363,28	6,89	2,00	0,50	8,43	20,01
8.4. Laminated pavement of HDF board	58,61	m^2	44,10	3,22	2,00	0,18	2,83	24,17

(*continued*)

Table 2. (*continued*)

Activities	Measure	Units	GE/unit		Workers (A5)			Cost/unit
			A1-2-3 (MJ)	A4-5 (MJ)	N° (units)	Wt/unit (h)	Lc/unit (€)	A1-5 (€)
8.5. Skirting of MDF board	29,30	m^2	1,66	0,17	1,00	0,09	1,37	8,33
8.6. Application manual of paint	236,51	m^2	9,17	0,18	2,00	0,34	5,68	10,80
8.7. Cement rendering in vertical surfaces	10,16	m^2	39,89	1,38	2,00	1,01	17,07	18,73
8.8. Suspended continuos false ceilings	36,80	m^2	113,96	4,73	2,00	0,45	7,14	22,58
9. Equipment								
9.1. Sanitary porcelain toilet	1,00	pc.	1.020,04	12,48	1,00	1,26	22,84	250,97
9.2. Sanitary porcelain basin	1,00	pc.	743,99	8,85	1,00	1,31	23,75	485,13
10. Waste management								
10. 1. Transport of mix of waste without sort	20,00	m^3	0,00	55,13	0,00	0,00	0,00	3,32
11. Health and safety								
11. 1. Pair of gloves against mechanical risks	2,00	pc.	0,00	0,00	0,00	0,00	0,00	3,41
11. 2. Pair of shoes of security	2,00	pcs.	0,00	0,00	0,00	0,00	0,00	89,38
11. 3. Mobile scaffolding tower	1,00	pc.	0,00	0,00	0,00	0,00	0,00	251,58
11. 4. Portable lamp	1,00	pc.	0,00	0,01	1,00	0,11	1,78	5,49
11. 5. General electrical board	1,00	pc.	6,64	0,33	2,00	1,12	18,10	321,51
11. 6. Powder extinguishers	1,00	pc.	754,44	3,45	1,00	0,11	1,78	44,48
12. Furniture								
12. 1. Warehouse furniture	5,00	pc.	79,08	4,77	2,00	1,06	18,03	175,73
12. 2. Sales space furniture	10,00	pc.	706,67	4,77	2,00	1,06	18,03	292,48

following data are indicated for each activity: measurement and units, energy consumption for the A1-3 and A4-5 phases, workers number required in the A5 phase (construction process), working time and labour cost of the workers in this phase and finally, the total cost of each activity. All data are expressed per unit of measurement.

3.3 Impact Assessment

Indicators defined in Table 1 are used to assess the sustainability performance. Results obtained for the activities involved in the service creation are summarized in Table 3. Global values are also indicated. A total of 7.298,02 $kgCO_2$-eq and 81.195,11 MJ are obtained for the GWP_{100} and GE environmental indicators respectively. In the economic dimension, the global BME is 27.979,40 € and the global EE is 3,83 €/$kgCO_2$-eq. In addition, 396,55 h and 6.018,47 € are, respectively, obtained for the W_t and L_c social indicators.

Table 3. Indicators obtained for each sustainability dimension.

Construction and assembly activities	Environmental		Economic		Social	
	GWP_{100} (kg CO_2-eq)	GE (MJ)	BME (€)	EE (€/kg CO_2-eq)	Tw (h)	Sw (€)
1. Dissasembly	0,17	2,12	19,87	119,70	1,15	19,48
2. Base	825,46	9.008,19	309,47	0,37	5,22	78,34
3. Loft fabric	496,98	5.327,47	475,65	0,96	14,05	250,27
4. Facades	1.047,95	14.353,20	3.298,28	3,15	20,77	297,30
5. Partitions	1.461,45	17.805,87	3.224,10	2,21	99,65	1.652,50
6. Facilities	1.338,24	9.475,65	8.860,37	6,62	74,74	744,20
7. Insulations	200,54	1.829,58	511,71	2,55	14,49	225,59
8. Coatings	1.317,60	14.567,37	5.865,31	4,45	145,40	2.412,00
9. Equipment	55,16	752,99	736,10	13,34	2,62	46,59
10. Waste manag.	11,00	275,68	66,40	6,04	0,00	0,00
11. Health and safety	34,47	263,34	808,64	23,46	2,46	21,70
12. Furniture	509,00	7.533,65	3.803,50	7,47	16,00	270,50
Total	7.298,02	81.195,11	27.979,40	3,83	396,55	6.018,47

3.4 Results Interpretation

In order to facilitate the analysis and interpretation of results, the percentage distribution of the impact generated by the activities involved in the service creation are shown in Fig. 3. GWP_{100}, BME and W_t indicators are analyzed. We observe that activities associated to the groups 5. Partitions, 6. Facilities and 8. Coatings have the most noteworthy impact in each sustainability dimension. These activities accumulate more than 60% of the GWP_{100} and BME indicators and more than 80% of the W_t indicator.

Additionally, those activities with the highest impact in each indicator are shown in detail. It is observed that the impact of specific activities denominated 5.1. Partitioning (13,5%); 6.5. Fixed and movable luminaries (18,3%) and 8.6. Application manual of paint (21,1%), is particularly elevated. Thus, activities that have the greatest effect on the sustainability of the service implementation stage are identified. These activities should be reviewed to achieve an effective improvement of the sustainability indicators. The sustainability assessment of the entire service life cycle as well as the development of strategies to achieve its improvement will be the object of future work.

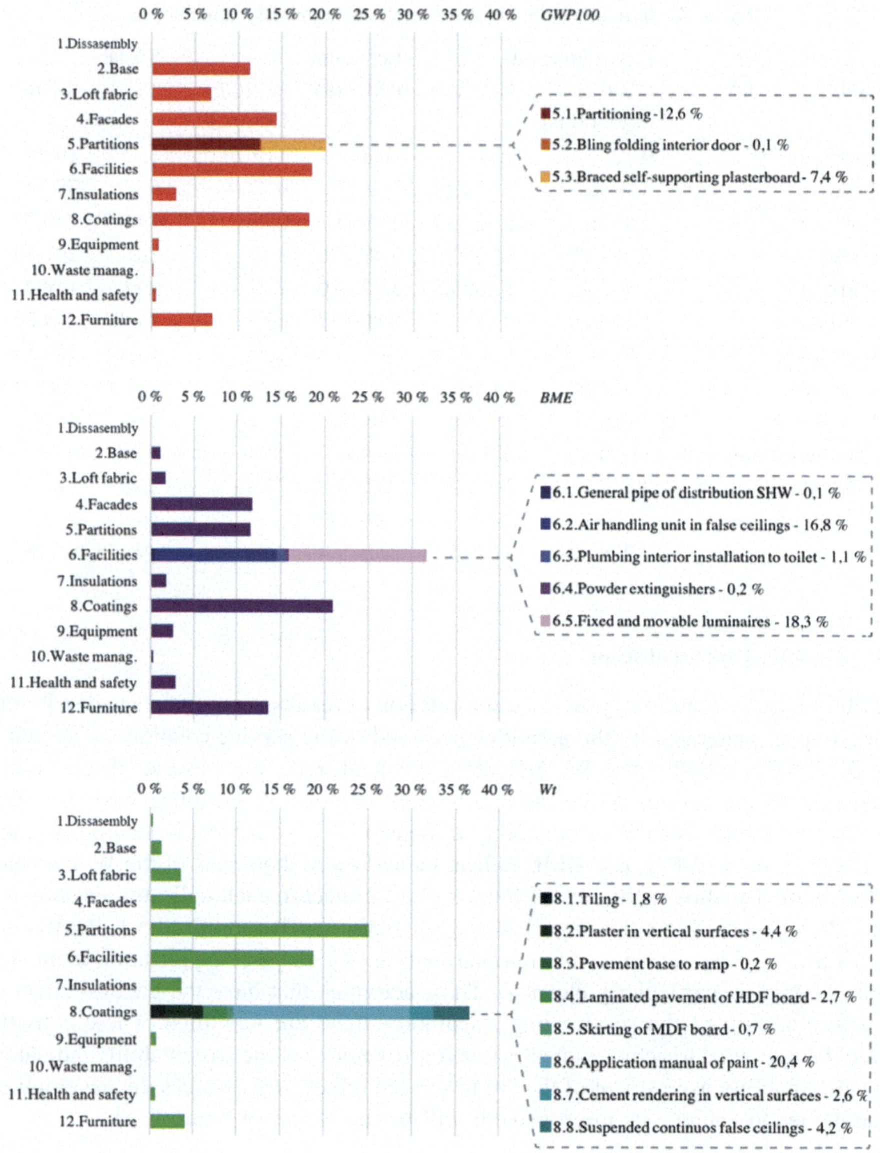

Fig. 3. Percentage distribution of the impact caused by the activities involved in the service creation.

4 Conclusions

In this work, the sustainability performance of a Product-Service System was addressed using LCSA approach. Links between life cycle stages of a clothing retail service and the clothing retail store in which service will be operated, were identified. The study was focused on the service implementation stage, which is coupled to the product and construction stages of the clothing retail store.

A set of appropriate indicators were selected to assess each sustainability dimension. Results showed that activities associated to Partitions, Facilities and Coatings accumulate the highest impact and should be conveniently reviewed to achieve an effective improvement of the PSS sustainability.

This method could be applied to assess the sustainability of other similar services in which a space is conveniently transformed in a store within the implementation stage. A wider list of indicators can be selected if a closer impact knowledge is required.

The activities and resources used during the implementation stage will be different depending of the sale product and the service provided, but is proposed the sequential development of stages which had been described in this work.

References

1. Brezet, J.C., van Hemel, C.G.: Ecodesign: a promising approach to sustainable production and consumption. UNEP, United Nations Publications, Paris (1997)
2. Mont, O.K.: Clarifying the concept of product-service system. J. Clean. Prod. **10**, 237–245 (2002)
3. Brezet, J.C., Bijma, A.S., Ehrenfeld, J., Silvester, S.: The design of eco-efficient services. Design for Sustainability Program. Delft University of Technology (2001)
4. Tukker, A., Tischner, U.: Product-services as a research field: past, present and future. Reflections from a decade of research. J. Clean. Prod. **14**(17), 1552–1556 (2006)
5. Sousa-Zomer, T.T., Cauchick, P.A.: Sustainable business models as an innovation strategy in the water sector: an empirical investigation of a sustainable product-service system. J. Clean. Prod. **171**, 119–129 (2018)
6. Székely, F., Knirsch, M.: Responsible leadership and corporate social responsibility: metrics for sustainable performance. Eur. Manag. J. **23**, 628–647 (2005)
7. Kloepffer, W.: Life cycle sustainability assessment of products (with comments by Helias A. Udo de Haes, p. 95). Int. J. Life Cycle Assess. **13**(2), 89–95 (2008)
8. Finkbeiner, M., Schau, E.M., Lehmann, A., Traverso, M.: Towards life cycle sustainability assessment. Sustainability **2**, 3309–3322 (2010)
9. Lacasa, E., Santolaya, J.L., Biedermann, A.: Obtaining sustainable production from the product design analysis. J. Clean. Prod. **139**, 706–716 (2016)
10. ISO: ISO 14040 International standard. In: Environmental Management - Life Cycle Assessment - Principles and Framework. International Organisation, Geneva (2006)
11. CYPE Ingenieros. http://www.cype.es

Comparative Analysis of Ecodesign in the Design and Manufacturing Methods for Mechanical Parts Made of Nylon PA6

José M. Paricio-Sánchez(✉), Ramón Miralbés-Buil, Juan A. Peña-Baquedano, and Alfonso Casas-Albiñana

Departamento de Ingeniería de Diseño y Fabricación, Escuela de Ingeniería y Arquitectura, Universidad de Zaragoza, Campus Río Ebro, C/María de Luna, 3, 50018 Saragossa, Spain
joparisa@unizar.es

Abstract. The life cycle analysis (LCA) is the tool that allows comparing products, processes or services in terms of the eco-design, that is, according to the environmental impact that said product, process or service has throughout all the phases of its life, both during its lifetime and after. In this study, three manufacturing techniques of mechanical parts in use in industry, such as machining, injection and additive manufacturing, are compared. This is made through the testing of a specific series of common mechanical pieces designed for that purpose, by studying through the tools of eco-design the different impact of each technique, and arriving at its quantification in terms of carbon footprint. From this obtained result, hypotheses are established about the modifications to the design that benefits each one of the techniques the most and allows a redesign of the most sustainable product.

Keywords: Eco-design · Life cycle analysis · LCA · Machining · Plastic injection · Additive manufacturing · 3D printer · CO_2 footprint · Design · Manufacturing

1 Introduction

The objective of this work is to make a comparison in terms of eco-design between three different manufacturing processes, by means of the Life Cycle Analysis (LCA) tool.

The three manufacturing techniques to be compared are machining, injection and additive manufacturing, and are used for mechanical parts in use in the industry. These three techniques have been developed at very different times, covering different needs and applied in their origin to different materials. Therefore, usually each process has been used with preference in certain designs, or with certain materials [1].

Due to this adaptation or preferential specialization of the processes to some designs or materials, it is not usual to find direct comparisons between the different methods. Since each manufacturing technique is applied to different types of parts, caution must be applied with the type of piece to be used. In order to make these

F. Cavas-Martínez et al. (Eds.): INGEGRAF 2019, LNME, pp. 40–48, 2020.
https://doi.org/10.1007/978-3-030-41200-5_5

comparisons, a possible approach is to apply the different processes to the same designs and materials, as far as possible, to identical parts.

The need to homogenize the result of each of the processes conditions the study approach. In the first place, it is necessary to choose materials that can be used interchangeably with the three manufacturing methods. In the second place, it is necessary to define designs that are suitable simultaneously for these methods [2].

Currently, the advance in techniques allows these processes to be applied to the same material provided that it is chosen appropriately, as well as that the part design is also adequate.

2 Materials and Methods

2.1 Material Selection

In order to select a material suitable for the work with each of the three desired processes, it is necessary to take into account the set of diverse materials with which each of them can work and choose one that is located at a common place.

Although the advance in the techniques allows a very wide margin for the selection, it is evident that certain materials lend themselves to some and not to the others. For example, wood is suitable for machining operations and not for injection. In the case of machining, the factor that limits our selection is the temperature reached in operations. The material must be able to withstand the temperature generated by chip removal. In the case of injection, in which plastics have been generally the selected material, for a use related to eco-design the use of thermoplastics is recommended as it allows the recycling of the supply conduits. Additive manufacturing requires a material that is susceptible for its use, resistant and that is to be supplied in the appropriate format.

By applying these criteria, the Nylon PA-6 material has been selected. This material is usable with the three manufacturing techniques and it allows to minimize the material as a source of discrepancies in the analysis. It is available in rods for machining, pellets for injection and filament for additive manufacturing.

2.2 Design Definition

As mentioned in the introduction, each design must be able to be manufactured without an, a priori, method enjoying a very specific advantage. But this is not entirely possible because each method is born by the appearance of concrete needs and is more suited to them. Any chosen design will have a small bias. What can be done is, first, to minimize the bias with a simple design that does not present difficulties and second, to test the variations produced in the results when using different designs. In the case of this work these two models of parts are selected, and named as conical part and L-part (Figs. 1 and 2).

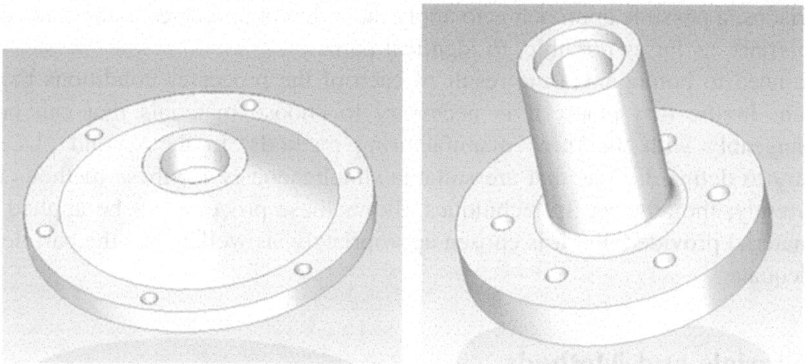

Fig. 1. Conical part (left) and L-part (right).

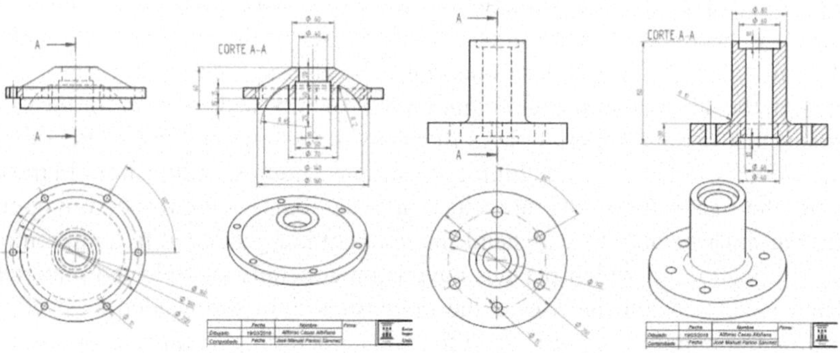

Fig. 2. Conical part (left) and L-part (right) drafts.

From these two different designs the functional units of the LCA are chosen, as the final parts made of the Nylon PA-6. These functional units of the analysis are chosen as an intermediate stage to the comparative analysis of the different methods, which is the final object of study.

2.3 Software

The simulation software is used to prepare the sheets of processes necessary to carry out each manufacturing process.

Table 1 lists the various software packages used.

Table 1. Software used: methods

Method	Software
Machining	Siemens NX 9.0
Plastic injection	Moldflow Adviser 2016
Additive manufacturing	Ultimaker Cura 3.3.1

Apart from the software for simulation, other programs are used to carry out the modeling of the parts and to carry out the life cycle analysis (Table 2).

Table 2. Software used: other uses

Method	Software
Parts design	Siemens NX 9.0
Life cycle analysis	Open LCA 1.7.0

2.4 Simulation: Machining

With the help of the selected software, Siemens NX 9.0, the sheets of processes necessary for the production of the two designs selected with the machining method are defined [3]. The input material selected for this process with the L-part is Nylon rod PA 6 dimensions 200 × 154 long, a usual presentation, commercial type, of this product. With the conical support, a 200 × 64 rod is used (Fig. 3).

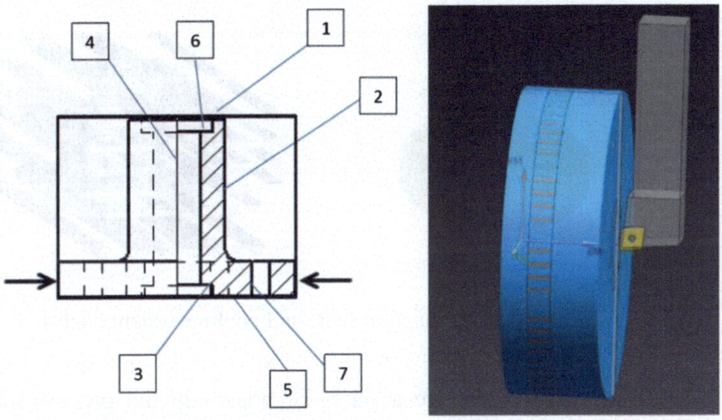

Fig. 3. Machining operations (left) and simulation (right)

The complete operation is also simulated and results tables are obtained that include: times, energy consumption, waste in the form of chips [4] (Table 3).

Table 3. Machining parameters.

Machining parameters	L-part	Conical part
Brute weight [Kg]	5.515	2.292
Part weight [Kg]	1.482	0.835
Waste [%]	73.1278	63.5689
Required time [seg]	382	839
Energy [kWh]	0.500408518	0.441288532
Energy [MJ]	1.801470665	1.588638713
Total consumption [MJ]	2.401960887	2.118184953

2.5 Simulation: Injection

Using the Moldflow Adviser 2016 software, the injection system is structured. For this process PA 6 is selected as material in the form of pellets, and the injection machine must be capable of producing an injection temperature of 250 °C and a mold temperature of 80 °C (Fig. 4).

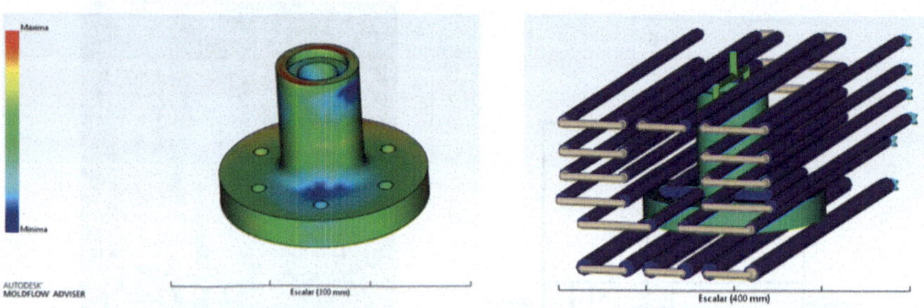

Fig. 4. Flow resistance diagram (left) and cooling system (right)

In the case of injection, there are a series of phases in the process that require energy supply, such as the pre-heating of the material, pressure, post-pressure, closing force and cooling for expulsion. All contributions are taken into consideration when performing the combined calculation [5] (Table 4).

Table 4. Injection parameters

Injection parameters	L-part	Conical part
Brute weight [Kg]	1.501309	0.842005
Part weight [Kg]	1.482	0.835
Waste [%]	1.286	0.8319
Required time [seg]	645.85	345.38
Energy [kWh]	1.1294	0.576991
Energy [MJ]	4.0659	0.88767846
Total consumption [MJ]	6.2553	3.19564246

2.6 Simulation: Additive Manufacturing FDM

Simulations are carried out with two simulation packages: Ultimaker S5 and Raise 3D n2 plus single. This is done in order to compare the results of both and verify the reliability and variations from one machine to another. When observing that the results are consistent, a weighted measure of them is used for the LCA [6] (Fig. 5 and Table 5).

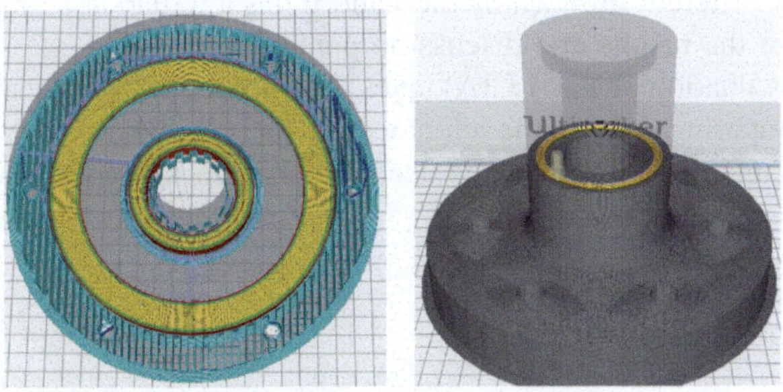

Fig. 5. Support (left) and layer generation (right)

Table 5. Additive manufacturing parameters.

Additive manufacturing parameters	L-part weighted	Conical part - weighted
Brute weight [Kg]	1.4955	0.890
Part weight [Kg]	1.482	0.835
Waste [%]	0.9027	6.1798
Required time [seg]	139650	90660
Energy [kWh]	18.43	11.974
Energy [MJ]	66.347	43.10

2.7 Life Cycle Assessment

The results of the previous phases are transferred to the OpenLCA program. The methodology of the ISO 14040 standard is applied, detailing the object, scope, function, functional unit, product system, always bearing in mind that the field of application must leave out everything that is foreign to the manufacturing processes. Not the pieces themselves, but the processes that lead to their manufacturing are evaluated. Since the result of the different processes is the same piece, we can discard any aspect that we would consider in a normal evaluation. As a result, several stages of the LCA, as defined by the standard, have to be minimized.

The goal and scope phase, as detailed in the introduction and materials and methods paragraphs, defines the functional unit as the two mechanical parts, what become an intermediate objective to compare the different manufacturing methods. The system boundaries are defined as the methods themselves, leaving to the software and databases used the determination of the external influence caused by the differences in the raw material inputs. The specific working parameters of the actual machines used, have an impact in the comparison, and the software is able to take into account this to a certain extent. But this is a limitation of the study. This study is not taking into account each and every possible machine used.

The inventory phase (LCI) is limited to take into account the raw materials and energy needed, as inflows, and the waste and emissions, as the outflows. The geometry of the parts, and the output of the different simulation software is used to calculate this. The waste is very different for each method and it has a remarkable influence, which is to be discussed in the results and discussion paragraph.

The impact evaluation phase (LCIA) takes the output of the software and selects the relevant data, comparing them to the references. This is shown in results and discussion paragraphs. The interpretation phase is accomplished in a limited way within results and discussion paragraph also.

3 Results and Discussion

For the application of the LCA, the software OpenLCA 1.7.0 version is used with the CML (baseline) evaluation method [v4.4, January 2015]. The evaluation with this program shows 11 impact categories. The category on which attention is focused is Climate Change (CO2 emissions). This category is chosen because of its special relevance and because there is good correlation between the other categories and this one.

Database "elcd_3_2_greendelta_v2_17" is used, which is appropriate for the scope of this work.

The results of the calculation of the inflows and outflows are those indicated in Table 6.

Table 6. CO_2 emissions.

Method	L-part - kgCO_2	Conical part - CO_2
Machining	51.1175	21.3786
Injection	14.5878	8.14403
Additive man.	21.7488	13.37723

By performing a critical analysis of these results we can see that they correspond to the data found in the literature [7].

The machining method generates the greatest emissions of the three, and that is in part caused because it generates the biggest proportion of waste. If it is taken into account the recycling of the waste, it is possible to reduce the emissions (Table 7).

Table 7. Recycling parameters. Machining.

Parameter	L-part	Conical part
Weight (kg)	1.482	0.385
Brute weight (kg)	5.515	2.292
Waste (kg)	4.033	1.457
Recycling weight (kg)	3.456	1.248
Recycling iterations (n)	7	7

The recycling of waste without losing its properties can only be achieved a limited number of times [8, 9]. The number of iterations is used to calculate the proportion of the waste that can be reintegrated to the loop.

The new results for the outflow of the machining with the recycling are shown in Table 8.

Table 8. CO_2 emissions comparative. Machining.

Method	L-part - $kgCO_2$	Conical part - CO_2
Not recycling (kg)	51.1175	21.3786
Recycling (kg)	19.9162	10.1065

The reduction in emissions is significant and brings the results near the level of the additive manufacturing. The recycling of the waste for the other two methods has less impact because of the less proportion of waste present in them, and it is not evaluated.

It can be reached greater gains from the redesign of the parts, applying specific design optimizations to each method [10]. For the injection, a preliminary study of a new design of the parts with similar mechanical properties but better thermal behavior, allows a reduction of heating and cooling which translates in 30% less emissions. For the additive manufacturing, the new design only shows a reduction of around 15%, but adding a change in the filling proportion to 60%, increased the reduction to around 27%. A line of research for the future is to improve the designs and compare the results when they are optimized for each manufacturing process.

For the models of parts studied, and for the material nylon PA 6, the less emissive of CO2 manufacturing process is the injection.

4 Conclusions

In this work the comparative analysis of eco-design of three manufacturing methods has been carried out. Materials and designs that were feasible in any of the three have been selected. Its manufacturing has been simulated by software packages of proven validity that have provided the necessary results to be able to establish a comparison through a specialized package in LCA. In carrying out the analysis, it has been taken into account that the material is highly recyclable, although the waste that is obtained (chips, burrs, supports) are of very different type for each of the methods, which leads to diverse results.

Aspects that can be incorporated in other studies have not been taken into account in this analysis. For example, the emissions necessary for the manufacture of injection machines, molds, machine tools or 3D printers. Also, aspects of costs or runs have not been taken into account. Given that there is disparagement in manufacturing times by one method or another, there is an obvious economic impact if long runs are to be carried out through additive manufacturing.

The results show that the manufacture, by means of the injection of plastics for these geometric designs and this material, -per unit of product-. It is the manufacturing method, not taking into account the mold, which is the most eco-sustainable taking into account the CO2 emissions.

References

1. Morbidoni, A., Favi, C., Germani, M.: CAD-Integrated LCA tool: comparison with dedicated LCA software and guidelines for the improvement. In: Hesselbach, J., Herrmann, C. (eds.) Glocalized Solutions for Sustainability in Manufacturing. Springer, Heidelberg (2011)
2. Kellens, K., Dewulf, W., Overcash, M., et al.: Int. J. Life Cycle Assess. **17**, 69 (2012). https://doi.org/10.1007/s11367-011-0340-4
3. Schudeleit, T., Züst, S., Wegener, K.: Methods for evaluation of energy efficiency of machine tools. Energy **93**(Part 2), 1964–1970 (2015). https://doi.org/10.1016/j.energy.2015.10.074. ISSN 0360-5442
4. Dietmair, A., Verl, A.: Energy consumption forecasting and optimisation for tool machines. MM Sci. J. **62**, 62–67 (2009)
5. Telenko, C., Seepersad, C.: A comparison of the energy efficiency of selective laser sintering and injection molding of nylon parts. Rapid Prototyp. J. **18**(6), 472–481 (2012). https://doi.org/10.1108/13552541211272018
6. Karania, R., Kazmer, D.: Low volume plastics manufacturing strategies. ASME J. Mech. Des. **129**(12), 1225–1233 (2007). https://doi.org/10.1115/1.2790978
7. Tang, Y., Yang, S., Zhao, Y.F.: Sustainable design for additive manufacturing through functionality integration and part consolidation. In: Muthu, S., Savalani, M. (eds.) Handbook of Sustainability in Additive Manufacturing. Environmental Footprints and Eco-design of Products and Processes. Springer, Singapore (2016). https://doi.org/10.1007/978-981-10-0549-7_6
8. Camañes, V., Elduque, D., Javierre, C., Fernández, Á.: The influence of different recycling scenarios on the mechanical design of an LED weatherproof light fitting. Mater. (Basel) **7**(8), 5769–5788 (2014). https://doi.org/10.3390/ma7085769
9. Lozano-González, M.J., Rodriguez-Hernandez, M.T., Gonzalez-De Los Santos, E.A., Villalpando-Olmos, J.: Physical-mechanical properties and morphological study on nylon-6 recycling by injection molding. J. Appl. Polym. Sci. (2000). https://doi.org/10.1002/(SICI)1097-4628(20000509)76:6%3C851::AID-APP11%3E3.0.CO;2-D
10. Cicconi, P., Landi, D., Germani, M.: An ecodesign approach for the lightweight engineering of cast iron parts. Int. J. Adv. Manuf. Technol. **99**, 2365 (2018). https://doi.org/10.1007/s00170-018-2649-7

Co-creative Experiences for the Achievement of the Sustainable Development Goals

Jorge Sierra-Pérez(✉) and Ignacio López-Forniés

Department of Design and Manufacturing Engineering, University of Zaragoza, Calle María de Luna 3, 50018 Saragossa, Spain
jsierra@unizar.es

Abstract. The Sustainable Development Goals (SDGs) have been established by the United Nations to end poverty, protect the planet and ensure that all people enjoy peace and prosperity. The challenges to address in the achievement of these objectives, due to their complex and abstract nature, will require the search for specific solutions to local problems in a framework of collaboration. This collaboration will have to take place between different fields of knowledge, social strata or countries. For this reason, co-creation methodologies have been identified treated in a generic way, as a group of promising tools in the achievement of the SDGs, since they are successfully applied in complex and multidisciplinary work environments. This study reviews and analyses various cases of how co-creation methodologies have addressed the SDGs so far; among them, the study has focused on Design Thinking, Living Labs, Citizen Participation and Collective Intelligence. The results of each case are presented and its common applications are discussed for all the SDGs or as a discrimination of methodologies based on the selected objective.

Keywords: Co-creation · Sustainable development · Creative methodologies · Eco-design

1 Introduction

The United Nations (UN) adopted "The 2030 Agenda for Sustainable Development" including 17 Sustainable Development Goals (SDGs) and their 169 targets, being a set of priorities and aspirations to guide to the most urgent challenges in the world. These needs include ending poverty and hunger; protecting the planet from degradation and addressing climate change; ensuring that all people can enjoy prosperous, healthy and fulfilling lives; and fostering peaceful, just and inclusive societies free from fear and violence [1].

All countries are responsible about their implementation and success through their own sustainable development policies, plans and programs. The Sustainable Development Goals (SDGs) will be a compass for aligning countries' plans with their global commitments.

One key distinction between the SDGs and their predecessor, the Millennium Development Goals (MDGs), is that SDGs require the involvement of all stakeholders: the governments of rich and poor countries, the private sector, civil society

F. Cavas-Martínez et al. (Eds.): INGEGRAF 2019, LNME, pp. 49–57, 2020.
https://doi.org/10.1007/978-3-030-41200-5_6

organizations (CSOs), and, perhaps most importantly, people [2]. The challenges within the SDGs are set in an environment characterized by complexity and multi-disciplinary. On the one hand, high number of interlinkages within the system and with systems of higher orders [3]. On the other hand, multi-stakeholder partnerships have been identified as an important strategy that seek to mobilize all stakeholders around the new agenda, within a range of sectors including business, sub-national and local governments, youth, civil society, philanthropic organizations and development banks. Leading successful projects in multi-stakeholder settings requires a broad skill set in the area of dialogue and collaboration [4].

In addition, complexity is not only affected by each of the challenges and objectives, but an interconnection between the SDGs that increase this complexity and mix of disciplines is observed. Complexity is seen as a great impediment; however, it may be the principle of opportunity. Because when complexity is high, the capacity to adapt, evolve, coordinate, innovate, and change is also high [5]. Collaboration is the strategy to reach a solution to the goals of each objective. The collaboration has to take place vertically, between different social strata and with different functions as indicated by the quadruple helix. And horizontally, where there are multiple relationships between goals in varied areas of knowledge. Solutions must take place integrating knowledge.

All the co-creation methodologies described in this work are characterized by being participatory, generating knowledge, with collective creation and focused on the user or beneficiary.

2 Materials and Methods

2.1 Objectives

This study aims to review some relevant co-creation methodologies and how they have approached the achievement of the SDGs by means of the analysis of various case studies.

2.2 Co-creation Methodologies. Selection and Analysis of Case Studies

The co-creation methodologies analysed in this work are living labs (LL), design thinking (DT), collective intelligence (CI) and public participation (PP).

By means of the study of several cases carried out with the LL, DT, CI and PP methodologies, a comparative table is made in which the following factors that affect the co-creation processes are analysed: objectives, team composition, communication, distribution (people and information), design approach, problem nature, area of application [6]. The selection of the information and case sources is based in governmental organizations web pages like United Nations (UN) or European Network of Living Labs (ENoLL), among others, and then to go deeper into sources that detail more each case. The selection of case studies has been made based on the relevance of the results obtained, a geographically representative selection, as well as the relevance in the scope of study of the participating entities. The selected cases are not comparable in objectives, territorial, economic and social dimension, so it is not possible to find indicators or a metric to compare them quantitatively. But the present study makes a qualitative description of each one.

Living Lab (LL) is an open innovation ecosystem based on a systematic user co-creation approach. They are integrated by public and private entities, developing research and innovation activities in the community, placing citizens at the center of innovation with the help of various approaches, instruments, methods, and tools [7].

The concept of Living Labs is being developed constantly. There exist different types, yet all Living Labs are learning configurations, which refer to a 'quadruple helix' innovation process [8]. This refers to Living Labs as a gathering of stakeholders from public institutions (at the level of cities, regions & local, regional, national & European policy), private organizations (start-ups, SMEs, corporations), as well as academia (researchers, universities, research organizations) and citizens. This leads to the inclusion of representatives from each sector in innovation processes, creating results from which all involved stakeholders can benefit [9].

Design thinking (DT) refers to the cognitive, strategic and practical processes by which design concepts are developed by designers or design teams [10]. Designers take charge of the process of co-designing and co-creating solutions to problems, whether complex or simple problems. On the other hand they are responsible for showing management the processes that allow them to be more productive with a different approach by applying the strategy of the organization.

Design thinking uses abductee thinking, the abductive thought, allows after an observation, to describe a hypothesis as an approach to a possible open and divergent solution as the beginning of the experimentation towards the final solution. Design and designers will form multidisciplinary teams to work abductively in the co-creation of more meaningful and prosperous solutions to current wicked problems [11].

Collective intelligence (CI) is the term to describe the ability of large groups of individuals doing things collectively that seem intelligent by applying knowledge to adapt to a changing environment, being the result more than the sum of their parts [12]. It is linked to consensus decision-making and group behavior in harmony that prioritizes collaboration in which everyone wins against competition [13]. Initiatives as Wikipedia and Linux using collective intelligence to develop new solutions through online communities show that collaborative online community could do the same for sustainable development [14].

The basic insight of collective intelligence is that there will always be value in mobilizing latent sources of intelligence outside the system, rather than relying only on what's inside, or on the contribution of consultants. The general idea is associated with open innovation, allowing outsiders to approach the group and establish mutually beneficial relationships [15].

In order to reach collective intelligence, ICTs are used to reach more people and faster [16] and get the necessary collective consciousness for collaboration, content creation, participation in decision making process, more complex and multi-media applications created, new approaches for dealing with privacy and other issues of governance. This manifestation of CI can be seen as an extension of the Open Innovation, Crowdsourcing, Wisdom of Crowds, Wikinomics, and Service Dominant Logic [17].

Public participation (PP) Public participation encompasses a group of procedures designed to consult, involve, and inform the public to allow those affected by a decision to have an input into that decision [18]. They are tools that allow the transfer of information and opinion between the citizens and the holders of the projects, which facilitate their communication. It is a participatory, inclusive and open process, fostering tolerance and respect for others. Normally it is applied in projects that are linked to a territory, in which a sense of belonging is generated by the neighbors, as if they themselves had designed it, a form of co-creation. Cases in vogue is the participation of people of all types in citizen science, from schools to senior centers. Citizen science refers to the participation of the public in scientific research activities in which citizens actively contribute, either with their intellectual effort or with knowledge of their environment or by contributing their own tools and resources. The citizens make experiments.

Social collaboration refers to processes that help multiple people or groups interact and share information to achieve common goals. In this regard, one of targets of SDGs (SDG16.7) refers to "Ensure responsive, inclusive, participatory and representative decision-making at all levels", that fits perfectly with this co-creation methodology.

Table 1. Comparison between co-creation methodologies

	Complexity	Collaborative/Cooperative	Resources	Participating entities
LL	Medium	Collaborative	Low. Only people and spaces, occasionally economic and materials	Public + private + citizen
DT	Medium	Collaborative	Low. Only people and spaces	Public + private
IC	High management and low in participation	Collaborative in management cooperative in participation	High in programming and economic needs. Low. Only people and spaces	Public
PP	Low	Cooperative	Low. Only people	Public

Table 1 shows the comparison between the four methodologies. The complexity of the environment is compared, if a cooperative or collaborative work is proposed, to the necessary resources and to the type of participating entity. The complexity of the environment is considered high when the number of participants in the Project is high (more than 15) and diverse (more than 5 disciplines), medium when the number of participants or disciplines is reduced, and low when the number of participants is low (less than 6) and not very diverse (less than 3 disciplines) [19]. It is considered a cooperative work if the objectives are group and individual and to achieve them several actions are developed independently; and collaborative thinking when the objectives are common and group and a single action is developed among all participants. The resources can be personal, space, material and economic, giving various combinations. Finally public or private entities or associations can lead the projects in each methodology between both.

3 Results and Discussion

Table 2 comprises all selected case studies for co-creation methodologies under study. In general, the majority of case studies are public initiatives, so the participants in teams are civil servants, as teachers or politicians. Due to the lack of knowledge in co-creation methodologies, some of these initiatives include a public-private collaboration, generating a relevant interrelation in both directions towards the SDGs. Moreover, all of these co-creation methodologies include the participation of citizens.

Regarding the field of action, the majority of studies has an environmental focus because it is established, and more deeply inserted, in the institutional agenda than social an economic vectors. The care of nature is the most common goal but with an important social burden, because they address the accessibility to quality water and healthy living environments. With respect social issues, the democratizing access to education and health focuses this dimension. Moreover, the reduction of inequalities, from improving the integration of rural areas to increase the safety of women.

In the case of Living Labs, its level of action is very focused on the urban level, due to policies that have promoted its creation. Therefore, its characteristics are very focused on the achievement of SDGs in this field, more specifically in the Circular Economy, with applications in city planning, sustainable mobility and waste management, among others.

In Design Thinking initiatives, there has been collaboration between private entities with a design profile for their introduction in public institutions, which has caused them to undergo a design process. It is necessary the application of protocols to replicate in other cases. In the majority of cases, a multidisciplinary action has characterized the process, facilitating the interrelation of knowledge among the different participating profiles. Also note the low complexity and level of abstraction of this experiences.

In the analyzed cases of Collective Intelligence, ICT is fundamental for the access to platforms open to the data to make comments and proposals. For that, combining collective human brain power and modern information technology has the potential to form just such an innovative channel for the sustainable development projects [14].

Public participation is the participation methodology most established and most accepted by citizens, but it is important to indicate the need for a facilitator to design the process to help obtain acceptable results. The majority of identified study cases have a great social burden, because the actions where citizen participation is developed are those that most affect the citizens, and they demand their involvement.

Table 2. Summary of selected case studies

	Case study	Objectives	Team composition	Ref.
Living lab	First solar school in Brazil	Improvement of school lighting and thermal comfort	Teachers, students and local community	[20]
	University of South Africa: open & distance learning	Transformation towards e-learning and the development of the associated information and communication infrastructure	Academia, business, government and citizenry	[20]
	Helsinki city plan and lighthouse projects	Apply environmental policies at urban level	Collaboration with residents, companies, city officials and other stakeholders	[7]
	Urban Wins. 8 European cities	Improvement of waste management	Different sources of fundings	[7]
Design thinking	SmartLife. Kenia	To ensure the accessibility to pure drinking water and wellness products	Designed by IDEO.org, Water & Sanitation for the Urban Poor (WSUP) and Unilever + GAIN	[21]
	Kundikunta lake restoration	To sensitize communities around the importance of preservation and restoration of the city's lakes	Kaava Design Studio, Government of Telangana, NGOs	[22]
	Family Justice Court	To reduce disputes and anxiety for child custody disputes	Singapore Mediation Centre, Ministry of Social and Family Development, Law Society	[23]
	H&M and textiles environment design	Provide sustainable education to design teams to generate ideas for new products and processes	The Head of The New Development Team and its team from H&M and textile design researchers	[24]
Collective intelligence	Open education information in Tanzania.	Improve school success avoid abandonment	National Examinations Council of Tanzania	[25]
	ACLÍMATE Colombia, Open Data to Improve Agricultural Resiliency	Help farmers to understand how to better navigate shifting weather patterns	International Center for Tropical Agriculture (CIAT with private-sector)	[26]
	The Open Public Services Network	Improve social services, achieving equal opportunities	Professional, civil servants and user	[27]
	Renewable Energy, Biomass Gasification	Provide electricity to rural areas to stimulate economic development	Asia Energy Group with the Earth Group & Green Planet Association	[28]

(*continued*)

Table 2. (*continued*)

	Case study	Objectives	Team composition	Ref.
Public participation	INVOLVE	Drive up standards, provide guidance and facilitate partnerships in health	Researchers, research commissioners and the public	[29]
	Harassmap	To engage all of Egyptian society to create an environment that does not tolerate sexual harassment	Tech partners, NGOs, politicians and volunteers	[30]
	Madrid Repara (MADRE)	Urban regeneration from mobility, energy, planning, environmental approach	Civil servants and citizens	[31]
	Water and Health in Romania	Ensuring participation of public in water and sanitation related matters in Romania	Water Company of Bucharest and citizens	[32]

4 Conclusions

The environmental aspect is more established and more deeply inserted in the institutional agenda than social an economic vectors. Also the accessibility to quality water and healthy living environments are significant acting fields. There are very heterogeneous examples, by thematic, as by the design of the co-creation process, which makes its comparison difficult. Despite having the right label, the co-creation process is not always followed. Exploratory research approaches like open observation or participative co-creation are rarely used. There is an opportunity for these processes to be given more frequently due to the complexity of the SDGs and the need for co-creation. Therefore, it is essential to understand the problem to be solved as a system, where the multiple parties involved see the structures, patterns and cycles, instead of seeing only one element. This will allow identifying solutions that address different areas, of the different participants involved, which also improve simultaneously within the system.

References

1. United Nations (UN) General Assembly: Transforming our world: the 2030 agenda for sustainable development. A/RES/70/1 (2015)
2. Sriskandarajah, D.: Toward an accountability revolution? Citizen participation and the SDGs. In: Kharas, H., Kato, H., McArthur, J., Desai, R. (eds.) From Summits to Solutions: Innovations in Implementing the Sustainable Development Goals. Brookings Institution Press, Washington, DC (2018)
3. Schick, A., Hobson, P.R., Ibisch, P.L.: Conservation and sustainable development in a VUCA world: the need for a systemic and ecosystem-based approach. Ecosyst Heal Sustain. **3**, e01267 (2017). https://doi.org/10.1002/ehs2.1267
4. Kuenkel, P., Schaefer, K.: Shifting the way we co-create. How we can turn the challenges of sustainability into opportunities (2013)

5. Kauffman, S.A.: At Home in the Universe: The Search for the Laws of Self-Organization and Complexity. Oxford University Press, New York (1995)
6. Ostergaard, K.J., Summers, J.D.: Development of a systematic classification and taxonomy of collaborative design activities. J. Eng. Des. **20**, 57–81 (2009)
7. European Commission: Cities as living labs – increasing the impact of investment in the circular economy for sustainable cities, Brussels (2017)
8. Ståhlbröst, A.: A living lab as a service: creating value for micro-enterprises through collaboration and innovation. Technol. Innov. Manag. Rev. (2018). https://doi.org/10.22215/timreview/744
9. Vaittinen, I., Vervoort, K.: Guidelines for living labs in CS (2017). http://eu-macs.eu/outputs/livinglabs/panelmanagement/
10. Brown, T.: Change by Design: How Design Thinking Transforms Organizations and Inspires Innovation, vol. 31. Harper Collins, New York (2009). https://doi.org/10.1017/cbo9781107415324.004
11. Watson, M.: Sustainable development goals through the design thinking lense. Design Project (2017)
12. Mulgan, G.: Collective intelligence and achieving the sustainable development goals (2018). https://www.nesta.org.uk/
13. Center for Collective Intelligence: Handbook of Collective Intelligence (2012). http://scripts.mit.edu/~cci/HCI/index.php?title=Main_Page
14. Musa, E.: Crowdsourcing social innovation, pp. 54–77 (2017). https://doi.org/10.4018/978-1-5225-0568-6.ch004
15. Bogers, M., Zobel, A.K., Afuah, A., Almirall, E., Brunswicker, S., Dahlander, L., et al.: The open innovation research landscape: established perspectives and emerging themes across different levels of analysis. Ind. Innov. (2017). https://doi.org/10.1080/13662716.2016.1240068
16. Jacovella, L., Lio, P.: Speeding up the transition to collective awareness. In: 2013 IEEE International Conference on Communications (ICC), pp. 220–224 (2013). https://doi.org/10.1109/iccw.2013.6649232
17. Wise, S., Paton, R.A., Gegenhuber, T.: Value co-creation through collective intelligence in the public sector. VINE **42**, 251–276 (2012). https://doi.org/10.1108/03055721211227273
18. Smith, L.G.: Impact Assessment and Sustainable Resource Management. Routledge, Harlow (1983)
19. Yesersky, P.T.: Concurrent engineering: your strategic weapon in today's jungle. C. Manag. Acc. Mag. Hamilton **67**, 24–27 (1993)
20. Filho, W., Andrade Guerra, J.B., Mifsud, M., Pretorius, R.: Universities as living labs for sustainable development: a global perspective **26** (2017)
21. Smart Life (2017). www.designkit.org, http://www.designkit.org/case-studies/4
22. Water Series (2018). https://wdo.org
23. Family Justice Courts. Family justice: strengthening the fundamentals (2017)
24. Earley, R., Reitan Andersen, K.: Design thinking for sustainability: a case study of a research project between Hennes & Mauritz and textiles environment design. In: Resilience – The New Research Frontier. ISDRS (2014)
25. McMurren, J., Verhulst, S., Young, A., Sangokoya, D.: Open Education Information in Tanzania (2016)
26. Ceschin, F.: Critical factors for implementing and diffusing sustainable product-service systems: insights from innovation studies and companies' experiences. J. Clean. Prod. **45**, 74–88 (2013). https://doi.org/10.1016/J.JCLEPRO.2012.05.034
27. Hogge, B.: The Open Public Services Network (2016)

28. Kirwan, C.G.: Smart Asia: a new platform for collective intelligence. In: Rau, P.L.P. (ed.) Cross-Cultural Design. Lecture Notes in Computer Science, vol. 8528, pp. 299–308. Springer, Cham (2014)
29. Involve (2019). https://www.invo.org.uk
30. Wael, R.: #FirstTimeIwasHarassed (2018)
31. Madrid Repara (2019). https://planmadre.madrid.es/
32. Regional Forum on Sustainable Development: Regional Forum on Sustainable Development. Cases Studies. Geneva (2018)

A Swift Clench-and-Release Device for Bicycles' Wheels with Disk Brakes

Giovanni Berselli[1], Matteo Bonomi[2], and Roberto Razzoli[1(✉)]

[1] Department of Mechanical, Energy, Management and Transportation Engineering, University of Genova, Via all'opera Pia 15, 16145 Genova, Italy
roberto.razzoli@unige.it
[2] Via Paverano 83, 19123 La Spezia, Italy

Abstract. The paper describes a novel Clench-and-Release Device (CRD), which can be used to quickly replace the rear wheels of bikes equipped with disk brakes, while also being suitable for front wheels. Since a quick wheel replacement is a pressing demand during competitions, the CRD reduces time wasted in situations like flat tire replacement. At the same time, it can be useful also for amateur bikers, who desire a simple way to lock/release a wheel without the risk of getting dirty or being unable to properly align the brake's disk and pads. In fact, thanks to the CRD, both brake and pinion packs are kept in place at all times and the chain remains engaged with the gears, even after wheel removal. Although similar devices can be found on the market, the proposed CRD provides an improved design since: *(i)* pinions and brake are placed on the same side with respect to the wheel symmetry plane perpendicular to the wheel axis of rotation; *(ii)* the front and the rear wheel share the same hub; *(iii)* manual operations are definitely easier; *(iv)* the wheels are compatible with different types of transmission; *(v)* the brake discs are more protected in case of crashing. A patent for the Italian region has been filed.

Keywords: Computer-aided design · User centered design · Wheel quick release · Through axle shaft · Wheels' fast removal

1 Introduction

As it may be self-evident within the bike industry, the design of the mechanical coupling between the wheels and the power transmission (i.e. chain/sprocket and braking systems) plays a crucial role. In particular, existing marketed solutions have been conceived so as to possibly increase the ease and, therefore, the speed, in wheels' replacement. These mechanical systems are mostly based on the separation of the hub from the power transmission rotors, so as to maintain chain, sprockets, and brakes in place at all times (even when the wheel is actually disconnected from the bike itself). In addition, modern designs aim at speeding-up the insertion of the pin that fixes the wheel hub.

Within this scenario, over the last few years, several patents have been filed, in order to propose wheel coupling systems with enhanced properties. These devices are often rather similar, although differences can be underlined for what concerns the

F. Cavas-Martínez et al. (Eds.): INGEGRAF 2019, LNME, pp. 58–67, 2020.
https://doi.org/10.1007/978-3-030-41200-5_7

connection of the wheel hub to the bike's body. Moreover, to date, no patented solution can be found on the market share dealing with those particular needs of racing gear; even high performing bikes, mostly equipped with disc brakes, employ traditional pin coupling for the wheel hub.

Owing to these considerations, the aim of the present discussion is to evaluate the feasibility of designing a fixing system to be used for bike's rear wheels equipped with a disc brake and up to a 13-speed transmission. Such a novel device shall be able to guarantee the necessary safety and stiffness provided by modern fastening systems, while improving rapidity and simplicity of wheel's replacement tasks.

In particular, note that the introduction of the disc brake, in bicycles, represented an important innovation as compared to the old "Caliper" technology [1]. It is indeed important to highlight the difficulty encountered by the end-users (e.g. the team mechanics) when they are asked to work, without protective gloves, in the proximity of steel discs that can reach very hot temperatures (especially when brakes are frequently used in challenging descents).

The design solution presented in this work, and hereafter referred to as Clench-and-Release Device (CRD), will provide a system in which the transmission of power, both during direct and inverse power flow (i.e. braking), is realized by means of a single shaft that combines safety of fastening with straightforwardness of wheel replacement.

In addition, note that the teams and the organization that assist the riders during cycling races and competitions, are required to promptly intervene when a wheel needs to be replaced. This action must be obviously performed as quickly as possible. At present, vehicles following the race must carry two types of spare wheels, i.e. front and rear, possibly complicating the team's tasks. Therefore, the proposed CRD provides a solution in which the wheels' hubs are identical, so that a wheel can be indifferently used on a bike's front or back, thus simplifying the replacement and limiting the number of spare parts to be carried around.

The rest of the paper is organized as follows: Sect. 2 describes the state-of-the-art concerning past patented solution; in Sect. 3, conceptual design alternatives are provided, whose embodiments are compared in Sect. 4. In Sect. 5, a physical prototype of the system is presented and the selected solution is critically evaluated. Section 6 finally draws the concluding remarks.

2 State of the Art

One of the first companies that dealt with the issue of speeding up wheel changing tasks, before the introduction of disc brakes in the bicycle industry, was Cinelli [2]. In the late 60 s, the company presented the idea of a hub that frees the sprocket set from the wheel hub, along with the very first prototype allowing identical wheels for the front and the rear. This project, despite the early technology, was based on a long pin, hereafter referred to as "Through Axle" (TA) system, actually smaller than modern ones, whose characteristics were comparable to a conventional quick release joint. Figure 1 (left and right) are published on the website of the Italian Cycling Federation [3] and they clearly depict the conceptual idea employed in Cinelli's prototype.

The correct position of the pin is guaranteed by a semi-cylindrical seat that forces frame and hub axles to coincide, while the pinion pack transmits the motion to the hub through a splined shaft, whose insertion may become tricky in critical conditions, like in competitions. Another patent dealing with the same subject has been filed by Angelo Morelli [4]. In this case, the rear and front hub share the same shape and dimension, so that wheels are fully interchangeable. In Fig. 2, a cross section of the Morelli's mechanism is shown. It exploits a TA system like the Cinelli's concept.

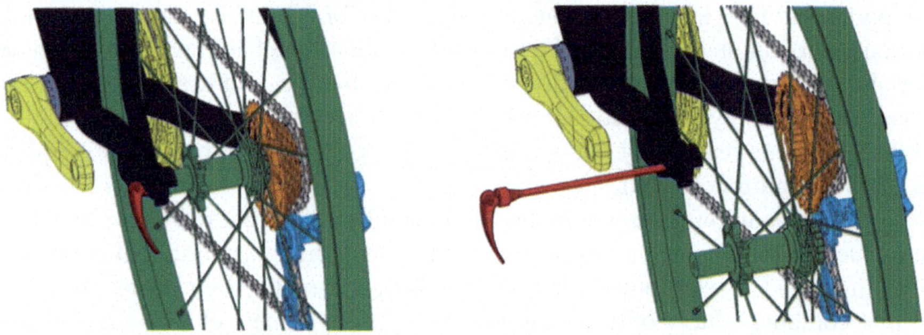

Fig. 1. The hub in the Cinelli's system [2].

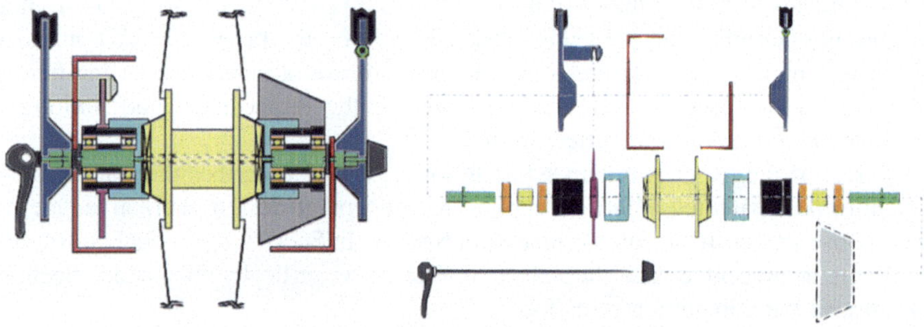

Fig. 2. Angelo Morelli's assembly (left) and exploded (right) sectioned view [3].

3 Comparison Among Possible Alternatives

The two cited patents can be summarized, in a very rough manner, as in Fig. 3. The upper part of the picture depicts both mechanisms in their locked configuration, whereas both the unlocked configurations, i.e. the ones allowing the hub removal, are shown in the bottom part of the picture. Note that, in both systems, a TA is required. For what concerns the Morelli device (on the right), the hollow shaft (2) hosts the pinions, whereas the hollow shaft (1) hosts the brake's disk. For Cinelli's design (on the left), the brake's disk is simply not present.

After an in-depth investigation of the past literature, here including patents [2–12], two different conceptual schemes are proposed hereafter, as shown in Fig. 4. In both designs, pinion and disk are held by the element (2), the only difference being either the presence (as in the left picture) or absence (as in the right picture) of the TA. Naturally, supposing that the same cross-sections are employed in both designs, the leftmost solution allows to provide a higher stiffness thanks to the presence of the TA (i.e. the deformation of the forks' free ends are constrained by the TA itself). In the rightmost solution, the joining action of a through pin is lost, so the forks must be reinforced so as to possibly grant a good flexural rigidity. Nonetheless, as a conceptual solution, either design may perform the required task, so that more detailed studies have been carried out to select the best one. In summary, the novelty as compared to the state of the art may be summarized as follows:

- Interchangeability between front and rear wheel, along with possibility of having equal wheels to be employed on different transmission groups and chassis;
- Chains, gears and brake's disk placed on a single side, i.e. on shaft (2) in Fig. 4, thus possibility to insert a single safety cover on the rotors (for protecting the athletes), which does not require to be disassembled during wheel replacement;
- Possibly, absence of TA (feasibility to be further explored), for what concerns the rightmost conceptual solution;
- Overall, increased ease and speed of the wheel replacement task.

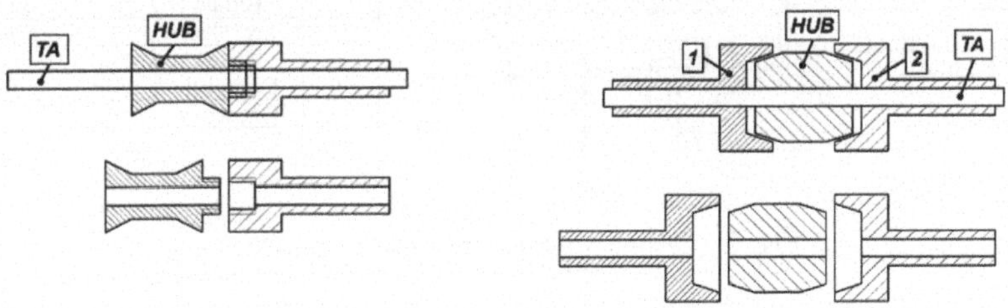

Fig. 3. Schematic of Cinelli (left) and Morelli (right) system. In both designs, the TA has to be removed in order to release the hub.

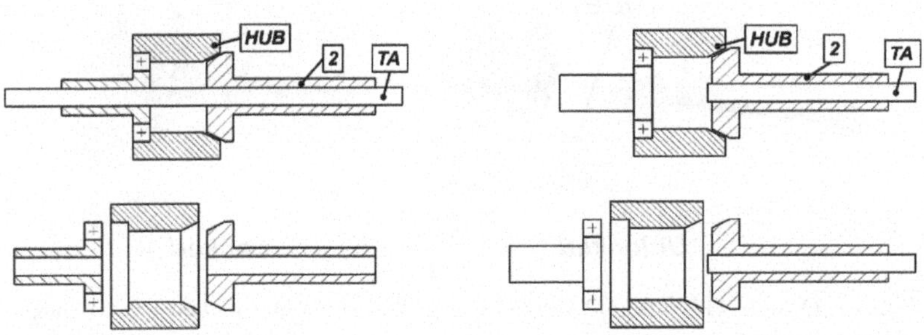

Fig. 4. Sketches of the authors' proposed alternatives; left: with a through axle, right: without through axle.

4 Discussion About the 3D Assemblies for the Two Options

The design of the conceptual alternatives in Fig. 4 has been carried out to evaluate advantages and drawbacks of each solution. Specifically, attention has been focused on the rear wheel, since it is more difficult to replace during a race as compared to the front one. With reference to Figs. 5 and 6, a common part of both alternatives is the transmission (primary) shaft (16), which needs to be modified as compared to components which are already on the market, the reason being that braking and power transmission are placed on the same (right) side. These solutions need the manufacturing of purposely designed hubs, which can be equal for both the rear and the front. As said, specific advantage is the possibility for the team to carry a single type of spare wheel for all cyclists; this is a very important feature because could allow team and service race's car to have on-board a single kind of wheels for all the cyclists. In addition, a relevant characteristic of the present design is that the new hubs can be used by different derailleur makers as only the primary shaft (16) dimensions have to be modified.

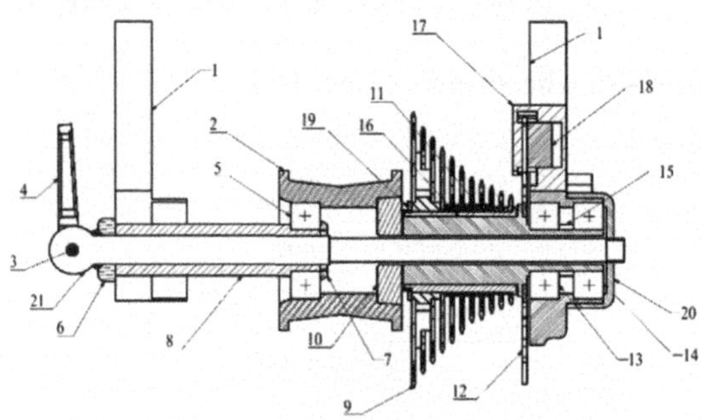

N.	Element	N.	Element
1	Frame	12	Brake disk
2	Hub	13	Bearing
3	Pin	14	Bearing
4	Lever	15	Shim
5	Bearing	16	Primary Shaft
6	Shim	17	Brake Caliper
7	Ferrule	18	Brake Piston
8	Sec. Shaft	19	Ferrule
9	Pinions	20	Threaded cap
10	Conical el.	21	TA
11	Free wheel		

Fig. 5. Design with TA, assembly X-section.

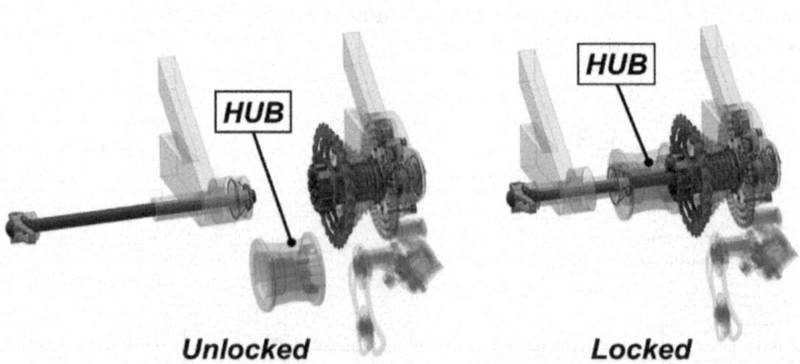

Fig. 6. Design with TA. 3D views in unlocked (left) and locked (right) states.

In Fig. 5 the structure of the mechanism is illustrated; it was built on a standard wheel, with rocket pinions, free wheel (11) and rolling bearings, but all these components being places on the right side, i.e. connected the primary shaft (16).

The transmission of power (also during braking) to the hub is granted by a toothed conical coupling (10), linked in the primary shaft through a form joint and a threaded pin for the sake of safety: the coupling cannot be realized as a single part along with the primary shaft, since such solution would not allow the placement of pinions in the correct position.

The left side of the mechanism is simpler as compared to the right one, this part being used to put in tension the entire system and for guaranteeing a good hub rotation. In practice, it consists of two parts only: a roller bearing (5), placed in the left inner side of the hub, thus allowing its rotation, and a secondary shaft (22), which acts as structural support. As for this architecture, it is necessary to employ a standard TA, whose function is to load the whole system in tension via a threaded connection placed on the frame (20).

Therefore, by having the TA placed on the right side, the user does not need to pay particular attention to the parts' correct placement, since the presence of the secondary shaft (with the bearings) basically forces the axially alignment of the entire system. In summary, the user will simply fix the axis by means of a screwed connection.

The second design alternative (Figs. 7 and 8) shares several similarities with the design proposed in Fig. 5, although, as previously reported in Sect. 3, a TA is not employed. Owing to the absence of the TA, this latter solution may possibly maximize assembly simplicity, thus further decreasing the time necessary for a manual wheel change. Seemingly, not having an axis which compresses the entire system, it is necessary to compute/optimize the mechanical pre-stresses so as to avoid excessive displacements of the frame ends, that may result in the hub getting off its site. Without providing numerical details, it can be stated that this solution can be effectively applied only to the rear wheel because the intrinsic characteristic of the front fork makes it impossible to offer a suitable stiffness. In any case, for the front fork, it is possible to employ the through pin design (see Fig. 5), which matches with a front hub with identical shape/dimensions as the rear one (thus allowing the use of undifferentiated wheels).

The two alternatives only differ on the way tension loads are applied to the hubs: in fact, for what concerns the second design characterized by the absence of the TA (see Fig. 7), it is necessary to provide a compression load by means of frame deformation (e.g. pre-compression of the whole system) and fixing mechanism, that need to be designed and built accordingly. As compared to the previous design solution, the secondary shaft (2) is different only in proximity of the frame; in fact, it is fixed through a sliding structure. The right side of the system is compressed by an axis (7), placed between the toothed conical coupling and the last roller bearing (9). As said, in case this concept would prove feasible, it would be possible to lock the hub without having the axis running inside. This architecture would allow for eliminating the TA and the time necessary for its centering during wheel change. Two seats of jaws (3) are built within the structure of the secondary shaft. These jaws allow an instantaneous lock or unlock of the entire system (by means of a mechanical or an electronic command), thus applying the required tension that is commonly generated by the pin traction in standard designs.

Concluding, the proposal of a second architecture is not a declaration of improvement with respect to the first one, although it seems more innovative with respect to the state of the art and more interesting in terms of ease of use and speed of operation.

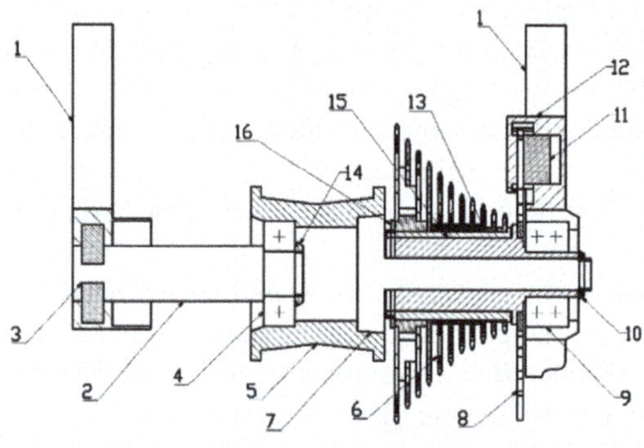

N.	Element	N.	Element
1	Frame	9	Bearing
2	Sec. Shaft	10	Seeger
3	Lock System	11	Brake Piston
4	Bearing	12	Brake Caliper
5	Hub	13	Free Wheel
6	Pinions	14	Seeger
7	Conical el.	15	Primary Shaft
8	Brake disk	16	Seeger

Fig. 7. Design without TA, assembly X-section.

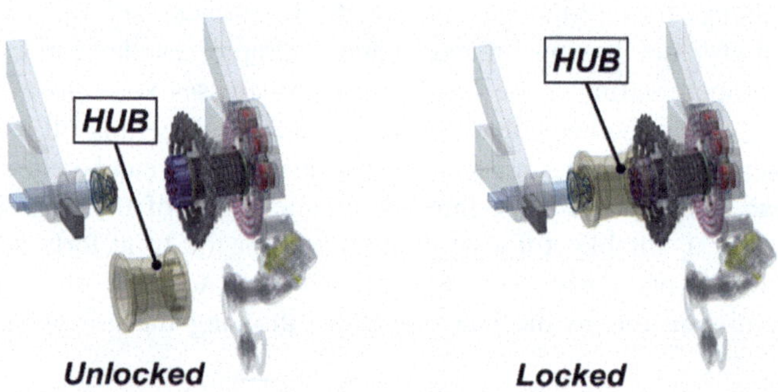

Fig. 8. Design without TA. 3D views in unlocked (left) and locked (right) states.

5 Strength and Weakness Points of the Selected Architecture

The proposed CRD can heavily reduce the time necessary for changing a wheel, since it makes it possible to avoid paying attention to the correct placement of the chain around chain-rings and pinions, and to the exact brake disc alignment between the brake pads. Notably, this procedure is neither difficult nor excessively time consuming for an amateur and for a cycle-tourist. Nonetheless, when dealing with a professional race, even a slight reduction of time wasted replacing a flat tire becomes crucial. In any case, regarding the cycle-tourism world, the CRD can allow the riders to change the wheel without touching transmission and brake, so it actually decreases the possibility of getting dirty and/or coming into contact with greasy fluids.

In order to adapt the CRD to standard frames, it is necessary to build specific hubs complying with the UCI rules about dimensions [13]. On the other hand, the adaption of the cycle's frame to the new position of brake and pinions, would allow to reduce the so-called Q-factor. With reference to Fig. 9, the Q-factor is defined as the distance between the external faces of cranks and it represents an important biomechanical factor: as proven in the past literature [14], since a low Q-factor allows the cyclist's legs to work with natural movements and in a more efficient way.

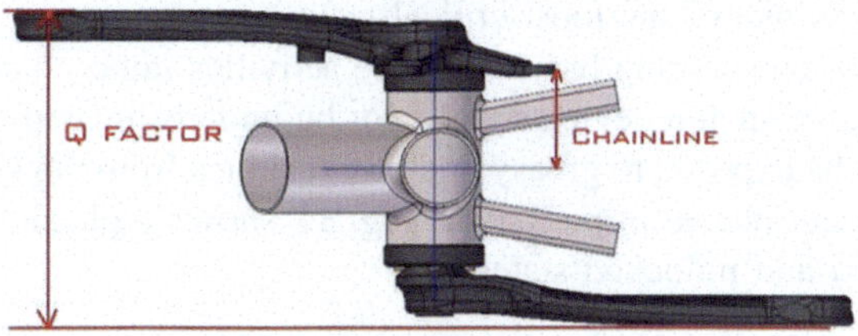

Fig. 9. The "Q factor" [15].

The main challenges, in order to produce a CRD design that would be effectively employed, are about the compatibility between brake system and transmission; in fact, the brake disc must be placed outside the derailleur working area, thus requiring a purposely designed transmission.

In Fig. 10, on the right side, the blue area identifies the normal working area of a standard derailleur: the depicts layout cannot grant a good shifting caused by a large distance from pinions, for such reason, it is mandatory to build a dedicated cage, operating in the red area.

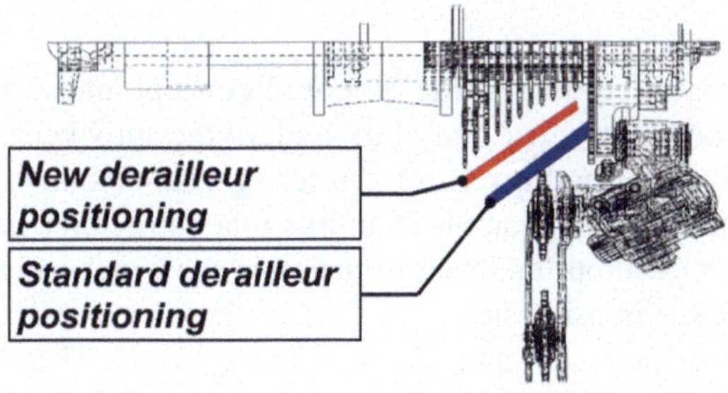

Fig. 10. Assembly and X-section with derailleur works area.

The frame design increases slightly the cycle's total weight. Nonetheless, this issue may be regarded as a minor drawback, since the weight limit for competition bikes has been set to 6.8 kg by the "Union Cycliste Internationale" (UCI) [13]. In practice, competition bikes are usually properly ballasted (still respecting the weight limit) to lower their center of mass. As for a cycle-tourist, weight increase surely represents a negligible disadvantage, whereas ease of wheel replacement may be seen as an advantageous feature, even in light of increased potentials for marketing.

6 Conclusion

In the present study, a novel Clench-and-Release Device (CRD) allowing to ease and speed-up bicycle's wheel replacement has been proposed. The system has been primarily conceived for usage in bike competitions, where a quick wheel substitution is always considered as one of the most critical tasks.

In addition, even considering bicycle leisure activities rather than the racing world, the potential impact of such a device should not be underestimated, since it allows the average user not to be exposed to greasy and oily residues from the transmission during the wheel replacement phase. In particular, Fig. 11 shows a physical prototype of the system in its locked and unlocked states.

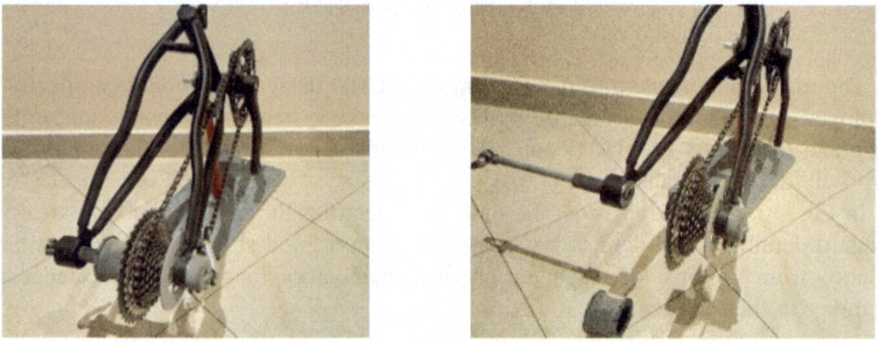

Fig. 11. Real prototype (with a TA): locked and unlocked state.

For what concerns safety issues, the proposed concept allows to place a cover for the exposed part of the brake's disc and, if desired, of the sprockets, so as to prevent the athletes, but also any other end-user, from suffering cuts or burns.

The CRD main innovation guarantees, unlike other structures presented up to now, the fixing of the wheel without the need to maintain the synchronism between the two areas involved in power transmission.

In fact, in terms of precision and stiffness, the presence of a double transmission shaft, as in the case of the Morelli's patent (see Fig. 1), is more accurate but involves difficulties in the assembly step, since it is not always possible to provide a proper alignment between both the grooves.

In conclusion, the prototype derived from the design with a through axle seems simple and efficient and it may represent a starting point for innovation in the cycling field. The authors are aware of the limits about frame dimensions imposed by the UCI rules for competitions (the interested reader may refer to [13, 15]), but they rely on the assumption that a CRD mass production will eventually allow for respecting these rules.

References

1. Orthwein, W.C.: Clutches and Brakes-Design and Selection, 2nd edn. Marcel Dekker Inc, New York (2004)
2. Derven, J.: Cinelli Bivalen Hubs. http://www.classicrendezvous.com/Italy/Cinelli/Cinelli_BiVal_B-Guide.htm. Accessed on 23 May 2019
3. Morelli, A.: Brevetto d'invenzione di Morelli Angelo, Ufficio Brevetti Giambrocono & Co., A28333/GE (2014)
4. Feng, B.-J.: Quick-release structure. J D Components Co Ltd., TW 20070109621 20070320 (2007)
5. Buder, J.: Quick release device of bike. Tokenproducts Inc., TW20160208571U 20160607 (2016)
6. Lin, M.-L.: Improved bike quick-release device. Ful Chee Ent Co Ltd., TW20100216086U 20100820 (2010)
7. Liu, W.-T., Jang, Jr.-Y.: Quick release device for bicycle wheel. Lee Chi Entpr Co Ltd., US201715611291 20170601 (2017)
8. Lin, J.-C.: Bicycle wheel axles having quick release structure, TW20100218659u 20100928 (2010)
9. D'Addario, L., D'Addario, S., Senezergues, S.: Bicycle with rear wheel quick mounting/dismounting mechanism, Mgop, WO2017EP80548 20171127 (2017)
10. Cheng, H.-L., Teng, C.-C.: Rear wheel quick release structure of bicycle. Motive Power Industry Co Ltd., JP20160094738 20160510 (2016)
11. Takachi, T.: Bicycle quick release structure. Shimano Kk, US20070680770 20070301 (2007)
12. Watarai, E.: Bicycle quick release adapter and bicycle fork using the same. Shimano KK, US20070676052 20070216 (2007)
13. Union Cycliste Internationale (UCI): Cycling sport constitution and regulations, F0319 (2019)
14. Disley, B.X., Li, F.X.: The effect of Q Factor on gross mechanical efficiency and muscular activation in cycling. Scand. J. Med. Sci. Sport **24**(1), 117–121 (2014)
15. Italian Cycling Federation. https://www.federciclismo.it/it/. Accessed 23 May 2019

Design of Urban Furniture for the Rehabilitation of Spaces

Silvia Doblas Expósito, Óscar D. de-Cózar-Macías(✉), Francisca J. Castillo-Rueda, and Patricia Mora-Segado

Dpto. de Expresión Gráfica, Diseño y Proyectos,
Escuela de Ingenierías Industriales, Universidad de Málaga,
c/Doctor Ortiz Ramos s/n, 29071 Málaga, Spain
odecozar@uma.es

Abstract. The work presented here is based on one of the strategic lines proposed by the Vice Chancellor of the Smart-Campus (VSC) of the University of Malaga (UMA), specifically within the line "Nature and environment" with the project called "Islands and Green Routes" (IGR). The main objective of the project is, through the work of a multidisciplinary team, to develop a space for leisure, recreation, rest, and outdoor study that is sensorized and sustainable, meeting the requirements of users and making these areas the focus of great social activity. The aim of the present work is to design urban furniture that will be located in the outer area of the School of Computing and the School of Telecommunication of the UMA. The methodology will follow the phases of the design and development of products, beginning with research on markets and competition, establishing user requirements regarding the product, development and discussion of the idea and, finally, the definitive proposal through a detailed technical design where the product is completely defined in terms of form, functionality, materials and mechanical strength. The results of the project reveal the development of a virtual prototype that meets the established specifications, obtaining 1st prize in the 2nd edition of the Islands and Green Routes contest.

Keywords: Sustainability · Urban furniture · Design

1 Introduction

The work presented here is included within the framework of the theme proposed by the Vice-chancellor of the Smart-Campus project of the University of Malaga within the second edition of the "Islands and Green Routes" contest. The purpose of the latter is to train multidisciplinary teams of students—with the support of teaching staff—to contribute ideas for adapting disused outdoor spaces of the various faculties and schools that could meet the needs of users. This is one of the actions included in the strategic line referred to as "Nature and environment" in the Smart-Campus Master Plan, and is included within the following six strategic lines: emissions, energy and water, nature and environment, health and wellbeing, mobility, research, and teaching and innovation and ICT. The intention is to develop the university campus into a Smart City, marking new lines of action at academic level that will allow the project to

F. Cavas-Martínez et al. (Eds.): INGEGRAF 2019, LNME, pp. 68–76, 2020.
https://doi.org/10.1007/978-3-030-41200-5_8

become an international reference in sustainability. The project is developed in the School of Computer Engineering and Telecommunications. This faculty has an external transit area located between the cafeteria, the parking lot, the library and the assembly hall, which is where the green island will be located, see Fig. 1.

Fig. 1. Area of operation.

The project involves a total of 17 people whose areas of expertise cover disciplines such as architecture, biology, geography, computing, telecommunications and industry. Meetings were held weekly that allowed for the development of the project by establishing new objectives and milestones.

The idea was to design a space that includes areas for rest, study, sports, leisure, and walking, combining vegetation, furniture, structures, and technology, with the provision of systematized solutions for each Green Island [1]. In addition, the space must be kept in alignment with the Boulevard Landscape Project during the construction process. The vegetation zone must be the heart and the main engine of the project, creating a magnificent visual impact within the area. A specific project developed by the industrial engineering department formed the proposal for the design of urban furniture for the projected area. Participants from other areas of expertise developed the urban planning of the area (architecture), the design of the vegetative space (biology) and finally, the development of the sensor and communications network in the external environment was developed by IT and telecommunications personnel.

The methodology followed was based on that provided by Pugh, [2] and included the following steps:

- Investigation of market needs and demands.
- Confirming Product design specifications.
- Generation and screening of ideas and definition of alternatives.
- Evaluation and selection of the solution.
- Conceptually defining the solution.
- Finalizing the technical design & prototype.

2 Market Research and Design Specifications

The objective was to obtain a rigorous perspective of the market [3–5], the needs and demands of the public [6], users [7–10], and those involved in technical maintenance. This analysis allows for having a complete view of the requirements that must be met by the product, as well as an initial approximation of ideas in reference to materials and mechanisms. A detailed study of the competition was carried out by analyzing such elements as forms, styles, dimensions, materials, assembly system, exterior/interior, fixed, or removable. Examples of the competition in outdoor street furniture are provided by the images shown in Figs. 2 and 3.

Fig. 2. Competition. Lorenz, modular seat.

Fig. 3. Competition. As you want, couple chair.

In addition to the market study, an analysis of the needs of the users was carried out through surveys [11] with a battery of questions that helped the system to direct and channel the product design towards the detected needs. The results thereof are shown in Fig. 4.

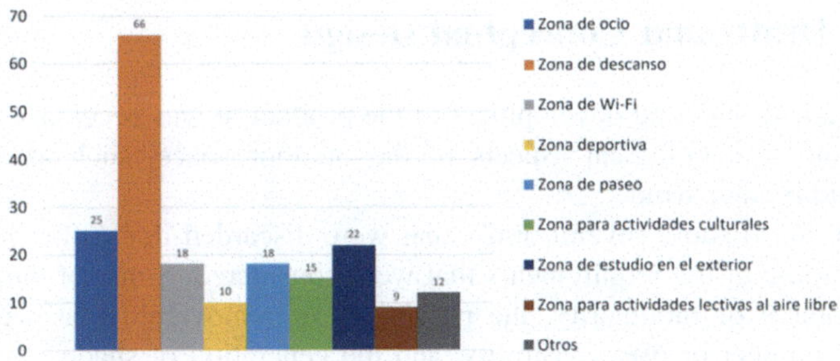

Fig. 4. Results of user surveys.

Among the conclusions reached, those worth noting include the tendency of users to gather in groups of four people, and there is also some tendency to form groups of five or six people, a number that will be taken into account for the final design. The vast majority of users carry electronic devices such as laptops and tablets, and thus it will be necessary to have charging points. A high percentage of users eats lunch at the school more than one day a week and would like to do so in outdoor areas. Another issue to note is the preference for a multifunctional table (for the use of devices, notes, books, and lunch). The responses to the open questions included indications for the need to install seats and tables, along with the provision of power outlets and WIFI coverage. With respect to sports and leisure, table tennis tables and a basketball hoop appeared to be the most requested facilities.

Another point worth noting is the ergonomic study of the design, taking into account human anthropometric measures [12] and access for users with functional diversity [10], predominantly wheelchair users.

Once the aforementioned studies had been conducted, it was necessary to list the set of specifications and requirements [8, 13, 14] to be met by the design of the product according to the Vice chancellor's office, the market, the users, and the specified regulations [15–18].

The resulting specifications were as follows:

- To provide ecological solutions that include self-sufficient energy, respect, integration of the existing vegetation, and reuse and recycling of materials or use of those with a lower environmental impact [19].
- Technological specifications included the integration of intelligent sensors for the control of temperature, humidity, and luminosity through an open source platform that is available through the university's mobile application [20, 21].
- Integration with the Boulevard landscape project.
- Active i.e. meeting the needs of users.

- Accessible. Can be navigated by people with functional diversity.
- Easy construction and maintenance, with cost-effective assembly.
- Resistant and durable; able to withstand exposure to inclement conditions.
- Resistant to vandalism, since it is an open space that is used by the public.

3 Rough Drafts and Conceptual Design

For the adequate design and development of the product, it was necessary to take in-to account various climatological aspects of the outdoor space, such as temperature, rainfall, humidity, and wind.

Figures 5 and 6 show several drafts that were discarded due to the lack of compliance with some of the requirements that were considered. Some of these sought to solve the problem of modularity, the possibility of removability, as well as multi-functionality, number of users, visibility, and the generation of shade.

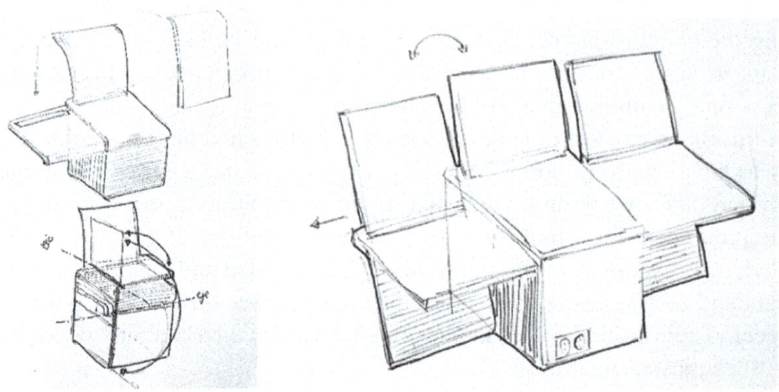

Fig. 5. First draft of suggested design.

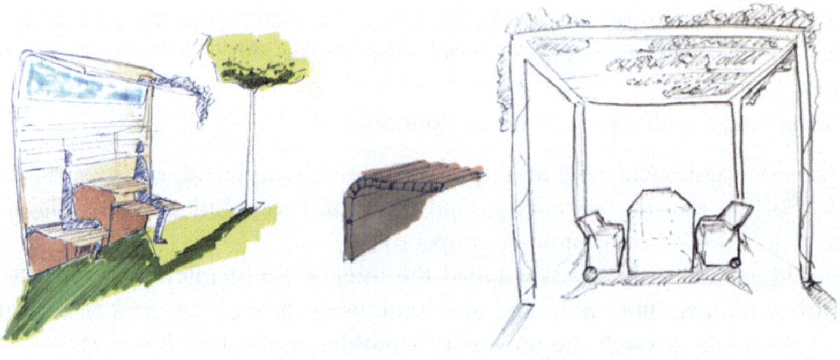

Fig. 6. Second draft of suggested design.

4 Detail Design

The final design (Fig. 7) allowed for the creation of a cubicle with space for 5 people along with space for one person with functional diversity (special needs). The table has three charging points, an outdoor LED light, an information beacon and a motion sensor that allows activation of the area. In addition, this meets the majority of the desired specifications and requirements.

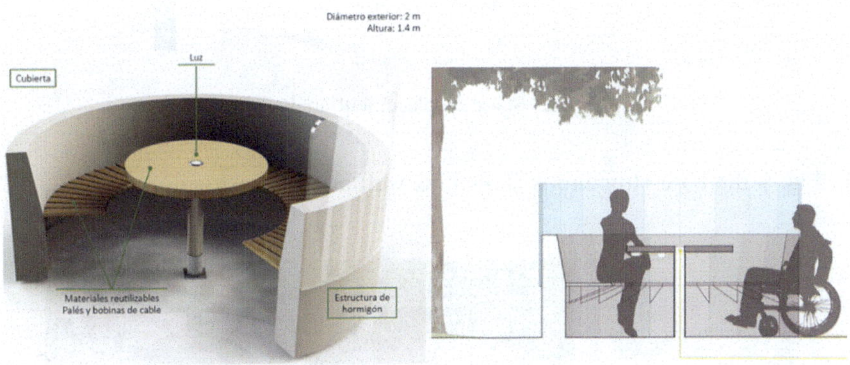

Fig. 7. Final design.

Regarding the materials used, the wood of reused pallets was used for the seats and the wood of cable reels was recycled for the table to which an external treatment was applied to prolong its durability and resistance. Concrete was one of the most used materials for the exterior component of the main structure and backrest. Cellular polycarbonate or methacrylate served as a wind protection element that makes it possible to visualize the level of occupancy of the area due to its high resistance to weathering and changes in temperature.

In reference to the electrical installation and protection of the users, it is necessary to provide, for the whole unit, a spike for adequate ground connection. Similarly, mechanical simulation was carried out using computer-aided engineering programs (SolidWorks) to check the mechanical strength of the furniture in its most stressed points (hotspots), such as seat and table (See Fig. 8), but without generating any problematic situation and achieving a factor of safety of 2.5 for standard stresses.

For the case of the seat, the study of stresses was developed according to different loads considering the actual materials used, that is, reused wood for the planks of the seat, steel brackets and slats, and the concrete structure. The maximum load hypothesis, taking into account that the average weight of a person is around 76 kg, was set at 200 kg, thereby achieving a safety factor of approx. 2.6.

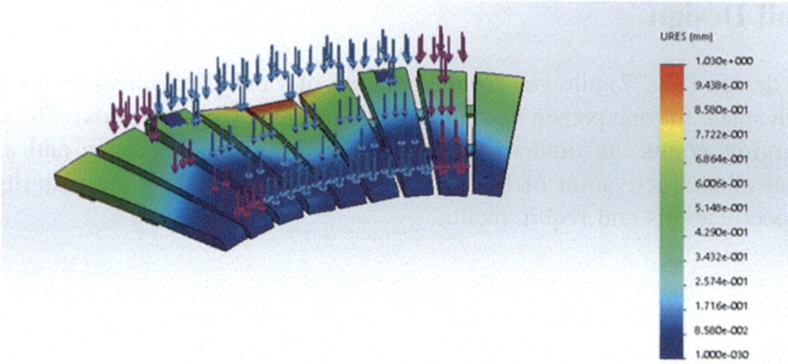

Fig. 8. Static displacement analysis.

In Fig. 9 the base structure is shown, as well as the construction process.

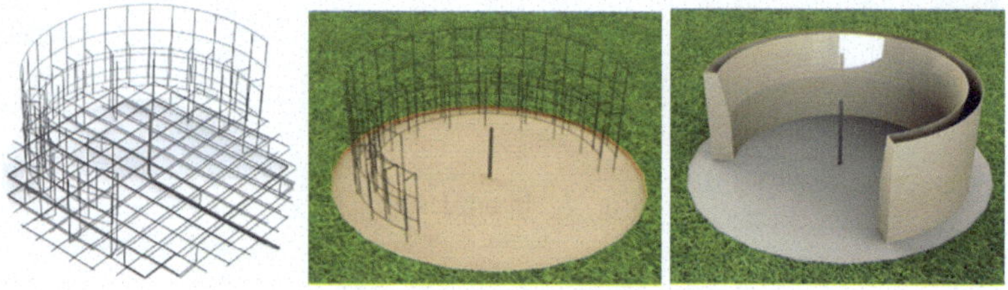

Fig. 9. Construction process.

Figure 10 shows the finished product and Fig. 11 displays the distribution of the cubicles.

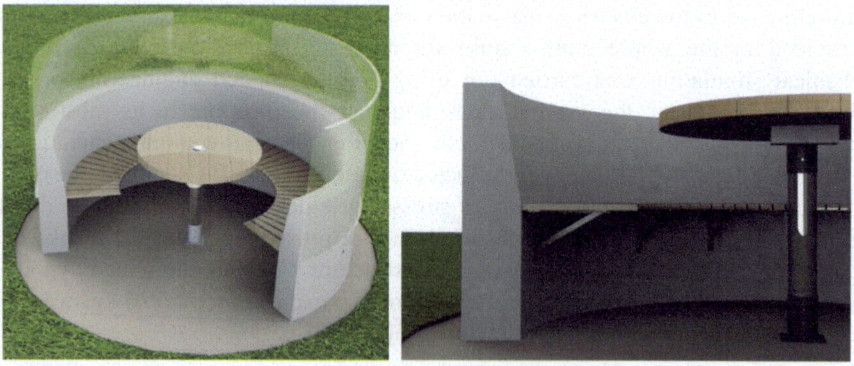

Fig. 10. Finished product.

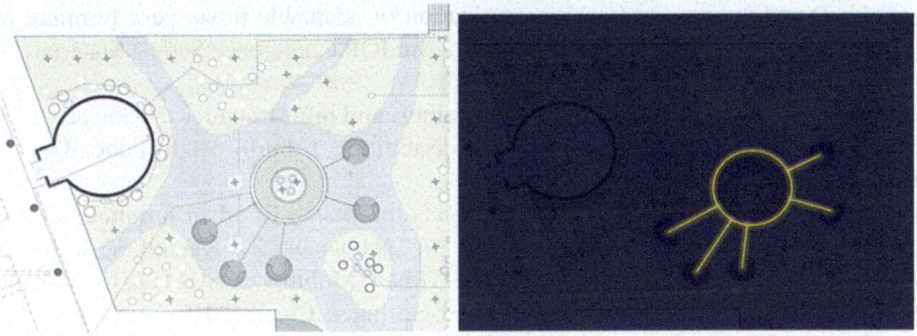

Fig. 11. Distribution of the cubicles around the school. Daytime and evening simulation.

5 Conclusions

This paper describes the process of design and development of a product, specifically urban furniture, as part of a multidisciplinary project developed by a team of students and faculty personnel. This work has resulted in a design that will be implemented as part of one of the strategic lines of the Master Plan put forward by the Vice Chancellor of the University of Málaga for its Smart-Campus project. This project has been awarded the first prize of the 2nd Islands and Green Routes contest, convened by the aforementioned Vice-Chancellor.

Thus, the island is both a technological and natural experience, where users can enjoy outdoor spaces using their electronic devices, transmitting a message of sustainability and responsibility for energy consumption through the information beacon.

References

1. de Paiva, R.B.F.: Phenomenology and emotional design: the conceptual synergy between architecture and design for urban furniture. Adv. Intell. Syst. Comput. **483**, 361–373 (2017)
2. Pugh, S.: Total Design: Integrated Methods for Successful Product Engineering. Addison-Wesley (1991)
3. Das, S., Rijas, M.P., Das, A.K.: DOT: design of a space-saving furniture with prototype-driven innovation approach. Smart Innov. Syst. Technol. **134**, 745–755 (2019)
4. Toso, T.M.: Lorenz asiento componible. http://www.metalco.it/prodotto/lorenz-2/?lang=es. Accessed 21 Apr 2019
5. Bae, S.: As You Want Couple Chair by Bae Se-hwa. https://mocoloco.com/as-you-want-couple-chair-by-bae-sehwa. Accessed 21 Apr 2019
6. Arruda, A., Moroni, I., Bezerra, P., Silva, P., de Paiva, R.B.F.: Practical urban: the urban furniture and its relationship with the city. Adv. Intell. Syst. Comput. **486**, 413–423 (2017)
7. Magdziak, M.: Flexibility and adaptability of the living space to the changing needs of residents. In: IOP Conference Series: Materials Science and Engineering, vol. 471 (2019)
8. Rinaldi, A., Caon, M., Khaled, O.A., Mugellini, E.: Designing urban smart furniture for facilitating migrants' integration: the co-design workshop as approach for supporting inclusive design. Adv. Intell. Syst. Comput. **824**, 461–470 (2019)

9. Thamrin, D., Mulyono, G.: Usability evaluation of adaptable urban park furniture product with cellular light-weight concrete as material. In: IOP Conference Series: Materials Science and Engineering, vol. 408 (2018)
10. Moriña, A., Morgado, B.: University surroundings and infrastructures that are accessible and inclusive for all: listening to students with disabilities. J. Furth. High. Educ. **42**(1), 13–23 (2018)
11. Şahin, S.H., Curaoğlu, F.: How big data affects the design of urban furniture: an approach from the perspective of industrial design. In: Nathanail, E., Karakikes, I. (eds.) Data Analytics: Paving the Way to Sustainable Urban Mobility. CSUM 2018. Advances in Intelligent Systems and Computing, vol. 879. Springer, Cham (2019)
12. Panero, J., Zelnik, M.: Las dimensiones humanas en los espacios interiores. Estándares antropométricos. G. Gili, Mexico (1989)
13. Pizzato, G.Z., de Macedo Guimarães, L.B.: Emotional attributes of urban furniture. Adv. Intell. Syst. Comput. **824**, 2087–2097 (2019)
14. Fusaro, G., D'Alessandro, F., Baldinelli, G., Kang, J.: Design of urban furniture to enhance the soundscape: a case study. Build. Acoust. **25**(1), 61–75 (2018)
15. UNE-EN 13198:2004: Productos prefabricados de hormigón. Mobiliario urbano y productos de jardín. AENOR, Madrid (2004)
16. UNE-EN 581-3:2017: Mobiliario de exterior. Asientos y mesas de uso doméstico, público y de camping. Parte 1: Requisitos generales de seguridad. AENOR, Madrid (2017)
17. UNE-EN 581-2:2016: Mobiliario de exterior. Asientos y mesas de uso doméstico, público y de camping. Parte 2: Requisitos mecánicos de seguridad y métodos de ensayo para asientos. AENOR, Madrid (2016)
18. UNE-EN 581-3:2017: Mobiliario de exterior. Asientos y mesas de uso doméstico, público y de camping. Parte 3: Requisitos de seguridad mecánica para mesas. Madrid: AENOR (2017)
19. Barcic, A.P., Oblak, L., Motik, D., Govedic, T.B., Kopljar, A., Vukovic, A.: Street furniture in Croatia and Slovenia - Park users requirements for (WOOD) urban equipment. In: Increasing the Use of Wood in the Global Bio-Economy - Proceedings of Scientific Papers, pp. 46–55 (2018)
20. Savio, L., et al.: Smart street furniture: innovation in the concept design process. In: Charytonowicz, J. (ed.) Advances in Human Factors, Sustainable Urban Planning and Infrastructure. AHFE 2017. Advances in Intelligent Systems and Computing, vol 600. Springer, Cham (2018)
21. Ciaramella, A., Bellintani, S., Savio, L., Carbonaro, C., Pagani, R., Pennacchio, R., Peretti, G., Thiebat, F.: Smart furniture and smart city. In: Conference Series: Materials Science and Engineering, vol. 408 (2018)

Application of New Systems for Positioning the Steering Wheel in Vehicles for the Improvement of Ergonomics in Autonomous Driving

Cristina Martin-Doñate[1(✉)], Antonio Gines-Alcaide[1], Jorge Manuel Mercado-Colmenero[1], Annalisa Di Roma[3], and Fermin Lucena-Muñoz[2]

[1] Department of Engineering Graphics Design and Projects, University of Jaen, Campus Las Lagunillas, s/n, 23071 Jaén, Spain
cdonate@ujaen.es

[2] Department of Management, Marketing and Sociology, University of Jaen, Jaén, Spain

[3] DICAR Department, Politecnico di Bari, Bari, Italy

Abstract. The current location of the steering wheel in the car makes accessibility to the driver's seat difficult for people with physical or mobility problems. Additionally, the external visibility of the steering wheel when the vehicle is parked could be a factor of lack of security in car theft. The automotive industry is currently facing the challenge of achieving completely autonomous driving, with this challenge being a new line of research. In light of these problems this paper presents a new concept that surpasses the state of the art by improving the user's experience in the decision of autonomous or guided driving of the vehicle. Similarly, the proposed system allows several functionalities such as extending the driver's room and improving comfort, especially for people with reduced mobility, or as an anti-theft system due to lack of visibility of the steering wheel making it difficult to steal the car.

Keywords: Industrial design · Ergonomics · CAD · Industry 4.0 · Entrepreneurship

1 Introduction

Technological development and improvements in the standard of living have made the car adapt to the new user requirements. Car bodies have undergone an aesthetic and functional revolution, becoming more personal and seeking to improve the user's experience. The customer looks for a new concept of car according to their style, rejecting designs related to high mass production. The shared use of the car by different drivers and with different ergonomic requirements causes time wastage originated by the car seat and steering wheel placement. People with reduced mobility or pregnant women require a greater mobility space inside the vehicle, with the steering wheel posing a problem when accessing the driver's seat. The automotive industry is

F. Cavas-Martínez et al. (Eds.): INGEGRAF 2019, LNME, pp. 77–85, 2020.
https://doi.org/10.1007/978-3-030-41200-5_9

currently facing the challenge of automating car driving. Autonomous driving presents multiple advantages [1]; fewer traffic collisions, due to elimination or minimization of human errors [2], smoother, more comfortable and less stressful rides [3], greater mobility freedom for the disabled, fatigued, drunk, inattentive, elderly and children [4], less demanding or unnecessary ownership for individuals and feasible automated event handling in the case of vehicle failure or incapacitated users. The adoption of self-driving cars into mobility services is in a position to not only transform the transport service sector but also the automotive industry. The use of renewable energies and sustainable materials poses a strong influence on design [5, 6] and therefore on autonomous driving [7]. Regarding the design and positioning of the steering wheel, some commercial brands have proposed some initial concepts mainly applicable to the use of the autonomous car. TRW [8] presents a design in which the steering wheel is embedded in the dashboard. Jaguar [9] presents a steering wheel concept separated from the steering axle, driving the vehicle like a tablet. BMW [10] has designed a small-sized steering wheel that can easily be installed on the dashboard, allowing the driver to drive independently and enabling the rotation of the co-driver seat. There are some patents applied to the field of design of steering wheels for autonomous driving [11, 12]. Unfortunately, none of them have managed to present a solution adapted to customers′ current requirements. This paper presents a new solution to the challenge of obtaining a system that improves the experience of the driver regarding his own decision to drive the vehicle autonomously or not, thus improving the driver's experience. It also improves access to the driver's compartment for people with physical disabilities by increasing the space available. Finally, the proposed mechanical design improves the safety of the car and hampers theft, since the steering wheel would only be incorporated into the axle with the original key.

2 Materials and Methods

In this section, the geometry of the new mechanical design is presented and described. The proposed mechanical design (see Fig. 1) consists of a system for positioning the mechanism of the steering wheel, improving the driver experience. This new device is composed of an automatic electric linear guide, a main mechanism formed by two linear electric actuators joined to the automatic linear guide, a coupling system, a mechanism composed of a set of elements for controlling and moving the steering wheel, a set of supports and joining elements and a set of newly designed elements such as a dashboard and a coupling element to the car steering axle.

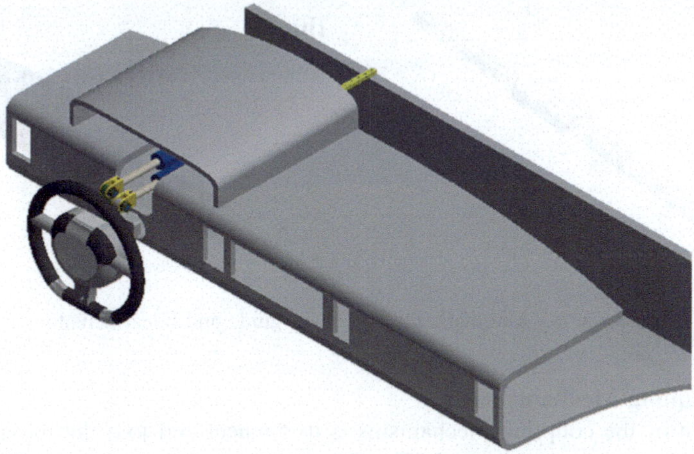

Fig. 1. Designed device positioned for manual driving

2.1 Mechanism for the Displacement and Positioning of the Steering Wheel

2.1.1 Electric Linear Actuators and Automatic Electric Linear Guides

The movement control system consists of an automatic electric linear guide Fig. 2a responsible for performing the translation movements of the mechanical device from the dashboard. Additionally, the system includes two electric linear actuators located parallel to each other with capacity to handle loads of up to 150 kN Fig. 2b. The electric linear actuators are attached to the automatic electric linear guide by a coupling mechanism that has been specially designed for this purpose. The movement control system of the steering wheel mechanism is activated by means of an electronic system that sends a start signal when the car keys are inserted into the dashboard, causing the adequate positioning of the steering wheel with respect to the driver. The return of the steering wheel to its resting position occurs when removing the car key. During the translation movement of the device, the mechanism of the steering wheel remains in a horizontal position. The electric linear actuators start to operate when the guide reaches its final position, generating a movement that allows the rotation of the steering wheel from a horizontal position (position of introduction of the steering wheel in the dashboard) to a vertical position (position for coupling the steering wheel to the steering axle).

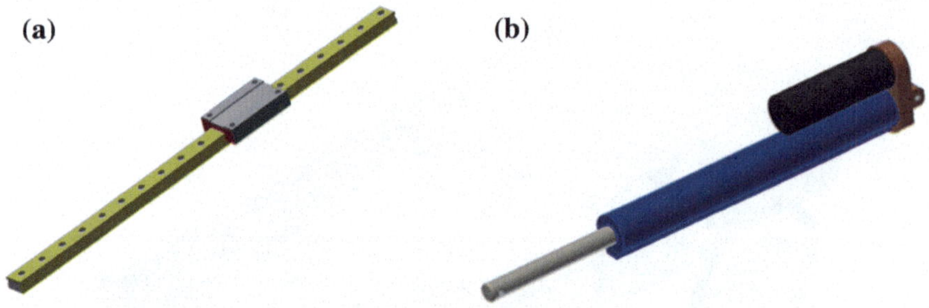

Fig. 2. a, b - Automatic electric linear guide and linear actuators

2.1.2 Coupling Mechanisms

The function of the coupling mechanisms is to connect and join the movement generator elements (linear electric guide Fig. 2a and electric linear actuators Fig. 2b) with the movement receptor element (steering wheel). The coupling mechanisms are classified into two groups; the coupling mechanism between the electric linear guide and the electric linear actuators (see Fig. 3) and coupling mechanisms between the linear actuators and the adapted bearing Fig. 4a, b and c. The coupling mechanism between the electrical linear guide and the electrical linear actuators consists of an upper coupling element characterized by connecting the electrical linear guide with the electrical linear actuators by means of a detachable union. The geometry of the upper coupling element is adapted to the geometry of the linear guide and the linear actuators. The upper coupling element comprises two grooves that fit into the bases of the linear actuators, with two rear tabs that allow the actuators to be fixed to the automatic electric linear guide by means of a removable union. In this way, it is possible to apply a translation movement to the electric linear actuators from the electrical linear guide.

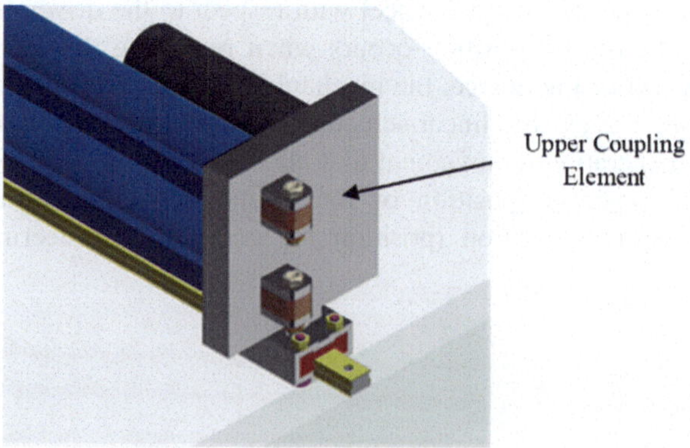

Fig. 3. Coupling mechanism between the electric linear guide and the electric linear actuators

The coupling mechanism between the electric linear actuators and the adapted bearing Fig. 4a, b, c comprises a slotted longitudinal support Fig. 4a and an L-shaped support element Fig. 4c, both connected to the electric linear actuators and the adapted bearing. The slotted longitudinal support joins the electric linear actuator located on the upper position to the upper union rod of the adapted bearing Fig. 5. The slotted longitudinal support is characterized by having a cylindrical geometry at one extreme including slots as stops. The function of the slotted stops is to limit the rotation movement of the element in a range of between 0° and 90° with respect to the Z axis.

The L-shaped support element Fig. 4c comprises one end with two tongue-like surfaces with through holes designed to house the lower union rods of the adapted bearing by means of a detachable connection. The upper part of the L-shaped support element includes a cylinder with a central through hole connected to a union adapter. The union adapter (Fig. 4b) is characterized by having two parallel faces with separate through holes. The union adapters include several limiting surfaces that connect with the surfaces of the slotted longitudinal supports and with the L-shaped support element. The function of the slotted stops is to limit the rotation movement of the element in a range between 0° and 90° with respect to the Z axis and therefore limiting the rotation movement of the steering wheel in a range of 0º to 90º. The rotation movement of the steering wheel is achieved from the translation movement of the linear actuators which push the slotted longitudinal support connected to the adapted bearing. In this way the slotted longitudinal support transmits a rotation movement to the L-shaped support element, placing the steering wheel in a vertical position as indicated in Fig. 7. In Fig. 5 it is indicated an assembly of the elements slotted longitudinal support, union adapter, L-shaped element.

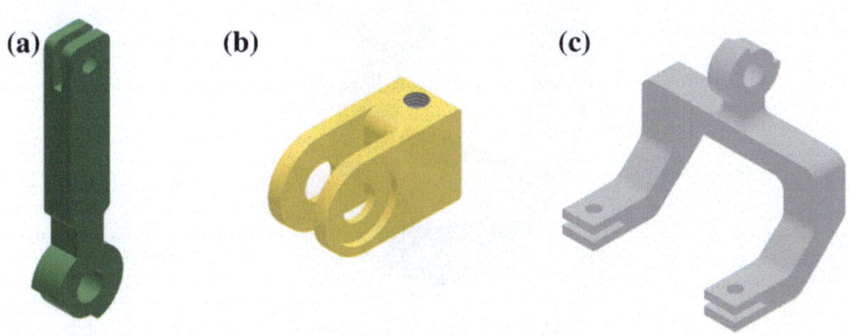

Fig. 4. a, b, c - From left to the right: slotted longitudinal support, union adapter, L-shaped element.

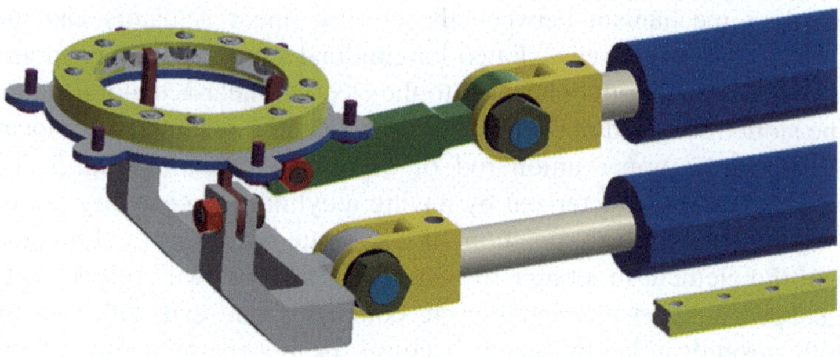

Fig. 5. Assembly of the elements slotted longitudinal support, union adapter, L-shaped element

2.1.3 Adapted Bearing

The adapted bearing see Fig. 6 is an element whose main function is to join the adapted steering wheel to the coupling mechanisms Figs. 4a, b, c. The adapted bearing consists of two housings in charge of housing the rolling elements, these being spherical steel balls. The adapted bearing allows the independent rotation of the steering wheel with respect to the steering axle adaptor. The adapted bearing includes three arms specially designed for connecting the coupling system composed by the slotted longitudinal support and the L-shaped element to the protective casings.

Fig. 6. Adapted bearing

2.2 External Elements

The external design elements of the mechanical device are those that are in contact with the end user being these, the adapted dashboard, the adapted steering wheel Fig. 7 and an adapter to the car steering axle. These external elements are adapted to the specifications and requirements of the mechanism for the displacement and positioning of the steering wheel.

2.2.1 Adapted Steering Wheel

The adapted steering wheel of the device presented Fig. 7 is characterized by its mobility with respect to the steering axle of the car. The rear part of the steering wheel has a concave recess whose function is to house the adapted bearing that connects the slotted shaft with the steering wheel. The steering wheel incorporates a rear cover with a circular shape so as to fasten all the components that are located inside and a grooved shaft that connects to the slotted steering axle adapter as indicated in Fig. 8.

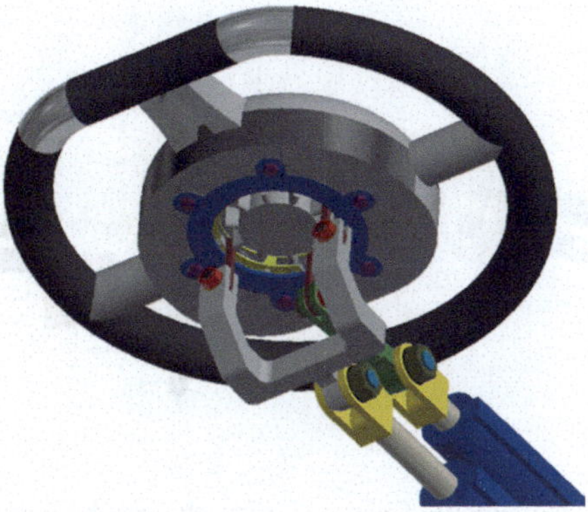

Fig. 7. Adapted steering wheel

2.2.2 Dashboard

The dashboard is specially designed to meet the geometric and functional requirements of the mechanical device presented in this paper. The geometry of the dashboard is characterized by free-form surfaces including a set of holes to house different accessories of the car.

3 Results

In Fig. 8 the movements performed by the mechanism for the displacement and positioning of the steering wheel in the driving position are shown. In Fig. 9 is represented the movements for positioning the mechanical device in autonomous driving.

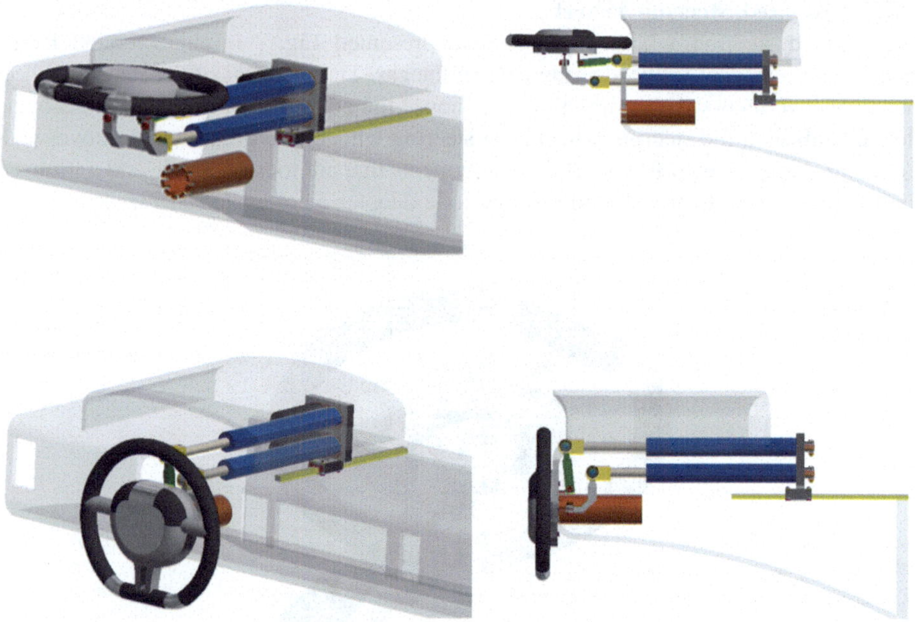

Fig. 8. Movements for positioning the mechanical device

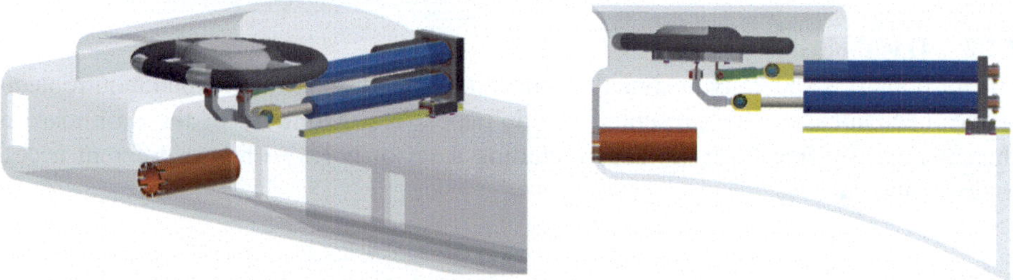

Fig. 9. Position of the mechanical device in autonomous driving

4 Conclusions

This paper presents a new concept that improves the user's experience in the decision of autonomous or guided driving of the vehicle. Similarly, the system allows several functionalities such as extending the driver's room, improving comfort, especially for people with reduced mobility, or as an anti-theft system due to the lack of visibility of the steering wheel making it difficult to steal the car. The design presented in the paper would be integrated with the car's safety system (airbag, etc.), thus guaranteeing the safety of the driver. The device has as an additional advantage in its adaptability to the ergonomics of the user in that it can adapt to their ideal position at the wheel in the case of non-autonomous driving by adding to the design an electronic system capable of analyzing the user's measurements in an automated way.

Acknowledgments. The research work was supported by the University of Jaen with the Plan de Apoyo a la Investigación 2017-2019- ACCION1 PAIUJA 2017-18: TIC-159.

References

1. Andreasson, H., et al.: Autonomous transport vehicles: where we are and what is missing. IEEE Robot. Autom. Mag. **22**(1), 64–75 (2015)
2. Sepúlveda-Arróniz, V., Savaresi, S.: Survey on biomctry for cognitive automotive systems. Cogn. Syst. Res. **55**, 175–191 (2019)
3. Dixit, V., Xiong, Z., Jian, S.: Risk of automated driving: implications on safety acceptability and productivity. Accid. Anal. Prev. **125**, 257–266 (2019)
4. Fagnant, D.J., Kockelman, K.: Preparing a nation for autonomous vehicles: opportunities, barriers and policy recommendations. Transp. Res. Part A: Policy Pract. **77**, 167–181 (2015)
5. Mercado-Colmenero, J.M., Rubio-Paramio, M.A., Martin-Doñate, C.: A numerical and experimental study of a new Savonius wind rotor adaptation based on product design requirements. Energy Convers. Manag. **158**, 210–234 (2018)
6. Mercado-Colmenero, J.M., Rubio-Paramio, M.A., la Rubia-Garcia, M.D., et al.: Int. J. Adv. Manuf. Technol. (2019). https://doi.org/10.1007/s00170-019-03626-0
7. Lim, H., Taeihagh, A.: Autonomous vehicles for smart and sustainable cities: an in-depth exploration of privacy and cybersecurity implications. Energies **11**(5), 1062 (2018)
8. https://www.adandp.media/blog/post/trw-the-disappearing-steering-wheel
9. https://www.abc.es/motor/novedades/abci-sayer-volante-futuro-201709041051
10. https://www.latercera.com/mtonline/noticia/bmw-sorprende-con-un-modelo-futurista-sin-volante/354224/
11. Schwarz, F.: Arrangement of a steering wheel in a motor vehicle suitable for autonomous driving (2018). US2018/0037248A1
12. Joyce, J.: Removable steering wheel and pedals for autonomous vehicle (2017). MX 2017001582A

Conceptual Design of Foldable and Stackable Furniture for Preschool Classrooms

J. S. Velázquez-Blázquez, R. G. Silva-Quituisaca, J. Nieto-Martínez, F. L. Sáez-Gutiérrez, D. Parras-Burgos, and F. J. F. Cañavate(✉)

Department of Graphical Expression, Technical University of Cartagena, 30202 Cartagena, Spain
francisco.canavate@upct.es

Abstract. Currently, school furniture used in preschool classrooms counts with a solid and functional design that has stayed practically invariant for the last 20 years, although it does not invite children to interact much with it. This furniture fills a great part of the classroom's functional space, and it is normally destined to a specific use, which avoids that it was used at other moments of the day, or in other activity zones. Moreover, last tendencies relative to preschool classroom's space organization demand more flexible layouts that allow a quick adaptation, which is impossible with traditional furniture. This work shows the result of the conceptual design of a new line of modular, foldable and stackable preschool furniture, destined for its use indoors in schools, kindergartens, or even at home, composed by a highchair, a desk and a chair. The design has been made following ergonomic, functional and easiness of assembly criteria, as well as the ones of health and safety. Finally, optimal materials have been selected, and the manufacturing cost of each product has been studied, in order to evaluate its competiveness.

Keywords: Modularity · Layout organization · Kindergarten · Ergonomics · 3D modelling

1 Introduction

In every context, furniture design is tried to be optimized for the role it must perform, and school is not an exception [1]. Along the centuries, the concept of school has been changing, and consequently, it has also done the furniture design used in it [2]. With the disappearance of traditional school at the end of XIXth century and the appearance of modern school at the beginning the XXth, furniture design is centralized. This transformed, through subtle and progressive modifications, the rigid tandem wooden banks of fixed dimensions, into single desks composed of laminated wooden table and chair, with a steel structure, which, have remained practically unaltered in our country since their appearance in the 70's [2] (Fig. 1).

F. Cavas-Martínez et al. (Eds.): INGEGRAF 2019, LNME, pp. 86–93, 2020.
https://doi.org/10.1007/978-3-030-41200-5_10

Fig. 1. Wooden bank of XIXth century (left) vs. present typical chair and table (right) [2].

Some studies [3, 4] highlight that sometimes, mainly in the first stages of schooling, boys and girls give more value to emotional implication aspects rather than ergonomics, this is, they prefer furniture whose themes, colours and designs appeal to them, even if they are less comfortable than other models.

However, the design of school furniture has a great importance, from an ergonomic point of view, as learning resents every time a student have to modify his position because of the furniture he is using [5], which can also cause incorrect stances that may lead up to low back pains or other more serious pathologies [6, 7].

Sometimes, this design based in anthropometric parameters is difficult to achieve, as population shows a great variety of dimensions depending on the age range, and even between groups of the same age [8]. Furniture manufacturing companies try to redress this deficit offering different sizes that adapt, in some way, to the dimensions of the pupils [9], but this forces the schools to store a great quantity of desks and chairs of different sizes, which implies a loss of useful space, as can be seen in Fig. 2, given the fact that even if the chairs allow some stacking, tables cannot be stacked in an efficient way, and much less be folded.

Fig. 2. Typical way of storing tables and chairs in a school.

If the attention is focused in the first stage of preschool, it must be also considered that latest tendencies in space organization inside the preschool classroom (learning corners, Frabboni or High Scope) [10–12] demand flexible spaces [13], that allow their quick adaptation, which is impossible with traditional furniture.

2 Objectives

This work shows the result of the conceptual design of a new line of foldable and stackable furniture for children, destined to be used in the classrooms of schools and kindergartens, and even at home, composed of a highchair, a desk and a chair, which aims to achieve the following main objectives:

- Making a functional system, foldable and stackable, capable of saving at least 60% of storing space, and therefore giving more flexibility to the classroom when doing activities.
- Reaching a design as solid, safe and stable as the one of the traditional furniture, but at the same time more eye-catching and attractive for children.
- Using ergonomic criteria to adapt the design to the dimensions of the kid, responding to the growth that takes place during the early childhood.

Also, the following secondary objectives are pursued:

- Quick assembly (less than 1 min) and folding (less than 10 s).
- Possibility of coupling several similar elements in a modular way.
- Reduce the weight of conventional solutions as much as possible.

3 Materials and Methods

From the school furniture manufacturing segment, as well as from a wide range of other different productive sectors and technological areas, different working methods aimed to develop new products have been stablished with a single goal: creating them according to a traceable and repeatable scheme that optimises its finishing [14].

Many of these developing methods have generated tools that proved to be very useful for designers, when planning their work, detecting their customers' needs and desires or proposing solutions.

In this case, we have opted for a User Centered Design (UCD) approach: once the context of use has been identified (the classroom) and after an in-field observation, the user requirements were stabilised as our main design objectives. Then, a conceptual design of the furniture line was created, with the idea of presenting these design concepts to both teachers and children in a future Testing and Verification stage. The process will work in an iterative way, as the initial design will be corrected as many times as necessary considering user's feedback. Therefore, our products design is separated in the following stages [15]:

1. Planning
2. Design
3. Testing and Verification

4. Manufacturing
5. Marketing

Present communication is centred just in the first two stages, and regarding the Design stage, it is focused only in Conceptual Design, leaving the definition of the coupling gaps between pieces, screws and fixation elements for the legs of the desk, for later development stages.

During this Conceptual Design phase, several technical solutions that improve existing furniture models available at market have been proposed, basing on ergonomic, functional and easiness of assembly criteria, as well as the ones of health and safety, being modelled later with CAD/ CAM software.

3.1 Software Used and Rules, Standards and Regulations Followed

For the design, modelling and later calculation of each element of the collection, Solidworks® 2017 software has been used, as it is widely used in other investigation fields, [16–19], while CES EduPack® 2010 was the software used for selecting the manufacturing materials.

Also, for Conceptual Design phase, UNE-EN 1729 European and National families of standards were taken into account [20–22].

4 Results and Discussion

All the designed elements have in common the presence of a handle that eases their transport, the use of vivid colours to result more eye-catching to children, and that their legs are telescopic and foldable, allowing them to acquire a flat form that facilitate their stacking, achieving a volume saving of at least 80%, as five folded elements can be stored in the space of an unfolded one.

Also, as the most common mismatches between children measures and furniture dimensions usually appear in seat height, seat depth, and desk height [23], our design has included telescopic legs in all cases to adapt to the variation, as children grow, of the 13 anthropometric mean measures that define the seated position [24].

All elements were designed with flat borders to avoid gaps between them when used as modules of a bigger formation. Also, all corners were rounded to increase safety. The particular characteristics of each design are described hereunder:

4.1 Highchair

The highchair counts with a total number of twelve pieces: the handle, the platform, the back, the tray, two tray holders, two lateral covers and four telescopic legs. These latter allow the adaptation of height to three different levels, simply adding or removing segments. It has been designed taking into account typical anthropometric measures for children between 1 and 3 years [24], as well as safety and stability criteria set by UNE-EN 14988:2018 [22]. The final result of the design and detail of the process of folding can be seen in Fig. 3.

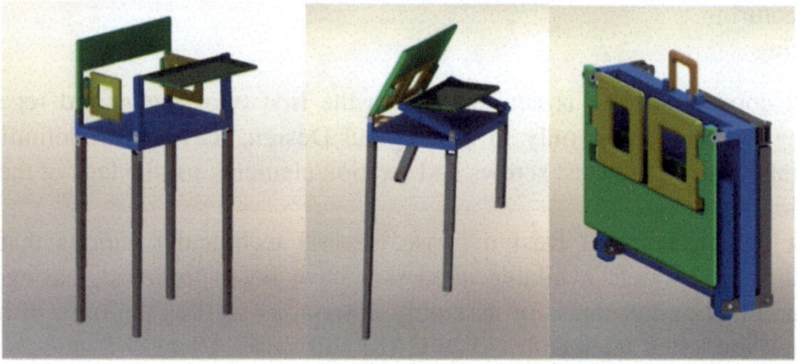

Fig. 3. Final design and process of folding of the highchair.

4.2 Chair

The design of the chair (Fig. 4) has been conceived with a total number of seven pieces: the handle, the seating, the back and four telescopic legs. It has been sized taking into account typical anthropometric measures for children between 3 and 6 years [19], as well as safety and stability criteria set by UNE-EN 1729-2:2012 [21]. Attaching laterally some of them, a bank could be formed, so it has some modularity.

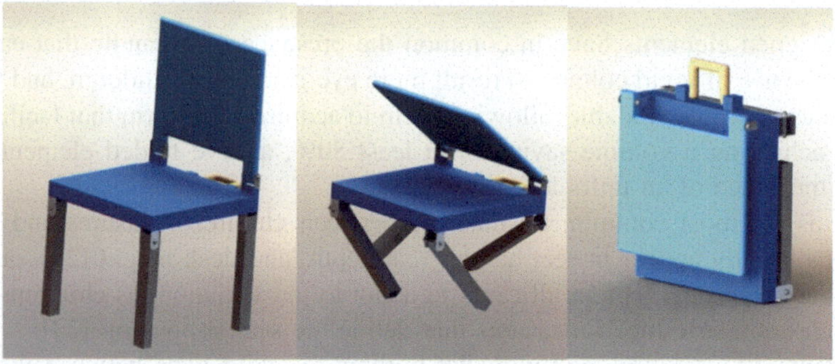

Fig. 4. Final design and process of folding of the chair.

4.3 Desk

The desk is the element which counts with a higher modularity among all the designed, as its trapezoidal shape allows putting it together with up to another five to form a regular hexagon, a quite common structure in work teams in preschool [25]. The legs located in the longest side would be mounted on a slider rack to allow that the chair can be inserted under the board.

The desk is composed of six pieces in total: the handle, the board and four telescopic legs. As in the case of the chair, it has been designed taking into account typical anthropometric measures for children between 3 and 6 years [19], as well as safety and stability criteria set by UNE-EN 1729-2:2012 [21]. Details of the final design and the process of folding can be observed in Fig. 5.

Fig. 5. Final design and process of folding of the desk.

4.4 Materials selection

Once finished the Conceptual Design stage, a selection of the materials to be used for all the designed products was made by means of software CES EduPack 2010.

The starting premise was that material should show good mechanical properties, be easy to be cleaned, non-toxic and easy to manufacture and mechanize, stablishing that it should therefore be a plastic, as it is a common use material when manufacturing all kinds of furniture [6, 26, 27].

In this way, selection parameters stablished were: tensile strength – elastic limit, price – elastic limit, price under 2 €/kg, and good environmental behaviour, concluding that the ideal material would be polyethylene.

The backs, trays, seatings and boards have been designed in rotomoulded polyethylene, as it reduces weight in up to a 12%, keeping structural stability with static loads of up to 500 kg (displacements of less than 0.2 mm), and its smooth finishing is as easy to clean as laminated wood.

5 Conclusions

In this work, the concept of a new foldable and stackable school furniture line for children of ages between 0 and 6 years, composed of a highchair, a chair and a desk has been developed, which achieves:

- Reducing the problem of furniture storing in schools, as it saves more than 80% of the volume occupied by traditional furniture.
- Making classroom more flexible, as its quick folding and unfolding (less than 10 s according to simulations) allows it to be withdrawn from its location, when this space wants to be used for other activities.
- At first, being more eye-catching and comfortable to children, as it has been designed following their mean anthropometric dimensions for each age, and basing in more appealing colours than traditional school furniture. This will be assessed in later stages gathering teachers' and students' feedback by means of tests and personal interviews.
- Being as sturdy as traditional furniture, allowing weights of up to 500 kg with displacements of just 0.2 mm, reducing total weight in a 12%.
- Having a certain modularity, as several chairs can be put together to form a bank, and several desks can be joined to form a bigger hexagon with a central gap.

All the elements that have not been totally defined in present communication, such as leg locking system and its fixation to the plastic surfaces, the fitting gaps between backs and seatings, or the screws to be used, are planned to be specified in the Detailed Design stage.

Finally, it has been considered ending the design with a future prototyping stage, in order to test and validate the products in a real school environment.

References

1. García, C., Moraga, R., Page, Á., Tortosa, L., Verde, V.: Ergonomía y mueble.Guía de recomendaciones para el diseño de mobiliario ergonómico (1997)
2. Jácome, L.R., Torre, P.L.G.: La evolución del mobiliario escolar. Técnica Ind. **295**, 64–69 (2011)
3. Salvador, C., Vicente, J., Martins, J.P.: Ergonomics in children's furniture–emotional attachment. In: Proceedings of the 5th International Conference on AHFE, pp. 5478–5485 (2014)
4. Salvador, C.: Adapting furniture to the child – ergonomics as a main tool in a design project. In: Proceedings of the 20th Congress of the International Ergonomics Association (IEA 2018), pp. 1294–1305. Springer, Cham (2019)
5. Parcells, C., Stommel, M., Hubbard, R.P.: Mismatch of classroom furniture and student body dimensions: empirical findings and health implications. J. Adolesc. Health Off. Publ. Soc. Adolesc. Med. **24**(4), 265–273 (1999)
6. Knight, G., Noyes, J.A.N.: Children's behaviour and the design of school furniture. Ergonomics **42**(5), 747–760 (1999). https://doi.org/10.1080/001401399185423

7. Gonçalves, M.A., Arezes, P.: Analysis of the mismatch between school furniture and children. In: Proceedings of the 5th International Conference on AHFE, Advances in Ergonomics in Design, Usability and Special Populations: Part I, pp. 268–274 (2014)
8. Aparicio, E.Q., Nogueras, A.M., Rodríguez, J.O., González, J.R., Pedraz, L.S., García, R.D. J.F.: Estudio del mobiliario escolar en una población infantil **26**(1), 3–12 (2004)
9. Rosal López, G.A.: Elaboración de metodología basada en la ergonomía de producto y ecodiseño aplicada al mobiliario escolar. Validación metodológica del product (2011)
10. Piatek, A.: El trabajo por rincones en el aula de Educación Infantil. Ventajas del trabajo por rincones. Tipos de rincones. Innovación y experiencias educativas **15**, 1–8 (2009)
11. Frabboni, F.: La educacion del nino de 0 a 6 anos. Cincel (1984)
12. Wiltshire, M.: Understanding the HighScope Approach: Early Years Education in Practice. Routledge, New York (2018)
13. Gamero EM (2008) Organización de los espacios en la Educación Infantil.
14. Walker, D.H.: Reflecting on 10 years of focus on innovation, organisational learning and knowledge management literature in a construction project management context. Constr. Innov. **16**(2), 114–126 (2016)
15. Page, A: Nuevas técnicas para el desarrollo de productos innovadores orientados al usuario. Instituto de Biomecánica de Valencia (2001)
16. Cavas-Martínez, F., Fernández-Pacheco, D.G., De La Cruz-Sánchez, E., Nieto Martínez, J., Fernández Cañavate, F.J., Vega-Estrada, A., Plaza-Puche, A.B., Alió, J.L.: Geometrical custom modeling of human cornea in vivo and its use for the diagnosis of corneal ectasia. PLoS ONE **9**(10) (2014). https://doi.org/10.1371/journal.pone.0110249
17. Cavas-Martínez, F., Bataille, L., Fernández-Pacheco, D.G., Cañavate, F.J.F., Alio, J.L.: Keratoconus detection based on a new corneal volumetric analysis. Sci. Rep. **7**(1) (2017). https://doi.org/10.1038/s41598-017-16145-3
18. Cavas-Martínez, F., Fernández-Pacheco, D.G., Cañavate, F.J.F., Velázquez-Blázquez, J.S., Bolarín, J.M., Alió, J.L.: Study of morpho-geometric variables to improve the diagnosis in Keratoconus with mild visual limitation. Symmetry **10**(8) (2018). https://doi.org/10.3390/sym10080306
19. Cavas Martinez, F., Garcia Fernandez-Pacheco, D., Fernandez Cañavate, F.J., Velázquez Blázquez, J.S., Bolarin, J.M., Tiveron, M., Alio, J.L.: Detección De Queratocono Temprano Mediante Modelado 3D Personalizado Y Análisis De Sus Parámetros Geométricos. Dyna Ingenieria E Industria **94**(1), 175–181 (2019). https://doi.org/10.6036/8895
20. UNE-EN: Mobiliario. Sillas y mesas para centros de enseñanza. Parte 1: Dimensiones funcionales, UNE-EN 1729-1:2016 AENOR (2016)
21. UNE-EN: Mobiliario. Sillas y mesas para centros de enseñanza. Parte 2: Requisitos de seguridad y métodos de ensayo, UNE-EN 1729-2:2012+A1:2016 AENOR (2016)
22. UNE-EN: Tronas para niños. Requisitos y métodos de ensayo, UNE-EN 14988:2018. AENOR (2018)
23. Agha, S.R.: School furniture match to students' anthropometry in the Gaza Strip. Ergonomics **53**(3), 344–354 (2010)
24. Motmans, R., Ceriez, E.: Body dimensions of the Belgian population (2005). https://www.dinbelg.be/anthropometry.htm
25 Lei, J.O.W.: Learning environments: design solutions for elementary schools (2011)
26. Turner, S.: Multi-orientation modular furniture having an energy releasable design (2016)
27. Leibovies, D., Shimelmits, I., Weiss, S.: Furniture and storage set. Google Patents (2017)

Computer-Aided Design and Interactive Design

The Inclined Plane Rosary Pump of Benito Bails' 'Elements of Mathematics' (1790): Geometric Modeling and Virtual Reconstruction

José Ignacio Rojas-Sola[1(✉)], María de las Mercedes Catena-Ibáñez[2], and Eduardo De la Morena-De la Fuente[3]

[1] Department of Engineering Graphics, Design and Projects, University of Jaén, Campus de las Lagunillas, s/n, 23071 Jaén, Spain
jirojas@ujaen.es
[2] University of Jaén, Campus de las Lagunillas, s/n, 23071 Jaén, Spain
[3] Research Group 'Engineering Graphics and Industrial Archaeology', University of Jaén, Campus de las Lagunillas, s/n, 23071 Jaén, Spain

Abstract. This paper shows the geometric modeling and virtual reconstruction of the inclined plane rosary pump included in the work 'Elements of Mathematics' by Benito Bails of 1790, specifically in Volume X, Tome IX, Part II on Hydraulic Architecture. The software Autodesk Inventor Professional has been used, which has allowed us to obtain its geometric documentation (plan of the ensemble, isometric and exploded views and 3D digital restitution). The material used in this research has been downloaded from the Biblioteca Digital Hispánica of the National Library of Spain. However, and as a drawback, it is necessary to indicate that the only available sheet does not have a graphic scale, nor does there exist expressly in the specification any reference measure that would allow obtaining said graphic scale, which is insufficient for obtaining a reliable 3D CAD model. For this reason, a graphic scale has been adopted from an estimated dimension of a known element (yoke), as well as a series of geometric, dimensional and movement constraints (degrees of freedom) in order to obtain a coherent model.

Keywords: Benito Bails · Inclined plane rosary pump · Geometric modeling · Virtual reconstruction · Industrial heritage · Industrial archeology

1 Introduction

Benito Bails was a Spanish mathematician and architect, a relevant figure of the Spanish Enlightenment. He was born in San Adrián de Besós (Barcelona) in 1730 and died in Madrid in 1797 [1]. A prolific author, he was noted for his well-known work 'Elements of Mathematics' composed of 11 volumes. One of these volumes, Volume X, Tome IX, Part II, referring to hydraulic architecture is dated to 1790.

In this volume, a compendium of historical inventions is presented, reflected in 51 sheets that cover concepts from the founding of hydraulic works, hydraulic works in

F. Cavas-Martínez et al. (Eds.): INGEGRAF 2019, LNME, pp. 97–104, 2020.
https://doi.org/10.1007/978-3-030-41200-5_11

rivers, leveling, navigable rivers and navigation channels, irrigation canals, of the channels for drainage of the marshy lands, and of internal navigation of Spain in general.

In particular, regarding the first aspect on the basis of hydraulic works, it is worth highlighting the description that he gives of machines used to extract water, specifically the inclined plane and vertical plane rosary pumps, whose purpose was to extract the water enclosed in a flooded area.

This paper centers its attention on the inclined plane rosary pump because it is a historical invention of relevance in its historical context.

Therefore, the main objective of this paper is to present the digital restitution obtained from a reliable 3D model, as well as its geometric documentation. For this, the objectives set out in the document Principles of Seville [2] on virtual archaeology have been followed, as well as the London Charter [3] on computer-aided visualization of cultural heritage. Thus, we seek to obtain a digital restitution based on research, documentation and dissemination that emphasizes the importance of authenticity, historical rigor and scientific transparency.

2 Materials and Methods

The material used in this research has been downloaded from the Biblioteca Digital Hispánica of the National Library of Spain [4], specifically Volume X, Tome IX, Part II. A brief description from said volume is reproduced below, together with a single sheet without graphic scale that is the only graphic information available (Fig. 1), for initially understanding the operation of the invention.

"Having made the drawer and enclosed the water we just need to remove it, for which different pumps have been invented called rosaries or mills, or rosary pumps, having two types, namely the inclined pump and the vertical, whose construction and use will become clear with the description that will be given of the first one.

The inclined plane AB is the floor of a channel (lower figure) whose end A is inserted into the water that is to be removed, and the other end B is directed to another channel BC that is on the outside of the drawer; the pump channel is served by another DE plane with two flanges, which form a channel.

The beads of this rosary pump are small boards, called vanes, linked with each other so that they form an endless chain.

The beads pass over two small gears, F and G, partly inserted into the channel, whose vanes push the water upward, with the gears moving in the corresponding direction; the other beads go down and over the support to collect water.

When this pump is horse-driven an IK shaft is added, which serves as the axle to two small gears: the first, G, above which the rosary passes, corresponds to the channel where the water falls; the other H, which in the front view cannot be seen because the first covers it, is fixed to the LM wheel that guides horses with the N shaft. But when it is men who operate the pump there is no other gear than G, to which winch handless are added. When the water to be drained does not have to be removed at a great height the drainage is performed with simpler devices."

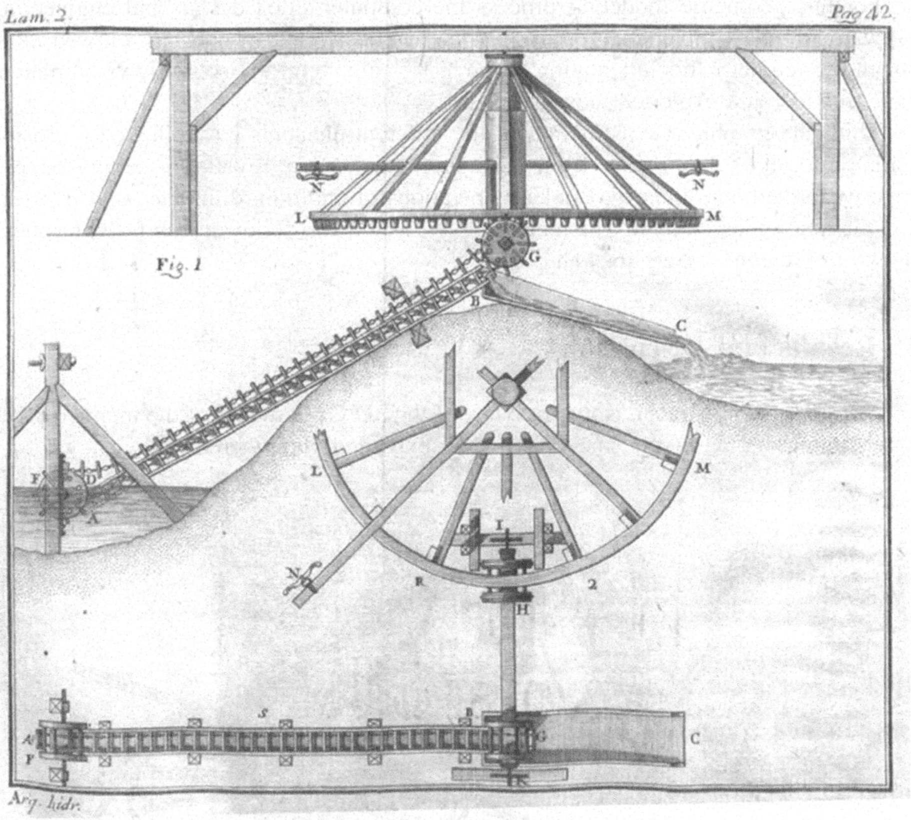

Fig. 1. Inclined plane rosary pump [4].

The methodology used has been based on knowledge of descriptive geometry and of the empirical techniques of direct measurement of the dimensions appearing in the figure, in order to obtain a solid modeling of it.

The geometric modeling process has been complex due to the lack of information since the only original illustration available (Fig. 1) is not drawn to scale, and it has therefore been necessary to take a reference measurement in order to obtain the graphic scale and be able to obtain a reliable 3D model. Specifically, we have taken as a reference measure a yoke with a length of 1.30 m, from which we have determined the graphic scale of the illustration that has turned out to be 1:100 on a proportional impression in format A3.

Given that the purpose of the sheet of the invention was not to provide accurate data for the purpose of material realization but rather to be a conceptual design that conveyed the idea of its operation, it has been necessary to establish dimensional, geometric and movement constraints (degrees of freedom) so that its design would be coherent and ensure the correct functioning of the invention.

For this geometric modeling process the computer-aided design and engineering software Autodesk Inventor Professional [5] has been used, which has allowed us to obtain the geometric documentation of the historical invention (isometric view, plan of the ensemble, and exploded view).

Thus, the graphic reconstruction of the different elements has followed a process common to all of them, since from a 2D profile extrusion, revolution, emptying, etc. were performed, and then by Boolean operations of addition, difference and intersection the final set assembled. The methodology, which was followed carefully, has been shown in previous papers [6–11].

3 Results and Discussion

Figures 2, 3 and 4 show an isometric view of the 3D CAD model of the inclined plane rosary pump, a plan of the ensemble and an exploded view thereof.

Fig. 2. Isometric view of the 3D CAD model.

For a better understanding of its operation, a description based on the plan of the ensemble with an indicative list of different elements is given (Fig. 3).

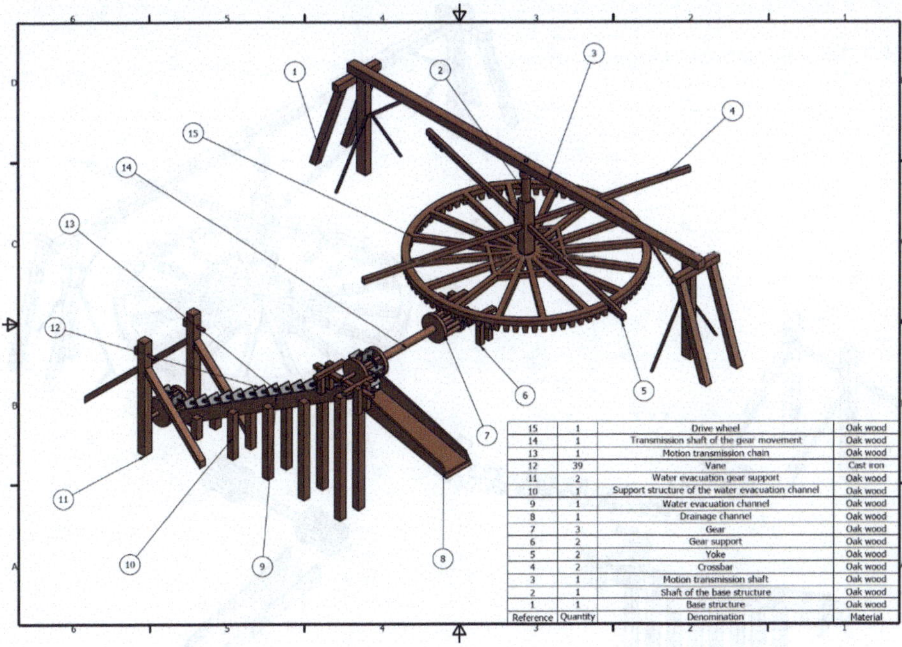

Reference	Quantity	Denomination	Material
15	1	Drive wheel	Oak wood
14	1	Transmission shaft of the gear movement	Oak wood
13	1	Motion transmission chain	Oak wood
12	39	Vane	Cast iron
11	2	Water evacuation gear support	Oak wood
10	1	Support structure of the water evacuation channel	Oak wood
9	1	Water evacuation channel	Oak wood
8	1	Drainage channel	Oak wood
7	3	Gear	Oak wood
6	2	Gear support	Oak wood
5	2	Yoke	Oak wood
4	2	Crossbar	Oak wood
3	1	Motion transmission shaft	Oak wood
2	1	Shaft of the base structure	Oak wood
1	1	Base structure	Oak wood

Fig. 3. Plan of the ensemble with an indicative list of different elements.

The invention has a fixed base structure (1), providing the assembly with rigidity and stability, and thanks to the motion transmission shaft (3) present in the shaft of the base structure (2), the movement of the drive wheel is allowed (15).

The movement of the shaft of the base structure (2) is driven by animal traction thanks to two yokes (5) located at the ends of a crossbar (4), so that when the drive wheel (15) is actuated, said movement is transmitted to a small gear (7) which in turn drives a second gear, both gears supported by supports (6) and joined by a motion transmission shaft (14). This last gear drives a motion transmission chain (13) formed by links, rings and vanes (12) that drag the water from the flooded area where it is enclosed (thanks to a third small gear partially submerged and sustained thanks to some supports (11)), by an evacuation channel (9) supported by a support structure (10), to the drainage channel (8) through which it is expelled.

Figure 4 shows an exploded view for a better understanding of the order of assembly of the different elements.

As mentioned previously, there arose several complications throughout the geometric modeling process. The first, as already mentioned, is the absence of any graphic scale or reference measurements, since the planimetry found was only intended to facilitate understanding of the operation of the invention, but not that of its reliable reconstruction. But in addition to the absence of the graphic scale there are inconsistencies between the different orthographic views, which do not have an exact geometric or dimensional relationship.

Fig. 4. Exploded view of the 3D CAD model.

The constraints used in the final assembly have been, mainly, the opposition or alignment of faces (allowing freedom of movement in two directions), used for example in the transmission chain; insertion constraint (allowing the rotation of one element with respect to another), used for the rotating axes of several elements; constraint of rotation movement, used in the contact between the gear and the drive wheel; and finally, tangency constraint, used between the vanes and the gear teeth.

On the other hand, most of the elements are quite simple, so the geometric modeling process is no more difficult than the usual operations for obtaining solid modeling.

However, the most difficult parts to model were the chains, due to their complex geometry and the difficulty of defining their relationship with the surrounding elements. The main obstacle appears when designing and placing the different elements that make up the chain, that is, the links, the rings and the vanes that make up the water evacuation system, since this implies the existence of numerous constraints of movement which sometimes conflict with each other.

For the assembly of the transmission chain, the following process has been carried out specifically: a work point has been defined on the internal surface of the link and then another point on the inner surface of the ring. Next, one element has been restricted by the opposition with respect to the other, taking this point as a reference. The last step has been to rotate these elements so that they have the proper orientation forming the chain of links (Fig. 5).

Fig. 5. Internal points in links and rings for the final assembly of the transmission chain.

4 Conclusions

This paper out shows the result of the complex process carried out to obtain the reliable geometrical modeling and virtual recreation of the inclined plane rosary pump of the work 'Elements of Mathematics' by Benito Bails from 1790, thanks to Autodesk Inventor Professional software.

The methodology employed is based on knowledge of descriptive geometry and the use of empirical techniques for measuring dimensions on orthographic views. However, since the only original illustration is not drawn to scale it has been necessary to take a reference measure in order to obtain the graphic scale and later establish dimensional, geometric and movement constraints (degrees of freedom) in order to ensure the correct operation of the historical invention and obtain a coherent 3D model of it.

From this 3D model the geometric documentation of the invention has been obtained (isometric view, exploded view and plan of the ensemble with indication of the list of elements and materials), which makes it easier for the reader to interpret its operation and composition.

Moreover, the 3D model will be the starting point of a subsequent static analysis through Computer-Aided Engineering techniques where the von Misses stresses, displacements and deformations will be analyzed, as well as the safety coefficient of the model under operating conditions.

Acknowledgments. The research work reported here was made possible by the Extracurricular Practices Program with the Research Group 'Engineering Graphics and Industrial Archaeology' of the University of Jaén.

References

1. Biography of Benito Bails. https://es.wikipedia.org/wiki/Benito_Bails. Accessed 16 May 2019
2. Principles of Seville. http://smartheritage.com/wp-content/uploads/2015/03/FINAL-DRAFT.pdf. Accessed 16 May 2019
3. London Charter. http://www.londoncharter.org. Accessed 29 Mar 2019
4. Bails, B.: Elementos de Matemática. Volumen X. Tomo IX. Parte II. Que trata de la Arquitectura Hidráulica. Imprenta de la Viuda de D. Joaquín Ibarra, Madrid (1790)
5. Shih, R.H.: Parametric Modeling with Autodesk Inventor 2016. SDC Publications, Mission (2015)
6. Rojas-Sola, J.I., De la Morena-de la Fuente, E.: Agustín de Betancourt's plunger lock: approach to its geometric modeling with Autodesk Inventor Professional. In: Eynard, B., Nigrelli, V., Oliveri, S.M., Peris-Fajarnes, G., Rizzuti, S. (eds.) International Joint Conference on Mechanics, Design Engineering and Advanced Manufacturing (JCM), pp. 757–765. Springer, Heidelberg (2017). https://doi.org/10.1007/978-3-319-45781-9_76
7. Rojas-Sola, J.I., De la Morena-de la Fuente, E.: Geometric modeling of the machine for cutting cane and other aquatic plants in navigable waterways by Agustín de Betancourt y Molina. Technologies **6**(1) (2018). https://doi.org/10.3390/technologies6010023. Article 23
8. Rojas-Sola, J.I., De la Morena-de la Fuente, E.: Agustin de Betancourt's wind machine for draining marshy ground: approach to its geometric modeling with Autodesk Inventor Professional. Technologies **5**(1) (2017). https://doi.org/10.3390/technologies5010002. Article 2
9. Rojas-Sola, J.I., De la Morena-de la Fuente, E.: Digital 3D reconstruction of Agustin de Betancourt's historical heritage: the dredging machine of the port of Krondstadt. Virtual Archaeol. Rev. **9**(18), 44–56 (2018). https://doi.org/10.4995/var.2018.7946
10. Rojas-Sola, J.I., Galán-Moral, B., De la Morena-de la Fuente, E.: Agustín de Betancourt's double-acting steam engine: Geometric modeling and virtual reconstruction. Symmetry **10**(8) (2018). https://doi.org/10.3390/sym10080351. Article 351
11. Rojas-Sola, J.I., De la Morena-de la Fuente, E.: The Hay inclined plane in Coalbrookdale (Shropshire, England): Geometric modeling and virtual reconstruction. Symmetry **11**(4) (2019). https://doi.org/10.3390/sym11040589. Article 589

The Model Based Definition Features - MBD Applied to the Technical Computer Aid Design and Drawing

Ramón Miralbes-Buil(✉), Juan Antonio Peña-Baquedano, Jose Manuel Paricio-Sánchez, and David Ranz Angulo

University of Zaragoza, 50018 Zaragoza, Spain
miralbes@unizar.es

Abstract. The graphic drawing of the industrial products, that is mainly associated with the technical drawing, will probably evolve towards the progressive abandonment of 2D representation and paper-based support due to the new possibilities that the computer aided design (CAD) tool have developed; these new features are the 3D representation of these objects using computers, tables and other digital media that provide a better visualization and new features and, in another way, allow the use of the augmented reality technologies, the collaborative work. These tools are currently very mature and have been included in most of the technical offices of the companies. Another advantage of these tools are that they allow the measurement of dimensions, the representation of the pieces and mechanical assemblies, the generation of cuts, etc.

However, in order to correctly represent all the necessary graphic information, it is essential to include in the 3D model some aspects such as main dimensions or the representation of the tolerances, surface finishing, welding, material bill tables, gear data, review tables, etc. Consequently, main CAD tools have included the Model-Based Definition tools or MDF to include such information. This tools in conjunction with the range of possibilities that 3D pdf documents provide, will be determinant in the way of how the technical drawing will evolve.

The aim of this article is to deal with this topic and to analyze the MDF tools, their advantages and disadvantages and their functionality for two of the main 3D CAD programs: Solid Works and Autodesk Inventor.

Keywords: MDF · Technical drawing · CAD · 3D

1 Introduction

Technical drawing in the field of the mechanical engineering deals with the representation of parts and assemblies and continuously evolve in parallel with the CAD drawing tools; these programs have begun to include new functionalities to manage, incorporate and show the necessary information of the parts and assemblies directly in the 3D model. These new tools are called the "Model-Based Definition" (MBD) [1] and enable to add annotations, dimensions, surface finished notes, welding indications,

F. Cavas-Martínez et al. (Eds.): INGEGRAF 2019, LNME, pp. 105–114, 2020.
https://doi.org/10.1007/978-3-030-41200-5_12

dimensional and geometric tolerances, etc. that are essential for the adequate representation of these elements and, later, to build them.

Traditionally, these data have indicated thus the use of a 2D drawing of the part or the assembly using necessary views and a scale, but this has several draw-backs [2, 3]; in this drawing it is included some indications like the dimensions, positions of the elements, the tolerances, the ribbed areas, the type and dimensions of the welding, the holes and screwed zones, the material of the piece and, in the case of the assemblies the material bill table that include the necessary in-formation of each piece (designation, quantity, regulations, material, measurements, etc.) [4]; for this purpose. It is necessary the study of the associated regulations because there are some elements that are commercial and some machining zones are subjected to a regulation like the ribbed ones. In the same way, some key information must be included in the title box table, such as the scale, the name of the company, the drawer, the responsible, the name of the piece/assembly, etc.

This information, which usually is represented in 2D, can be visualized either digitally through dwg or pdf files and/or printed on paper. However, the development of the 3D Pdf files has allowed the generation of digital files that allows the visualization of parts and mechanical assemblies in three dimensions without the need to install 3D modelling programs. Additionally, these programs have some tools with great applicability in the industrial sector such as the possibility of generating cut and section views, measure distances, angles, etc., obtaining additional information, modifying the appearance and transparency of the piece or the lights, indicating some data model, hide pieces of assembly, etc. These functionalities were included in the CAD 2D drawing but, in addition, 3D pdf files allow the possibility of use 3D glasses and the augmented reality that will be the cornerstone in the industry 4.0.

It has been carried out a study of the current regulations [5] and the main conclusion obtained is that there is a lag between the regulation and the current functionalities that 3D CAD tools provided; then there is no regulation related to this type of spatial representations that allows the ordering and defining a way to generate these representations.

This article is focused on the MBD functionalities of two of the most common 3D CAD tools in the field of the mechanical engineering: Autodesk Inventor 2018 and Solid Works 2018 (Fig. 1).

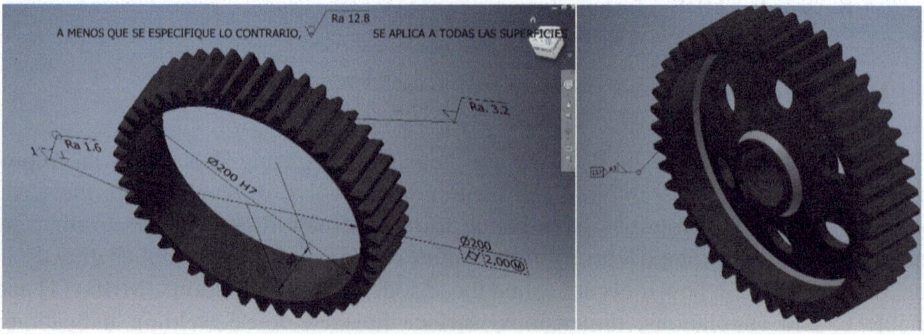

Fig. 1. Examples of Inventor MBD part (left) and welded assembly (right)

Fig. 2. Examples of Solid Works MBD part (left) and welded assembly (right)

2 Analysis of Autodesk Inventor 2018

The main tools that offer the annotate tab [6] can be seen in Fig. 3:

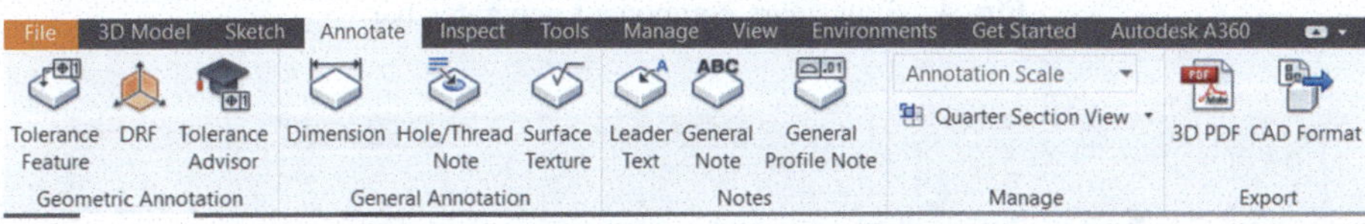

Fig. 3. Annotate bar of Autodesk Inventor

In addition, Autodesk Inventor is able to include the information contained in the title block table and much more additional information and even customize it through the "Iproperties" tab that can be later transferred to the 3D pdf and visualized.

One of the main weak point of this tool is the impossibility to generate MDF tables for the gear teeth and for the holes, so it is necessary to generate a 2D draw, to obtain and export these tables. However, the data of the gear can be visualized inside the "Design Accelerator" bar and exported to a html format (Fig. 4).

Analyzing the capabilities of representation of assemblies, during the assembly the list of materials (see Fig. 5) is automatically generated and it can be modified in "List of materials"; this table includes all the necessary information (name, quantity, designation and measurements, standards, material, etc.) It is possible to include more additional information, as well as to export it to Excel formats, modify its structure, change the order, etc. The main problem of the program is that it is not possible to include it in the 3D representation, to export it to the 3D pdf file or to generate the balloons to mark each sub-assembly or piece. In the same way, the "annotation" bar is not available for assemblies so it is impossible to include any of its options.

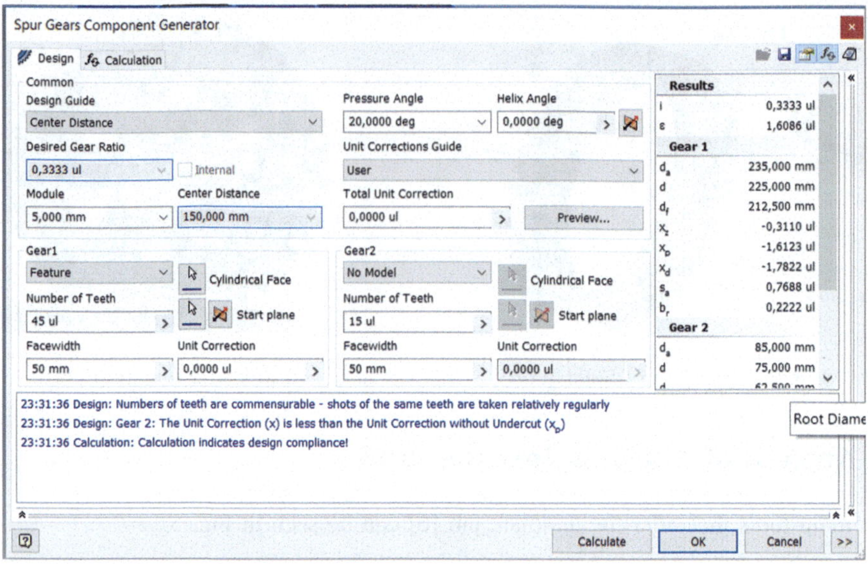

Fig. 4. Spur gears component generator bar

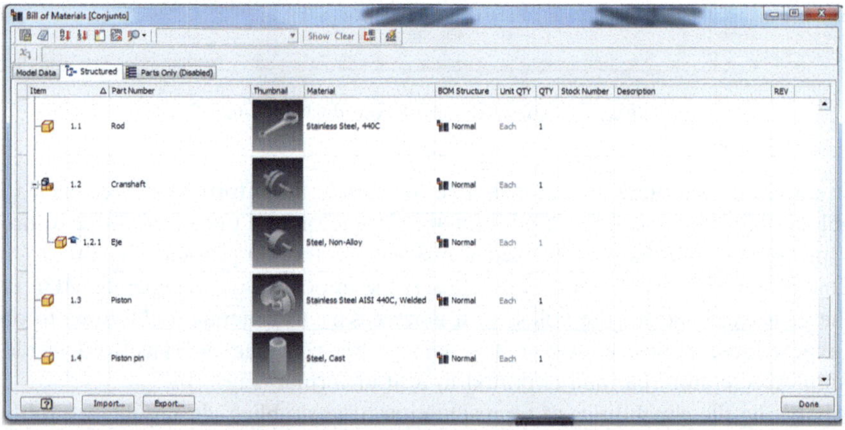

Fig. 5. Bill of materials of Autodesk Inventor

Besides that, after analyzing some of the generation resources of Inventor to obtain shafts, axes, gears, belt drivers, cams, springs, ribbed zones, etc., it must be said that it is possible to generate perfectly any of this type of elements/zones and it includes all the associated regulation, therefore to obtain the data to machining them; however, these information is impossible to translate to the MBD model using a table and so to the 3D pdf file and, in some cases like the teeth of the gears, it can only be exported to a html file. In some cases, like the ribbed zones, it can be noted with a leader text but not automatically. As a conclusion, although the CAD model contains all the information, its visualization is not possible through a 3D PDF and, part of the information (like spring generation data), is not possible to be exported easily to Excel or to another file.

In the case of the sheet metal parts, Inventor generates the flat Pattern, the centre line through "cosmetic centreline" and the annotations for a punch tool. Additionally, it contains all the necessary data such as sheet metal rules, K factors, the data of the corners and bends (e.g.: relief shape, relief data, 2 and 3 bend intersection) but it is difficult to represent it as a table and export it to excel, as well as its visualization in 3D pdf. Neither is it possible to generate the folding tables or export them to Excel, so the 2D draw must be used to obtain and visualize this information.

It should be pointed that, although it is possible to generate the flat pattern and visualize in it the folds and folding order, the MBD module cannot note these order or the annotations of the fold. In the same way, it is impossible to generate a 3D pdf file that shows the flat pattern and so, only the final shape can be visualized.

Analysing the tools for the welded assemblies (see Fig. 6), it must be indicated that the program allows the generation of the "previous preparation" model of each piece, the welded model and the post-welding machining model. In the case of the welding model, it is possible to generate the graphic representation (limited to fillet and grove welds) and the symbolic one using the cosmetic weldments.

Additionally, the welding bar allows to generate MDF indications for any type of welding using the "symbol", it is possible too, to generate a report of the welds by mean of the Bead Report button, as well as calculating the welds by means of the Weld Calculator.

However, it should be pointed that, after generating the 3D pdf file, it does not allow to visualize any of these aspects, so the generated 3D model lacks welds, previous and/or subsequent machining, welding annotations, welds, etc.

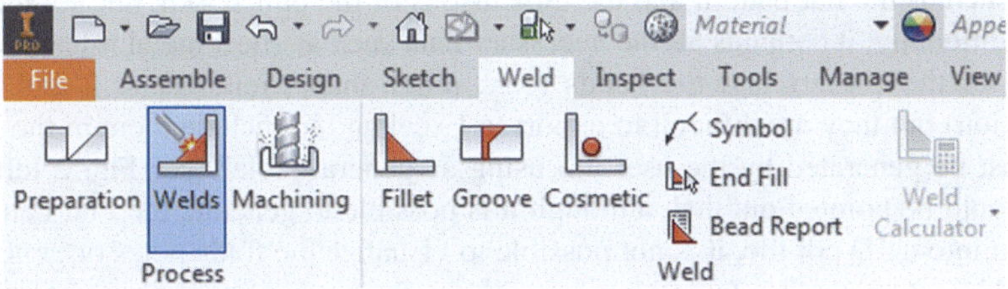

Fig. 6. Weld bar of Autodesk Inventor

Once again, it must be indicated that, since it is an assembly, it presents the same functionalities and weak points.

Finally, it must be highlighted that the program does not allow to indicate in the MDF representation some aspects related to moulded parts such as the partition line, maximum burrs, feeders, etc.

3 Analysis of Solid Works 2018

The other analyzed program is Solid Works that include the tab "Solid Works MBD" that is shown in Fig. 7 [7]:

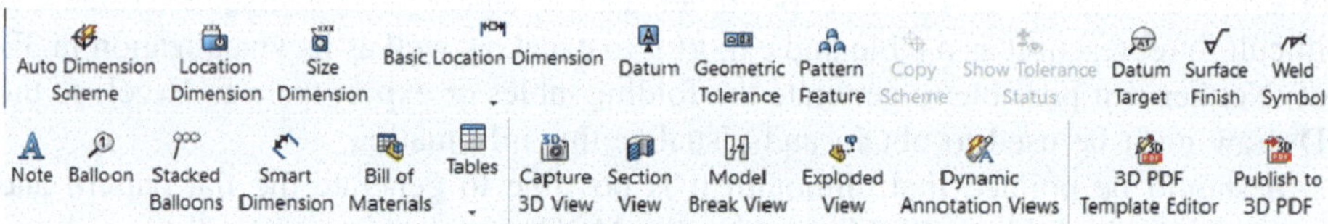

Fig. 7. Solid Works MBD bar

Analyzing the main functionalities in this toolbar, it can be observed that there are more tools than in Inventor.

It should be noted that, in all dimension type annotations, it is possible to indicate all types of dimensional tolerances (deviations, maximum and minimum values, class, etc.).

Analyzing the possibilities offered by Solid Works related to the assemblies, it allows to add the list of materials (see Fig. 8) and the marks of parts in the model. Therefore it is possible to export this information to Excel and to a 3D pdf file.

In the case of the generators of elements, Solid Works allows to generate any kinds of gears and, by using "general tables", to show the data of the teeth of these elements in the 3D pdf file and in the model. However, Solid Works has limitations to generate elements such as shafts, axes, cams, springs, ribbed zones, etc.

Analyzing the tools related to the sheet metal parts, the program allows to include aspects such as the flat pattern and the dimensions of the punch tool, but not the centre line. Additionally, it contains all the necessary data such as sheet metal rules, K factors, the data of the corners and the bends (e.g.: relief shape, relief data, 2 and 3 Bend Intersection) but they are difficult to export to Excel and to include them in the pdf file, so it must be generated by the user too using a "general table" (see Fig. 2 left).

It should be pointed out that, although it is possible to generate the Flat Pattern and to print it into a 3D pdf file, it is not possible to visualize the folds notes or generate the folds' table automatically. In another way, although it is possible to generate elements such as "cross-break", "forming tools" and "Vents", its MBD dimension cannot be noted.

In the case of the welded assemblies, it should be noted that the program only allows the model the final representation, which makes difficult to differentiate between the previous preparation stage of the pieces, the representation with welds and the representation with the post-welding machining. Solid Works is capable to note both, graphical and symbolic representation of the welds, as well as the complete annotation of any type of weld. Additionally, this information can be included in the 3D pdf file.

Finally, it must be noted that Solid Woks does not allow to indicate in the MDF representation aspects related to molded parts such as the partition line, maximum burrs, feeders, etc.

Fig. 8. Bill of materials of Solid Works and marks for the welded gear

4 Comparative Analysis of Solid Works and Inventor

Figure 9 shows the same part with MBD annotations for a metric thread, a geometric tolerance, a dimensional tolerance and a surface texture.

Table 1 compares Solid Works and Inventor and shows their advantages and disadvantages and summarizes all the previous observations. This table could be really useful to select the adequate program depending on the type of part, assembly, etc.

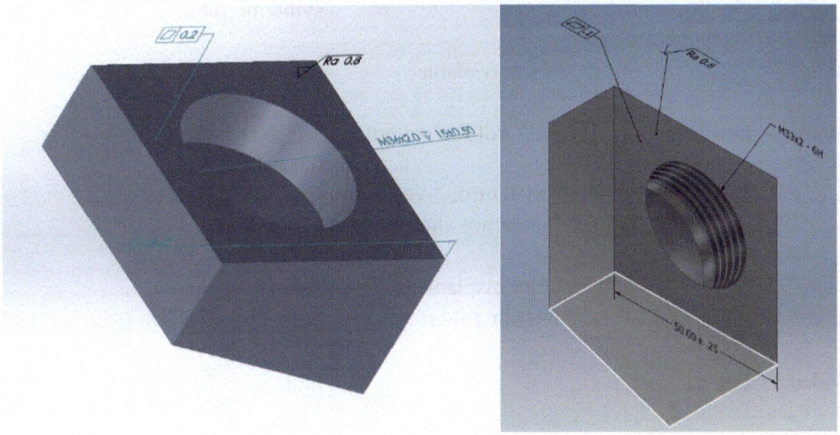

Fig. 9. MBD annotations with Inventor (left) and Solid Works (right)

Table 1. Comparative analysis of Solid Work and Inventor.

	Autodesk Inventor 2018		Solid Works 2018	
	Pros	Cons	Pros	Cons
Dimension note	Full		Full	
Dim tolerances	Full		Full	
Notes	Full			General Tol and Surface finish
References	Full		Full	
Aut pattern feat		Not available	Available	
Holes, threated	Full		Full	
Holes table		Not available		Not available
Geometric toler	Most types	Partial Ref.	Full	
G. tol. automat	Available		Available	
Surface finish	Full		Full	
Welding notes	Full		Full	
Welding drawing	Prep., weld., machining Graphic and symbolic rep	3D Pdf	3D PDF Graphic and symbolic rep	Prep., weld., machining
Bending notes		Not available		Not available
Bending draw	Full	3D Pdf flat pattern	Full, 3D Pdf flat pattern	
Data of the draw	Any type. Included in 3D PDF	MBD title block table not allowed	Any type. MBD Title block table	
Balloons		Not available	Available	
Material bill	Full	MBD Table	Full, MBD Table	
Exploded view	Full		Full	
Gears	Full-PDF 2D	MBD Table Excel	Full-Excel-MBD Table	MBD table not automatic
Ribbed elements	Full	MBD not auto		Not available

(*continued*)

Table 1. (*continued*)

	Autodesk Inventor 2018		Solid Works 2018	
	Pros	Cons	Pros	Cons
Cams	Full	Excel-MBD tab.		Not available
Shaft and axes	Full	Excel		Not available
Belt drivers	Full	Excel		Not available
Springs	Full	Excel-MBD tab.		Not available
Molded parts		Notes not avail		Notes not avail
STEP 304	Full			
PDF 3D	Available	Section view Break view	Available	
PDF templ tool		Not available	Available	

5 Conclusions

The main conclusion that can be obtained from this article is that, currently, Solid Works and Autodesk Inventor, have enough MBD tools to replace the 2D drawings for most of the pieces in the industrial field through representations in 3D that will contain all the necessary information for the manufacture and assembly of the parts. Therefore, it is possible to indicate any type of tolerance and almost any type of annotation, as well as the treatments and surface finish. At this point it should be noted a weakness of Inventor that cannot include partial references.

However, for other types of parts, it should be noted that both parts must improve some aspects in MBD representation of sheet metal parts, as well as in the automatic generation of associated tables by both softwares. In addition, both CAD tools are not able to generate annotations related to molded parts, which is a weak point for both.

Related to the generation of MBD tables, in one hand, only Solid Works allows this possibility, but in another hand, Inventor is clearly superior in the generation of shafts, axes, gears, belt drivers, cams, springs, ribbed zones, etc., but the process to export and to draw the data necessary for mechanization are insufficient.

Analyzing the features related to export to a 3D pdf file, Solid Works is clearly superior and allows to include more information in file and to generate templates (an accessory previously owned by Inventor through Inventor Published and that has disappeared).

Finally, in the representation of the welds, Inventor is the only one that allows the representation of the different phases of the welding process in a simple way, although these representations cannot be exported to a 3D pdf file (see Fig. 10).

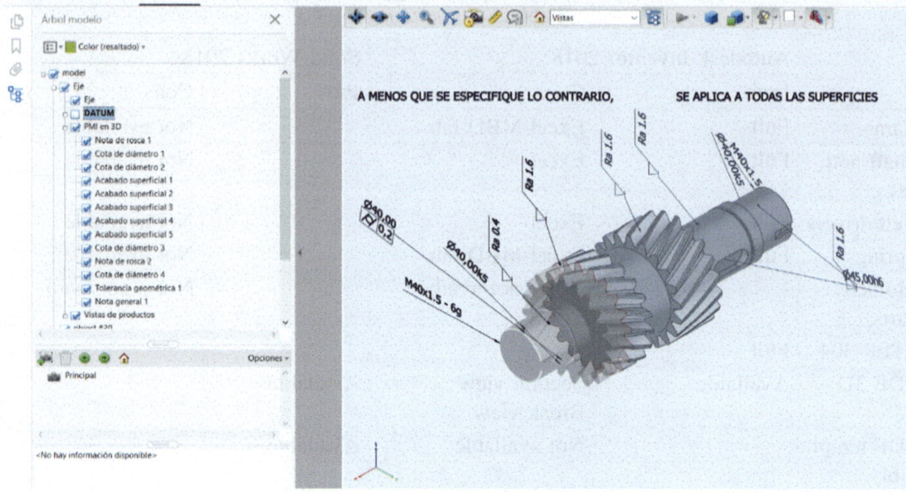

Fig. 10. PDF 3D generated with Inventor

References

1. PTD. https://www.ptc.com/en/products/cad/creo/model-based-definition. Accessed 03 Apr 2019
2. Bijnens, J., Cheshire, D.: The current state of model base definition. Comput.-Aided Des. Appl. **1**(1), 308–318 (2019)
3. Fernandez-Vazquez, A.: Son aún necesarios los programas CAD 2d en el ámbito de la ingeniería y la arquitectura. In: JORNADA INGEGRAF 2016 "Ingegraf+D+i" Contribución de la ingeniería gráfica a la sociedad Universidad de Zaragoza, pp. 165–169. Universidad de Zaragoza, Zaragoza (2016)
4. Barnard, A.: Promoting model-based definition to establish a complete product definition. In: ASME 2016 International Manufacturing Science and Engineering Conference, pp. 1–12. ASME (2016)
5. AENOR regulations. https://www.aenor.com/normas-y-libros/buscador-de-normas. Accessed 03 Apr 2019
6. Inventor help. http://help.autodesk.com/view/INVNTOR/2018/ENU/. Accessed 03 Apr 2019
7. Solid works help. http://help.solidworks.com/. Accessed 03 Apr 2019

Geometrical Study of the Barrel as an Element of the Wine-Making Productive Process. Parameterization and Modelling

Adriana Conde-Fernández(✉), Eduardo Zurita-De La Vega, Pablo Vila-Lameiro, and Patricia Eva Tato-Sánchez del Valle

Dept. of Agroforestry Engineering, Escuela Politécnica Superior de Ingeniería, University of Santiago de Compostela (USC), St/Benigno Ledo s/n, 27002 Lugo, Spain
adriana.conde@usc.es

Abstract. The barrel in the winemaking process is more than a recipient, it also intervenes in its maturation and fermentation. Its geometry determines the exchange of tannins and micro oxygenation, but so far, there is no research that studies the influence of geometry with the stored volume. Geometrically, the free surface according to the barrel filling level determines the alcoholic and malolactic fermentation, which influences the oxygenation of the must. The engineering of the productive process should study the optimal form of the barrel, as a resistant element and connecting it with the inner processes. The geometry influences the productive processes, the cost of the barrel and its storage and consequently, the added value of the wine. The methodology employed in this research is the following: 3D modelling of commercial barrels, measuring inner volume and surfaces with 3D-CAD, comparing results with traditional formulas, parameterizing the geometry and simulating the model with the parametric CAD tools. We conclude that the 3D model enables precise measures and delimits mistakes using the traditionally used empirical formulas. Also, the parameterization of the models allows the functional and economic optimization of the barrel, connected with the ratio of the inner exchange area and the storage volume.

Keywords: Parametrical CAD · 3D modelling · Optimization · Barrels · Winemaking process

1 Introduction

The aim of this paper is to study and optimize the parameterization of the barrel's geometry. The problem with it is that the mounting forms used for parameterization do not use any curve, with the most commonly used being the circumference arcs. Nevertheless, the deformation of the staves that build a barrel should be close to the deformation of a beam with central load and supported on two points. That is to say, an elastic deformation adjusted to a third grade equation and closer to a parabola. Obviously, a stave from the heartwood of a trunk is not an isotropic environment, the layers of phloem and xylem that are generated are not uniform in thickness or radially or

F. Cavas-Martínez et al. (Eds.): INGEGRAF 2019, LNME, pp. 115–122, 2020.
https://doi.org/10.1007/978-3-030-41200-5_13

longitudinally. We can approach the behavior of the material, but each stave is a unique case.

Red wines are aged in barrels because of the barrels contribution of wood notes [...] that quality wines should have. But beyond the flavoring, the wood is a container with the capacity of profoundly modifying the composition and quality of wine (Vivas 2005).

If the barrel is a key element in the maturing process of wine, the relation of the barrel form with the maturing process should be studied as well as, consequently, the geometrical parameters that determine the relation between the storage volume and the wood contact surface. In cooperage, the recipients are defined by the diameter of the head hoop (hereinafter called *d*), the bilge diameter (hereinafter called *D*), the length of the barrel (hereinafter called *L*) and the thickness of the stave. With these dimensions, it could be considered that the barrel, as in Fig. 1, was generated by a curve that passes through points A, B and C when rotating around the axis. This curve could be closer to a parabola and, in case of parameterization of the geometry, a circumference arc that passes through A, B and C should be used.

This communication shows the calculations of the revolution volume and the inner surfaces of both geometrical models as well as the reduction of the margin of error when choosing the circumference arc model confronting the parabolic arc, used in the most popular commercial barrels. Finally, the circumference arc model will be parameterized and the evolution of the capacity volume will be studied in connection with the inner surface. Thus, it will be possible to assess the storage capacity in relation to the wood surface in contact with wine.

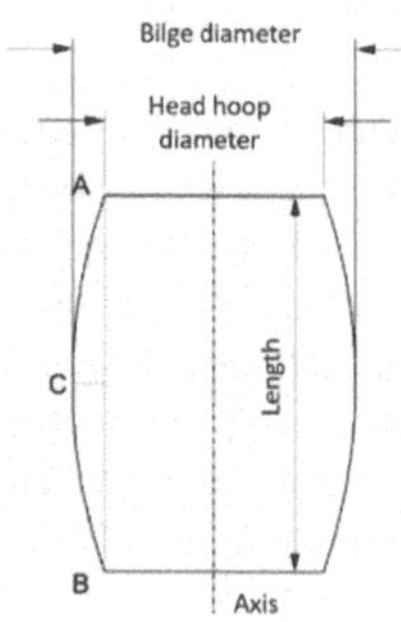

Fig. 1. Parts of the barrel

2 Traditional Empirical Formulas

Until today, tradition has passed the use of empirical formulas that relate the volume to parameters of shape that are the following: bilge diameter, head hoop and length or even the distance from the bunghole to the bottom corner of one of the bottoms. As an example, below (Table 1) are some of the expressions[1] from the book "Tratado de Enología" de Hidalgo-Torres (1997).

For a commercial barrel of 350 l we obtain, with L dimensions L = 1020 mm, D = 830 mm, d = 670 mm and stave thickness of 27 mm, the values in Table 1 (subtracting the thickness to work with the previous dimension).

[1] Where: D = bilge diameter, d = head diameter, L = length and A = distance from upper bunghole to the bottom corner of a bottom.

Table 1. Empirical formulas of barrel capacity.

Volume formulas	A (dm)	L (dm)	D (dm)	d (dm)	Volume (l)
$V_1 = 0.635 \cdot A^3$	8.531	9.66	7.76	6.16	394.25
$V_2 = 0.8 \cdot L \cdot D \cdot d$		9.66	7.76	6.16	369.41
$V_3 = 0.268 \cdot L \cdot (D^2 + D \cdot d + d^2)$		9.66	7.76	6.16	377.89
$V_4 = 0.3927 \cdot L \cdot (D^2 + d^2)$		9.66	7.76	6.16	372.38
$V_5 = 0.694 \cdot L \cdot D^2$		9.66	7.76	6.16	403.70
$V_6 = ¼ \cdot \pi \cdot L \cdot (d + (D - d) \cdot 0.56)^2$		9.66	7.76	6.16	377.73
$V_7 = 0.087 \cdot L \cdot (2 \cdot D + d)^2$		9.66	7.76	6.16	395.02
$V_8 = 0.15 \cdot L \cdot (d + 1.27 \cdot D)^2$		9.66	7.76	6.16	371.65
$V_9 = 0.012 \cdot L \cdot (5 \cdot D + 3 \cdot d)^2$		9.66	7.76	6.16	380.33
$V_{10} = \pi \cdot L/60 \cdot (8 \cdot D^2 + 4 \cdot D \cdot d + 3 \cdot d^2)$		9.66	7.76	6.16	397.95

Taking into account this distribution based on observations and measurements through history, this implies the following statistical results, in Table 2.

Table 2. Statistics on empirical formulas.

Min (l)	369.41
Max (l)	403.70
Geom. mean. (I)	383.850207
Average (I)	384.03
Interval error (I)	34.29

If hypothetically the commercial barrel has a capacity of 350 L, using these expressions the error can be higher than 10%.

3 Delimitation of the Error When Considering a Circular Stave or Parabolic Barrel and the Relation of Inner Surface over Storage Volume

With the 3D modelling tools and according to a commercial barrel catalog (Table 3, Ibérica de Tonelería 2019) a modelling of the volumes has been carried out. Using measuring tools of the CAD modelling, the inner volume and the inner surface of the models have been measured through the revolution of a circumference arc and a parabolic arc, bearing in mind the thickness of the stave used.

Table 4 shows the error results when using a circular model instead of the parabolic. There is an excess error delimited between 8 and 14% in relation to volume while when considering the inner surface the default error is between 8 and 11%.

When it comes to the maturing process and the exchange of tannins for red wines, the barrels with less volume are best suited because the ratio surface per exchange liter favors these sizes.

Table 3. Commercial dimensions of barrels

Barrel type	Weight (kg)	Height (mm)	Diameters (mm)		Thickness (mm)
			Head-hoop	Bilge	
300 l	58	1008	610	780	27
350 l	63	1020	670	830	27
400 l	67	1050	690	850	30
500 l	82	1092	780	950	32

On the contrary, barrels with a longer length have a smaller exchange relative surface (Table 4).

Table 4. Length and surfaces of the circular and parabolic models and error delimitation in connection with the parabolic model.

Barrel type	Circular model			Parabolic model			% vol. error	% inner superior error
	Inner surface	Vol.	m^2/l	Inner surface	Vol.	m^2/l		
300 l	2.15890	0.35700	0.00605	2.15650	0.35670	0.00605	0.08410	−0.11129
350 l	2.35300	0.42080	0.00559	2.35100	0.42020	0.00559	0.14279	−0.08507
400 l	2.46690	0.45000	0.00548	2.46500	0.44940	0.00549	0.13351	−0.07708
500 l	2.89020	0.59290	0.00487	2.88800	0.59210	0.00488	0.13511	−0.07618

This circumstance leads us to consider how the geometry of the barrel, increasing the relative surface of exchange between the wine and wood, shortens the maturing process and what the storage and productive processes would cost. Thus, it would be possible to favor price competition with equal quality or even to increase the industrial revenues.

4 Experimentation in the Variation of the Barrel Geometry with the Help of Parametric CAD

If the geometry of the section of the barrel is parameterized and with these parameters, the volume and the inner surface of the barrel are introduced in the CAD, it will be possible to obtain the m^2 ratio of the inner surface as opposed to an exchange volume. All of these values will be regarding the parameters of the head hoop diameter, bilge diameter and barrel length. The deformation arrow of the stave that is connected with these parameters is the consequence of the experience in the stave deformation and is connected to the thickness of it and with the nature of the materials used. If this value is admitted without exceeding the wood fatigue in the deformation, carrying out a simulation will be possible by increasing the head hoop diameter or increasing the length of the barrel. In both cases, the deformation of the stave is reduced if the bilge diameter is maintained.

With the parametric CAD, once all of the formulas have been introduced, if the parameter head hoop diameter is introduced and the mousse is automatically moved, the parameter value changes, simultaneously changing the section shape and the values of the rest of the parameters, including the ratio inner surface - exchange volume. This can be repeated simulating the variation of the barrel length and, with the results of the simulation, the optimal geometry for the barrel can be chosen.

The parameterization of the section can only be done with circumference arcs, therefore, to carry out the parameterization and simulate the model, the volume and surface functions depending on the head hoop diameter, bilge diameter and length of the barrel are needed.

5 Volume and Surface of the Barrel Considering the Revolution of a Circumference Arc Through the Infinitesimal Calculus

Obviously, since the emergence of the infinitesimal calculus, the use of empirical formulas does not make sense. Considering that the stave arc is a circumference arc, we could calculate the volume, which will be calculated below, with data from Fig. 2.

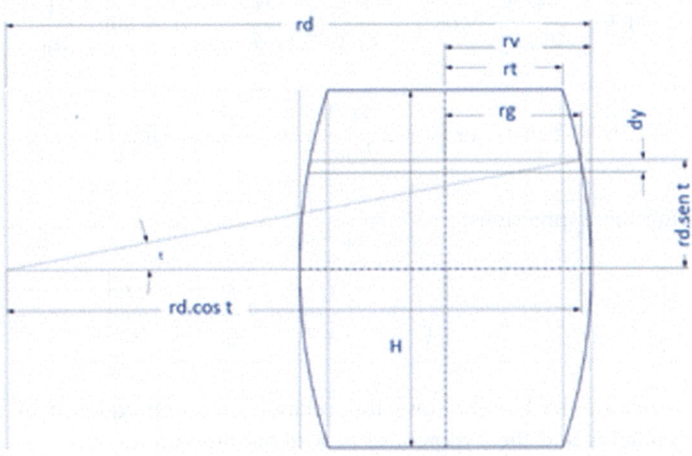

Fig. 2. Model components

The parametric equation of the circumference would be:

$$\begin{cases} x = r_d \cos t \\ y = r_d \sin t \end{cases} \qquad \begin{cases} dV = \pi r_g^2\, dy \\ dy = r_d \cos t\, dt \end{cases}$$

$$r_g = r_d \cos t - (r_d - r_v) = r_d(\cos t - 1) + r_v \tag{1}$$

The volume of half of the barrel when varying t from 0 until t_1, with this last value being the angle that makes the horizontal line that crosses the center of the stave arc with the ratio that goes to the end of the head hoop:

$$dV = \pi r_d^3 cos^3 t\, dt + \pi\left(2r_d^2 r_v - 2r_d^3\right) cos^3 t\, dt + \pi\left(r_d^3 - 2r_d^2 r_v\right) \cos t\, dt \tag{2}$$

Half of the volume of the barrel will be:

$$V = \pi r_d^3 \int_0^{t_1} cos^3 t\, dt + \pi\left(2r_d^2 r_v - 2r_d^3\right) \int_0^{t_1} cos^2 t\, dt + \pi\left(r_d^3 - 2r_d^2 r_v\right) \int_0^{t_1} \cos t\, dt \tag{3}$$

Substituting, we would have the following expression:

$$\frac{V}{2} = \frac{\pi r_d^3}{3} \sin^3 t_1 + \frac{\pi}{4}\left(2r_d^2 r_v - 2r_d^3\right) \sin 2t_1 + \pi\left(r_d^3 - 2r_d^2 r_v\right) \sin t_1 + \pi\left(r_d^2 r_v - r_d^3\right) t_1 \tag{4}$$

According to the geometry of the barrel:

$$\sin t_1 = \frac{H}{2r_d} \qquad \cos t_1 = \frac{1}{r_d}\sqrt{r_d^2 - \frac{H^2}{4}} \qquad t_1 = \sin^{-1}\frac{H}{2r_d}$$

$$\sin 2t_1 = 2 \sin t_1 \cos t_1 = 2\frac{H}{2r_d}\frac{1}{r_d}\sqrt{r_d^2 - \frac{H^2}{4}} = Hr_d^2\sqrt{1 - \frac{H^2}{4r_d^2}} \tag{5}$$

Substituting the expressions:

$$V = \pi H r_d \left[\frac{H^2}{12 r_d} - r_d^2(r_d - r_v)\left[\sqrt{r_d^2 - \left(\frac{H}{2}\right)^2} + 2\frac{H}{r_d}\sin^{-1}\frac{H}{2r_d}\right] + (r_d - 2r_v)\right] \tag{6}$$

The geometry of the barrel shows this relation as a consequence of the relation between the cathetus and the hypotenuse of a right-angled triangle:

$$r_d = \frac{\frac{H^2}{4} + (r_t - r_v)^2}{2(r_t - r_v)} \tag{7}$$

The complexity of this expression justifies the use of empirical formulas bearing in mind, on one side, an artisan industry of barrels, dominated by a century-old tradition.

But even using the infinitesimal calculus, bear in mind that the first digital calculator does not appear until 1965 and its implementation is in the decade of the 70's. Until that moment, carrying out a complex calculus of trigonometric and exponential expressions had to be done with log tables or with the ability and the limitations of the accuracy of the calculation rule. Today we can think about spreadsheets, but remember that Microsoft Excel did not appear until 1987. That is to say, we started using it only 30 years ago. Seeking easier expressions at that historical time was understandable, but nowadays, if the objective is to optimize the design, it is not possible to continue with that kind of dynamic.

We have analyzed the volume that will affect the stock management but for the purpose of micro oxygenation through the staves that influence the fermentations and the tannins exchange, the inner surface of the barrel should be taken into account. Interestingly, in the references this piece of data is not available because the study of the volume has been only oriented towards stock management.

According to the maturing process, the exchange of tannins should be minimal for fruity wines. However, in the case of red wines, if the objective is to shorten the maturing process with the same tannin level, it is interesting to have a greater contact surface with the barrel wood. In any case, the study will have to do with the ratio between the volume and the exchange surface of tannins or micro oxygenation.

It is necessary to know the inner surface of the barrel. If we consider revolution surface, it can be determined if we have the parametric equation of the stave arc, through the following expression if the revolution axis matches the OY axis and the axis ratio is the x coordinate:

$$\begin{cases} x = f(t) = r_d \cos t - (r_d - r_v) \\ y = g(t) = r_d \sin t \end{cases} \qquad \begin{cases} x' = -r_d \sin t \\ y' = r_d \cos t \end{cases}$$

$$S = 2\pi \int_{t_2}^{t_1} f(t)\sqrt{[f'(t)]^2 + [g'(t)]^2 dt} \tag{8}$$

Substituting and simplifying:

$$S = 4\pi r_d \left[\frac{r_d H}{\sqrt{4r_t^2 + H^2}} + (r_d - r_v)\tan^{-1}\frac{H}{2r_t} \right] \tag{9}$$

With this last expression, in parametric CAD, it is possible to simulate by changing the values of the parameters and seeing the evolution of the inner surface ratio/storage volume.

Bearing in mind that these parameters affect the deformation arrow of the stave, the simulation trial will have to adjust to the following criteria:

- Maintaining *D* constant, slightly increase or decrease *d*
- Maintaining *D* and *d* constant, slightly increase or decrease *L* of the barrel.
- Maintaining *L* constant, increase *D* and *d* by the same amount.

6 Conclusions

1. 3D CAD provides accuracy when taking decisions in relation to the process of barrel elaboration as well as an evolution in the decision-making process, in comparison to the artisan process.
2. The parametric CAD is an unavoidable tool in the research on optimization of the geometry related with the design function.
3. Maintaining *D*, if *d* is increased, maintaining *L*, the ratio inner surface/storage volume is reduced, as well as the tannin exchange, being a more convenient geometry for fruity and white wines, whisky or beer or to store wine that needs slower maturing.
4. Maintaining *D*, if *d* is decreased, maintaining *L*, the ratio inner surface/storage volume is increased, as well as the tannin exchange, being a more convenient geometry for red wines.
5. Maintaining the *L* constant, increasing equally *D* and *d*, does not increase with the same speed facing the variation of the parameters that in the essays mentioned in the conclusions 2, 3 and 4 and might be a criteria to bear in mind for stock optimization.

References

Vivas, N.: Manual de tonelería. Ediciones Mundi-Prensa (2005)

Hidalgo-Torres, J.: Tratado de enología. Volumen I-II. Ediciones Mundi-Prensa (1997)

Martel-Moreno, J.: Fórmulas generales para la determinación de áreas y volúmenes. Universidad de Gran Canaria. Rev. El Guiniguada Nº 8/9, pp. 308–331 (1999–2000)

Website: Ibérica de Tonelería (Formato de barrica y capacidades). http://www.ibericadetoneleria.com/es/dargaud-jaegle-es/. Accessed 03 June 2019

Walking Mechanisms: Study and Design for Use in Sandy Environments

Gregorio Romero Rey(✉), M. Luisa Mtz. Muneta, Antonio Carretero Díaz, Juan de Dios Sanz Bobi, and Jorge Gonzalez-Onieva

Grupo de Ingeniería Gráfica, Departamento de Ingeniería Mecánica, ETSII, Universidad Politécnica de Madrid, C/José Gutiérrez Abascal, 2, 28006 Madrid, Spain
gregorio.romero@upm.es

Abstract. Since the beginning of time, the human being has tried to adapt to the environment and has had the ability to overcome obstacles in a comfortable and effective way. There exist mechanisms with the ability to move on sandy or rough terrain, where it is very difficult to move or even practically impossible for systems that move by using wheels. Therefore, it is considered vital to study and analyze these kinetic structures in order to provide a series of guidelines aimed at improving their design for application in sandy or desert areas. In this work, and after analyzing the possible advantages of the walking mechanisms, as well as studying other existing ones, we will undertake the design of one of them that can be easily transported. I will be low cost and simple to assemble so it will be able to manufacture the specimens that are needed in the various places where you want to use it. All this taking into ac-count different fundamental parameters.

Keywords: Mechanisms · Design · Parameterization

1 Introduction

Since the invention of the first machines, developing mechanisms that have the ability to "walk" [1], animals have been in the mind of designers. These mechanisms have evolved in just a few years trying to adapt to the environment and with the ability to overcome obstacles in a comfortable and effective way. One of the main advantages of these mechanisms is the ability to move on sandy or abrupt terrain, where it is of great difficulty or impossible for mechanisms that are moved by wheels.

There are many applications and areas where these mechanisms could be applied, such as agriculture, the displacement of people with disabilities on land as common as the beach [2], or even for the movement of devices on the moon. For this reason, it is considered an opportunity to analyze these structures and be able to advance in their design. In this work we will focus on the design of a walking mechanism that can be transported easily, with low cost and a relatively simple assembly to be able to manufacture numerous specimens in different places where you want to use.

F. Cavas-Martínez et al. (Eds.): INGEGRAF 2019, LNME, pp. 123–131, 2020.
https://doi.org/10.1007/978-3-030-41200-5_14

In this work, these mechanisms will be studied in order to design a walking structure whose parts are cut in plywood with an automatic milling machine, in predetermined planks and saving both material and production in order to reduce the environmental impact that it may have.

To do this, in a first stage the main advantages of the mechanisms that have the ability to "walk" through the terrain will be analyzed, in order to understand the possible advantages and utilities in different areas. The proposed design will be based on the proportions of Theo Jansen's mechanism [3–5] and it will be a structure capable of "walking" accumulating the energy of the wind.

In a second stage, the design of the structure will be carried out, where the different parts of the mechanism will be designed and its placement will be defined in the final assembly, for which a series of fundamental parameters such as dimensions or cutting conditions for its manufacture will be taken into account.

Next, we will proceed to the study of different key points in the structure, which will allow us to know in which terrains the mechanism designed can be moved, which obstacles can be overcome and the kinematics thereof will be analyzed.

Finally, possible uses of the designed mechanism will be analyzed, having been adjusted the pieces as much as possible in order to reduce the amount of waste and the cutting path of the milling machine. In such a way that not only is reduced the wear of the milling machine but also the energy consumption and the environmental impact of the project.

2 Analysis of Movement by Wheels and Legs

The main advantage of the mechanisms that move with "legs" is found in the irregular terrains since this mechanism allows moving over obstacles without causing many friction losses, as is the case of the one that can occur in the case of the wheels (Fig. 1).

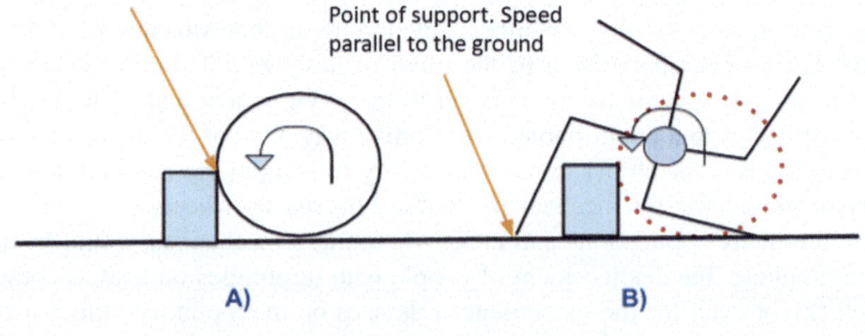

Fig. 1. Reactions in front of an obstacle. (A) rolling mechanism. (B) walking mechanism.

As can be seen by analyzing case (A), it is observed that when the displacement direction hits, it begins to have a vertical component to the terrain, braking the mechanism, while in case (B) the 4 legs that are part of the mechanism they would avoid the obstacle while preserving the horizontal displacement.

Another point of view is the damage that occurs to the ground, since while the wheel moves horizontally having contact with all the terrain through which it moves, the mechanisms with "legs" only require certain supports at certain distances, just like the animals when walking, which prevents damage to the entire floor. In addition, the wheels are less stable in the contact against changes of terrain since they can cause detachments in the same (Fig. 2).

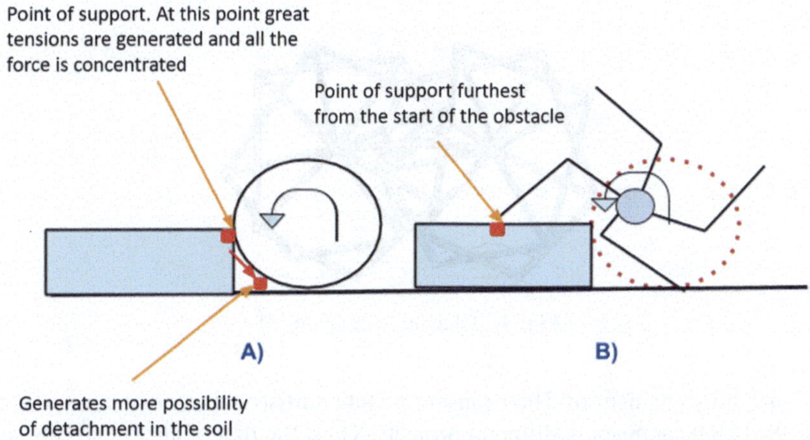

Fig. 2. Support Point. (A) rolling mechanism. (B) walking mechanism

3 Walking Mechanisms

There are different mechanisms capable of reproducing movements similar to those necessary for walking, although, nevertheless, most of these structures do not have much stability or require many legs to be stable. Some of the most important mechanisms in this field are the Klann mechanism, designed in 1994 using the Burmester theory [6], in which it is studied the kinematics of a flat mechanism that attempts to imitate the movement of animals with several legs through the movement of a crankshaft (Fig. 3).

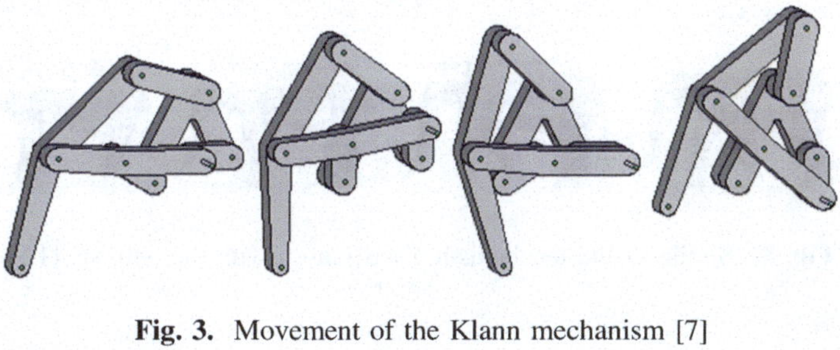

Fig. 3. Movement of the Klann mechanism [7]

The Klann mechanism offers many of the benefits of the most advanced vehicles to walk without some of its limitations and can overcome curbs, climb stairs, or move through areas where it would be impossible to do it with wheels.

Similarly, there is the Ghassaei mechanism [8], which is inspired by the Theo Jansen mechanism and consists of 2 pairs of legs, as shown in Fig. 4, where it´s possible to see the two trajectories of the parts lower of the mechanism or points of support, as well as the red trajectory associated with the displacement in contact with the ground and the blue with the elevation of the point of support to overcome possible obstacles.

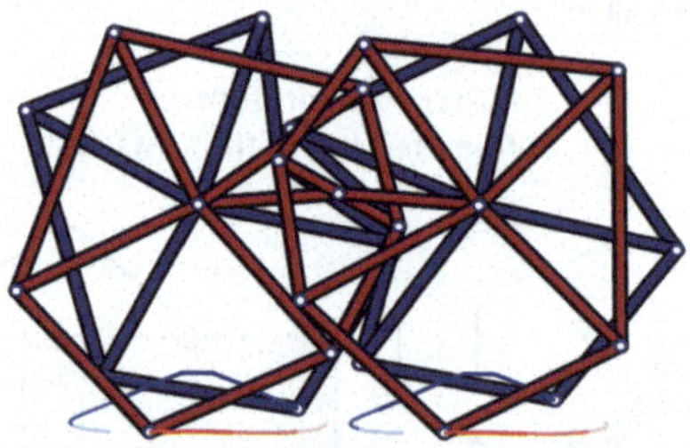

Fig. 4. Ghassei Mecanism [8]

As for the mechanism of Theo Jansen, a Dutch artist with a great connection with physics [9–11], he appears in different projects. Thus, the first of all, "the flying saucer" was a project that began in 1974 building a PVC-like structure similar to a flying saucer and which later filled it with helium and in 1980 let it fly on a cloudy day over Belgium.

Fig. 5. Simile of the mechanism Theo Jansen with the animals [12]

Another of his inventions was a machine that painted according to the change of light it received. Its main attribution was the work called "strandbeest" - beach animal in Dutch-, able to move on sand with the push of the wind and composed of different legs that simulate the movement of four-footed animals when walking (Fig. 5). The movement of the legs is transmitted by a circular movement transmitted by a crankshaft, which makes it can be simply connected to a rotary engine on the crankshaft and walk by itself.

4 Mechanism Design

From the kinematic study of the movement of the mechanism, where it´s possible to see the route of its different parts, a leg has been designed since its operation is symmetrical and the same in the 5 pairs of legs considering a walking structure of approximately 500 × 500 × 500 mm volume, as shown in the following figure (Fig. 6).

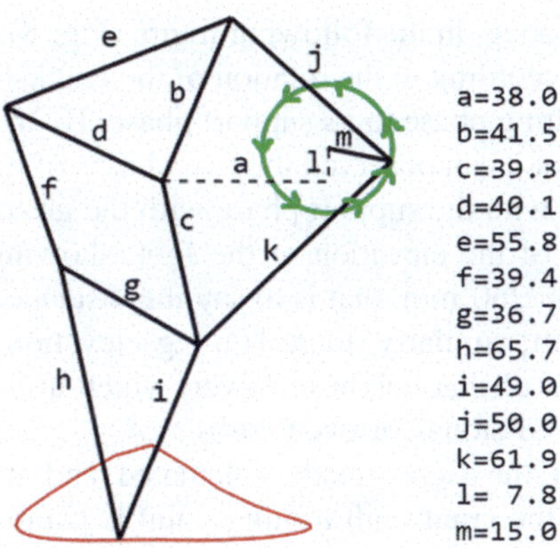

Fig. 6. Crankshaft movement in the Theo Jansen structure [2]

Thus, to simplify the dimension of the element "c" it has been modified, making it equal to that of the element "f", which would be considered the same piece and reduces the number of different parts in the assembly. On the other hand, the *e-d-b* and *g-h-i* assemblies have been taken as non-deformable triangles, so it can be formed as two triangular pieces, showing in the following figure the final scheme of one of the pairs of legs in a characteristic position, specifically at the moment of raising the point of support (Fig. 7).

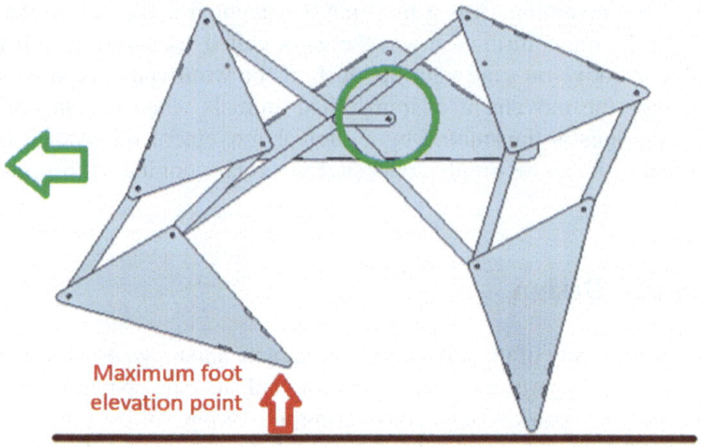

Fig. 7. 90° position of the crankshaft. Point of maximum elevation of the "foot"

In this way, isolating a leg, in the following figure (Fig. 8) the different positions of the leg can be observed according to the rotation of the crankshaft, being able to see the path of the leg from its lifting phase to its support phase. By tracing the trajectory of the point of support of the foot it can observed the two phases of its route: the lower line (in green) would correspond with the support phase with the ground, while the second one (in red) defines the path of the elevation of the foot, showing a step of the designed structure of approximately 260 mm, that is to say the distance of support in the ground in a turn of the crankshaft; similarly, there is a leg elevation of around 95 mm, so it could overcome obstacles of that height or lower, which allows the structure to move on land with medium-sized stones or even curbs.

With the distance of the step already calculated and its elevation, it has been verified that the route of the crankshaft turning point is parallel to the ground in order not to produce elevations or instabilities to the machine, having observed that the support is parallel to the ground and which one has a smooth progressive elevation. Similarly, it has been seen that the triangle *e-d-b* performs an exclusively rotary movement since it has a fixed point, and an angular path of 40–50°.

Based on a DC electric motor for turning the crankshaft [13], for example with specific characteristics of 2500 rpm, 12 V and a torque of 0.93 MNm, a speed between 18 and 20 km/h will be obtained depending on the degree of existing slip on the ground, so it is considered optimal for use in smooth movements.

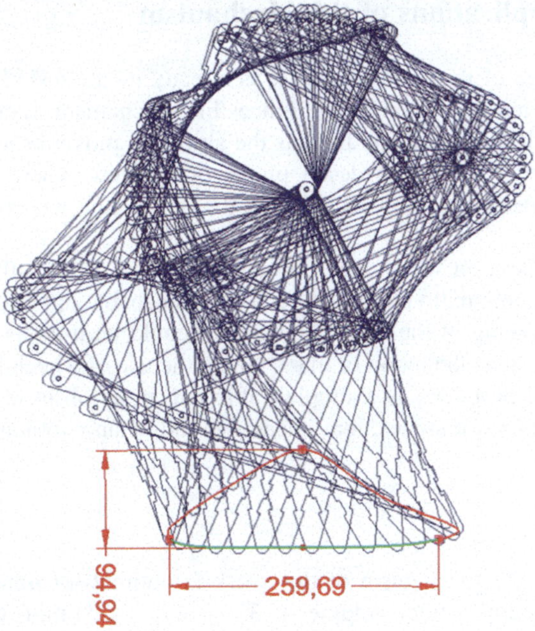

Fig. 8. Leg motion sequence in the Theo Jansen structure [2]

5 Prototype Construction

A prototype has been built in plywood. The different parts except the crankshaft have been milled and assembled in scale 1:1. The Fig. 9 shows the wood parts and the final object.

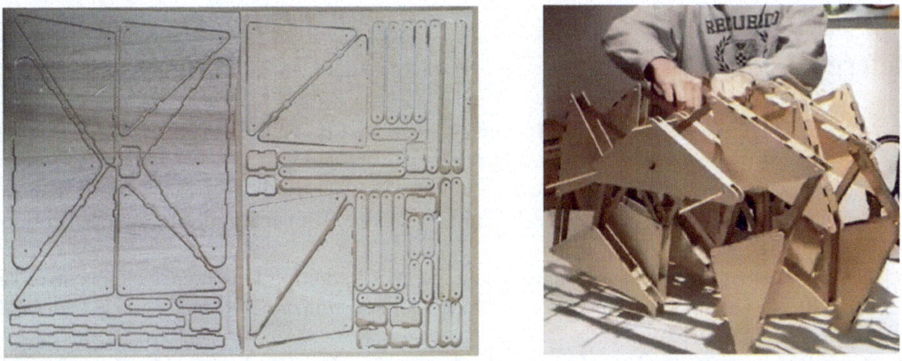

A) B)

Fig. 9. Prototype construction. (A) Parts of the mechanism. (B) Final prototype

6 Possible Applications of the Mechanism

The main advantage of this mechanism is, as already mentioned above, the displacement by complex terrains of any type, since this mechanism manages to overcome obstacles close to 10 cm in height and has the ability to move through sandy soils, so be used as a transport trolley in desert places or beaches, where it would allow to transport, in a lighter way, the load to avoid the jam of the wheels of the traditional trolleys in the sand.

In addition, an area in which this mechanism could evolve is the field and everything related to agriculture, to be able to get a design capable of performing work in the field such as the sowing of the land. Similarly, it could be used as a rover for exploration in terrains such as the moon or mars, where the use of wheeled vehicles causes a series of associated problems as a simple traffic jam in an obstacle along the way, or being in front to a row of stones, may mean the end of an exploratory mission [14].

7 Conclusions

It can be said that the mechanism of this work is optimal for transport since all the pieces together occupy a total volume of 33 × 400 × 600 mm, besides being plywood, so it is a light structure. Therefore, it would allow it to be easily sent to places where mechanisms of this type are necessary, such as the desert, a beach, abrupt field or even the moon, being at the same time, an economic product accessible to any person.

Although this work tries to be a step towards the advance in the design of these structures and their future applications, both of transport of merchandise and its application in agriculture or sandy land, its validity can be observed as a substitute design to other types of vehicles that are traditionally used in such places.

The future lines of work in this mechanism are oriented to the incorporation of motorization, -possibility of flexing the crankshaft-, improvement of the supporting point and some aspects related to the transport capacity (mass, speed, etc.).

References

1. Avello, J.: Teoría de Máquinas. Tecnun (Universidad de Navarra) (2014). ISBN 978-84-8081-250-4
2. Mohsenizadeh, M.: Kinematic Analysis and Simulation of Theo Jansen Mechanism. Birmingham 2015 Early Career Technical Conference, University of Alabama at Birmingham (2015)
3. Borges, J.S.: Aplicação dos mecanismos de locomoção do tipo Jansen em cadeiras de rodas. 2016. 193 f. Dissertação (Mestrado em Modelagem e Otimização) - Universidade Federal de Goiás, Catalão (2016)
4. Asok Raj Kumar, M., Ramalingam, N., Ajithkumar, B., Arunkumar, M., Boopalan, M., Jeyamurugan, K.: Design and Fabrication of Spider by Using Klann's Mechanism for Remote Sensing (2018)
5. Sánchez Ron, J.M., Artola, M.: Ciencia: lo que hay que saber. Editorial Planeta (2017). ISBN 9788467050110

6. Bustamante Tellez, J.: Diseño de mecanismo de locomoción andante con cambio de dirección. TFG. Universidad Pública de Navarra (2016)
7. Kulandaidaasan Sheba, J., Martinez-Garcia, E., Mohan, R.E., Le, T.-P.: Design and evaluation of reconfigurable Klann mechanism based four legged walking robot, pp. 1–5 (2015). https://doi.org/10.1109/icics.2015.7459939
8. Ghassaei, A.: The Design and Optimization of a Crank-Based Leg Mechanism. Thesis. Pomona College Department of Physics and Astronomy (2011)
9. Herzog, L.: Strandbeest. The Dream Machines of Theo Jansen. https://www.taschen.com/pages/es/catalogue/art/all/05782/facts.lena_herzog_strandbeest_the_dream_machines_of_theo_jansen.htm
10. Patnaik, S.: Analysis of Theo Jansen Mechanism (Strandbeest) and its comparative advantages over wheel based mine Escavation system. IOSR J. Eng. (IOSRJEN), **05**(07), 43–52 (2015). ISSN 2278-8719
11. IMMA. Vehículos sin ruedas basados en el mecanismo de Theo Jansen (2014). http://immf.uma.es/diseno_maquinas.php?PHPSESSID=ec773d1813d31cb3331d934da869ac73&accion=ver&linea_sel=10
12. Herrera, I.: Muestra Theo Jansen en el Espacio Telefónica (2016). https://espacio.fundaciontelefonica.com/evento/theo-jansen/
13. Elmeq. Product catalogue (2018). https://www.elmeq.es/
14. Xiao, L.: China's touch on the Moon. Nat. Geosci. **7**(6), 391–392 (2014). https://doi.org/10.1038/ngeo2175

A Proposed Methodology for Calculating the Rigid Body Natural Frequencies of EPDM Rubber Fixed Supports with the Finite Element Method (FEM)

Eduardo Jiménez-Ruiz[1](✉), Fátima Somovilla-Gómez[2], Saúl Iñiguez-Macedo[2], Carlos Berlanga-Labari[1], Marina Corral-Bobadilla[2], and Rubén Lostado-Lorza[2]

[1] Public University of Navarre, 31006 Pamplona, Navarra, Spain
Eduardo.jimenez@bshg.com
[2] University of La Rioja, 26004 Logroño, La Rioja, Spain

Abstract. This paper proposed a methodology for calculating the rigid body natural mode shape and frequencies of a piping system that are formed by a compressor and rubber supports with the aim to improve its reliability and durability. The rubber supports are manufactured by Ethylene Propylene Diene Monomer (EPDM) hyperelastic material. The methodology combines the Finite Element Method (FEM), experimental modal analysis (EMA) and ANSYS Workbench parameter evaluation software. First and using SpaceClaim Software, the geometry of the EPDM support was generated and imported in the ANSYS Workbench Finite Element Analysis Software. Then, a 3D parameterized finite element model was created in ANSYS Workbench considering tetrahedral elements with linear formulation. An experimental modal analysis was performed on said rubber supports, and using Brüel & Kjaer BK Connect Modal Analysis and BK Connect Geometry Software, the rigid body natural mode shape and frequencies were obtained. Subsequently, these natural mode shapes and frequencies obtained experimentally were compared with those obtained from the ANSYS Workbench Finite Element simulation varying Young's Modulus of rubber supports. An agreement between the natural mode shapes and frequencies obtained from the experimental analysis and the FEM demonstrates that the proposed methodology could be valid for improve the design process of these type of products.

Keywords: Ethylene Propylene Diene Monomer (EPDM) · Natural frequencies · Finite Element Method (FEM) · Finite Element Modal Analysis (FEMA) · Experimental Modal Analysis (EMA)

1 Introduction

In the design process of a piping system, for applications where compressors are needed, one of the main problems that need to be solved is the calculation of its natural frequencies. To avoid the effects of dynamic amplification, the compressors must always avoid working on the natural frequencies of the pipe system. In order to achieve

F. Cavas-Martínez et al. (Eds.): INGEGRAF 2019, LNME, pp. 132–141, 2020.
https://doi.org/10.1007/978-3-030-41200-5_15

enough precision in the calculating these frequencies, it is very important to correctly characterize the boundary conditions of the whole system. The main components in this type of applications are the compressors, playing an important role in the characterization of boundary conditions of the whole system. In addition, the system formed by grommets (rubber supports) and compressors, has its own natural frequencies, which are the rigid body natural frequencies of the whole system. There-fore, having a good characterization of natural frequencies is essential. A typical material with which the rubber grommets are manufactured is the Ethylene Propylene Diene Monomer (EPDM). The calculation of the natural frequencies by means of a modal analysis (linear calculation) with the Finite Element Method (FEM) of a hyper-elastic material (non-linear mechanical behavior) is complicated, and the data obtained may be incorrect. However, when the EPDM is subjected to a low load and undergoes small strains, its mechanical behavior can be considered linear elastic, and therefore, Young modulus (E) could be assumed also as linear elastic [1]. For EPDM, E is not linear with the frequency, but its value can be assumed as linear for the nor-mal operating range of the whole system (0 to 50 Hz) [2, 3]. Some authors [4] already performed similar investigations, obtaining a good approach for determining the rigid body natural frequencies for simple geometries for the rubber supports. Other authors [2] made a simplification of the grommets with a radial and axial stiffness for static and dynamic loads, but the rotational stiffness was not characterized. In a more recently research work [5], the low frequency vibration of rotary compressors were studied. In this case, a good correlation between a numerical method, FEM and measurements of rigid body natural frequencies were found. In the current paper, a Finite Element Modal Analysis (FEMA) is proposed for calculating the rigid natural frequencies, assuming a linear behavior for the grommets. An Experimental Modal Analysis (EMA) was developed for validating the linearity considerations assumed in the FEA. An agreement between the natural mode shapes and frequencies obtained from the FEMA and the EMA were found.

2 Finite Element Model Proposed

The FEMA was performed in this study assuming that the mechanical behavior of EPDM supports was linear elastic with small strains. The system simulated was formed by a rigid body (compressor), which was simulated considering its center of gravity and its principal inertia moments supported by three EPDM grommets. ANSYS Space-Claim Software [6] was used for designing the geometry of grommets (See Fig. 1).

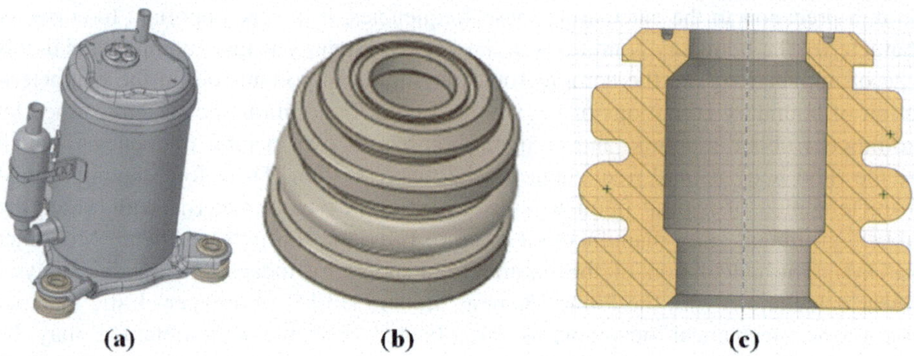

Fig. 1. ANSYS SpaceClaim design: (a) Compressor; (b) 3D Grommet image; (c) Grommet detail.

After the design process was finalized, the geometry was imported to ANSYS Workbench to perform the FEMA. This process was developed in the following steps:

- A material with linear mechanical behavior was created considering and elastic Young's Modulus (E).
- A Tetrahedral mesh was selected for modeling the grommets (Fig. 2a).
- The rigid body was modeled as a point mass, which was attached to the upper surface of the grommets (Fig. 2b).
- The Inertia properties on the centre of mass of the rigid body was considered.
- A fixation boundary condition on the bottom surface of grommets was defined (Fig. 2c).

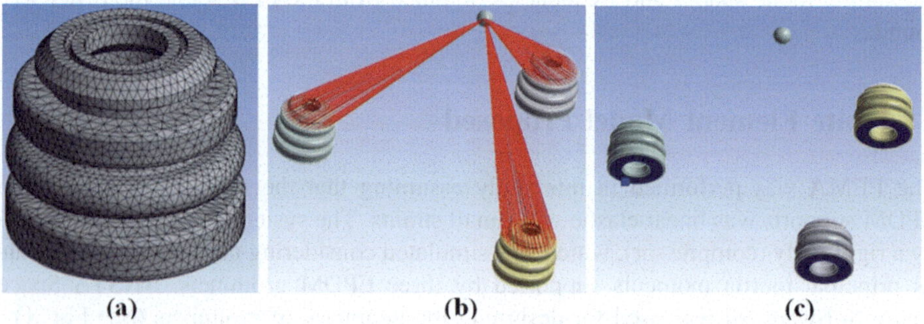

Fig. 2. ANSYS Workbench steps: (a) Fine mesh of a grommet; (b) View of the mass connectors; (c) Boundary condition for fixing the bottom surfaces of the grommets.

As previously mentioned, a tetrahedral element mesh was selected instead of hexahedral mesh. Tetrahedral elements are usually used in modal analysis instead of the hexahedral element because the FEM models with complex geometries converge more quickly if this type of elements is selected. A comparison of number of elements needed and the natural frequency reached assuming tetrahedral and Hex dominant meshes are shown in Table 1.

Table 1. Ansys workbench modal analysis summary results

Natural frequency	Coarse mesh (Tet./Hex.)	Medium mesh (Tet./Hex.)	Fine mesh (Tet./Hex.)
1st [Hz]	8.3/8.2	8.1/8.1	8.1/8.0
2nd [Hz]	8.3/8.2	8.1/8.1	8.2/8.0
3rd [Hz]	21.7/21.5	21.3/21.2	21.3/21.1
4th [Hz]	24.0/23.6	23.5/23.2	23.4/23.2
5th [Hz]	29.3/28.8	28.7/28.4	28.6/28.4
6th [Hz]	29.9/29.5	29.3/29.0	29.1/28.9

From this table it is observed that the differences between tetrahedral and hexahedral results are less than 1 Hz, which can be considered as a valid error for this analysis. In addition, it is recommended to work with higher number of elements in order to have more precision in the obtained results. However, a greater number of elements also require a higher computational cost, besides working with a smaller number of elements facilitates the work with BK Connect Structural Dynamics [7]. For this study, the type mesh chosen for the FEMA were fine mesh with tetrahedral elements. In addition, the E of EPDM grommets was estimated through Shore (A) hardness test and Eq. (1) according to [8]. The calculation of E is approximate, but it gives a starting point to iteratively find through FEMA the most appropriate values of E that correspond to the grommets

$$E = \frac{1-\mu^2}{2R}\frac{C_1 + C_2 Sh(A)}{C_3(100 - Sh(A))} \tag{1}$$

Where:

C_1 = 0.549 N (Offset constant between indentation Force and Shore hardness)
C_2 = 0.07516 N (Linear constant between indentation Force and Shore hardness)
C_3 = 0.025 N (Linear constant between indentation depth and Shore Hardness)
Sh_A = Shore (A) hardness
μ = Poisson's ratio
R = Indenter's Radius (0.395 mm)
E: Young's Modulus in N/mm^2

According to the theory of elasticity, it is possible to deduce an indentation depth (w) of shore indenter, calculating its value as the depth of a uniform load distributed over the area of a circle [9]. In that case, the maximum indentation depth that is produced at the centre on the indenter could be calculated as:

$$w_R = \frac{1}{2}\frac{F}{R}\frac{1-\mu^2}{E} \tag{2}$$

In addition and based on ISO 868:2003 [10], it is assumed linearity between indentation force and Shore-A hardness (Eq. 3) as well as between indentation depth and Shore-A hardness (Eq. 4)

$$F = C_1 + C_2 Sh(A) \tag{3}$$

$$w = C_3(100 - Sh(A)) \tag{4}$$

Then, several FEM simulations, just changing E, in the range of 1 MPa to 10 MPa were carried. This range was selected based on Eq. (2), and values from literature [11–13].

3 Experimental Modal Analysis

In order to validate the results obtained from the FEMA, an EMA with impact hammer was performed. For a correct development of this test, it is highly recommended to create a geometry describing the position and orientation of accelerometers and hammer impacts. For this task and using BK Connect, the geometry was imported from the FEM mesh previously created in Ansys software. This process, called "decimation" (Fig. 3) is able to create a geometry based on mesh's nodes, which allows to maintain the original position of the nodes (important data for the experimental analysis) while saving time in creating a new geometry.

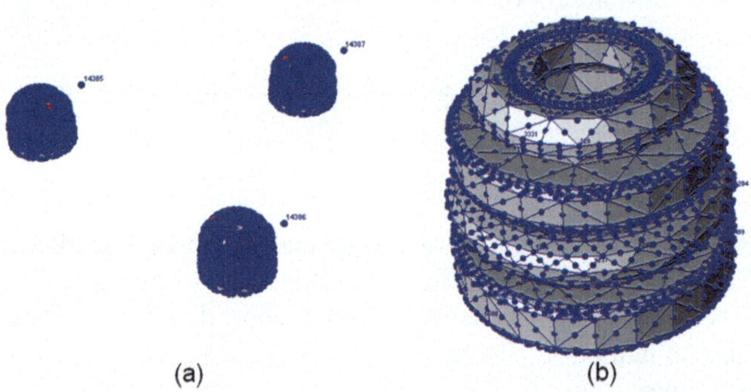

Fig. 3. Decimation process (a) General view. (b) Detail of decimation process and mesh.

The natural frequencies calculated from the FEMA were also imported with BK Connect, which allow minimizing the number of accelerometers needed to avoid losing any natural frequency during the EMA. For that, BK Connect check the selected locations and the amount of accelerometers and plots the confidence to distinguish modes. The results are plotted in MAC style, and when the diagonal elements of this matrix show values equal to "one" and the other elements of the diagonal have values close to "zero", the natural frequencies are measured experimentally without spatial aliasing. Figure 4a shows the results obtained from the proposed location and number of accelerometers. Also, and in order to validate the results obtained, an EMA of a selected body was performed to compare results with the FEM simulation. The main properties for the tested body (compressor) in this case were: Weight = 5.38 kg;

$I_x = 7098$ kgmm^2; $I_y = 15604$ kgmm^2; $I_z = 16328$ kgmm^2. The locations of the accelerometers in the body are shown in Fig. 4b.

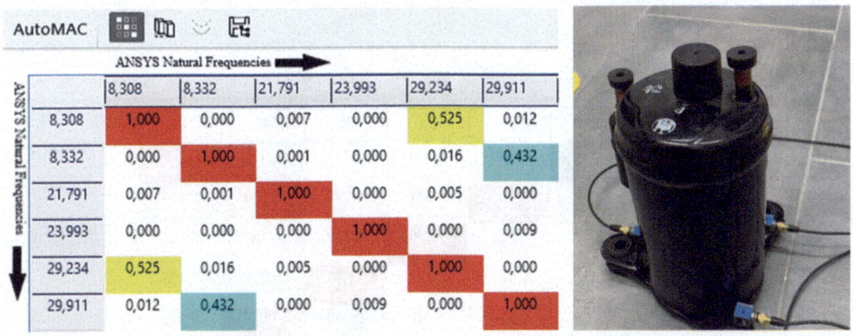

(a) (b)

Fig. 4. Decimation process (a) General view. (b) Detail of decimation process and mesh.

For validating the measured data, the following points were checked:

- Deviation between Measured and Synthesis modal response.
- Complex Mode Indicator Functions (CMIF).
- Stability Diagram.
- Similarity within measured mode shapes (AUTOMAC).
- Comparison of animations between FEMA and EMA Mode Shapes.

Stability diagram, Complex Mode Indicator Functions (CMIF) and Rational Fraction Polynomial-Z (RFP-Z) were used to determine modal the parameters (Frequency, damping, and mode shape). Both, stability diagram and CMIF can be observed in Fig. 5a. Also, the error between actual (measured) and synthesis (calculated by RFP-Z) can be obtained and graphically represented in Fig. 5b, which can be considered as an acceptable error for the estimated parameters that have been considered in this analysis.

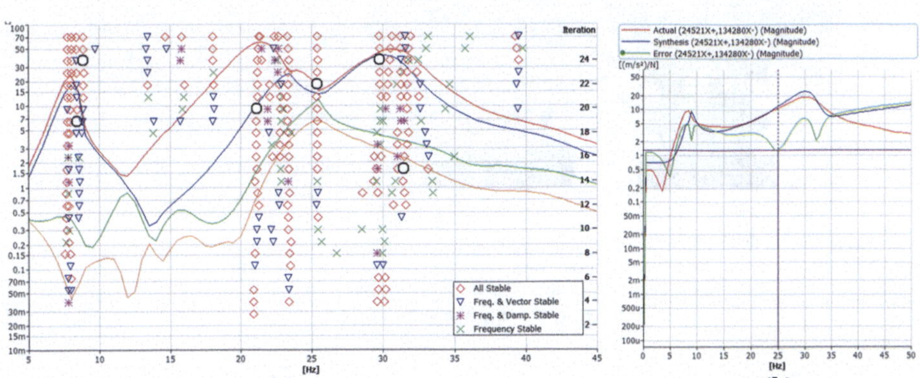

(a) (b)

Fig. 5. Validation: (a) Stability diagram and CMIF lines. (b) RFP-Z error.

Additionally, in order to validate the obtained results, Modal Assurance Criterion (MAC) was performed. With MAC it is possible to determine the similarity among obtained mode shapes. The more different are mode shapes; the closer is value to zero. Figure 6 shows the MAC, where all obtained mode shapes differs each other.

AutoMAC: Mode Table

	8,402	8,859	21,186	25,396	29,743	31,214
8,402	1,000	0,533	0,002	0,133	0,459	0,839
8,859	0,533	1,000	0,001	0,087	0,519	0,464
21,186	0,002	0,001	1,000	0,197	0,147	0,131
25,396	0,133	0,087	0,197	1,000	0,238	0,149
29,743	0,459	0,519	0,147	0,238	1,000	0,520
31,214	0,839	0,464	0,131	0,149	0,520	1,000

Fig. 6. AutoMAC matrix.

Due to nature on the movements (translations and rotations), to assure that FEMA and EMA mode shapes for each frequency are the same, a comparison of animations were made. Figures 7 and 8 shows the Z-axis translational mode and rotational mode shapes respectively. Figure 7a shows a grommet at its maximum and its minimum displacement in the Z-axis translational mode. Figure 7b is an overlap of three images obtained from BK Connect animation of the same Z-Axis translational mode. From both figures, it is observed that the shape mode is the same either obtained from the FEM or obtained experimentally.

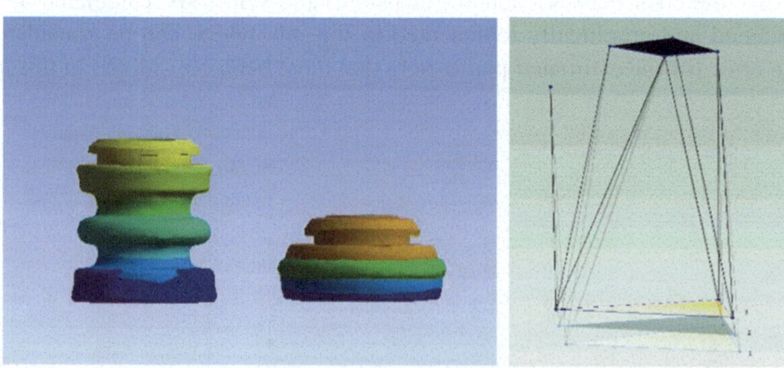

(a) (b)

Fig. 7. AutoMAC matrix.

Similar results can be observed for the Z-Axis rotational mode (Fig. 8a and b).

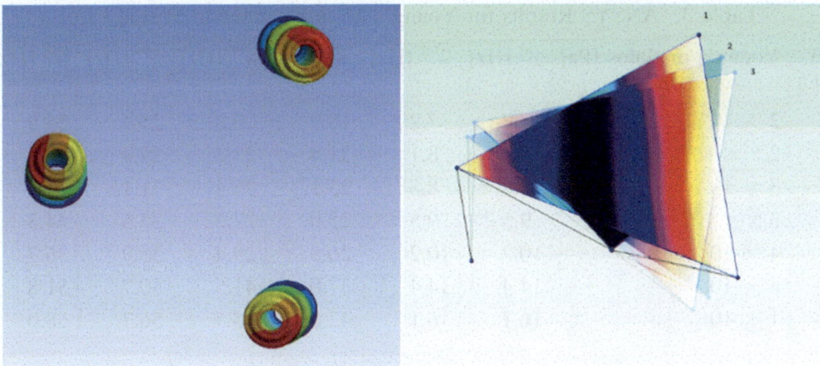

(a) (b)

Fig. 8. AutoMAC matrix.

Summarized results obtained from BK Connect are shown in Table 3. The analysis of the results should be focused on the frequency because FEMA doesn't calculate any damping ratio. Values of complexity also help to validate the measurements because some causes why complexity is not close to cero are: non-proportional damping, bad measurements or poor modal parameter estimation, inconsistent data due to e.g. time variant conditions. The value of complexity should be close to 0 as it is shown in Table 2.

Table 2. Result summary for Body 1.

Mode damped frequency	[Hz] damping	[%] complexity	[-]
1st [Hz]	8.402	3.764	0.129
2nd [Hz]	8.859	4.997	0.003
3rd [Hz]	21.186	14.207	0.032
4th [Hz]	25.396	4.770	0.119
5th [Hz]	29.743	13.710	0.383
6th [Hz]	31.422	7.830	0.223

4 Results

After the validation of the results of Table 3, they were used as a target for the iteration on EPDM Young's Modulus (E) in the FEMA. Table 3 summarized the results for these iterations, where the design point 2 (DP2) with E = 2.5 MPa, is the design point that has the best correlation to predict Rigid Body natural frequencies (Maximum of deviation 2 Hz).

Table 3. ANSYS Results for Young's Modulus iteration of Body 1.

Design point [-]	Young's modulus [Pa]	1st [Hz]	2nd [Hz]	3rd [Hz]	4th [Hz]	5th [Hz]	6th [Hz]
DP1	2×10^6	7.2	7.2	18.9	20.8	25.4	25.9
DP2	2.5×10^6	8.1	8.1	21.4	23.2	28.4	29.0
DP3	3×10^6	8.8	8.8	23.1	25.4	31.1	31.8
DP4	3.5×10^6	9.5	9.5	25.0	27.5	33.5	34.3
DP5	4×10^6	10.2	10.2	26.8	29.4	35.9	36.7
DP6	8×10^6	14.4	14.4	37.8	41.5	50.7	51.8
DP7	1×10^7	16.1	16.1	42.3	46.5	56.7	58.0

5 Conclusions

This paper proposed a methodology that combines FEM, EMA and ANSYS Workbench parameter evaluation software for calculating the rigid body natural mode shape and frequencies of a piping system composed by a compressor and rubber supports manufactured by EPDM hyperelastic material (non-linear mechanical behavior). The FEM considers the compressor as a rigid body, and used the mass, centre of mass, inertia properties and geometry of supports for the FEMA. For the EPDM hyperelastic material, in spite of having a non-linear mechanical behavior, for the lower deformation value to which it is subjected, it can be considered to have a linear behavior. In addition, from the good results obtained, it can be concluded that:

- Equation (1) could be considered as an estimator of EPDM Young's Modulus (E). However, more comparison for different Shore (A) hardness should be performed to extent valid range.
- EPDM behavior could be considered linear for low strains (from 1.46×10^{-6} to 3.14×10^{-6})
- EPDM Young's Modulus (E) can be considered constant from 0 to 50 Hz.

Because the natural mode shapes and frequencies obtained from the experimental analysis and the FEM are very similar, these results would demonstrate that the proposed methodology with the considered hypotheses could be valid to improve the product design process through an increase in its reliability and durability of these types of products.

References

1. Schaefer, R.J.: Mechanical properties of rubber. In: Harris' Shock and Vibration Handbook, 6th edn. McGraw-Hill, New York (2010)
2. Koblar, D., Boltežar, M.: Evaluation of the frequency-dependent Young's modulus and damping factor of rubber from experiment and their implementation in a finite-element analysis. Exp. Tech. **40**(1), 235–244 (2016)

3. Levent, M.E.: Characterisation of vibration isolators using vibration test data. In: 10th International Congress on Sound and Vibration, Stockholm, Sweden (2003)
4. Vendrami, C.E., Pellegrini, C.: Analysis of vibration isolators for hermetic compressors. In: International Compressor Engineering Conference, Indiana, USA (2016)
5. Wen, Z., Zhang, R., Zhang, Y., Gu, H., Hu, Y.: Research on low frequency vibration of rotary compressor. In: International Compressor Engineering Conference, Indiana, USA (2016)
6. SpaceClaim 3D Modeling Software. ANSYS SpaceClaim 19.0
7. BK Connect Structural Dynamics. Brüel & Kjær Sound & Vibration Measurement A/S
8. Kunz, J., Studer, M.: Determining the modulus of elasticity in compression via Shore-A hardness. Kunstst. Int. **96**(6), 92–94 (2006)
9. Timoshenko, S., Goodier, J.N.: Theory of Elasticity, 2nd edn. McGraw-Hill, New York (1951)
10. ISO 868:2003. Plastics and ebonite - Determination of indentation hardness by means of a durometer (Shore hardness)
11. Matweb. http://www.matweb.com. Accessed 14 May 2019
12. Gehrmann, O., Kröger, N.H., Erren, P., Juhre, D.: Estimation of the compression modulus of a technical rubber via cyclic volumetric compression tests. Tech. Mechanik **37**(1), 28–36 (2017)
13. Gent, A.N.: On the relation between indentation hardness and Young's modulus. Rubber Chem. Technol. **31**(4), 896–906 (1958)

The Importance of 3D Scanned Mesh Processing in FEM Simulation Results

Saúl Iñiguez-Macedo[1](✉), Eduardo Jiménez-Ruiz[2], Fátima Somovilla-Gómez[1], José Manuel Valle-Melón[3], Marina Corral-Bobadilla[1], María Ángeles Martínez-Calvo[1], and Rubén Lostado-Lorza[1]

[1] University of La Rioja, 26004 Logroño, La Rioja, Spain
saul.iniguez@unirioja.es
[2] Public University of Navarre, 31006 Pamplona, Navarra, Spain
[3] University of the Basque Country (UPV/EHU), 01006 Vitoria-Gasteiz, Álava, Spain

Abstract. There are many times when working with mesh elements is unavoidable, and what is more, necessary to obtain a reliable geometry. One of the most common situations of mesh working is reverse engineering, especially if the parts are composed of complex surfaces. When this happens, the reproduction of the product is not an easy task. In those scenarios, it is useful to follow a solid methodology in order to generate the most accurate model from 3D scanned meshes. In this paper, the main differences between the use of parametric models (CAD models) and scanned 3D meshes are discussed. For the parametric design, a complete footwear sole made in Rhinoceros was used. The sole was divided in two well differentiated parts: an outer part made in Nitrile Butadiene Rubber (NBR), and an inner part made in Expanded Polyurethane (E-PUR). The 3D scanned model was obtained with a structured light scanner from the complete sole once it was manufactured. The final scanned model was exported from IDEA software after combining 19 scans, and then modified with Geomagic to separate both parts (outer and inner parts). Both meshes (from parametric models and scanned 3D) were imported in a Finite Element Analysis (FEA) software in order to compare their mechanical behavior. In this case, the FEA was used for determining the stress levels into the complete footwear sole during a slip test in both meshes. Stress results showed no significant differences between the two FEA for the same load values. It can be confirmed therefore, that, the use of a 3D scanned model instead of the original CAD is not a cause of significant error if mesh geometry has a minimum quality.

Keywords: 3D · Scanner · Mesh · Accuracy · FEM · Simulation · Complete footwear sole

1 Introduction

When a company designs products with complex geometries, sketches and dimensions are not usually revealed and its reproduction become difficult or even impossible. This same problem is present in nature, due to the high amount of complex geometries, like

F. Cavas-Martínez et al. (Eds.): INGEGRAF 2019, LNME, pp. 142–150, 2020.
https://doi.org/10.1007/978-3-030-41200-5_16

stones, trees, animal bones, etc. The reproduction of such products only by drawing lines and surfaces in Computer-Aided Design (CAD) software would take months and the results would not be accurate at all. The 3D scanning technology is usually used to solve these problems. The development of new technologies has propitiated the appearance of new and cheaper scanners, making possible multiple studies about topics that years ago were unthinkable. Normally, each scanner is supplied with its own scanning software, but most of the time, this software is very limited, and it is not possible to obtain a good model with it. In many cases, accuracy is not the main objective, as can be the rendering models used for animation or video games. However, if the meshes obtained are going to be used later in FEA software, the accuracy of these meshes must be very important. An example of a basic program for use with 3D scanners is the IDEA software [1]. With this software, only mesh alignment and retouch operations can be done. In addition to these basic software, there are more advanced solutions such as Geomagic Design [2–4] or Rhinoceros [5] software. Geomagic is able to operate with mesh elements in a more detailed way, allowing even drawing over them. On the other hand, Rhinoceros is a software that is usually used for surface drawing and it is widely used in the footwear industry. In this paper the use of an original CAD complete footwear sole design is compared with the use of a 3D scanned model from the same complete sole after being manufactured. That comparison is made by two Finite Element Analysis (FEA) of a slip test in order to show the complete footwear sole stress levels. For acquiring the sole geometry in the 3D scanning process, a structured light scanner was controlled with IDEA software. After getting a 3D model, Geomagic Design plus Rhinoceros software were used to divide the outer part made in Nitrile Butadiene Rubber (NBR), and the inner part made in Expanded Polyurethane (E-PUR) (Fig. 1). Finally, both models were imported to MSC Marc [6] FEM software to carry out the FEA.

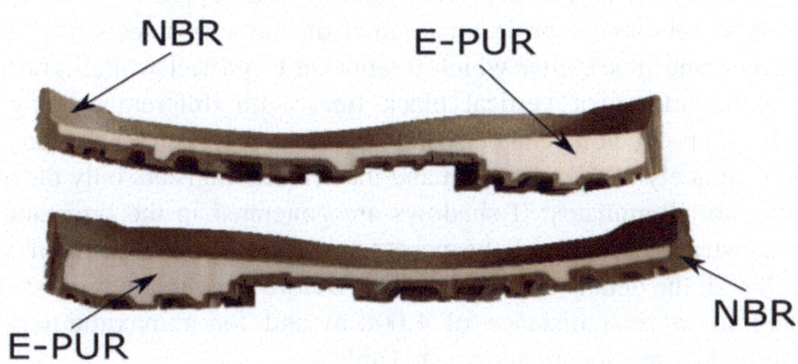

Fig. 1. Detail of the materials that composed the complete footwear sole.

2 Materials and Methods

Every 3D scanning process should start with an appropriate analysis of the product under study. In this case, the product is a footwear complete sole, and the desired geometry must be a closed mesh including the outer and the inner parts. That needs make necessary to take more than one scans and combine them to obtain the complete sole. The selected scanner was based on the structured light technology, concretely the Scan in a Box model [7], which is composed by a light pattern projector and two cameras. The structured light technology is based in the projection of several line patterns on the part surface. When the patterns are projected, the first camera is acquiring the information in order to compose a 3D model based on the line's deformation. The second camera favors the process of acquiring this information, also reducing the number of scans to obtain a better mesh quality. Having a second camera does not make the scan accuracy increase, but the two viewpoints help to reduce irregularities in the final scan. Scan in a Box can provide 0.04 mm accuracy, which is enough for most applications. Nowadays, several 3D scanning technologies are available, but structured light is by far the cheapest. Its accuracy, as previously said, is good enough for the requested applications. This technology has also some limitations. For example, as it is based in light projection, if the product has reflective surface, it must be painted. Another inconvenient that should be considered during the selection of the 3D scanning technology is that if technology is portable or not. Despite of the structured light has small equipment, it is not usually known by its portability. A fact that is evident when compared to portable laser scanners that can be operated using only one hand.

2.1 A Scanning Process

For the case studied, the complete sole had an outer part (shell) made of NBR. This material has a low reflective index. For achieving a good definition, the light conditions were controlled placing the structured light equipment in a laboratory. The control of the scanner was managed by a laptop with IDEA software. IDEA performs the calibration process as well as the data collection. Once the structured light scanner equipment was mounted and the laboratory lighting was adjusted, the scanning process was performed as follows: A projector emitted on the sole a sequence of three light colors (red, green and blue), after which it went on to project white light followed by white light with patterns of vertical black lines with different spacing. After the acquisition, the 3D point cloud can be seen in the IDEA software. The scan area of each record is approximately 300 × 400 mm and the scanner registers only the area that the scanner's projector illuminates. If shadows are generated in the projected area, they will not be registered, and it will be necessary to change the point of view of the projector so that all the details can be recorded. Before starting the process, the scanner was calibrated for a scan distance of 1.008 m and for a maximum scan area of 450 × 360 mm. The results are shown in Table 1.

Table 1. Mean Squared Error (MSE) of calibration and X, Y, Z waste in mm, 400 × 400 template.

ΔX	ΔY	ΔZ	Mean Squared Error (MSE)
0.070	0.062	0.058	0.110

The complexity for achieving a good 3D model not only remains in the accuracy of the mesh but also in the good combination of various meshes to generate a complete closed geometry. The selected product to be scanned is shown in Fig. 2 that corresponds to a complete footwear sole (view for the outer part).

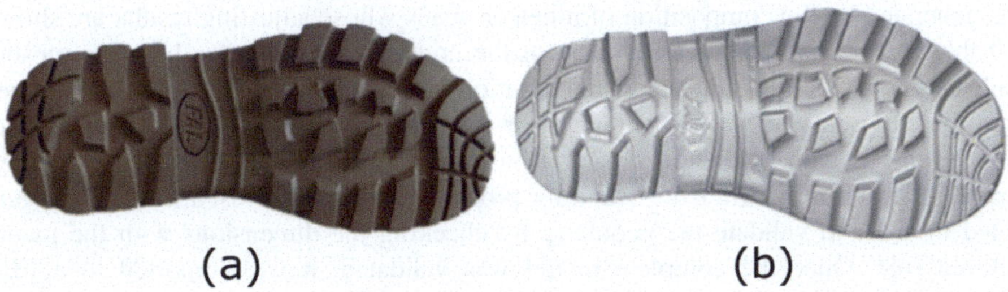

Fig. 2. (a) Complete footwear sole to be scanned (outer part). (b) Scanned complete footwear sole (outer part).

To obtain the complete scanned footwear sole, the upper side with the entire contour was scanned firstly and then the bottom side, adding again information to the contour of the model. By doing these two scans, two meshes were obtained (Fig. 3). The upper mesh contains the geometry of the cavity where the rest of the boot will be sewn, and the bottom mesh, where is shown the detail of the footprint.

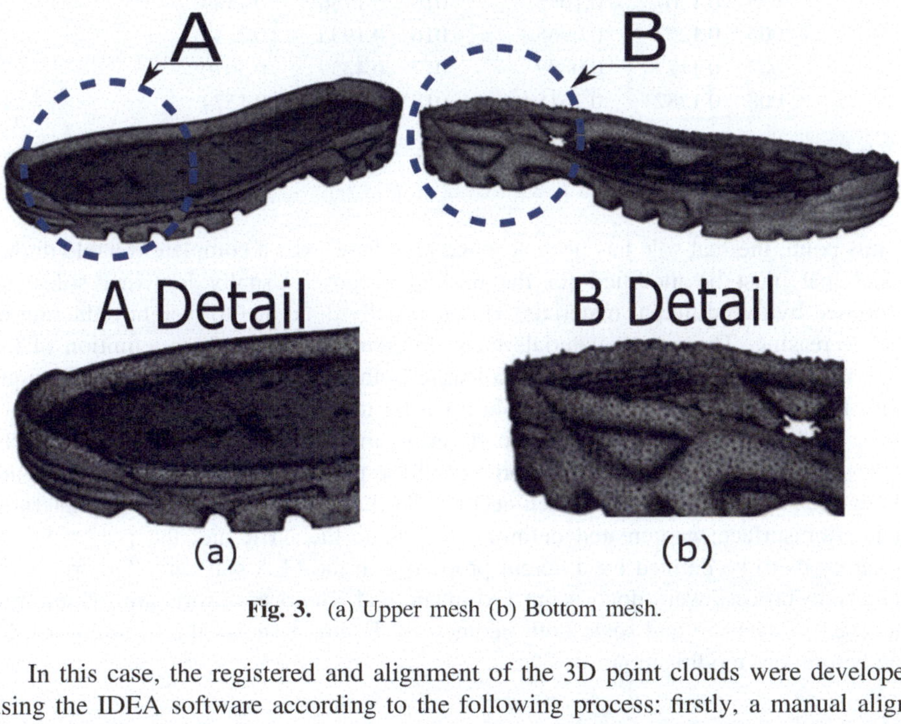

Fig. 3. (a) Upper mesh (b) Bottom mesh.

In this case, the registered and alignment of the 3D point clouds were developed using the IDEA software according to the following process: firstly, a manual alignment was made by looking for at least three homologous points of the point clouds resulting from each record and taking as reference one of the clouds of registered points of the trace. The aligned point clouds form a reference group from which the rest of the missing point clouds must be aligned. Once all registered point clouds were manually aligned, the IDEA software offers automatic alignment for a better fit. The final model was generated by the composition of nineteen scans whose adjusting results are shown in Table 2. The dist. parameter stands for the node average distance between meshes while std. dev. represents the standard deviation of those measurements. As can be seen from the exposed values, the average error is always lower than 0.19 mm except for scan seventeen with 0.43 mm error. In the other hand, standard deviation is always smaller than 0.3 mm. After the scanning process, a complete closed model was generated in order to validate the geometry by checking the dimensions with the manufactured sole. Once the complete model was validated, it was imported to a FEA software in order to simulate the slip test.

Table 2. Model alignment results for nineteen point clouds using IDEA software.

Scan	dist. [mm]	std. dev [mm]	Scan	dist. [mm]	std. dev [mm]
001	0.1563	0.1270	011	0.1382	0.1537
002	0.1650	0.1482	012	0.1315	0.1670
003	0.1014	0.0727	013	0.1137	0.1270
004	0.1413	0.1305	014	0.1262	0.1670
005	0.1207	0.1085	015	0.1186	0.1489
006	0.1207	0.0965	016	0.1935	0.2148
007	0.1127	0.0889	017	0.4359	0.3070
008	0.1782	0.1615	018	0.1746	0.1574

2.2 Preprocessing Process

At this point, the real sole has been scanned and there exist a complete reliable digital model that must be modified for the desired purpose. Usually footwear soles are composed by two different materials. This is usually done due to weight reduction or cost decreasing. These two materials have different properties. The definition of the correct material properties in the FEA software cannot be done if the model is a unique solid. For this reason, the scanned sole must be modified to separate the outer part (shell) made in NBR, and the inner part (infill) made in E-PUR. The scanner technology only captures the external geometry of the parts, being a task of the de-signer adding the interior information when needed. To study the influence of both materials, an interior surface was created defining two parts, the NBR and the polyurethane, which needs to be defined by different properties in the FEA software. The trimming and gluing process were done using Geomagic and Rhinoceros software, which also was used to measure and scale both geometries. Figure 4 shows the trimming of the sole before any modification.

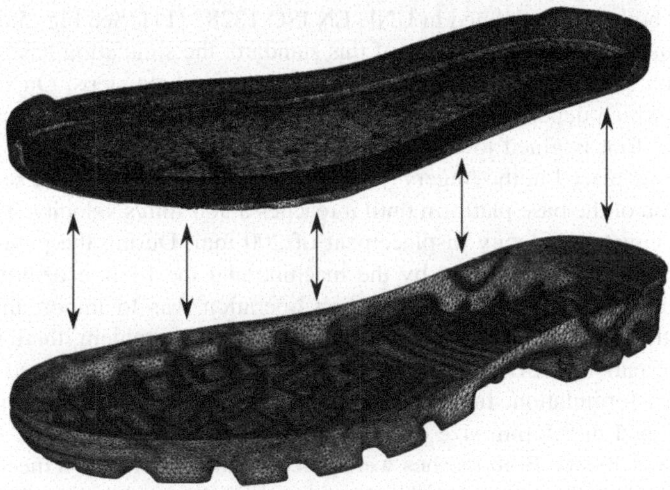

Fig. 4. Upper and bottom meshes before infill definition

Geomagic Design X is a 3D software specialized in working with mesh geometry. This software has the possibility to repair mesh holes, combine meshes by gluing them or even create CAD elements from scanned geometry. Geomagic has an assistant that makes the gluing process easy. The operation is simple; two meshes are positioned by selecting the same points in the upper and bottom mesh, like in the IDEA software. In this case, the upper side was trimmed from the CAD model and the lower side from the scanned model. The final model was generated by gluing the two meshes, by selecting points in the contour area. This combination of a 3D scanned mesh was the selected geometry to be compared with the CAD. The final model has a soft surface, without the appearance of holes or scratches, forming a closed mesh that makes possible the application of this geometry in any other software like Cura for 3D printing [8]; Autodesk Inventor [9] for a mold generation or MSC Mentat-Marc [10] for a FEM simulation.

2.3 FEM Model

Two FEM models were created, one using the CAD geometry and another using the scanned model. Both models are equal; same material properties, same boundary conditions, same contact bodies, etc. In this case, FEM models were created using MSC Mentat-Marc for simulating the footwear sliding test according to UNE-EN ISO 13287. This test is one of the most expensive requirements for manufactures, since there is necessary the creation of aluminum molds to vulcanize NBR and then obtain the injection E-PUR sole. Until the sole is tested, there is no way to know if it will obtain or not the approve. If the design does not pass, the mold must be remanufactured with the associated extra costs. A FEM simulation model was built from the 3D scanned mesh, and another one from an original surfacing CAD model. The new designed shoe models must prove their coefficient of friction under two different floors, steel and tile, both clearly defined in UNE-EN ISO 13287 [11] (See Fig. 5a). As steel is about to be discarded in new editions of this standard, the simulation has been defined for tile friction. The test machine operation consists in three steps. On first step the normal load, which depends on the sole size, is applied using a mechanical foot. The bottom of this foot is glued to the sole surface. This device di-vides the load in two round areas, one placed in the fingers' joint and another in the heel. The second step is the acceleration of the base platform until it reaches a 300 mm/s velocity. The third and last step is a constant velocity displacement of 200 mm. During the process, vertical and horizontal loads are recorded by the machine and the friction determined by its ratio. To define the whole problem the first operation was to import the generated geometry to the Marc software. As the simulation is a 3D problem, there is a need of remeshing the part with 3D elements. Hexahedral elements were selected using complete Hermann formulation. In order to reduce the computational cost only the footprint was meshed in 0.5 mm size elements where the rest of the model was meshed using 1.0 mm elements. Both meshes were glued by the addition of a mechanical contact constraint. Then the material properties were applied using Mooney-Rivlin material behavior models. The material constants that define its behavior were previously tested and adjusted by model updating methodology combined with MRS (Multi Response Surface) optimization [12]. The tile was modeled as a non-deformable surface applying the Coulomb Arctan friction model. The coefficient of friction was defined for the coupling of Tile-NBR with dependence of the part Von-Mises Stress, (See Fig. 5b). This function was also obtained by a process of testing the slip of several specimens under different loads. Then the specimen slip test was modeling with Mentat-Marc to obtain the real stress levels in the contact area.

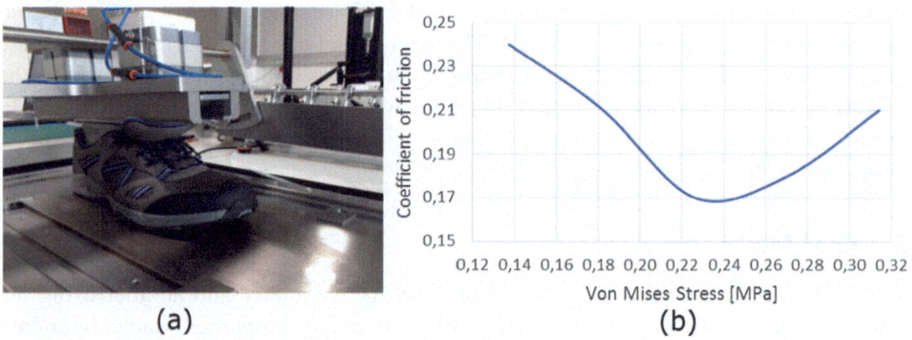

Fig. 5. (a) Sliding test according to UNE-EN ISO 13287. (b) NBR Material coefficient of friction dependent of Stress levels

After the definition of every material constant, the boundary conditions were applied as two groups of vertical point loads in the in the fingers' joint and heel. Furthermore, every node in contact with the mechanical feet was constrained preventing them to move in the horizontal plane. The last operation before the launch of the simulation was the definition of the floor velocity and load application following the three steps of the test machine. The final FE model can be seen in Fig. 6.

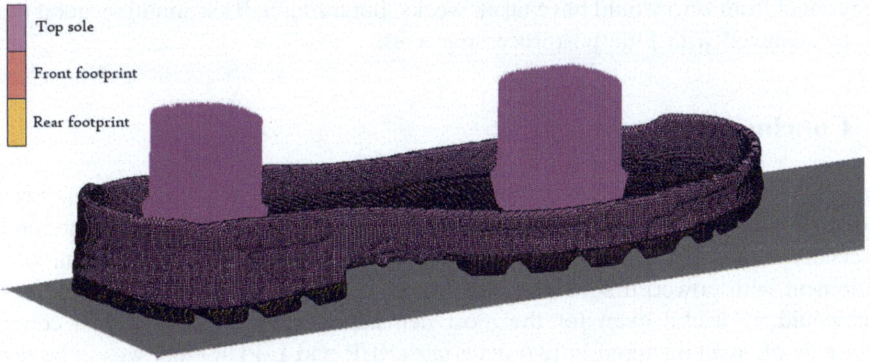

Fig. 6. Final slip test finite element model

3 Results and Discussion

After simulating the both FE models, normal stress values (MPa) were obtained for the original surfacing CAD model (Fig. 7a) and the 3D scanned mesh (Fig. 7b) respectively. From both figures it is observed that the tensional state of the study area (lower part of the complete footwear sole) is practically the same. Only a small difference in the maximum stress obtained is appreciated from both FE simulations (5.87 MPa from the CAD and 8.85 MPa from the 3D scanned mesh).

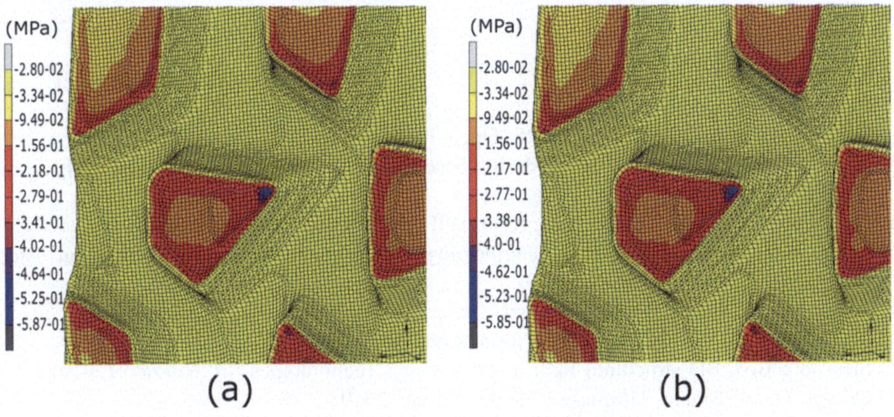

Fig. 7. Normal Stress obtained from: (a) CAD geometry (b) Scanned geometry.

Since the difference obtained in both models is very small, it can be deduced that either of the two models could be used to create an FE model and simulating the footwear sliding test according to UNE-EN ISO 13287. Fact that proves that through this methodology it is possible to obtain a reliable the geometry of a complex part. Drawing the product from zero would have taken weeks, but through 3D scanning, a good model can be achieved with little postprocessing work.

4 Conclusions

An agile methodology of 3D scanning can save a lot of time and money when it comes to obtaining complex parts models. Nowadays there are many economic scan models, especially those that use the structured light technology. The conjunction of this equipment with powerful software tools makes it possible to generate quality meshes that would be useful even for the most demanding jobs. In this case, a complete footwear sole manufactured in two materials (NBR and E-PUR) that was to be tested, was scanned with a Scan in a Box model based on structured light technology. Several software (trimming and gluing the mesh with Geomagic and Rhinoceros, IDEA for the automatic alignment of the mesh) a new mesh, complete, refined and able to differentiate the two parts of the sole was obtained to be imported into the FEM software. The results shows how there is hardly any difference between the results obtained from a FE model for simulating the footwear sliding test that has been created from parametric models and another that has been generated from a scanned 3D meshes. It can be confirmed therefore that, the use of a 3D scanned model instead of the original CAD is not a cause of significant error if mesh geometry has a minimum quality.

Acknowledgments. The authors wish to thank the La Rioja Government through Project ADER 2014-I-IDD-00162 as well as to the Institute of Riojan Studies (IER) for funding part of this research.

References

1. IDEA the Software Release 1.1 SR8. Open Technologies Srl, Rezzato (2019)
2. Hai, F.W.Z.: Reverse engineering design based on geomagic studio software. Tool Eng. **11** (2007)
3. Geomagic Design X. 3D Systems, Rock Hill (2019)
4. Chang, K.H., Chen, C.: 3D shape engineering and design parameterization. Comput.-Aided Des. Appl. **8**(5), 681–692 (2011)
5. Rhinoceros 6. Robert McNeel & Associates, Pacific (2012)
6. MSC Marc 2017. MSC Software Corporation, Newport Beach (2017)
7. Scan in a Box, 3D structured light scanner. Open Technologies Srl, Rezzato (2019)
8. Cura 4. David Braam, Ultimaker, Geldermalsen (2019)
9. Autodesk Inventor 2016. Autodesk, Inc., San Rafael (2016)
10. Javanbakht, Z., Öchsner, A.: Advanced Finite Element Simulation with MSC Marc. Springer, Cham (2017)
11. UNE-EN ISO 13287. Personal protective equipment - Footwear - Test method for slip resistance. AENOR, Madrid (2013)
12. Íñiguez-Macedo, S., Lostado-Lorza, R., Escribano-García, R., Martínez-Calvo, M.Á.: Finite element model updating combined with multi-response optimization for hyper-elastic materials characterization. Materials **12**(7), 1019 (2019)

Manufacturing and Industrial Process Design

Inspection of 3D Printing and Advanced Manufacturing Processes Using Hybrid 3D Metrological Technologies

L. Barrenetxea[1](✉), R. Minguez[1], O. Etxaniz[1], N. Ortega[2], and S. Plaza[2]

[1] Department of Graphic Design and Engineering Projects, University of the Basque Country UPV/EHU, Pl. Ing. Torres Quevedo. 1, 48013 Bilbao, Spain
lander.barrenetxea@ehu.eus

[2] Department of Mechanical Engineering, University of the Basque Country UPV/EHU, Pl. Ing. Torres Quevedo. 1, 48013 Bilbao, Spain

Abstract. The aeronautical industry has begun to use advanced manufacturing and 3D printing for functional parts. The Product Design Laboratory (PDL) has studied the feasibility of manufacturing a part by Additive Manufacturing and has compared the geometric quality of the resulting part with those currently manufactured by plate assembly. Traditionally, we carried out the manufacture of these elements by stamping their different parts and subsequent welding. The objective of this project is to analyse the resulting part using different methods; contact technology (CMM), non-contact based on structured blue light (triangulation technology) and X-rays (CT), determining the suitability and limitations of each inspection technology. The manufacture of this part is not analysed. We studied this component from different points of view: geometric and dimensional verification and analysis of internal defects. We carried out the first of the macro geometric dimensional control stages using a Coordinate Measuring Machine (CMM) and acquiring a point cloud with structured light, while the defect analysis activity was carried out using an X-ray machine. There are certain detail features that we processed by Computational Tomography (CT) tracing the results to those obtained with an optical microscope with a micrometer table.

Keywords: 3D-printing · Advanced manufacturing · Inspection · CMM · 3D-scan · Computational Tomography

1 Introduction

Additive manufacturing has been in use for some time and multiple studies have been carried out in different areas such as medicine [1, 2]. However, in the aeronautical sector, where agility in adapting to change and reducing costs are critical factors, the additive is undoubtedly a technology that has many advantages, but which, in parallel, faces significant challenges in the viability of its industrialisation.

F. Cavas-Martínez et al. (Eds.): INGEGRAF 2019, LNME, pp. 153–165, 2020.
https://doi.org/10.1007/978-3-030-41200-5_17

Traditional methods provide design and manufacturing safety and there is an uncertainty of entrusting oneself to a new technology yet to be industrialised [3–5]; additive manufacturing entails profound challenges for all members of the aeronautical industry: from manufacturers to end users, including technological development centres and training centres.

This document presents a practical case developed within the 3Dx3D project. An aeronautical component has been selected, specifically an arc segment of the Rolls Royce Trent 1000 engine. For confidentiality reasons, we do not provide the data relating to the plan of the part. We provided the areas of interest for the measurement and the alignment procedure specified by design.

We used different technologies in this project to characterize a segment of Inconel 718 manufactured by Selective Laser Melting [6, 7]. This is a prototype used to study the geometry concordance between the manufactured part and the nominal model. We know that the part presents important deformations but they have not been quantified. We use a high precision CMM machine to obtain traceable measurements of the geometry and general shape of the part.

Subsequently, the arc segment is 3D scanned using structured light, capturing as much of the outer surface as possible. The aim is to compare these two traditional technologies and look for evidence of errors in the manufacturing method used.

The part has highly complex areas with a honeycomb stiffener and an EDM slot. We used this slot to assemble the ring segments together. These areas are impossible to measure with the previous techniques, so we used a computerized tomography (CT) system. This system is especially suitable for additive manufacturing parts because it allows us to analyze the internal structure and look for lack of fusion defects [8].

2 Measurement with CMM

The first measurements were carried out with the Mitutoyo CRYSTA APEX-S machine using a PH10MQ PLUS probe and Renishaw SP25 module [9, 10], which serves as an equipment with sufficient precision to obtain the necessary traceability. In order to obtain the information that allows us to know exactly the real geometry of the part and ensure that our clamping does not introduce major changes in it, the measurements have been made with the following procedure:

1. Tie the part and carry out manual alignment.
2. Execute the automatic alignment and perform the complete program two times, getting two measurements of circumferences at different heights.
3. Release the part, reattach it and perform the necessary manual alignment.
4. Re-execute the automatic alignment and obtain two more measurements.

We also measured five radial sections along the perimeter of the arc segment, equidistantly positioned from the central plane and with intervals of 3.5°. Its objective is to analyze possible deformations of the revolution contour of the part. Figure 1 shows the elements measured on the arc segment.

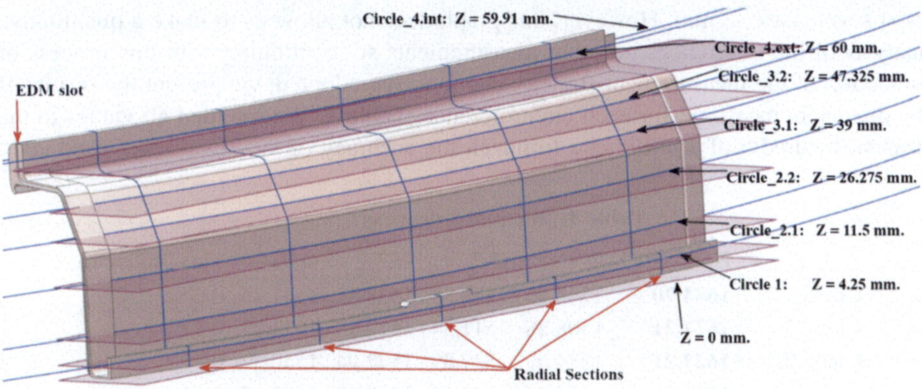

Fig. 1. Measured circles positions

2.1 CMM Measurement Results

In order to know the validity of the measurements, we analyzed their repeatability. We represented the results of the measured circumferences in two formats:

- In order to see the distance between the theoretical center of the circumference and the real one, we represented the X and Y coordinates of the centers of each circumference measured in each measurement, Fig. 2.
- Error of the measured diameter with respect to the nominal diameter

It begins by analyzing the position of the center of each circumference in the four measurements made. In the top view, all circumferences should be concentric.

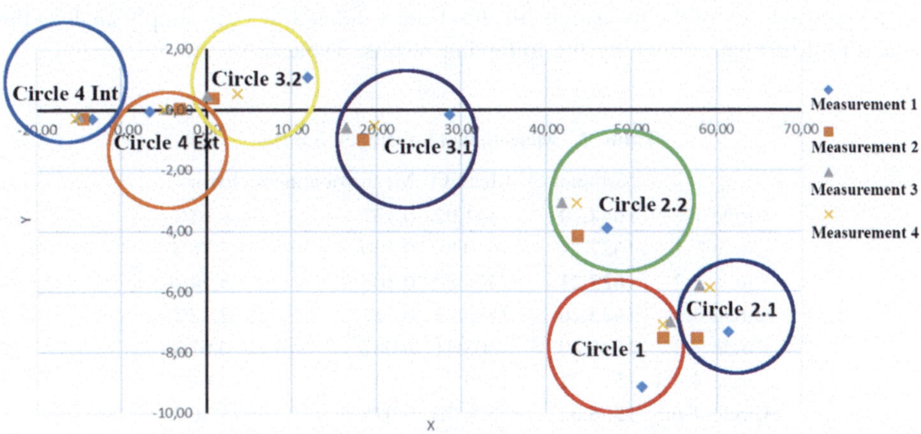

Fig. 2. Top view: Measured centers. Coordinates of the theoretical center (0.0)

We observe that the circumferences with lower Z coordinate have a greater error in their centers than the four upper ones. The lack of repeatability due to mooring is, in the most severe case, 2 mm. However, this graph does not allow us to make a quantitative analysis of the differences between measurements so, continuing with this process of validation of the measurement strategy and mooring adopted, we present the results of the diameters measured for each circumference, Table 1. We rounded all values to the nearest hundredth of a millimetre to match the accuracy of the 3D scanner used later.

Table 1. Measured diameters (mm).

	Nominal Ø	Ø 1	Ø 2	Ø 3	Ø 4	Max. difference
Circle 1	**1644.70**	1548.46	1538.85	1538.63	1541.75	0.64%
Circle 2.1	**1627.21**	1508.28	1511.54	1512.38	1511.25	0.27%
Circle 2.2	**1627.21**	1534.09	1537.89	1542.94	1540.37	0.57%
Circle 3.1	**1623.26**	1565.24	1584.13	1588.90	1582.67	1.51%
Circle 3.2	**1615.19**	1590.09	1611.37	1613.44	1606.74	1.47%
Circle 4 ext	**1579.91**	1593.27	1585.47	1588.99	1589.94	0.49%
Circle 4 int	**1568.02**	1594.64	1595.97	1597.62	1598.58	0.25%

Table 1 shows the differences between the values obtained from the diameter of each circumference in each of the measurements. This difference in no case exceeds 2%, a value that companies in the aeronautical sector usually take as a standard acceptance criterion when carrying out a repeatability analysis in a CMM measurement. Therefore, this validates the measurement criterion and the mooring system used.

2.2 Overall Results

For circumference measurements, we calculated the roundness of the circumferences. As the repeatability of the measurements has been validated, we can simply analyze the mean of all of them, obtaining the following results, Table 2.

Table 2. Measurement averages (mm).

	Nominal Ø	Mean Ø	Mean roundness	Error
Circle 1	**1644.70**	1541.92	0.137	6.67%
Circle 2.1	**1627.21**	1510.86	0.140	7.70%
Circle 2.2	**1627.21**	1538.82	0.108	5.74%
Circle 3.1	**1623.26**	1580.23	0.082	2.72%
Circle 3.2	**1615.19**	1605.41	0.062	0.61%
Circle 4 ext	**1579.91**	1589.41	0.045	-0.60%
Circle 4 int	**1568.02**	1596.70	0.075	-1.80%

In the mean roundness of Circle 4 Ext, we discarded measurement 1, because its result is noticeably different from that of the other measurements, and therefore we assumed that it was a measurement error. We analyzed separately the results of diameters and roundness obtained in the segment.

2.3 Diameters

As mentioned above, the zones in which the diameters have the greatest deviation is the lower zone of the part (zones 1 and 2), while zone 4, the upper zone, is the one that best adjusts to its nominal value. The Fig. 3 shows these results, comparing them with the nominal diameters.

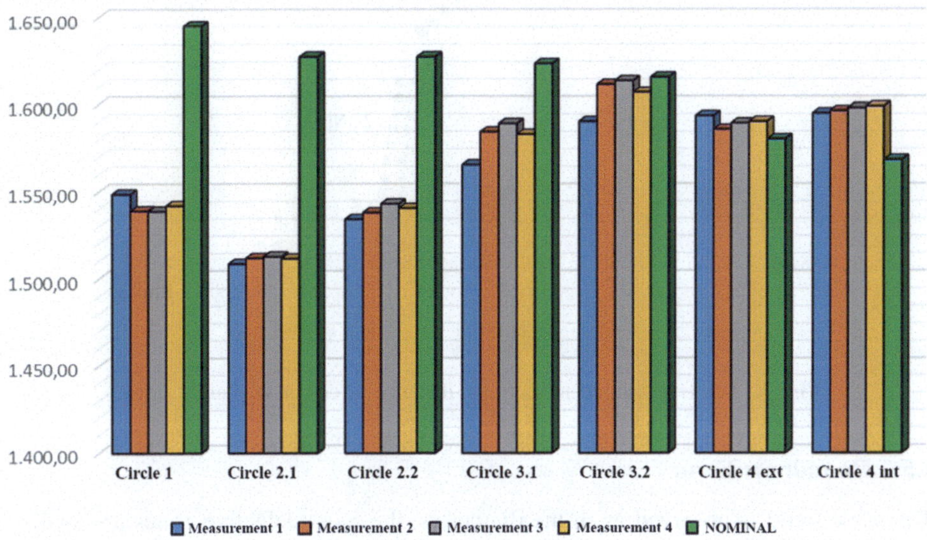

Fig. 3. Diameter bar chart (mm)

It is necessary to highlight a fact deduced from this representation; the ratio of diameters does not agree with the dictate the plans of the piece. In other words, the diameters that should be the largest in this piece are the smallest and vice versa. We appreciated this fact in the change of sign of the error available in the previous table (positive errors represent errors by default and negative errors by excess).

2.4 Sections

When analyzing the real profile of the part in each section where it has been measured, the coordinates of the probed points (distance to center vs. height) are represented, obtaining the profile of the part in that section, Fig. 4.

Observing the real profile of the piece, you can see the differences between the different sections that we measured. The upper zone of the piece (zone 4) is the one in which its dimensions remain with the least variation along the arc that forms this segment. The profile, therefore, in this zone, is practically constant in all the sections of the same one. The error increases from Sects. 1 to 5.

For its manufacture, the part rests on the side closest to Sect. 1. In other words, the greater the distance from the base of the part, the greater the deviation of its dimensions with respect to the first section of the part manufactured.

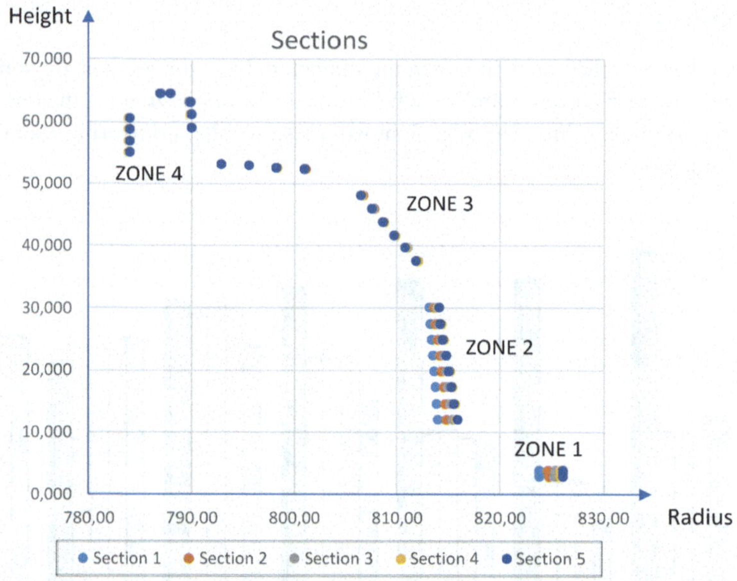

Fig. 4. Points measured in sections, distance to center vs. height (mm)

2.5 Roundness Error

The same trend is observed as in the diameters, the most defective zones are again 1 and 2. In addition, we can identify the discarded value in "Circle 4 ext" by its discordance with the rest of measurements. All this can be clearly seen in Fig. 5.

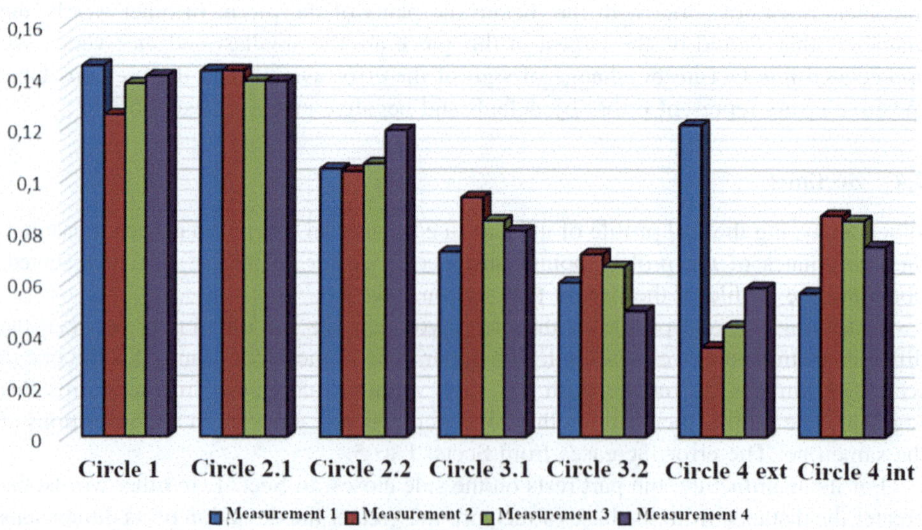

Fig. 5. Roundness bar charts

3 Measurement with Structured Light

We take the results obtained with CMM as the basis for evaluating the capacity of structured light systems. Although CMM can be used for generating three-dimensional meshes and reverse engineering [11], optical scanners are optimized for this task. The methodology used for the measurements with structured light passes, in the first place, through the selection of the scanning equipment. In this case, we chose a blue-light GOM ATOS Compact 5M scanner and GOM Inspect software [12] available in the Department of Graphic Design and Engineering Projects. We made the captures using lenses for small volume (150 × 110 × 110 mm) at an optimal distance of 590 mm. During scanning the system depends on the user's expertise that determines the amount of captures, angle to the piece, distance and exposure time. We did not simplify the mesh, we only eliminated noise points and elements foreign to the piece.

3.1 Measurement Results with Structured Light

The first results indicate that, as with CMM measurement, the part has smaller than nominal diameters. Figure 6 shows this effect perfectly. Aligning the revolution axes of the nominal part and the scanned one, the nominal diameter of Circle 1 is 1644.7 mm, while the measurement by structured light has been 1529.95 mm, which means a difference of 114.75 mm.

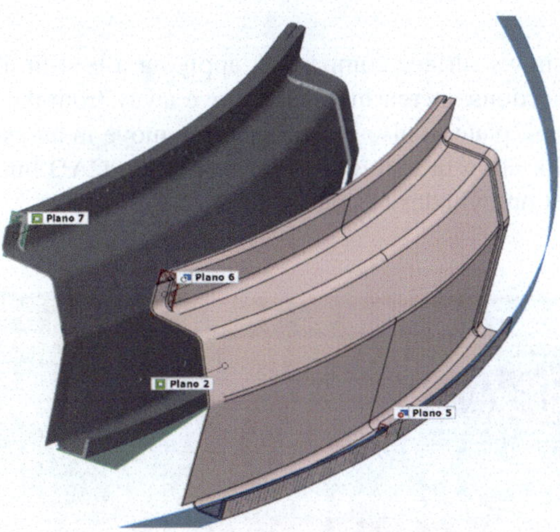

Fig. 6. Comparison between nominal model and scanned one aligning the revolution axes

In order to appreciate the effect of the deformation of the part with greater perspective, two processes will align the nominal and the measured parts.

First, we use alignment by geometric elements, as in the MCC. The diameters of the sections measured with structured light do not go further than 1% in the worst case with respect to the values obtained with CMM. The centres of these circumferences also show displacements similar to those of Fig. 2. Figure 7 shows the deformation of the part and the "cone" effect generated by this deformation. These deformations explain the inconsistencies in the diameter measurements made with CMM.

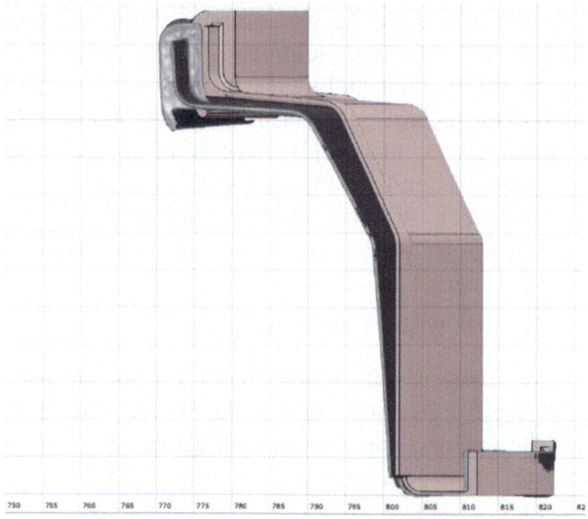

Fig. 7. Comparative graph of the nominal model and the real one measured by structured light

Second, Fig. 8 shows surface comparison applying a best-fit algorithm. The error varies in the radial sections, increasing as we move away from the right side joined to the manufacturing base plate. It also increases as we move in an axial direction, which implies not only a variation in radius with respect to the CAD but also in the type of surface, from revolution cylinder to conoid.

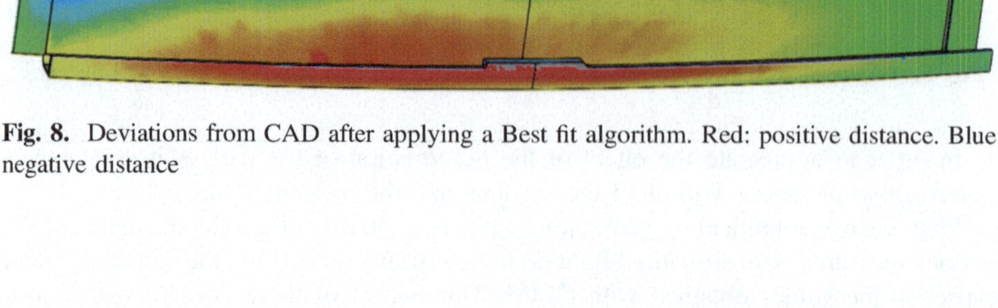

Fig. 8. Deviations from CAD after applying a Best fit algorithm. Red: positive distance. Blue: negative distance

Measuring with structured light does not only result in measurements similar to CMM. We can scan in 3D quickly the majority of the surface, and get an idea of the 3D deformation of a part with respect to its nominal. We have the freedom to align the elements after scanning according to different criteria and this makes it easier for us to discover faults in the manufacturing processes.

However, there are areas we could not analyze, such as honeycomb and the EDM slot. The scan focused on the theoretically flat, cylindrical and conical areas, which meant that we captured the other ones tangentially and with little detail. The reflective surface and the difficulty of covering these areas, as well as adjusting the orientation to capture the concavities with both cameras of the scanner, require an increasing number of captures. So many captures introduce noise and small misalignment errors that would negatively affect the rest of the geometries.

4 Computerized Tomography Measurements

As mentioned at the beginning of the document, we selected this piece as a practical case study in this project due to its special geometric difficulty. Among the features to be highlighted are:

1. Lack of representation of the complete circumference.
2. Honeycomb of thin wall in rear part.
3. Thin and deep EDM slot at one end.

Of the three types of characteristics mentioned, two are impossible to measure using the previously used methods (CMM and structured light). These are the thin-walled honeycomb and the EDM slot whose purpose is to materialize the union between segments.

The CT scans were carried out using a Seifert X-Cube Compact machine [13]. This type of machine works by throwing beams of rays from an X-ray tube to a receiving plate, through a previously placed piece in the path of these beams. The receiving panel will get more or less amount and intensity of rays depending on the parts that has passed through. The greater the thickness, in general, the less amount of rays will go through the piece and the darker the part will be seen. Figure 9 shows the EDM slot along the entire length of the segment in an image taken from an X-ray tomography.

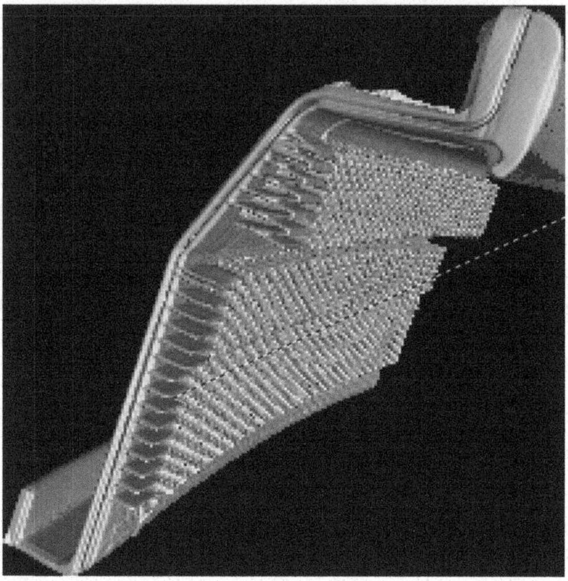

Fig. 9. EDM slot & honeycomb

4.1 X-Ray Inspection Results

The functionality of the software to visualize specific sections eases the realization of the measurements. Figure 10 shows how the sections of three perpendicular planes are generated in the part.

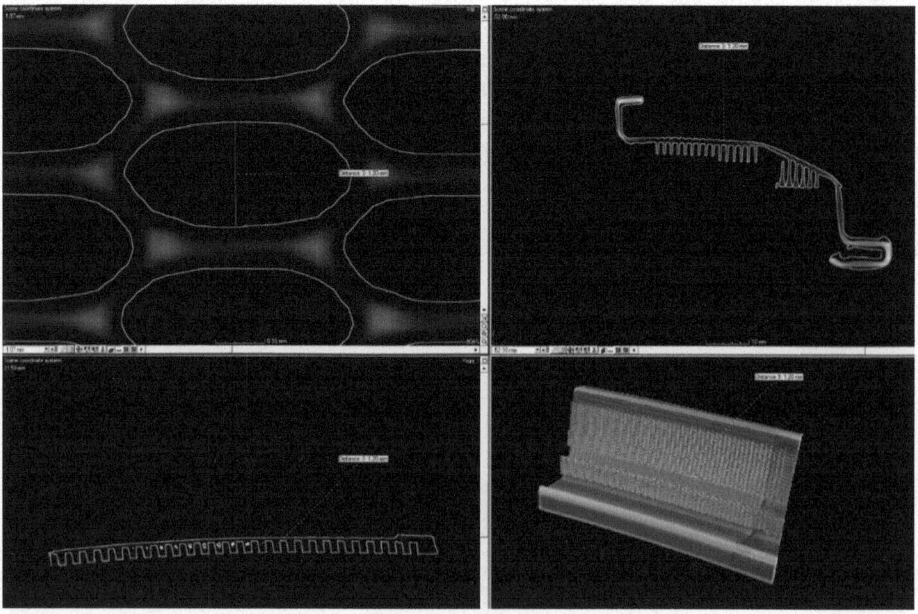

Fig. 10. Sections in the generated solid

Thanks to these sections we have obtained the levels corresponding to honeycomb (Height: 1,50 mm. Width: 2,74 mm. Thickness: 0,19 mm.), Figure 11. The theoretical values are; Height: 1,50 mm. Width: 2,75 mm. Thickness: 0,20 mm.

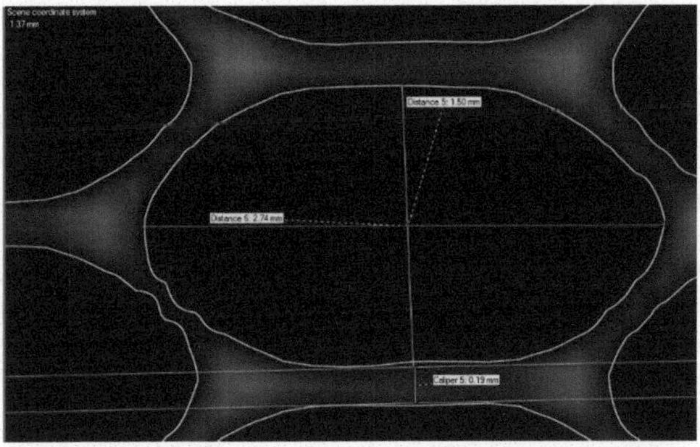

Fig. 11. Honeycomb height, width and thickness measurements

We obtained the thickness of the EDM slot in the same way, as shown in Fig. 12. The theoretical value is 0.6 mm.

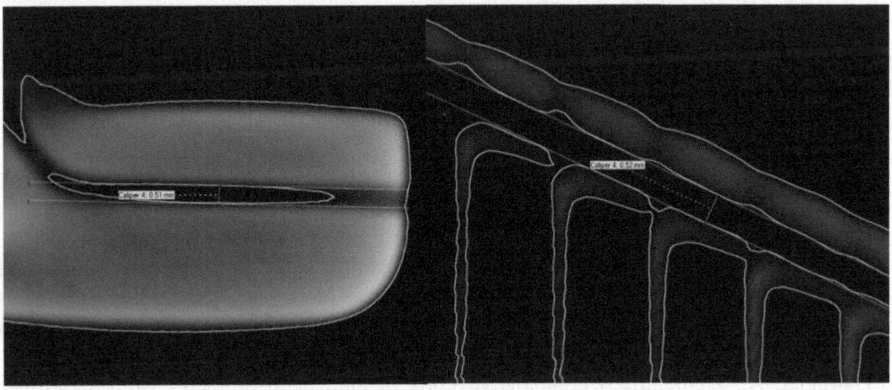

Fig. 12. EDM slot thickness measure: 0.51 mm.

5 Conclusions

The first obvious conclusion is that the part generated by additive manufacturing had serious geometric deformations due to thermal contractions associated with the process. Both the CMM and the structured light scanner have determined these deformations and the geometric deviations from the CAD file. However, although the 3D scanner loses in accuracy with respect to CMM and is more dependent on the user's expertise, its measurement process was faster and simpler. In addition, not relying so much on actual measurements for alignment, and doing so later, provides more information on deformations. On the other hand, the surface finish of the additive manufacture causes irregular reflections on the surface. This situation forces us to use a large amount of spray on the piece, which can distort the measurements made with the scanner. If the accuracy is not critical (more than 40 μ), scanning is the optimal system for warped and complicated positioning parts common in aeronautics.

None of these systems alone can perform a complete inspection of parts built by advanced manufacturing for the aeronautical industry. However, when we combine them in the right order, we can reduce time and cost.

The structured light and its different alignment possibilities quickly discriminate form errors that would be overlooked by other techniques. In this way, we can detect manufacturing errors in advance and optimize additive manufacturing processes by inserting stiffeners into the geometry or by compensating in the CAD for deviations inherent in any thermal process. We can then use the CMM to verify the parts more accurately.

However, both systems are ineffective against small thicknesses, slots and internal voids. In this situation, it is necessary to resort to computed tomography within the thickness and volume limits of this machine.

Acknowledgments. The research work reported here was made possible by the financial support from the Basque Government Industry Department under the ELKARTEK Program (KK-2017/00068 and KK-2018/00024).

References

1. El-Katatny, I., Massod, S.H., Morsi, Y.S.: Error analysis of FDM fabricated medical replicas. Rapid Prototyp. J. **16**(1), 36–43 (2010)
2. Bici, M., Cardini, V., Eugeni, M., Guachi, R., Bini, F., Campana, F., Marinozzi, F., Gaudenzi, P.: Digital design of medical replicas via desktop systems: shape evaluation of colon parts. J. Healthc. Eng. (2018)
3. Roberts, D., Zhang, Y., Charit, I., et al.: A comparative study of microstructure and high-temperature mechanical properties of 15-5 PH stainless steel processed via additive manufacturing and traditional manufacturing. Prog. Addit. Manuf. **3**(3), 183–190 (2018). https://doi.org/10.1007/s40964-018-0051-5
4. Hawaldar, N., Zhang, J.: A comparative study of fabrication of sand casting mold using additive manufacturing and conventional process. Int. J. Adv. Manuf. Technol. **97**(1–4), 1037–1045 (2018). https://doi.org/10.1007/s00170-018-2020-z
5. Subramanyam, B., et al.: Comparative analysis of additive manufacturing over conventional manufacturing. In: IOP Conference Series: Materials Science and Engineering, vol. 455, no. 1 (2018). https://doi.org/10.1088/1757-899X/455/1/012102
6. Yakout, M., Elbestawi, M.A., Veldhuis, S.C.: A review of metal additive manufacturing technologies. In: Solid State Phenomena, vol. 278, pp. 1–14, 14 p. (2018). https://doi.org/10.4028/www.scientific.net/SSP.278.1

7. Deng, D., Peng, R.L., Brodin, H., Moverare, J.: Microstructure and mechanical properties of Inconel 718 produced by selective laser melting: sample orientation dependence and effects of post heat treatments. Mater. Sci. Eng.: A **713**, 294–306 (2018). https://doi.org/10.1016/j.msea.2017.12.043
8. Gögelein, A., Bamberg, J., Zenzinger, G., Ladewig, A.: Process monitoring of additive manufacturing by using optical tomography. In: 14th Quantitative InfraRed Thermography Conference (2018). http://dx.doi.org/10.21611/qirt.2018.004
9. Mitutoyo. https://ecatalog.mitutoyo.com/Crysta-Apex-S-5007009001200-Series-191-Standard-CNC-CMM-C1812.aspx. Accessed 23 May 2019
10. Renishaw. https://www.renishaw.com/cmmsupport/knowledgebase/en/ph10m-plus-ph10mq-plus–22148. Accessed 23 May 2019
11. Dogariu, D.M., Andrei, C., Vieru, B.T., Plămădeală, R.A., Chiru, A.: A method for 3D geometry scanning of a combustion chamber. In: Proceedings of the European Automotive Congress EAEC-ESFA 2015 (2015). https://doi.org/10.1007/978-3-319-27276-4_53
12. http://www.henindo.co.id/home/ATOS-Compact-Scan_EN_RevA.pdf. Accessed 23 May 2019
13. Baker Hughes. https://www.industrial.ai/xcube. Accessed 23 May 2019

Conformal Cooling Systems Design and Dimensioning for Injection Molds

Abelardo Torres-Alba, Daniel Diaz - Perete, Cristina Martin-Doñate(✉), and Jorge Manuel Mercado-Colmenero

Department of Engineering Graphics, Design and Projects, University of Jaén, Campus Las Lagunillas, s/n, 23071 Jaén, Spain
cdonate@ujaen.es

Abstract. Plastic injection molding is one of the most versatile and widespread manufacturing processes of plastic parts manufacture. Obtaining the final plastic part from the melt thermoplastic material is divided into four main phases: filling, packing, cooling and ejection. The cooling phase is the most representative, since the time related to this phase is the longest. Current conventional machining processes only allow the manufacture of simple cooling systems based on straight drilled channels, which have a low cooling performance in regions of the plastic part with complex geometry. However, with the development of 3D additive manufacturing technologies, the cooling channels can adapt with greater versatility to the plastic part topology in order to achieve the high geometric and functional requirements that the industrial sector demands. In order to evaluate the performance of the conformal type cooling systems, three different configurations have been generated. For the case study the zig-zag, spiral and isocontouring pathing have been defined in order to model the conformal cooling channels that are adapted to its geometry. Through numerical simulations, the results of the cooling time and plastic part temperature after the cooling phase obtained for conformal configurations have been compared. The results obtained show that the conformal cooling channels provide a uniform cooling temperature, reduce the cooling time and improve the productivity of the plastic injection parts manufacturing process.

Keywords: Conformal cooling · Additive manufacturing · Injection molding · CAD

1 Introduction, Background and Related Work

In the plastic industry thermoplastic injection molding is widely used [1, 2]. Injection molding has many advantages, such as short production cycles [3], excellent final products surfaces and the easy way to mold complicated shapes, so it is the most popular molding process for manufacturing thermoplastic parts [4, 5]. The cooling phase is one of the most important in the molding process because of its influence on the productivity and quality of the plastic part. A large number of researchers have focused their work on the distribution design of the main elements that conform the

F. Cavas-Martínez et al. (Eds.): INGEGRAF 2019, LNME, pp. 166–174, 2020.
https://doi.org/10.1007/978-3-030-41200-5_18

injection mold cooling system and on the analysis of the geometric and technological parameters that determine the thermal performance and efficiency of the cooling phase. The first studies related to the cooling systems design were developed for conventional systems based on straight drilled channels [6, 7]. Li et al. [8, 9] developed a methodology based on the recognition of cooling features to design the layout of straight drilled channels. This conventional cooling channels design is based on the C-Space methodology. Using algorithms of genetic optimization and fuzzy logic, they established the design of the cooling system adapted to the other components of the plastic injection mold.

There are other hybrid systems that use auxiliary cooling elements, baffles and bubbles, to improve the thermal exchange between the melt plastic front and the coolant [10, 11]. These systems can be manufactured using conventional manufacturing techniques and allowing regions of the plastic part to be cooled where straight drilled channels lose efficiency. Furthermore, with the development of 3D additive manufacturing technology the design of these auxiliary cooling elements has been optimized in order to adapt it to the geometry of the plastic part [12, 13], being possible its implementation in the injection mold as an additional insert.

These 3D additive manufacturing techniques expand the possibilities of generating complex geometries and free surfaces in metallic materials with thermal and mechanical properties so that they allow their use in the manufacture of the main components of the plastic injection mold. The advantages of additive technology have allowed other researchers to focus their research on the design of conformal-type cooling systems adapted to the plastic parts topology. Wang et al. [14, 15] pro-posed conformal type cooling channel designs based on the Voronoi diagram and the Boundary-distance map associated with the topology of the plastic part. However, these methodologies are not fully automated. Because the process of dimensioning the technological variables of the cooling system is developed, in part, manually. Xu et al. [16] established an analytical model for the dimensioning of the main geometric and technological variables of the conformal type cooling system. This analytical model is based on the division of mold surfaces in discrete geometric regions designated as cooling cells. These comprise a prismatic volume that contains the cross section of the conformal type cooling channel and the geometric region of the plastic part corresponding to that volume.

Despite the development of different methodologies for the design of the main elements of the injection mold cooling system, their performance has not yet been evaluated and compared in order to establish a general methodology that allows its automated design and dimensioning. To resolve this problem, the present manuscript proposes the application of genetic optimization algorithms to the analytical model [9, 11, 14] which establishes the magnitudes of the geometric and technological variables that define the elements of the conformal refrigeration system type; along with the numerical analysis of the main pathing for the design of conformal type cooling channels: zig-zag, spiral and iso-contouring. By evaluating the numerical results obtained, the improvement in terms of thermal exchange of the conformal type solutions is determined and the design that best adapts to the geometric requirements of the case study under analysis is established, resulting in an improvement in the uniformity of the temperature map after the cooling phase as well as a reduction in the cycle time.

2 Parameterization and Design of the Conformal Cooling System

In order to establish a general and automated methodology for the design and dimensioning of conformal type cooling channels, a parameterization the main variables that determine the final geometry of the conformal type cooling system is proposed. As shown in Fig. 1, the geometry of the main elements of the conformal type cooling system is defined by: the separation distance between the conformal cooling channel and plastic part surface D_p [m], the diameter of the conformal cooling channel $Ø_c$ [m] and the separation distance between conformal cooling channels D_c [m].

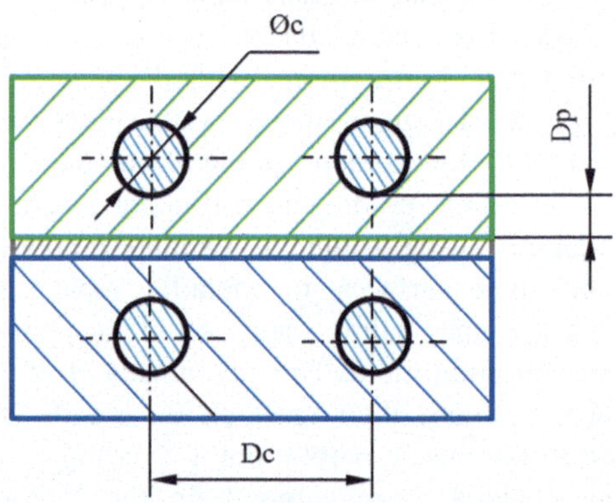

Fig. 1. Parameterization and design of the conformal cooling system.

In this way, the physical and analytical model [9, 11, 14] used for the dimensioning of the cooling system is established according to the present geometric parameterization. To evaluate the influence of the conformal cooling channels pathing on the temperature map of the plastic part after the cooling phase and the resulting cycle time, three main pathing are defined for the conformal cooling systems under study in the present manuscript: zig-zag, spiral and iso-contouring. As shown in Figs. 2, 3 and 4, the design of the conformal type cooling channels is adapted to the topology of the plastic part under study, using the magnitude of the geometric variables (see Fig. 1) obtained from the presented physical model, see (1) and (2).

3 Physical Model for the Analysis of the Cooling System

To define the physical model of analysis of the heat exchange between the plastic part and the coolant that runs through the conformal cooling channels, a set of criteria is first established in order to simplify its mathematical development:

- The mechanical and thermal properties of the material used for the additive manufacturing process remain constant.

- The geometric and technological variables resulting from the present physical model remain constant for the whole domain of the plastic part.
- The analysis of the thermal exchange is considered unidirectional, since the thickness of the plastic part is much lower than the rest of its dimensions on the plane perpendicular to the demolding direction.
- Only the thermal exchange produced by conduction between the coolant and the plastic part is taken into account, since this represents more than 95% of the total thermal exchange.
- It is assumed that the coolant regime inside the conformal cooling channels develops in a turbulent regime (Re > 4000).

According to Xu et al. [16], the steady state cycle mold cavity surface temperature, denominated as T_{mold} [°C], can be expressed as (1).

$$T_{mold} = T_{coolant} + \frac{\rho_p \cdot C_p \cdot t_p \cdot (2 \cdot \delta_s \cdot D_c + h \cdot \pi \cdot \varnothing_c \cdot D_p) \cdot (T_{melt} - T_{eject})}{2 \cdot h \cdot \pi \cdot \varnothing_c \cdot \delta_s \cdot t_{cycle}} \quad (1)$$

Where, $T_{coolant}$ [°C] represents the temperature of the coolant fluid, T_{mold} [°C] represents the temperature of the surface of the mold, ρ_p [kg/m^3] represents the density of the melt plastic flow, t_p [m] represents the average thickness of the plastic part, C_p [J/kg · °C] represents the specific heat of the molten plastic, δ_s [W/m · °C] represents the thermal conductivity of the metallic material of the mold, h [W/m^2 · °C] represents the transmission coefficient of heat between the mold and the coolant fluid, T_{melt} [°C] represents the temperature of the melt plastic flow, T_{eject} [°C] represents the ejection temperature of the plastic part of the mold and t_{cycle} [s] represents the time corresponding to the cooling phase. According to Menges et al. [17] the cooling time of the plastic part can be expressed as (2).

$$t_{cycle} = \frac{t_p^2 \cdot}{\pi \cdot \alpha_s} \cdot Ln\left(\frac{4 \cdot (T_{melt} - T_{mold})}{\pi \cdot (T_{eject} - T_{mold})}\right) \quad (2)$$

Where, α_s [m^2/s] represents the coefficient of thermal diffusivity of the metal material of the injection mold. In this way, the main geometric and technological variables dimensioned from the present physical model are: the coolant temperature $T_{coolant}$ [°C], the surface temperature of the mold T_{mold} [°C], the separation distance between the conformal cooling channels D_c [m], the diameter of the conformal cooling channels $\varnothing_c$ [m] and the separation distance between the conformal cooling channels and the plastic part D_p [m]. Through the application of genetic optimization algorithms to (1) and (2), the variables of the present physical model are optimized under the restrictions imposed by the manufacturer regarding the thermoplastic material and the metallic material for 3D additive manufacturing. Table 1 shows the recommended magnitude of the geometric and technological variables defined by the manufacturer of the 3D printing metallic and thermoplastic material, for the geometric and technological variables established for the cooling system dimensioning.

Table 1. Restrictions of geometric and technological variables defined by the manufacturer of the 3D printing metallic and thermoplastic material.

Description	Units	Value
Minimum distance between cooling channel and plastic part	mm	3
Minimum distance between cooling channels	mm	3
Minimum conformal channels diameter	mm	3
Maximum mold temperature	°C	71
Minimum mold temperature	°C	49
Maximum coolant flow rate	l/min	8
Maximum shear rate	1/s	50000

Furthermore, Table 2 shows the configuration of the genetic optimization algorithm for the proposed physical model. Otherwise, Table 3 shows the resulting magnitude of these variables, which in turn are used to model the conformal cooling systems studied in the present manuscript, see Figs. 2, 3 and 4.

Table 2. Definition of the technological optimization problem.

Genetic algorithm options structure	Defined set-up
Algorithm	NSGA-II
Reproduction	Two points
Mutation	2%
Crossover factor	80%
Migration	Direction: Forward Migration interval: 20 Migration factor: 30%

4 Numerical Analysis of the Conformal Cooling Systems

The numerical analysis of the plastic part cooling phase for each design of the conformal cooling system proposed has been carried out using the commercial software Moldex 3D R16. By means of the import to the numerical simulation software of the plastic part and conformal cooling channels 3D geometries, the discrete analysis mesh is generated. For each study case the average size of the elements edges is analogous and equal to 1.5 mm. The type of elements used to discretize the analysis geometries is an upper-order tetrahedron (Solid 186). In addition to improving the accuracy of numerical simulations, 5 boundary layers are defined in the mesh of the plastic part (from the contact surface between the plastic part and the injection mold) and 3 boundary layers in the mesh between the conformal cooling channels and the elements of the filling system. The type of elements used for boundary layers is upper-order prism (Solid 186) whose height is defined by the offset ratio (ratio between the average size of the mesh element and the size of the boundary layer element) equal to 0.1. Table 3 shows the technological and geometric parameters used to define the numerical simulation of the cooling phase for the case study presented. Table 4 shows the magnitude of the physical and thermal properties of the thermo-plastic material of the part (Cycolac MG 47) and the metallic material (Steel alloy 1.2709) of the main elements of the injection molds, which intervene in the physical resolution of the presented model, see (1) and (2).

Table 3. Setup of the cooling process simulations and solutions of the physical model parameters.

Nomenclature	Units	Description	Value
$Ø_c$	mm	Diameter of the conformal channels	6
D_c	mm	Distance between conformal channels	10
D_p	mm	Distance between conformal channels and plastic part	10
T_{melt}	°C	Melt temperature	238
T_{mold}	°C	Mold temperature	60
T_{eject}	°C	Ejection temperature	81
$T_{coolant}$	°C	Coolant temperature	55
Q	m^3/s	Inlet coolant flow rate	70
t_p	mm	Plastic part average thickness	2
t_{cycle}	s	Cooling time	11

Table 4. Thermoplastic and metallic material properties used in the physical model.

Nomenclature	Units	Description	Material	Value
α_s	m^2/s	Thermal diffusivity	Steel alloy 1.2709	$5.55 \cdot 10^{-6}$
h	$W/m^2 \cdot °C$	Heat transfer coefficient between mold and coolant	–	4500
ρ_p	kg/m^3	Density	Cycolac MG47	930.55
C_p	J/kg °C	Specific heat	Cycolac MG47	2314

5 Results

In this paper the design of three conformal refrigeration systems using different pathings to establish the geometry of the cooling channels: zig-zag, spiral and iso-contouring, is proposed. In addition, the present methodology performs an optimal dimensioning of the main technological and geometric variables of the cooling channels, which optimizes the thermal exchange in those regions of the plastic part where the conventional elements of the cooling system lose efficiency. The numerical simulations included in the present manuscript were carried out using the injection molding simulation Moldex 3D R16. For the validation of the proposed conformal cooling systems, the cooling temperature maps of the plastic part and the distribution of the cycle time along the geometry of the plastic part have been included (see Figs. 2, 3, 4 and Table 5). To evaluate the thermal performance of the proposed designs, see Figs. 2, 3 and 4, their cooling efficiency is included.

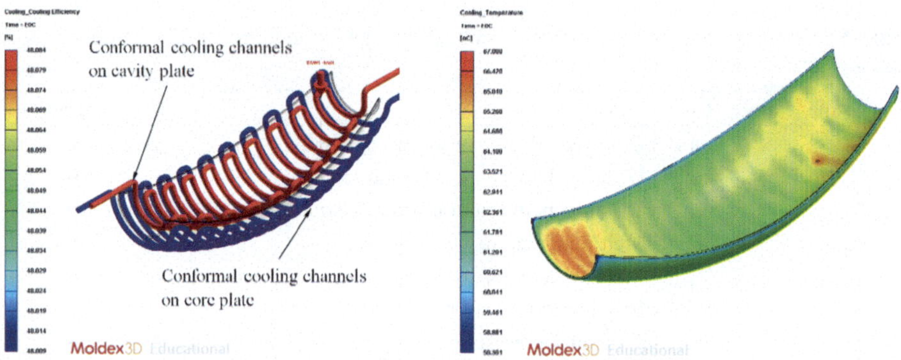

Fig. 2. Distribution of the plastic part temperature map after the cooling phase and efficiency of the conformal cooling system with zig-zag pathing.

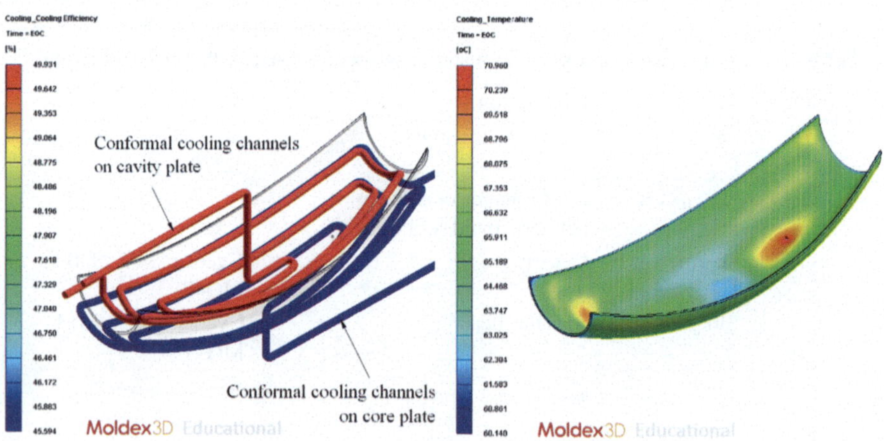

Fig. 3. Distribution of the plastic part temperature map after the cooling phase and efficiency of the conformal cooling system with spiral pathing

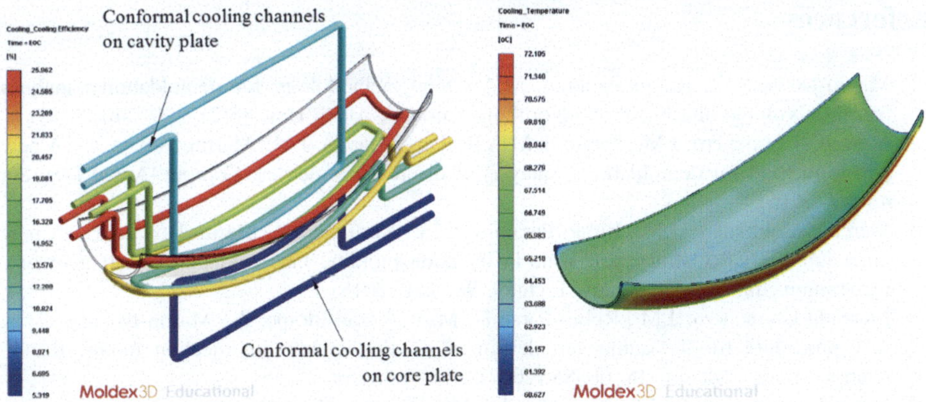

Fig. 4. Distribution of the plastic part temperature map after the cooling phase and efficiency of the conformal cooling system with iso-contouring pathing.

Table 5. Numerical simulation results of the cooling phase for each conformal pathing analyzed.

Variable	Units	Zig-zag	Spiral	Iso-contouring
Average/standard deviation of the part temperature	°C	63.51/1.33	64.68/1.68	65.57/2.80
Average/standard deviation of the cooling time	s	0.93/0.99	1.08/1.17	1.25/1.37
Average/standard deviation of mold temperature difference	°C	1.31/0.98	1.84/1.45	2.92/1.88
Cavity/core cooling efficiency	%	48.08/48.01	49.93/45.59	54.49/36.60

6 Conclusions

The results obtained show that the zig-zag pathing for the conformal cooling design channels is the one that best adapts to the plastic part topology under study, since it presents a lower average cooling time and lower standard deviations of the plastic part cooling temperature and mold temperature difference. Thus, it optimizes the plastic part′s final quality, since the temperature map after the cooling phase is uniform and the defects derived from the residual stress caused by the temperature gradients are reduced. Similarly, it is verified that the use of 3D additive manufacturing techniques improves the productivity and economy of the injection mold process, since it improves its cycle time as well as allowing for improvements in the cooling phase on complex plastic part regions and surfaces.

Acknowledgments. This research work was supported by the University of Jaen through Plan de Apoyo a la Investigación 2019-2020-ACCION1 PAIUJA2019-20: TIC-159.

References

1. Martin-Doñate, C., Rubio Paramio, M.A.: New methodology for demoldability analysis based on volume discretization algorithms. Comput.-Aided Des. **45**(2), 229–240 (2013)
2. Mercado-Colmenero, J.M., Paramio, M.A.R., Perez-Garcia, J.M., Martin-Doñate, C.: A new hybrid method for demoldability analysis of discrete geometries. Comput.-Aided Des. **80**, 43–60 (2016)
3. Mercado-Colmenero, J.M., Rubio-Paramio, M.A., Karlinger, P., Martin-Doñate, C.: A new procedure for calculating cycle time in injection molding based on plastic part geometry recognition. Int. J. Adv. Manuf. Technol. **98**, 441 (2018)
4. Mercado-Colmenero, J.M., Rubio-Paramio, M.A., Vizan-Idoipe, A., Martin-Doñate, C.: A new procedure for the automated design of ejection system in injection molds. Robot. Comput. Integr. Manuf. **46**, 68–85 (2017)
5. Mercado-Colmenero, J.M., Muriana, J.A.M., Paramio, M.A.R., Martín-Doñate, C.: An automated manufacturing analysis of plastic parts using faceted surfaces. In: Eynard, B., Nigrelli, V., Oliveri, S., Peris-Fajarnes, G., Rizzuti, S. (eds.) Advances on Mechanics. Lecture Notes in Mechanical Engineering. Design Engineering and Manufacturing. Springer, Cham (2017)
6. Opolski, S.W., Kwon, T.W.: Injection molding cooling system design. In: ANTEC 1986, pp. 264–268 (1987)
7. Tang, L.Q., Pochiraju, K., Chassapis, C., Manoochehri, S.: A computer-aided optimization approach for the design of injection mold cooling systems. Trans. ASME J. Mech. Des. **120**, 165–174 (1998)
8. Li, C.G., Li, C.L.: Plastic injection mould cooling system design by the configuration space method. Comput. Aided Des. **40**(3), 334–349 (2008)
9. Li, C.G., Li, C.L., Liu, Y., Huang, Y.: A new C-space method to automate the layout design of injection mould cooling system. Comput.-Aided Des. **44**, 812–823 (2012)
10. Park, H.S., Pham, N.H.: Design of conformal cooling channels for an automotive part. Int. J. Automot. Technol. **10**(1), 87–93 (2009)
11. Mercado-Colmenero, J.M., Rubio-Paramio, M.A., de Juanes Marquez-Sevillano, J., Martin-Doñate, C.: A new method for the automated design of cooling systems in injection molds. Comput. Aided Des. **104**, 60–86 (2018)
12. Crema, L., Sorgato, M., Lucchetta, G.: Thermal optimization of deterministic porous mold inserts for rapid heat cycle molding. Int. J. Heat Mass Transf. **109**, 462–469 (2017)
13. Mercado-Colmenero, J.M., Martin-Doñate, C., Rodriguez-Santiago, M., et al.: Int. J. Adv. Manuf. Technol. **102**, 1719 (2019). https://doi.org/10.1007/s00170-018-03235-3
14. Wang, Y., Yu, K.-M., Wang, C.C.L., et al.: Automatic design of conformal cooling circuits for rapid tooling. Comput.-Aided Des. **43**(8), 1001–1010 (2011)
15. Wang, Y., Yu, K.-M., Wang, C.C.L.: Spiral and conformal cooling in plastic injection molding. Comput.-Aided Des. **63**, 1–11 (2015)
16. Xu, X.R., Sachs, E., Allen, S.: The design of conformal cooling channels in injection molding tooling. Polym. Eng. Sci. **41**(7), 1265–1279 (2001)
17. Menges, G., Mohren, P.: How to Make Injection Moulds, 3rd edn. Hanser, Munich, Vienna and New York (2001)

Custom Elasticity Materials Through Mixing Thermoplastics with Extrusion 3D Printers

Ane Eguiazabal-Galán[1], Aitor Cazón-Martín[1](✉), María Isabel Rodríguez-Ferradas[1], Leire Frances-Morcillo[1], Paz Morer-Camo[1], and Luis Matey-Muñoz[1,2]

[1] University of Navarra, TECNUN, Paseo Manuel de Lardizábal 13, 20.018, San Sebastián, Spain
acazon@tecnun.es

[2] University of Navarra-CEIT-IK4, Manuel de Lardizábal 15, 20.018, San Sebastián, Spain

Abstract. Additive Manufacturing technology (AM) – or 3D printing – based on extrusion has experienced an undeniable boom due to the expiration of key patents. Nowadays many different companies are selling extrusion AM devices at very competitive prices bringing the advantages of AM closer to plain users, universities and companies dedicated to design or engineering. Parallel to the increase in the supply of these low-cost devices, the variety of materials available for 3D printing has become larger as well. However, the current availability of materials with an elasticity that lies between a rigid thermoplastic and a regular elastomer remains very limited. This paper aims to study how to produce custom elasticity materials by taking advantage of low-cost 3D printers capable of extruding the mix of two materials. Different materials will be created by combining a regular thermoplastic (PLA) with a Thermoplastic Polyurethane (TPU) that later will be mechanically characterized by means of tensile tests. The layer height was also introduced as a variable in the process. Results from this study show that a customized elastic modulus can be achieved for the printed part, and thus potential users can have the resources to cover their needs with regard to the choice of a mate-rial with certain properties.

Keywords: Additive Manufacturing · Dual printing · Extrusion · PLA · TPU · Tensile properties

1 Introduction

Additive Manufacturing (AM) is the process of creating a physical object based on its virtual representation by adding material layer by layer [1]. Its main advantage with respect to traditional manufacturing methods is that virtually any geometry can be built, offering almost unlimited and unconstrained geometric complexity. Among AM devices, material extrusion is one of the fastest, lowest-cost, most easy-to-use and widely available techniques available on the market for AM printers [1] due to the expiry of key patents which has led to economy-friendly printers. The working principle of this technique is to selectively deposit a semisolid thermoplastic material

F. Cavas-Martínez et al. (Eds.): INGEGRAF 2019, LNME, pp. 175–183, 2020.
https://doi.org/10.1007/978-3-030-41200-5_19

through a nozzle or orifice. The thermoplastic material, usually in filaments between 1 to 3 mm in diameter, is, first heated and, then, extruded through the nozzle in single paths to form each layer of the build. Current extrusion devices use a variety of either rigid thermoplastics such as Polylactic Acid Polymer (PLA) and Acrylonitrile Butadiene Styrene (ABS), or elastomers such as Thermoplastic Polyurethane (TPU) or Thermoplastic Elastomer (TPE) [2, 3]. In recent years, more exotic material has become available, such as ABS reinforced with short carbon fibre [4], iron particles in a nylon-type matrix [5] and electrically conductive materials [6].

Despite the wide range of available materials for low-cost extrusion devices, in terms of stiffness the user is still limited by the manufacturer's offer. This means that engineers can only 'play' with the part geometry in order to get a desired stiffness in the printed part. This drawback can be fixed with specific low-cost extrusion printers capable of mixing two materials in specific ratios, getting as a result a new material that can be normally deposited layer by layer to get the physical part. This current project aims to explore this possibility by studying how to produce custom elasticity materials by taking advantage of a low-cost 3D printer capable of extruding a mix of two materials. Several specimens were printed and then mechanically characterised by means of tensile tests. The reference materials used in the experiments are PLA and TPU. To ensure the homogeneity of the mixture, colour material was chosen: red for PLA and white for TPU.

2 Methods

2.1 Printing Specimens

The specimens for the tensile test were modelled in accordance with BS EN ISO 527-2:1996 as shown in Fig. 1 using CREO 5.0 CAD software from PTC (PTC Corporate Headquarters, Needham, USA). Later, the specimen was exported in the Stereolithograhy (STL) file format and loaded to Slic3r, an open-source 3D printing software application, to generate the necessary g-code.

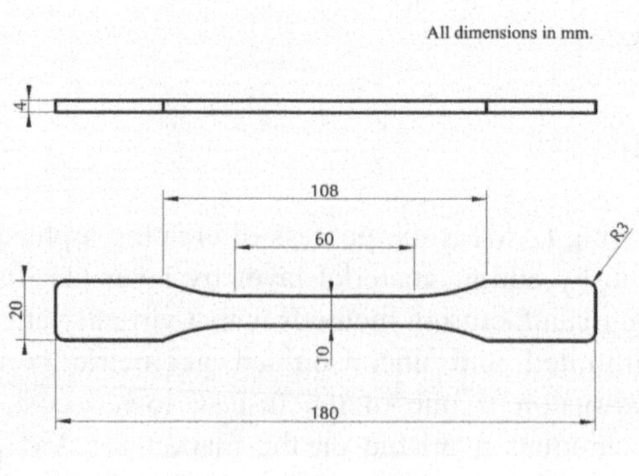

Fig. 1. Specimen dimensions in mm for tensile testing.

The extrusion printer selected for this study was the Colido X3045 Duo, whose cost is around 3,500 €. It is a dual material 3D printer that features a single extruder. The extruder is mounted on a head with movement in the X and the Y-axes. The extruder is just above a mixing chamber where the two materials get after being pushed by two Bowden extruders. Once mixed, the new material leaves the chamber through a nozzle that is 0.4 mm in diameter and is deposited layer upon layer to physically manufacture the part. A schematic drawing showing the basic working of the extruder is shown in Fig. 2 along with a real image.

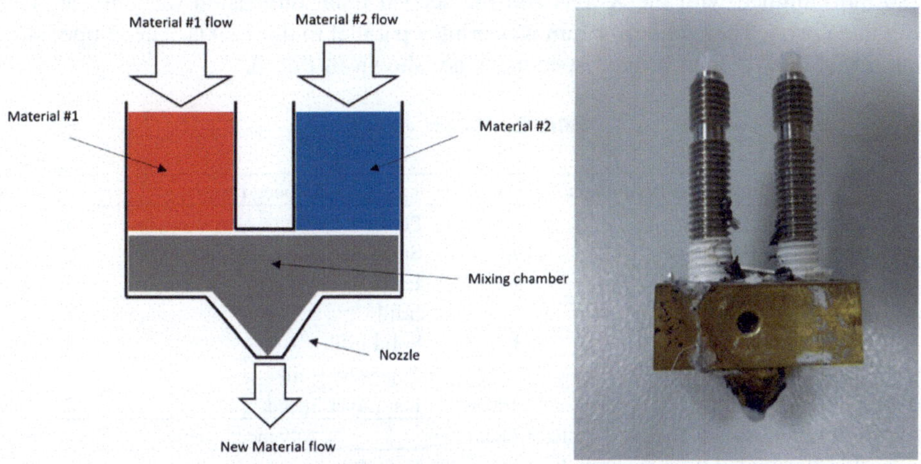

Fig. 2. Schematics (left) and real image (right) of the extruder.

Table 1. The 15 specimens according to 'mixing ratio' and 'layer height'.

Specimen number	Mixing ratio (%)		Layer height (mm)
	PLA	TPU	
1	100	0	0.1
2	100	0	0.2
3	100	0	0.4
4	75	25	0.1
5	75	25	0.2
6	75	25	0.4
7	50	50	0.1
8	50	50	0.2
9	50	50	0.4
10	25	75	0.1
11	25	75	0.2
12	25	75	0.4
13	0	100	0.1
14	0	100	0.2
15	0	100	0.4

A full factorial Design of Experiment (DOE) approach was followed by considering two main factors – mixing ratio and layer height – with five levels for mixing ratio and three levels for layer height. Two different polymers were used to print one specimen for each combination for a total of 15 specimens (Table 1). This strategy would allow for the evaluation of the main effect of both factors. However, as there was only one observation for each level combination, there were not enough degrees of freedom to investigate and test interaction effects, and so it was assumed that there was no interaction between the independent variables. The printing settings used within Slic3r are shown in Table 2. All specimens were printed with the first main dimension (180 mm) aligned with the X-axis and the second main dimension (20 mm) aligned with the Y-axis. The width of 4 mm was printed parallel to the Z-axis. The temperature was set to 230 °C. The printed specimens are shown in Fig. 3.

Table 2. Printer settings.

Layer and Perimeters	
Layer Height	LH
First Layer Height	LH
Perimeters	3
Solid Layer Top	3
Solid Layer Bottom	3
Infill	
Fill Density	100%
Fill Pattern	Rectilinear
Top/Bottom Fill Patterns	Rectilinear
Combine Infill Every	1
Solid Infill Every	0
Fill Angle	0
Skirt and Brim	
Loops	8
Distance from Object	5
Skirt Height	1
Maximum Extrusion Length	0
Brim Width	0

Speed (mm/s)	
Perimeters	5
Small Perimeter	5
External Perimeter	5
Infill	5
Solid Infill	5
Top Solid Infill	5
Fist Layer Speed	5
Advanced (mm)	
Default Extrusion Width	0.4
First Layer Height	0.4
Perimeters	0.4
External Perimeters	0.4
Infill	0.4
Solid Infill	0.4
Top Solid Infill	0.4
Overlap	0.4

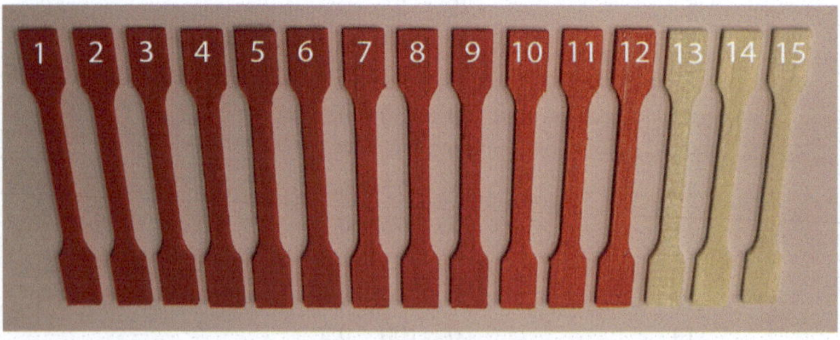

Fig. 3. The 15 printed specimens.

2.2 Testing Specimens

Tensile tests were performed on the specimens using an Instron model 4467 (Instrom Worldwide, Norwood, MA) tensile test machine that has a 30 kN load cell and is fitted with a long travel (50 mm) contact extensometer. Each sample was subjected to a cross-head speed of 1 mm/min until the failure of the specimen. The primary goal of the tensile testing was to generate force-extension data. To this end, a support computer running LabVIEW software (National Instruments, Newbury, Berkshire, UK) and connected to the tensile machine was used to visualize and record the force-extension curves of each specimen. The load-displacement outputs from the tensile test machine were normalized to engineering stress-strain data. Although engineering stress-strain approximation is more widely used, our data was transformed to true stress-strain curves to better describe the shape and volume change when deformation is not small, something that could happen with the more flexible material. Four mechanical parameters were then obtained: the Young's modulus (E), the Ultimate Tensile Strength (UTS), the Elongation at break and the Fracture stress. Young's modulus was calculated for each specimen by considering the stress values at the strains of 0.005 and 0.01.

3 Results

The results of the experiments are shown in Table 3 while the stress-strain curves are shown in Fig. 4. Specimens #1 to #12 were tested until fracture occurred in the neck region between the extensometer legs. The tests for the last three specimens, corresponding to 100% TPU, were stopped after 30 min before any of the specimens presented any sign of failure. Considering the Young's modulus, the Colido printer was able to deliver specimens with remarkable differences as a function of the selected mixing ratio.

Table 3. Printer settings.

Specimen number	E (MPa)	UTS (MPa)	Fracture stress (MPa)	Elongation at break (%)
1	3272	45.3	42.8	1.82
2	3199	48.0	47.7	1.73
3	2967	44.9	41.8	2.01
4	1720	23.0	21.3	1.18
5	1912	25.8	25.0	1.52
6	1534	22.0	20.8	1.97
7	871	13.3	8.8	2.02
8	976	15.2	12.0	2.46
9	1042	16.4	11.3	2.9
10	422	6.4	3.6	11.6
11	447	7.0	3.9	17.4
12	393	7.1	4.6	10.8
13	49	–	–	>57.2
14	42	–	–	>55.9
15	46	–	–	>59.5

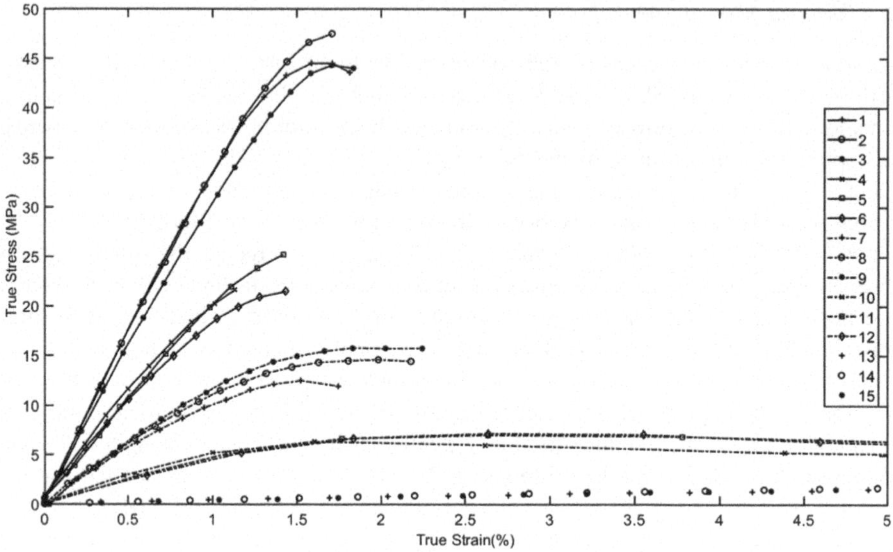

Fig. 4. Stress vs strain curves for all specimens for the first 5% of deformation.

A two-way between-groups analysis of variance was conducted to explore the influence of the two main factors on the elastic modulus, with three levels for 'layer height' and five for 'mixing ratio' at a confidence level of 95% (Table 4). ANOVA assumptions were met for E: residuals were normally distributed (p-value of 0.461 according to Anderson Darling's test) while groups presented equal variances according to Levene's tests of equality at a 95% level of confidence (p-value of 0.29 for 'mixing ratio' and p-value of 0.974 for 'layer height'). For elastic modulus and 'mixing ratio', there are statistically significant differences between the mean (p-value of 0.000) while there are no statistically significant differences for the 'layer height' (p-value of 0.304). The interaction plots of both variables for Elastic modulus are shown in Fig. 5. As can be observed, E curves have negative trends when more flexible material is in the specimen. However, the E modulus present flat curves for every mixing ratio, confirming the ANOVA results.

Table 4. Analysis of variance for E.

Source	DF	Sum squares	Mean square	F	P
Mixing ratio	4	18113200	4528300	352.24	0.000
Layer height	2	35687	17843	1.39	0.304
Error	8	102846	12856		
Total	14	18251734			

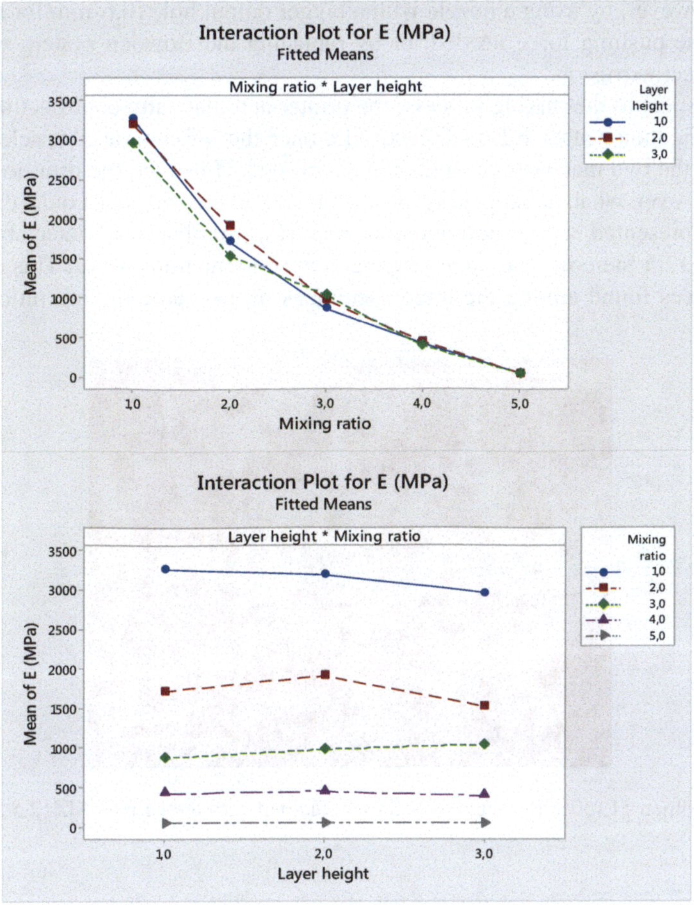

Fig. 5. Interaction plots of the Means for elastic modulus by 'mixing ratio' and 'layer height'.

4 Discussion and Conclusions

There are several topics for discussion beginning with the 3D printer selected for this study. Although there are numerous dual extrusion printers on the market, Colido was one of the few companies offering a Duo model that not only had the dual printing capabilities but also the option of mixing two raw materials to 3D print with a new third material. Although the principle of working was easy in theory, we faced several difficulties due to the fact of printing with two very different materials from the mechanical and material science points of view. Combining two materials based both on PLA or ABS is easy since the printing settings should be identical for both extruders. However, in our case, the PLA and the TPU needed considerably different settings, especially when dealing with the printing speed. In the end, the printing speed ended up being governed by the softer filament; otherwise, our feeder (Bowden extruder) developed clog due to the high flexibility of the TPU filament. This can be avoided, however, by using a nozzle with a bigger output hole (0.6 mm for instance) to minimize the pushing force needed, or by replacing the Bowden system with a head that has direct extrusion.

With respect to the mixing process, the printer automatically controls the activation of the electric motor of each Bowden extruder once the 'mixing ratio' is selected by the user. Then, the two materials are mixed and extruded. However, the user does not have any control over what is happening inside the chamber, and that could explain why specimens presented a non-homogenous mix (Fig. 6) that was detectable under a LEICA MZ12.5 Stereoscopic microscope and even to the human eye. This can explain the differences found among the three specimens on the same mixing ratio.

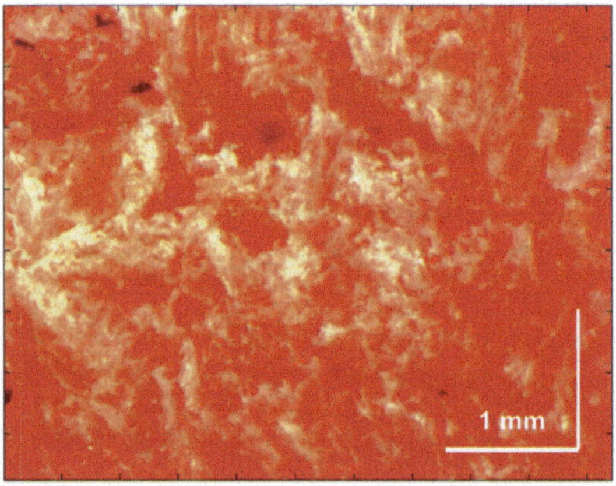

Fig. 6. Specimen 11 (50% PLA and 50% TPU) obtained with the LEICA MZ12.5 Stereoscopic microscope.

Finally, in terms of the results, they were more or less as expected with respect to the 'mixing ratio'. This variable greatly influenced the mechanical properties of the resulting material. In contrast, it seems that the layer height parameter does not affect the mechanical properties, so higher values can be used in order to save printing time.

In conclusion, we have demonstrated that commercially available AM devices can deliver custom elasticity materials by combining a regular thermoplastic (PLA) with a Thermoplastic Polyurethane (TPU). In addition, results indicate that, independently of the mixing ratio, the layer height does not play a significant role in the elastic modulus. Future research can analyse the influence of other printing parameters that are related to the printing process itself (printing temperature and printing speeds) rather than the geometry of the specimen (infill).

References

1. American Society for Testing and Materials International. ASTM F2792 - 12a Standard Terminology for Additive Manufacturing Technologies (2012)

2. Raasch, J., Ivey, M., Aldrich, D., Nobes, D.S., Ayranci, C.: Characterization of polyurethane shape memory polymer processed by material extrusion additive manufacturing. Addit. Manuf. **8**, 132–141 (2015). https://doi.org/10.1016/j.addma.2015.09.004
3. Lopes, L.R., Silva, A.F., Carneiro, O.S.: Multi-material 3D printing: The relevance of materials affinity on the boundary interface performance. Addit. Manuf. **23**(March), 45–52 (2018). https://doi.org/10.1016/j.addma.2018.06.027
4. Tekinalp, H.L., Kunc, V., Velez-Garcia, G.M., Duty, C.E., Love, L.J., Naskar, A.K., Blue, C.A., Ozcan, S.: Highly oriented carbon fiber–polymer composites via additive manufacturing. Compos. Sci. Technol. **105**, 144–150 (2014)
5. Masood, S.H., Song, W.Q.: Thermal characteristics of a new metal/polymer material for FDM rapid prototyping process. Assem. Autom. **25**(4), 309–315 (2005)
6. Flowers, P.F., Reyes, C., Ye, S., Kim, M.J., Wiley, B.J.: 3D printing electronic components and circuits with conductive thermoplastic filament. Addit. Manuf. **18**, 156–163 (2017). https://doi.org/10.1016/j.addma.2017.10.002

Wheelchair Mobility Aid Through the Adaptation of Electric Scooters

Bernardo Pajares Moreno(✉), Guillermo Peris-Fajarnés, Ismael Lengua, and María Moncho-Santonja

C.I.T.G. Universitat Politècnica de València, Camino de Vera s/n, 46022 Valencia, Spain
berpamo@etsia.upv.es

Abstract. This paper presents a design proposal for the use of an electric scooter to propel a conventional wheelchair. This adaptation will transform a wheelchair into a "handbike" using a scooter-chair hitch. The system allows access to these devices to a greater number of people with mobility assistance needs, mainly in the urban environment, thanks to the use of low-cost mobility devices. Keeping this is mind, the design of a scooter-chair hitch that respects the chair integrity is presented. The new set will allow the electric scooter to pull the chair analogously to the way modern handbikes do. The electric mechanism of the scooter together with the hitch mechanism has been designed under the premises of usefulness without complex ergonomic conditions, simplicity, efficiency, cost, and accessibility to parts, mechanisms and materials.

Keywords: Scooter · Hitch · Handbike · Electric · Attachment · Affordable · Transportable · Easy · Adjustable · Means of transport · Third wheel · Universal

1 Introduction

A "handbike" is a conventional wheelchair propelled by a driving wheel adapted to the front of it, in which both traction and steering control are performed by the user with a handlebar. At present, a great majority are electric and it can be stated that they are an evolution of those ones that were operated with cranks—hence their name "handbike". The great advantage of these devices is that they allow to propel the conventional wheelchair avoiding the need to change the chair or to "transfer" it.

This paper describes the hitch mechanism of a conventional wheelchair to an electric scooter in order to transform the wheelchair into a "handbike". The system has been designed to be used in adapted urban environments and for people with the minimum mobility capacity necessary for the use of an electric handbike with handlebars.

Although the article focuses on a specific type of aid mechanism, it should be noted that there are as many aid mechanisms as individuals need. They range from the aforementioned hitches to motors that push the chair from the back, or electric driving wheels directed with the hand rings. Some are more integrated than others in the chair

F. Cavas-Martínez et al. (Eds.): INGEGRAF 2019, LNME, pp. 184–190, 2020.
https://doi.org/10.1007/978-3-030-41200-5_20

assembly. They are more or less removable and transportable, and all of them are electric. But the main drawback is their excessive price.
Make a Table of the Different Types, Advantages and Cost Order

Model-type	Connection complexity Estimate the time for coupling and uncoupling… if the user can do it on his/her own	Manageability Maneuverability	Change in dimensions How the size is altered	User operability If the user can do it on his/her own	Cost Average price in stars…. For example, each star amounts to €500 approximately
BatecMini	4 s EASY-FIX	Good	+ 830 mm	Not so good	*********
Lomo Litio Pico	4 s	Good	+500 mm	Good	*******
Lomo 20"	15–20 s clamp	Good	+900 mm	Low	********
Quickie attitude	15–20 s clamp	Good	+930 mm	Low	******

For example, the *e-pilot* "handbike" is a traction device with an electric motor that, when installed in a manual wheelchair, transforms it into an electrically powered chair, which considerably increases the user's mobility and flexibility. Its price ranges are between €4,200 and €4,400.

On the other hand, we have the simplest ones, with the inconvenience of their size, which approximate prices are around €2,200 (see Fig. 1).

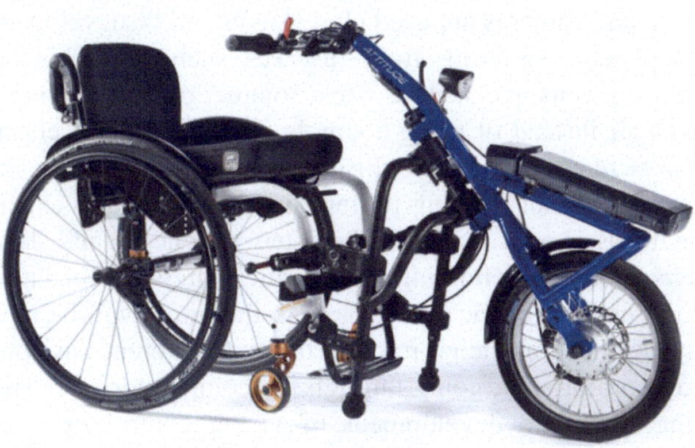

Fig. 1. Handbike image.

Therefore, this design aims to create an electric mobility aid that is as integrated as possible in the wheelchair assembly at a reduced cost. At the same time, it should serve as many chairs as possible, be electric and easily transportable. In any case, it is required not to alter very much the functionality of the handbikes existing in the market.

A review of the existing products in the market has been carried out by analyzing their technical characteristics and their prices. These data have been contrasted and studied in order to make decisions regarding the most appropriate design. Finally, we have opted for a "handbike" type design.

The reason for this decision is based on three basic points. The first one is the simplicity of the design as it is essential to reduce costs. The second one is the ease of hitching and unhitching the chair, and the third one is the possibility of creating a simple, lightweight and easily transportable model. These are basic characteristics for mobility.

2 Analysis of the State of the Art

In order to know the restrictions and requirements in the design of wheelchairs or the respective mobility aids, it is necessary to review bibliographies on this type of needs. This includes market research, in order to obtain information on what products are currently on the market. At the same time, it is important to analyse what types of projects are being developed. All these aims to obtain information on the products in which the industry and medical organizations are working.

2.1 State of the Art

In order to approach the current electric wheelchairs market, it has been necessary to carry out review of innovation projects.

In most cases, research papers published by universities have shown progress in Smart Wheelchairs (SW). These are wheelchairs in which modern technologies such as brain sensors, voice and cameras are used [1, 13]. SW can be developed to increase the independence of patients suffering from diseases such as Parkinson's and vision problems. Different movements, such as neck, tongue, or eye tracking can be used to control the wheelchair instead of using a simple joystick as most electric wheelchairs do. SW can also be equipped with collision detectors, Global Positioning System (GPS) technology and virtual tracking to improve the wheelchair experience and help the user. Research has shown that a further step in wheelchairs development is to design wheelchairs capable of climbing stairs, which offers obvious advantages [6, 7].

In general, the trend of electric wheelchairs in the future will be models capable of overcoming obstacles such as stairs, moving through different environments, such as traveling downhill comfortably, going through the countryside, etc. The user interface is also improving, with the development of vision control or voice recognition. Moreover, the functionality of wheelchairs is improving with the introduction of GPS and wall-following technology. New developments are taking place in the 'Supercapacitors' (also called Ultracapacitors) field, which are a hybrid between batteries and capacitors. They offer much faster charging times than lithium-ion batteries and are much more powerful than those used nowadays.

3 Conceptual Design. Initial Ideas

When a wheelchair user needs to travel medium and long distances, mainly in cities, he/she encounters a series of obstacles that are uncomfortable and sometimes insurmountable. Among others, slopes, both parallel and perpendicular, not adapted public transport, obstacles in the pedestrian roads, ramps, etc. For this reason, the need for the acquisition of a device that helps overcome them in a comfortable way and with minimal physical effort arises. There are many different aid models on the market, to a large extent electric. When purchasing one of them, it is necessary to choose between a wide variety of ranges and prices. The prices of all these aids depend on the type of propulsion and on their equipment but all of them have a drawback: their excessive price (above € 2,200). Furthermore, these types of mechanisms (such as the one shown in this article) are uncomfortable and heavy.

Therefore, the aim is to design an affordable electric handbike with a compact size, easy to transport, simple and universal.

It was decided to create an additional electronic device for wheelchairs taking a series of basic ideas as a starting point. They are the following:

- The use of a commercial electric scooter (see Fig. 2).
- Design of a new hitch mechanism wheelchair-handbike.
- Specific structural design for the use of several models of electric scooters.
- Parts design as universal as possible.

Fig. 2. Image of electric scooter.

These ideas were given shape as basic sketches using the CAD CAM PTC CREO PARAMETRICS 3.0 programme, for example in Fig. 3.

Fig. 3. Assembly design-wheelchair

The design of the "motorized front wheel" is an idea similar to that of the "hand-bikes" existing in the market, which feature a single traction motor incorporated in the front wheel. By means of the hitch, the front casters of the wheelchair are elevated and leant on the scooter-hitch assembly.

Thus, the motor, wheel, electrical components, handlebar, etc., of the commercial electric scooter are used. The commercial electric scooter placed in the front of the wheelchair intends to fulfil the traction and steering functions.

There are several components of the design that are essential to make the set universal. For this reason, the parts have been designed to be easily adjustable to different models of wheelchairs and scooters. For example, the assembly that works as a hitch and pulls the wheelchair has to be adjustable in width to each type of wheelchair (Figs. 4 and 5). Another essential element that must be adjustable too is the connection shaft that joins the wheelchair to the scooter. It needs to be mobile to allow adjustment in height (Fig. 6).

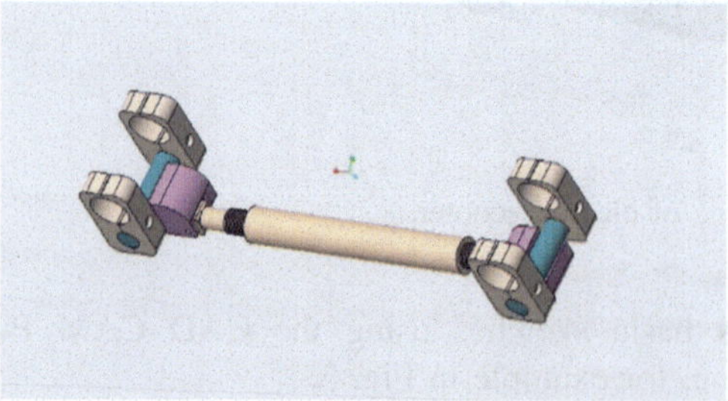

Fig. 4. Extensible assembly for coupling to wheelchair.

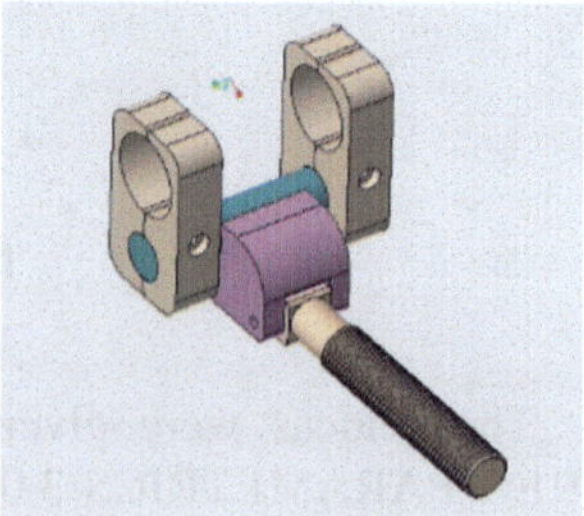

Fig. 5. Assembly detail.

Fig. 6. Swivel detail.

4 Conclusion

The objective of creating a prototype with propulsion system for wheelchairs using a conventional electric scooter, feasible and easy to use, has been achieved. The prototype's connection is adjustable to a large number of electric scooters and manual wheelchairs. It is easy to assemble and adjust. It is possible to sell only the scooter-wheelchair part, so that the price can become significantly more affordable. The use of components already existing in the market, such as electric scooters, should be noted.

On the other hand, since it is one of the priorities, the acquisition price of the proposed prototype can be found between a range of € 600 and € 700 depending on the price of the scooter purchased. The manufacture of the wheelchair-scooter hitch structure has an approximate cost of about € 250 which, added to € 350 (average price of electric scooters), which gives us a total price of € 600. It is much lower than the € 2,200 of the least expensive handbike model on the market.

Furthermore, we have designed an assembly that does not increase the dimensions of the wheelchair significantly, easy to transport thanks to its simplicity and with a similar performance to the one offered by the existing handbikes in the market.

The only exception is the maximum speed, which may be too high, and that can be easily changed in the configuration of the electric scooter.

References

1. Alcubierre, J.M., Minguez, J., Montesano, L., Montano, L., Saz, O., Lleida, E.: Silla de ruedas inteligente controlada por voz. In Memorias del Primer Congreso Internacional de Domótica, Robótica y Teleasistencia para todos, Madrid, España, November 2005
2. Birmanns, T.: España. N: E 16154456. Dispositivo de dirección delantero y tándem de silla de ruedas. Europa (2017)

3. Carrasco García, J.: ES 2584 477 B. Impulsor eléctrico trasero para sillas de ruedas. Murcia (España) (2015)
4. Cañete Aguado, J.: ES 1170 858 U. Dispositivo para la automatización de sillas de ruedas eléctrica. España (2016)
5. Conte, G.: Italy Patent No. EP 2 810 633 B1 (2014)
6. De Aguilar, J.M.D.: Silla de ruedas de bajo presupuesto capacitada para subir tramos escalonados (2012)
7. Escudero Mendoza, J.P., Ruiz Miranda, C.F.: Diseño y simulación de una silla de ruedas autónoma que permita subir y bajar escalones a su usuario (Bachelor's thesis) (2012)
8. European Committee for Standardization (CEN). DIN EN 12183:2014-06. Manual wheelchairs - Requirements and test methods. CEN - Technical Bodies, June 2014. https://standards.cen.eu/dyn/www/f?p=204%3A110%3A0%3A%3A%3A%3AFSP_PROJECT%2CFSP_ORG_ID%3A33939%2C6274&cs=12B13583893E1DCD34FFE2D9D93F1F3C6. Accessed 2019
9. European Comitee for Standardization (CEN). DIN EN 12184:2014-06. Electrically powered wheelchairs, scooters and their chargers - Requirements and test methods, June 2014. https://ec.europa.eu/eip/ageing/standards/healthcare/personal-autonomy/en-121842014_en
10. European Committee for Standardization (CEN). DIN EN 12182:2012-07. Assistive products for persons with disability - General requirements and test methods, July 2012. https://ec.europa.eu/eip/ageing/standards/healthcare/personal-autonomy/en-121822012_en
11. Leaman, J., La, H.M.: A comprehensive review of smart wheelchairs: past, present, and future. IEEE Trans. Hum.-Mach. Syst. **47**(4), 486–499 (2017)
12. Richard, M.A., Gentiletti, G.G.: Plataforma experimental de interfaz cerebro computadora orientada al control de sillas de ruedas. In: IV Latin American Congress on Biomedical Engineering 2007, Bioengineering Solutions for Latin America Health, pp. 1127–1130. Springer, Heidelberg (2007)
13. Sibai, M.: A Study on Smart Wheelchair Systems, December 2015. https://www.researchgate.net/publication/305699756_A_Study_on_Smart_Wheelchair_Systems. Accessed 1 Mar 2018
14. Tumboroller, S., Micro, H., Rodem Lomo Litio, H., Stricker City 7, H., Stricker City Compact, H., & Stricker City 11, H. et al. (2019). Rodem. https://www.dismovil.net/es/25_rodem?gclid=CjwKCAjwy7vlBRACEiwAZvdx9lemckhegEZw3byErbbQnh_vPZusEYTdtGBZxtHuxxAwtF2loe2RoCw7sQAv_BwE
15. Hsu, H.-L.: Conjunto de control auxiliar para sillas de ruedas (2010). Europa

Graphical Bioengineering

A Custom-Made Photogrammetry Scanner to Support Paediatric Surgery

Harkaitz Eguiraun[1,2(✉)], Lander Barrenetxea[1], Xabier Amezua[3], Oskar Casquero[4], Ruben I. Garcia-Fernandez[5], and Iñigo Tuduri[6]

[1] Department of Graphic Design and Engineering Projects, Faculty of Engineering in Bilbao, University of the Basque Country UPV/EHU, Bilbao, Spain
harkaitz.eguiraun@ehu.eus

[2] Research Center for Experimental Marine Biology and Biotechnology PiE, University of the Basque Country UPV/EHU, Plentzia, Spain

[3] Department of Graphic Design and Engineering Projects, Faculty of Engineering Gipuzkoa, University of the Basque Country UPV/EHU, San Sebastian, Spain

[4] Department of Systems Engineering and Automatic Control, Faculty of Engineering in Bilbao, University of the Basque Country UPV/EHU, Bilbao, Spain

[5] Unidad de Innovación ISS Biocruces Bizkaia/OSI EECruces, Barakaldo, Spain

[6] FEA Cirugía Pediátrica/OSI EECruces, Barakaldo, Spain

Abstract. Photogrammetry is a challenging technique used to digitalize objects and to build their surface points' meshes by connecting digital photographs taken from different positions. The work presented here is the first phase of a collaboration between the OSI EECruces Hospital and the Faculty of Engineering in Bilbao to develop a portable scanner to perform digital scans of paediatric patients' faces with enough accuracy to: (i) train surgery operations, (ii) analyse patients' evolution and (iii) perform virtual reality surgery guides. This first phase deals with the construction and calibration of a custom-made scanner. The scanner was built around 12 Raspberry Pi cameras and single-board computers, assembled on different positions in a custom-made structure inside an ad hoc designed and 3D printed container. The Raspberries were connected and synchronized via Ethernet to allow the 12 photographs to be taken simultaneously. Additionally, a projector was used to generate a particular light pattern to aid connecting the different photographs and generating the resulting two 3D points' meshes: (i) a real colour and texture of the test subject's face mesh and (ii) a mathematical mesh. Calibration was performed with a rule placed onto the test subjects' face. This first study was designed to test the accuracy of the photogrammetry scanner by comparing its performance to that of an industrial portable scanner. The promising results obtained indicate that the described scanner has a big potential to perform clinical studies.

Keywords: Photogrammetry · Paediatric · Surgery · Face · Texture

F. Cavas-Martínez et al. (Eds.): INGEGRAF 2019, LNME, pp. 193–201, 2020.
https://doi.org/10.1007/978-3-030-41200-5_21

1 Introduction

The interaction between engineering and medical sciences is becoming increasingly important and has led to a new knowledge area, namely, bioengineering. In short, bioengineering applies engineering techniques, including graphic expression, mechanics, cybernetics, automatic control, manufacturing and signal processing, to medical, biomedical and biological problems such as early detection of Alzheimer's disease [1, 2] surgical screws resistance [3], bone surface reconstruction [4], dentistry [5] and evaluation of positional plagiocephaly by means of a laser scan [6] among others.

Photogrammetry is a widely used technique in geosiences' applications [7] such as cartography [8] or topography [7]; scanning of full bodies [9] and it has demostrated a big potential in medical aplications [10, 11], for example in the detection of thoracic abnormalities [12] and cranial deformations [13]. It can be defined as a technique to digitalize objects and reconstruct them as 3D virtual objects by connecting digital photographs taken from different positions. Photogrammetry scans, unlike most industrial scans, have colour and texture, which opens up for new application areas, including clinical studies.

The work presented here is part of a collaboration between the OSI EECruces Hospital and the Faculty of Engineering in Bilbao to develop a portable scanner suitable to perform digital, colour and textured scans of paediatric patients' faces with enough accuracy to: (i) train surgery operations, (ii) analyse patients' evolution and (iii) perform virtual reality surgery guides.

2 The Photogrammetry Scanner

The photogrammetry scanner was built around a set of cameras placed in different positions surrounding the targeted object. The cameras were placed in a structure which has a LED light system used to illuminate the object directly. Additionally, a light

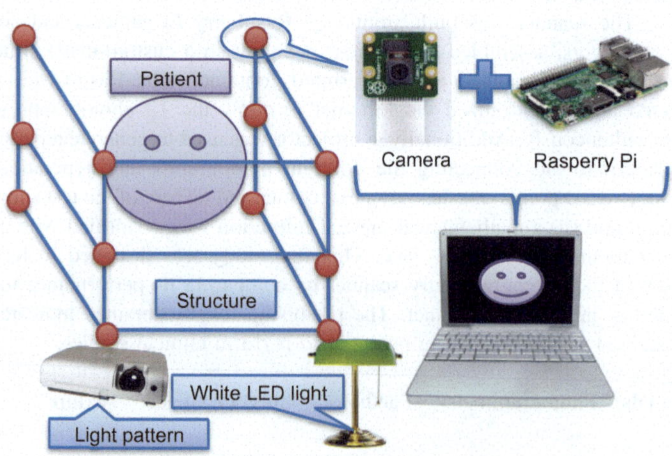

Fig. 1. General diagram of the photogrammetry scanner. The cameras and Raspberries were housed in a structure and connected to a computer. The scans were performed using white LED light and a projected specific pattern.

projector was used to project a specific light pattern which will aid in the reconstruction of the 3D object by linking the photographs. The photographs were taken simultaneously by the 12 cameras, stored locally and transferred to a computer for the reconstruction of the 3D object, which in this case was the test subject's face (Fig. 1).

2.1 Raspberries and Cameras

The camera system was based on the one developed by Pi3Dscan [9]. In the present work, 12 Raspberry Pi 3 Model B single-board computers with their respective camera modules (Camera Module V2 8 Mp), both of which were purchased from Raspberry Pi Foundation [14], were used to scan the test subject. Another Raspberry Pi 3 Model B single-board computer without the camera module was used to synchronize the light projector. Each Raspberry had a 16 Gb SD card.

The system was fed by a custom-made power supply built inside a specific rack where an AC-DC single output (5VDC 60A) power supply (MeanWell RSP-320-5) feeds the Raspberries and an AC-DC single output (24 VDC 10A) power supply (MeanWell SP-240-24) feeds the LED lights. The rack was machined to mount the required 13 USB and 2 DC ports.

2.2 Photographing, Light, Data Storage and Synchronization

All the photographs were taken simultaneously. Then, each photograph was stored locally in its corresponding single-board computers' SD card from where they were transferred to the computer for their post processing. The Raspberries were connected by Ethernet cables to an Ethernet Gigabit Switch (TP-LINK TL-SG1024 Switch 24) and synchronized with a laptop (Intel Core i7 and Nvidia Quadro).

The white light LED straps were placed alongside the structure to illuminate the test subject, and a light projector was used to project a special light pattern created by the software used. This special light pattern was designed to help joining together the different photographs to build the final 3D object. It has different background colors for left and right sides and showed a better performance than the black and white pattern from the Pi3Dscan software.

2.3 Image Acquisition and 3D Object Creation Software

The 3Dscan Management from Pi3Dscan software was used to project our custom build colour light pattern, shoot the photographs and join them together to create the 3D objects.

2.4 Housing

The Raspberries and the cameras were housed inside ad hoc boxes built from scratch for this project. The boxes were printed in ABS material and designed in Solid Edge ST10 from Siemens. They were subsequently mounted into a folding screen type wooden structure as shown in Fig. 2. This allowed us to place the Raspberries and the cameras in the required positions. Additional ABS material cones were used when necessary to move the cameras closer to the object.

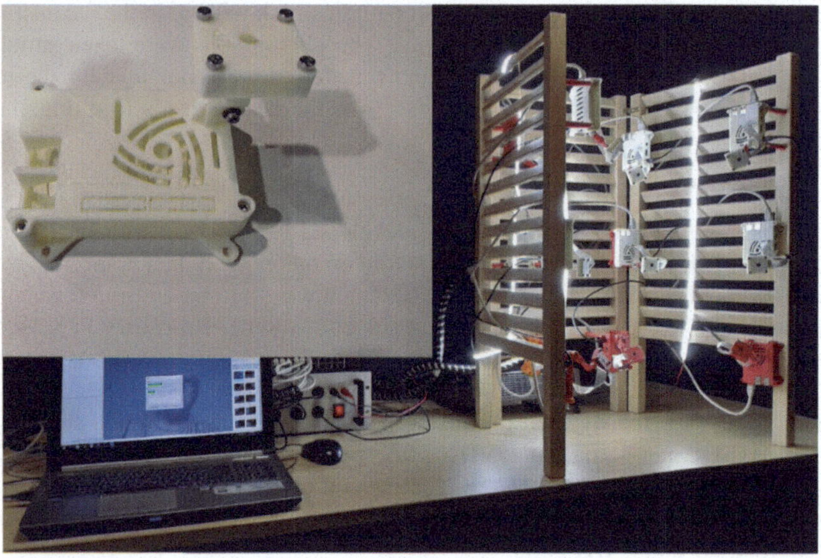

Fig. 2. The photogrammetry scanner. Note the wooden structure with the Raspberries and cameras' housings, the LED lights, the projector, the power supply, the Ethernet-switch and the computer. The upper left picture shows the custom made and 3D printed Raspberry and camera housing.

2.5 Calibration, Scaling and Test Scans

The scene is reconstructed using the structure-from-motion method, which reconstructs the scene by computing the camera's positions and geometry of the scene by matching repeated features in different images [7].

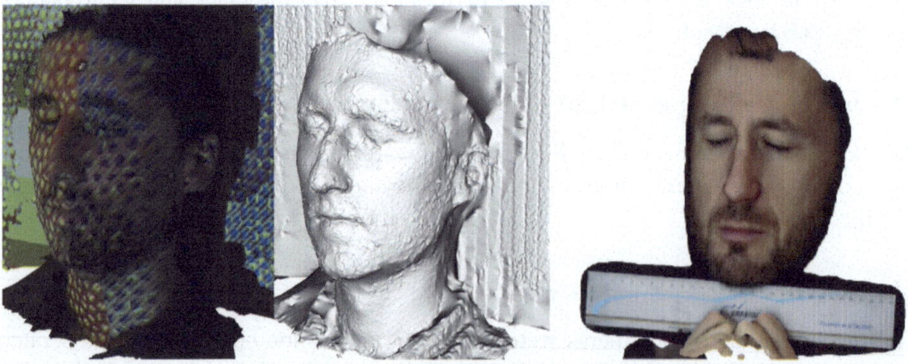

Fig. 3. Illustration of the calibration of the scanner. The left figure shows the test subject illuminated by the selected light pattern. The figure in the middle shows the resulting 3D object from that scan session. The picture in the right shows the resulting scan of the same test subject in a color-textured format with the rule used for scaling of the 3D object.

Calibration of the scanner and scaling of the scanned test objects was performed by placing a scale bar between two points of the scans and comparing the measured distance between those two points with the real distance measured by the scale bar, as shown in Fig. 3.

3 First Field Test

The first field study was conducted to test the accuracy of the photogrammetry scanner (P) in terms of precision and trueness. The scans performed with our methodology were compared to those obtained with the GO!Scan 3D G1 from Creaform, with a nominal accuracy of 0.1 mm, (G) on the same test subject. The experimental setup was as follows:

1. N = 5 scans from the photogrammetry scanner were compared to each other and named P1 for the reference scan and P2 to P5 the 4 additional scans.
2. N = 5 scans from the GOScan were compared to each other and named G1 for the reference scan and G2 to G5 the 4 additional scans.
3. The reference scans obtained in (a) and (b) were compared to each other, i.e., P1 versus G1.

Cleaning of the scans and alignment was performed by the GEOMagic studio 2013 from 3D Systems and the comparisons were carried out using the GOM Inspect 2018 from GOM software. Since the nose-mouth range was our area of interest, the accuracy of the P scanner was assessed in that region, and the nose of the test subject was used as the reference surface to perform the alignment. Moreover, we decided to test it in an unfavorable scenario by including hair in the face. To that end, a 38-year-old bearded male was used as the test subject (Fig. 3 right).

All the scans were performed consecutively in 1-h by a trained technician.

4 Results and Discussion

Figure 4 shows the results of the reference scans against the rest of the scans for the two scanners (photogrammetry and GoScan): P1 versus P2 to P5; and G1 versus G2 to G5. The color palette shows the deviation between the two compared meshes in a scale from 2 mm (red or beyond the reference) to −2 mm (blue or below the reference). The green color corresponds to the 0 deviation which means that there is no difference between the reference and the compared mesh.

Figure 5 shows the results for the comparison of the two reference scans named P1, for the photogrammetry scanner, and G1, for the GOScan. The upper figure shows the deviation between the two meshes in a scale for ±2 mm range, while lower figure depicts the deviation in a ±4 mm range. In both cases, the red colour means that the compared mesh will be beyond the reference mesh, while in the blue colour will be below the reference mesh. As expected, the lower ±2 mm range improved the resolution and detected more differences between the two scans.

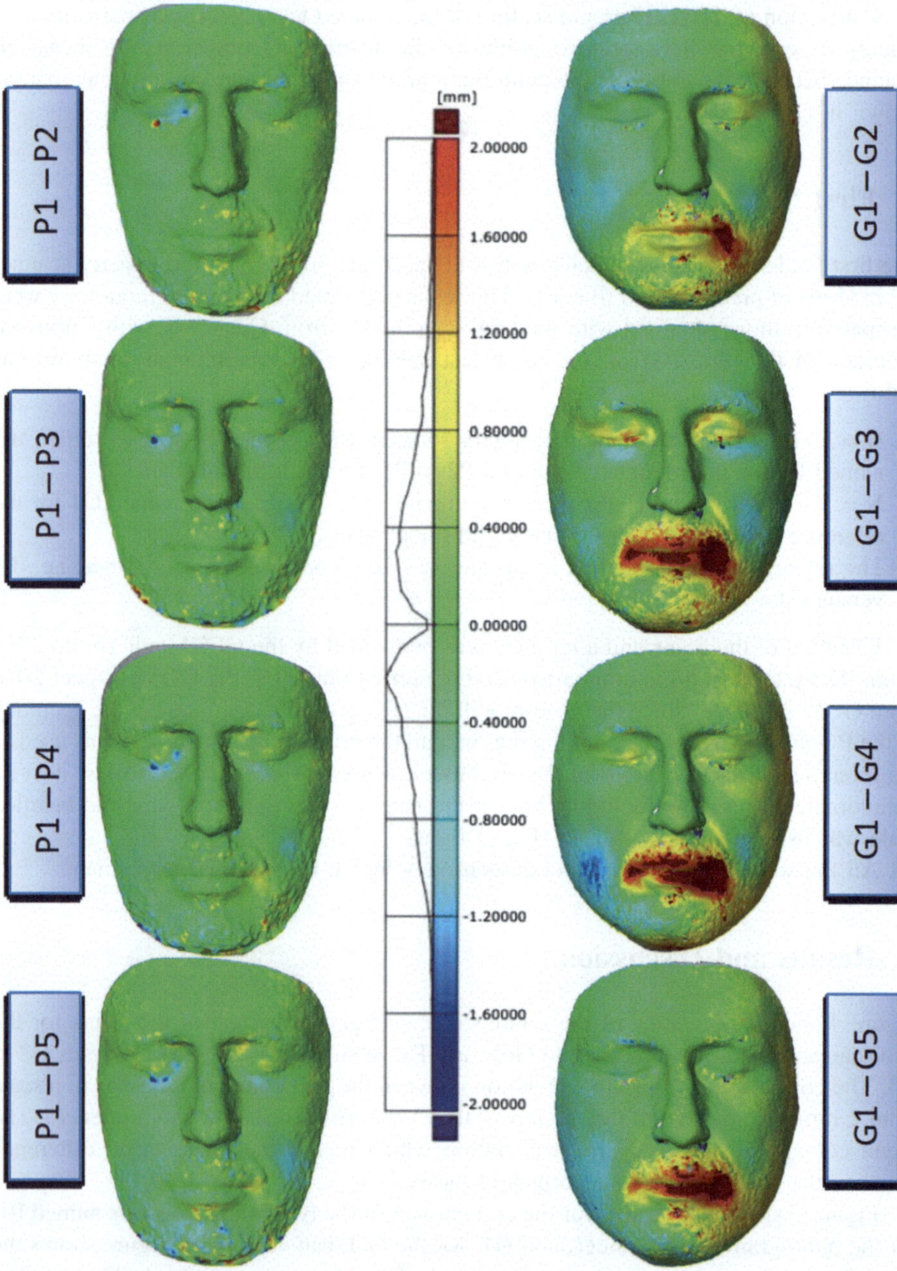

Fig. 4. Comparisons of the scans obtained from each of the scanners (Photogrammetry - P and GoScan - G) for the bearded test subject. P1 and G1 are the reference scans. The colour map shows the deviation between the two compared scans as indicated in the central scale.

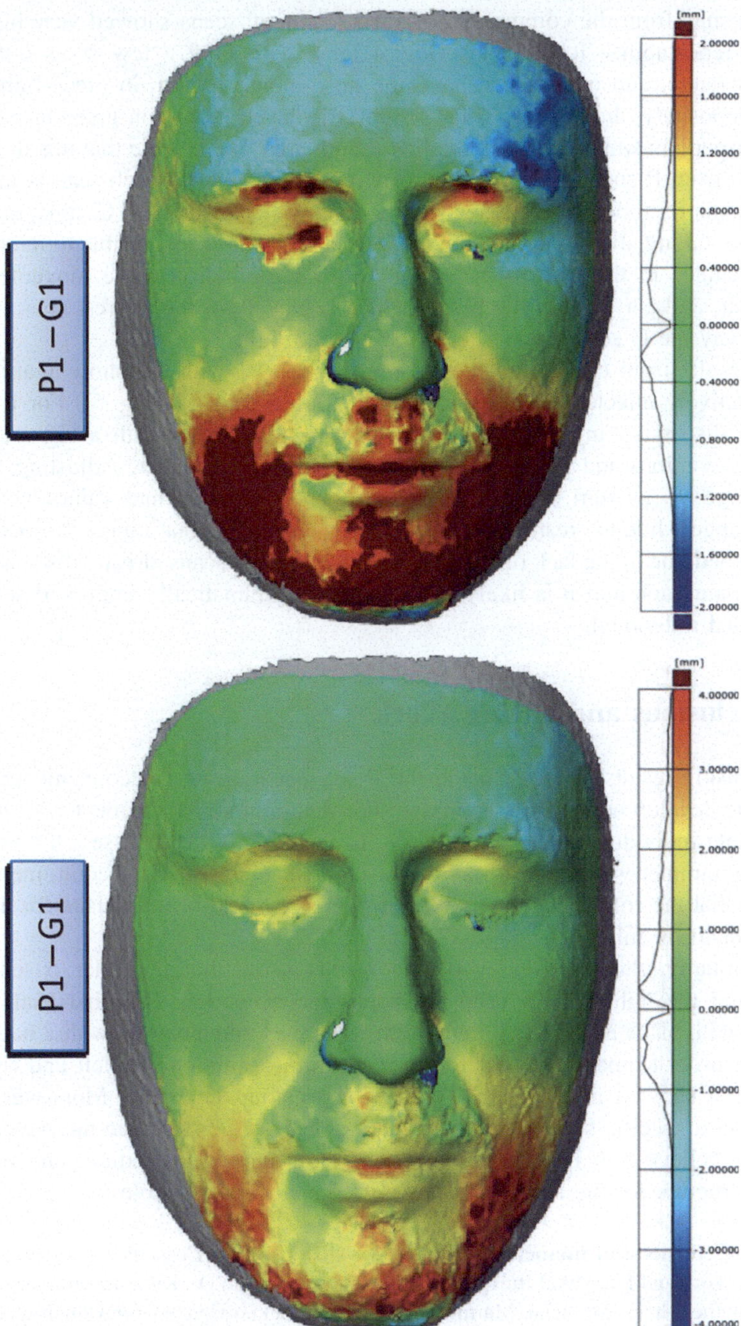

Fig. 5. Comparison of the reference scans of both scanners: P1 and G1. The colour map shows the deviation from the two compared scans. The upper figure is in ±2 mm range and the lower figure in ±4 mm range.

The results from the comparison of each scanners' scans showed very high accuracy and repeatability for P scans (Fig. 4 left), with only a few spots outside the ±0.4 mm range, and our interest area, the nose-mouth region, in green-light yellow colour. Oppositely, the G scans (Fig. 4 right) displayed more non-green areas and the biggest dispersion was detected in the lips-mouth area. We believe that this disparity in the results from P and G scans is due to two factors: (i) in P, each scan is taken in a given instant of time and it takes approximately 1 s, while in G it is more time consuming, taking about. 40 s, which means that G is more susceptible to interferences due to changes in the expression of the face and/or indiscernible movements; and (ii) the hair, and particularly the beard, which is exactly in the interest area, interferes with the way the G scan works, i.e. the structured light.

The results from comparing the reference scans P1 with G1 show again that the more negatively affected area is the one containing the hair (Fig. 5). For a ±2 mm range the dispersion in the bearded areas seems to be high with a big area in red (±2 mm deviation) and probably with some values even higher. Adjusting the measuring range to ±4 mm only one part of the bearded area had values beyond the ±2 mm range while the rest of the bearded area was inside that range. We believe that this happened due to the lack of accuracy related to the G scans already discussed in the previous paragraph and it is likely that it could be dramatically improved scanning a non-bearded individual.

5 Conclusions and Future Work

The promising results obtained using the P scanner indicate the convenience of continuing the development of this approach. For that, a larger experiment with more and more variate individuals included children, is needed in order to assess the validity of the photogrammetry scanner for real life clinical analysis. The requirement of the clinicians that the colour and the texture of the face should be a requisite for the scans, was satisfactorily fulfilled.

Additionally, the consistency of the results show that the ruler based scaling method and the light pattern used are appropietate enough. Our first trials showed problems with black and white light patterns thus we build a pattern with 2 background colours in order to make it easier to the algorithm to differenciate left and right side. This pattern showed no problem when reconstructing the scene. Moreover, we are working on a specific support structure with a scaled ruler to place the patient's face which we believe will make it much easier the scanning procedure; and on a new housing structure for the scanner to make it compact and portable.

Acknowledgements and financial support. We wish to thank Cesar Perez for his assistance during this work and Dr. Mikel Iturrate, Dr. Eneko Solaberrieta, Dr. Rikardo Minguez and Olatz Etxaniz for their help. Dr. Iciar Martinez is gratefully acknowledged for valuable comments, critical reading and editing of the manuscript.

The work was partially financed by contract no. TR40931 from Euskoiker Foundation and Unidad de Innovación ISS Biocruces Bizkaia/OSI EECruces.

The authors declare no conflict of interests and the founding sources have no involvement on the drafting of the manuscript.

References

1. López-de-Ipiña, K., Egiraun, H., Sole-Casals, J., Ecay, M., Ezeiza, A., Barroso, N., et al.: Feature extraction approach based on fractal dimension for spontaneous speech modelling oriented to alzheimer disease diagnosis. In: Drugman, T., Dutoit, T. (eds.) Advances in Nonlinear Speech Processing, pp. 144–151. Springer, Heidelberg (2013)
2. López-de-Ipiña, K., Alonso, J.B., Solé-Casals, J., Barroso, N., Henriquez, P., Faundez-Zanuy, M., et al.: On automatic diagnosis of Alzheimer's disease based on spontaneous speech analysis and emotional temperature. Cogn. Comput. **7**(1), 44–55 (2013)
3. López-Campos, J.A., Segade, A., Casarejos, E., Fernández, J.R., Vilán, J.A., Izquierdo, P.: Finite element study of a threaded fastening: the case of surgical screws in bone. Symmetry **10**, 1–13 (2018)
4. Gelaude, F., Vander Sloten, J., Lauwers, B.: Accuracy assessment of CT-based outer surface femur meshes. Comput. Aided Surg. **13**, 188–199 (2008)
5. Iturrate, M., Eguiraun, H., Etxaniz, O., Solaberrieta, E.: Accuracy analysis of complete-arch digital scans in edentulous arches when using an auxiliary geometric device. J. Prosthet. Dent. **121**, 447–454 (2019)
6. Nahles, S., Klein, M., Yacoub, A., Neyer, J.: Evaluation of positional plagiocephaly: conventional anthropometric measurement versus laser scanning method. J. Cranio-Maxillofac. Surg. **46**, 11–21 (2018)
7. Westoby, M.J., Brasington, J., Glasser, N.F., Hambrey, M.J., Reynolds, J.M.: 'Structure-from-Motion' photogrammetry: a low-cost, effective tool for geoscience applications. Geomorphology **179**, 300–314 (2012)
8. Colomina, I., Molina, P.: Unmanned aerial systems for photogrammetry and remote sensing: a review. J. Photogram. Remote Sens. **92**, 79–97 (2014)
9. Pi3Dscan. http://www.pi3dscan.com/index.php. Accessed 03 Apr 2019
10. Mitchell, H.L., Newton, I.: Medical photogrammetric measurement: overview and prospects. J. Photogram. Remote Sens. **56**, 286–294 (2002)
11. Ey-Chmielewska, H., Chruściel-Nogalska, M., Frączak, B.: Photogrammetry and its potential application. Adv. Clin. Exp. Med. **24**, 737–741 (2015)
12. Davidson, J., Dos Santos, A.M.N., Garcia, K.M.B., Yi, L.C., João, P.C., Miyoshi, M.H., et al.: Photogrammetry: an accurate and reliable tool to detect thoracic musculoskeletal abnormalities in preterm infants. Physiotherapy **98**, 243–249 (2012)
13. Barbero-garcía, I., Luis, J., Miranda, P., Marqués-mateu, Á.: Smartphone-based photogrammetric 3D modelling assessment by comparison with radiological medical imaging for cranial deformation analysis. Measurement **131**, 372–379 (2019)
14. Raspberry. https://www.raspberrypi.org. Accessed 03 Apr 2019

Best-Fit Alignment in the Digital Dental Workflow

Xabier Amezua-Lasuen[1(✉)], Mikel Iturrate-Mendieta[2], José Antonio Oriozabala-Brit[1], Xabier Garikano-Osinaga[1], Iñaki Martin-Amundarain[3], and Eneko Solaberrieta-Mendez[1]

[1] DEHI (Dental Engineering – Hortz Ingeniaritza), Department of Graphic Design and Engineering Projects, Faculty of Engineering Gipuzkoa, University of the Basque Country UPV/EHU, 20018 Donostia-San Sebastian, Spain
xabier.amezua@ehu.eus

[2] DEHI (Dental Engineering – Hortz Ingeniaritza), Department of Business Management, Faculty of Engineering Gipuzkoa, University of the Basque Country UPV/EHU, 20018 Donostia-San Sebastian, Spain

[3] DEHI (Dental Engineering – Hortz Ingeniaritza), Department of Graphic Design and Engineering Projects, Faculty of Engineering Gipuzkoa, University of the Basque Country UPV/EHU, 20600 Eibar, Spain

Abstract. In recent years, together with the development of specific software for inspection and analysis of 3D images, reverse engineering has been integrated in different kind of workflows. Among other applications, this kind of software is used to assess the accuracy of different scanners such as intraoral scanners. These scanners are used to obtain digital replicas of the dental arches and then diagnose, plan and design the customized treatment for each patient. In dentistry, these replicas are known as digital impressions. However, the accuracy of digital impressions is often called into question and there have been many research studies that measure precisely this accuracy. In these research works, reverse engineering has been widely used by using standard mesh aligning tools and measuring average deviations of aligned meshes. These measured deviations often have no correlation with the deviations that can influence the success or failure of a restoration in the mouth. A specific clinical case, such as the placement of a framework on multiple implants, requires an accurate digital impression in which the distances between implants and the angulation of the scanbodies do not exceed a certain error. This error cannot be inferred by measuring the deviation between two meshes that have been aligned using software automatic alignment tools. In this study, a procedure for measuring the accuracy of digital impressions acquired with intraoral scanners has been proposed.

Keywords: Best-fit alignment · Reverse engineering · Accuracy · Inspection software · Mesh processing · Digital impressions

F. Cavas-Martínez et al. (Eds.): INGEGRAF 2019, LNME, pp. 202–211, 2020.
https://doi.org/10.1007/978-3-030-41200-5_22

1 Introduction

In recent years, dentistry has made a huge impact on technological advances, and the methods used to date have resulted in new processes that have had a great impact on the functionality, safety and esthetics required by patients [1]. For this reason, computer-aided technology (CAD/CAM: Computer Aided Design/Computer Aided Manufacturing) is becoming increasingly important in this field of medicine. One of the computer-aided systems that have developed a wide range of applications in this field is reverse engineering. One of these applications is the digitization of dental arches to obtain digital impressions as a substitute for traditional plaster impressions (see Fig. 1). Thus, in recent years several intraoral scanners have been marketed for this purpose (Trios by 3Shape, True Definition by 3M, CS 3600 by Carestream, etc.).

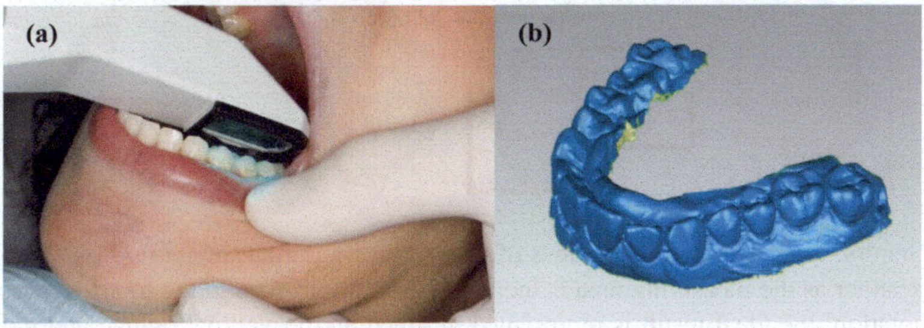

Fig. 1. Digital impression: **a** the scanning process and **b** the obtained file.

These intraoral scanners have introduced advantages such as reduction of patient discomfort [2, 3], reduction of time between the dental clinic and the dental laboratory [4, 5], simplification of the process [6] and improvement of communication flows [7]. But, despite these advantages, the accuracy of digital impressions obtained with these intraoral scanners is often called into question.

It has been demonstrated that one of the error causes of these digital impressions is a direct consequence of the technique that intraoral scanners use to match the different measurements [8]. That is because, using any 3D scanner, when the size or form of the surface to be measured is such that it cannot be covered in a single measurement stage, the surface must be measured in parts which are subsequently transformed into a common coordinate system to create a single 3D image of the object [9]. This process is called registration and in most cases it is performed by the specific software of each 3D scanner. For this purpose, depending on each company and some factors such as the field of application of the 3D scanner, different techniques are used: self-location, location by optical tracking, mechanical location, best-fit, etc. [9, 10]. The majority of intraoral scanners use the best-fit technique which relies on similarity calculations to determine the point of view of each measurement [10]. It means that the software analyses each measurement, selects the points of interest in each measurement, performs a similarity calculation that defines the coincidence of points of interest between different measurements and using them creates a unique three-dimensional representation of the object [10]. These points of interest can be found by detection of transition areas, such as strong curvatures, physical limits or differences of grey intensity ("Shape of Silhouette") [10]. Therefore, it has been demonstrated that the larger is the area to be scanned by the intraoral scanner, the greater is the error due to the registration of the different measurements, i.e., for restorations affecting 1 to 5–6 teeth intraoral scanners have an acceptable accuracy but not for restorations affecting the full arch (see Fig. 2) [2, 11, 12]. Also, in extreme cases of little variation in geometry and color, such as edentulous cases, sometimes intraoral scanners get lost even when scanbodies are used to facilitate the measurement process (see Fig. 3).

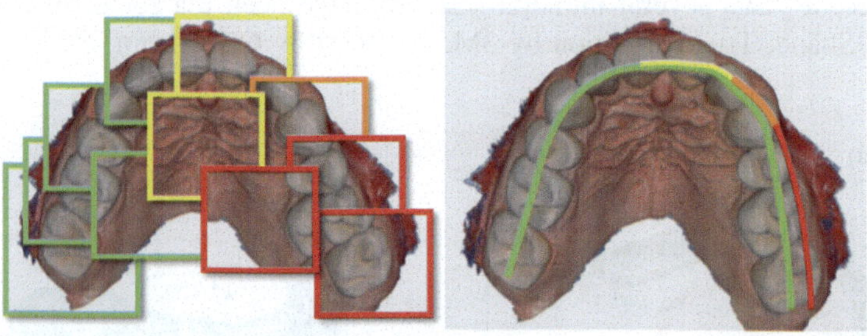

Fig. 2. Graphic representation of the loss of accuracy of a digital impression obtained with an intraoral scanner as the measuring area is increased. Starting to scan in the tooth 18 according to the FDI notation, the ideal result is represented in green and in yellow, orange and red the real result due to the error committed in the registration of different measurements.

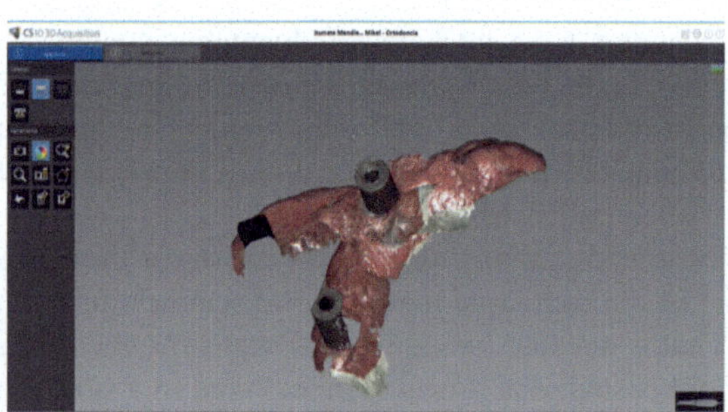

Fig. 3. Example of an error made by the software of the intraoral scanner due to the absence of geometric and colour variation. When arriving at the second scanbody the software interprets that it has returned to the first one and consequently performs the registration of the subsequent measurements in an erroneous way.

There have been many research studies that have analysed the accuracy of these digital impressions by using reverse engineering tools. In these research studies they have analysed the accuracy of these impressions aligning pairs of impressions obtained with different 3D scanners and then measuring the deviations between both of them. But in these research works, reverse engineering has been widely used by using standard mesh aligning tools and measuring average deviations of aligned meshes [13]. These measured deviations often have no correlation with the deviations that can influence the success or failure of a restoration in the mouth. A specific clinical case, such as the placement of a framework on multiple implants, requires an accurate digital impression in which the distances between implants and the angulation of the scan-bodies do not exceed a certain error. This error cannot be inferred by measuring the deviation between two meshes that have been aligned using software automatic alignment tools. This is because the software automatic alignment tools consider each mesh completely and perform the alignment minimising the distance error between both of them, thus, in the case of digital impressions obtained by intraoral scanners, among others, distributing the error due to the registration of different measurements. So, in these cases, it is necessary to use some other alignment tool. Depending on the software used and factors such as the objective of the alignment or the geometry of the meshes to be aligned, there are different tools which, in turn, yield different results (see Fig. 4): *Prealignment*, *3-Point Alignment*, *3-2-1 Alignment*, *Best-Fit by Reference Points*, *Local Best-Fit*, *Feature-Based Alignment*, etc. [14–16]. It is generally unknown in dentistry what alignment options each software offers and how each one works, which leads to alignments not suitable for the deduction of certain results.

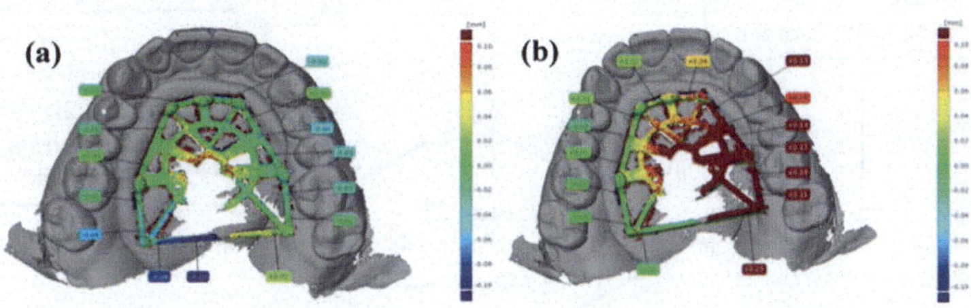

Fig. 4. Deviations obtained after aligning a pattern scanned with an intraoral scanner and a high accuracy industrial 3D scanner using two different alignment tools: **a** automatic *Prealignment* and **b** *Local Best-Fit* in the area in green in the left corner of the pattern.

Therefore, the objective of the study presented in this paper is to define and validate a procedure to measure in the digital impressions obtained with an intraoral scanner the deviations that can influence the success or failure of a restoration in the mouth, bearing in mind that one of the main causes of error lies in the way in which the software registers the different measurements and using for this purpose the appropriate alignment tools. This study has been carried out by the Product Design Laboratory (PDL, www.ehu.eus/PDL) and Dental Engineering – Hortz Ingeniaritza (DEHI, www.ehu.eus/dehi) research groups of the Department of Graphic Design and Engineering Projects of the University of the Basque Country (UPV/EHU).

2 Methodology

The procedure proposed in the study presented in this paper to measure in the digital impressions obtained by using an intraoral scanner the deviations that may affect in the success or failure of a restoration in mouth, is that represented in the flowchart in Fig. 5. As it can be appreciated, this procedure allows having a reference against which compare the digital impressions obtained with the intraoral scanners in real patients by using a pattern that can be extraorally digitalized with a high accuracy industrial 3D scanner and then attached to the patient's mouth to perform the intraoral digitalization. So, this procedure is valid for both in vitro (outside the patient's mouth) and in vivo (inside the patient's mouth) accuracy analysis.

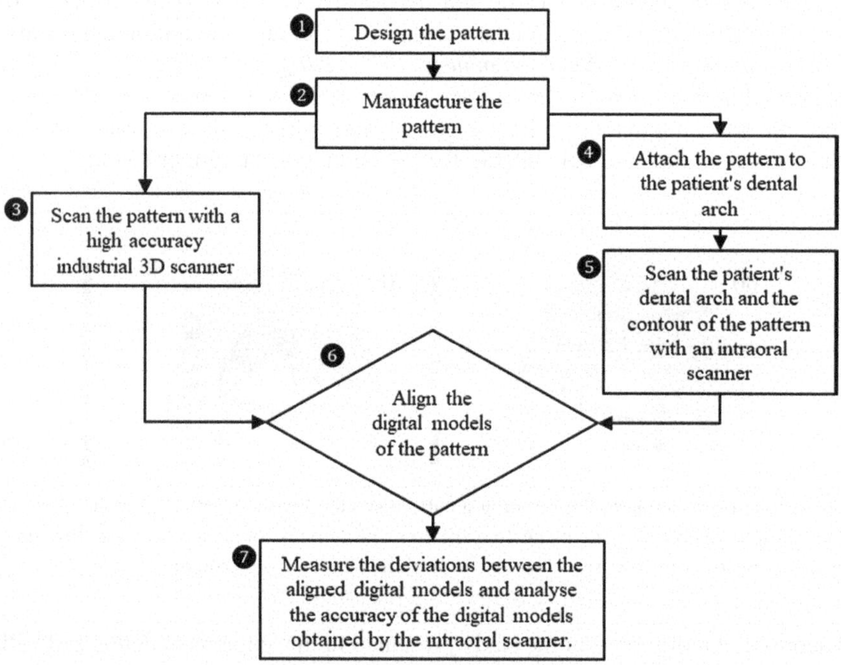

Fig. 5. Flowchart of the procedure that is valid to analyse the accuracy of a digital impression obtained in vitro like in vivo by using an intraoral scanner.

To validate the proposed procedure, an in vitro test has been performed. As this is the most critical case for intraoral scanning due to the absence of geometrical and colour variation, a plaster impression of an edentulous patient was used (see Fig. 6a). In order to reduce the error due to the registration of different measurements, a malleable element that provides geometric variation has been added to this model (see Fig. 6b). Finally, in order to validate the procedure, the pattern has been attached to this model (see Fig. 6c). Thus, from the comparison of the deviations obtained in the cases shown in Fig. 6a and b, it has been observed if the accuracy of the digital impression has improved by adding variation in the geometry; and from the comparison of the deviations obtained in the cases shown in Fig. 6b and c, the validity of the pattern for obtaining information about the magnitude of the deviations presented by the digital impression has been discussed. The following is a step-by-step description of the methodology followed to perform this in vitro test.

Fig. 6. The cases to be analysed: **a** the edentulous model with 4 scanbodies, **b** the edentulous model with 4 scanbodies and the malleable element and **c** the edentulous model with 4 scanbodies, the malleable element and the pattern.

Step 1. Digitization with a high accuracy industrial 3D scanner
This first step consists in digitizing with a high accuracy industrial 3D scanner (in this case with the ATOS Compact Scan 5 M by GOM) the models that will be used as a reference when measuring the deviations of the digital impressions obtained by using the intraoral scanner (see Fig. 7).

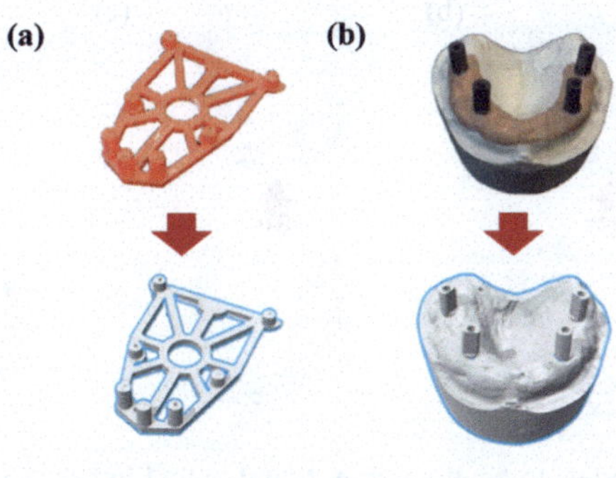

Fig. 7. Physical models (up) and their respective digital models (down) obtained with the high accuracy industrial 3D scanner ATOS Compact Scan 5M (GOM): **a** the pattern and **b** the edentulous model with 4 scanbodies. Note that **a** corresponds to step 3 of the flowchart in Fig. 5.

Step 2. Digitization with an intraoral scanner
This second step consists in digitizing with an intraoral scanner (in this case the Trios 2 by 3Shape) the models to be analysed (see Fig. 8). All models have been digitized starting from the position of the tooth 18 according to the FDI notation (right end of the arch in Fig. 8).

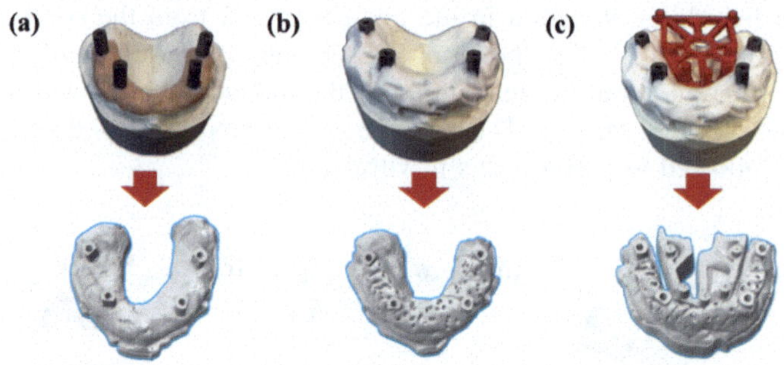

Fig. 8. Physical models (up) and their respective digital models (down) obtained with the Trios 2 scanner: **a** the edentulous model with 4 scanbodies, **b** the edentulous model with 4 scanbodies and the malleable element and **c** the edentulous model with 4 scanbodies, the malleable element and the pattern. Note that **c** corresponds to step 5 of the flowchart in Fig. 5.

Step 3. Alignment of the digital models
This third step consists in aligning the digital models obtained by the intraoral scanner with the corresponding ones obtained with the high accuracy industrial 3D scanner (see Fig. 9).

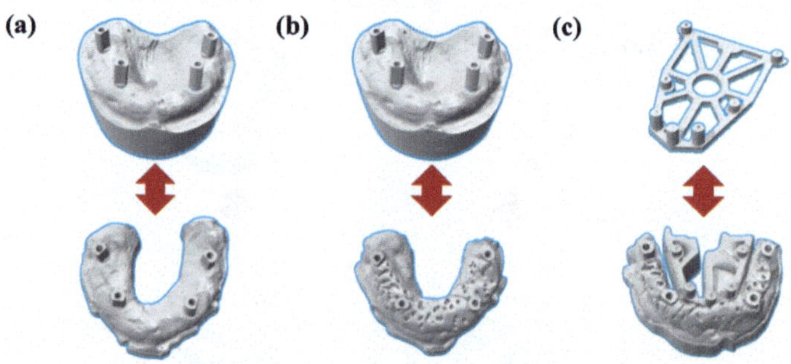

Fig. 9. Pairs of digital models to be aligned: **a** digital model in Fig. 8a with the digital model in Fig. 7b, **b** digital model in Fig. 8b with the digital model in Fig. 7b and **c** digital model in Fig. 8c with the digital model in Fig. 7a. Note that **c** corresponds to step 6 of the flowchart in Fig. 5.

This step is critical because, as mentioned above, depending on the alignment tool used, the deviations obtained will be very different. Taking into account the geometry of the models to be aligned, the error due to the registration of different measurements mentioned above and that in order to measure the deviations that can influence the success or failure of a restoration in mouth it is necessary to avoid the distribution of this error, it has been deduced that the most appropriate tool to align these models is the *Local Best-Fit* considering only the piece of the mesh where the intraoral scan has been initiated. In this way, in the area where the digital impression obtained with the intraoral scanner is more exact an almost perfect alignment is achieved (distance error between meshes practically null), and in the areas furthest away from there, i.e., in those areas that are more affected by the error due to the registration of different measurements, alignment errors are detected. Therefore, to perform these alignments it has been used the *Local Best-Fit* command offered by the reverse engineering software Geomagic Studio. A double alignment has been performed. For the first alignment, the pieces of the mesh that both models to be aligned have in common have been selected (for the alignments illustrated in Fig. 9a and b, as shown in the upper side of the Fig. 10a and b respectively, pieces of the mesh corresponding to the edentulous model; and in the case of the alignment illustrated in Fig. 9c, as shown in the upper side of the Fig. 10c, pieces of surfaces corresponding to the pattern). In the second alignment, these alignments have been adjusted by selecting only the piece of the mesh in common in each pair closest to the area where intraoral digitization has begun (see Fig. 10 down).

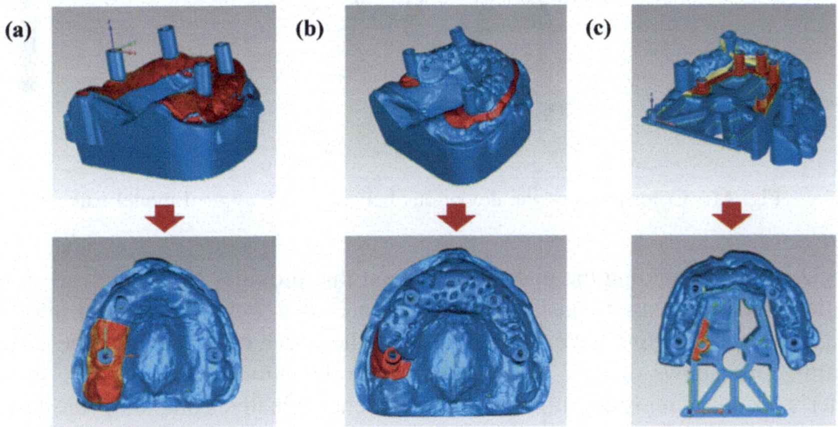

Fig. 10. Results obtained for each case in both alignments: first alignment up and second alignment down. In red the pieces of mesh selected to perform each alignment.

3 Results and Discussion

Importing the aligned model pairs to the inspection software GOM Inspect and measuring with it the deviations between both of them, the results in Fig. 11 have been achieved.

The deviations measured in this way are those that affect on the success or failure of a restoration in mouth and therefore are indicators of the accuracy of the digital impression (higher deviation, lower accuracy). Firstly, analysing the first two cases separately, it can be observed that, as expected, the value of the deviations increases as they move away from the area that was first digitized intraorally. It should be noted that in the area that was first digitized intraorally the deviations are almost null, which means that in small areas the accuracy of the intraoral scanner is acceptable. Furthermore, comparing the results of the first two cases (see Fig. 11a and b), it can be

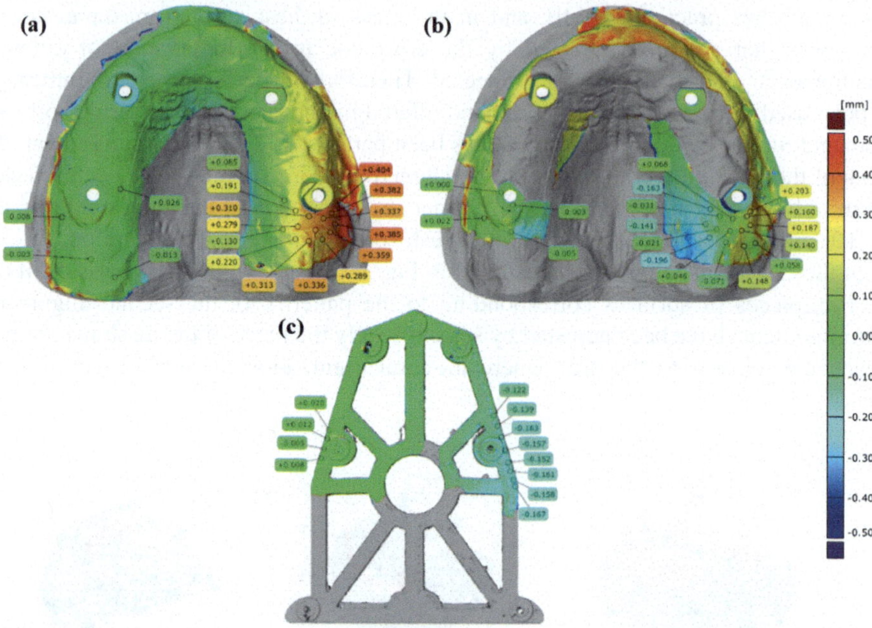

Fig. 11. Color maps of the deviations between each aligned model pair.

appreciated that by adding the malleable element that provides geometric variation the deviations have decreased considerably (in the area that was last digitized intraorally the maximum deviations have decreased from the order of 400 μm to the order of 200 μm). This shows that the main cause of the deviations is the way the intraoral scanner's software registers different measurements. Finally, comparing the last two cases (see Fig. 11b and c) it is observed that the deviations measured on the digital impression and on the pattern are from the same order (taking into account where the pattern is located, deviations of less than 200 μm in the area that has been last digitized intraorally). This means that the pattern can be used to evaluate the accuracy of digital impressions obtained by using intraoral scanners, i.e., that the procedure proposed in this study can be useful for the analysis of the accuracy of digital impressions obtained by using intraoral scanners.

4 Conclusions

From the results obtained in the study presented in this paper, the following could be concluded:

1. The selection criteria and the proper use of alignment tools are essential in order to achieve results that fit each application.
2. The procedure proposed in the study presented in this paper demonstrates that intraoral scanners can provide digital models of acceptable accuracy in small areas

but that due to the way in which the registration of different measurements is performed, for larger areas, accuracy goes decreasing.
3. The procedure proposed in this study can be useful for the analysis of the accuracy of digital impressions obtained by using intraoral scanners.

Acknowledgments. The authors thank the Faculty of Engineering Gipuzkoa for locating the DEHI (Dental Engineering – Hortz Ingeniaritza, www.ehu.eus/dehi) research group's Laboratory in their facilities and the University of the Basque Country UPV/EHU; and to the Country Council of Gipuzkoa to support this work (grant number: 70/19).

References

1. Solaberrieta, E.: Integración de la ingeniería en la odontología. Dyna **90**(1), 1–4 (2015)
2. Chochlidakis, K.M.: Digital versus conventional impressions for fixed prosthodontics: a systematic review and meta-analysis. J. Prosthet. Dent. **116**(2), 184–190 (2016)
3. Mangano, F.: Intraoral scanners in dentistry: a review of the current literature. BMC Oral Health **17**(1), 1–11 (2017)
4. Burhardt, L.: Treatment comfort, time perception, and preference for conventional and digital impression techniques: a comparative study in young patients. Am. J. Orthod. Dentofac. Orthop. **150**(2), 261–267 (2016)
5. Grünheid, T.: Clinical use of a direct chairside oral scanner: an assessment of accuracy, time, and patient acceptance. Am. J. Orthod. Dentofac. Orthop. **146**(5), 673–682 (2014)
6. Zimmermann, M.: Intraoral scanning systems - a current overview. Int. J. Comput. Dent. **18** (2), 101–129 (2015)
7. Patzelt, S.B.M.: The time efficiency of intraoral scanners: an in vitro comparative study. J. Am. Dental Assoc. **145**(6), 542–551 (2014)
8. Iturrate, M.: Accuracy analysis of complete-arch digital scans in edentulous arches when using an auxiliary geometric device. J. Prosthet. Dent. **121**(3), 447–454 (2019)
9. Luhmann, T., Robson, S., Kyle, S., Boehm, J.: Close-Range Photogrammetry and 3D Imaging, 2nd edn. De Gruyter, Berlin (2013)
10. Richert, R.: Intraoral scanner technologies: a review to make a successful impression. J. Healthc. Eng. **2017**, 1–9 (2017)
11. Patzelt, S.B.M.: Accuracy of full-arch scans using intraoral scanners. Clin. Oral Invest. **18** (6), 1687–1694 (2014)
12. Ender, A.: Accuracy of complete-arch dental impressions: a new method of measuring trueness and precision. J. Prosth. Dent. **109**(2), 121–128 (2013)
13. Zhang, F.: Validity of intraoral scans compared with plaster models: an in-vivo comparison of dental measurements and 3D surface analysis. PLoS ONE **11**(6), 1–10 (2016)
14. GOM: GOM Inspect Software (2018)
15. Creaform: VX elements 6.3 SR1 Software (2018)
16. 3D Systems: Geomagic Studio Software (2013)

Digitization of the Mechanical Articulator: Virtual Articulator

José Antonio Oriozabala-Brit(✉), Xabier Amezua-Lasuen, Xabier Garikano-Osinaga, Mikel Iturrate-Mendieta, Iñaki Martin-Amundarain, and Eneko Solaberrieta-Mendez

Grupo Dental Engineering/Hortz Ingeniaritza, Escuela de Ingeniería de Guipúzcoa, Universidad del País Vasco/Euskal Herriko Unibertsitatea, Pl. Europa, 20018 Donostia-San Sebastian, Guipúzcoa, Spain
joseantonio.oriozabala@ehu.eus

Abstract. This study presents the digitalization of the dental articulator, which is a device used by dentists as a replica of the patient's mouth. For this purpose, the digitalization has been carried out applying reverse engineering to obtain a three-dimensional model that totally simulates the behavior of the virtual model with the same measurements as the physical articulator Panadent PSH.

First, the physical dental articulator has been scanned using the ATOS Compact three-dimensional scanner. Then, the processing was carried out using the reverse engineering Geomagic software, where the necessary geometric elements were obtained. The parts have been modeled in 3D in the Solid Edge program and the assembly has been carried out. In this same CAD program, the parameterization of the adjustable parts, such as condylar pits, has been carried out. Finally, variables have been introduced that a physical dental articulator does not allow, for example the intercondylar distance.

By digitizing and parameterizing the dental articulator, the virtual articulator has been obtained. This virtual device allows the user to have greater control of the design, plan more accurately, enter more patient data, for example the 3D face of the patient, and data management is better.

Keywords: Reverse engineering · Motion simulation · Dental CAD/CAM

1 Introduction

In the industrial field, the last few decades have seen a process of digitization of processes related to product design, manufacturing and management. The advantages obtained are greater adaptability to demand, greater personalization of the product, shorter design, manufacturing and sales times, in addition, during the process a large amount of useful information has been generated for new designs or decision making.

As far as dentistry is concerned, it is a suitable field for applying CAD/CAM techniques since the design and manufacture of dental prostheses must always be made to measure for the client, in a personalized way to adapt the geometry of the prostheses

The original version of the chapter (23) has been revised. A correction to this chapter can be found at https://doi.org/10.1007/978-3-030-41200-5_68

F. Cavas-Martínez et al. (Eds.): INGEGRAF 2019, LNME, pp. 212–218, 2020.
https://doi.org/10.1007/978-3-030-41200-5_23

to that of the patient. The process of designing prostheses incorporating digital solutions is based on a digital model of the patient's mouth that is used for custom design of the prosthesis. It is then manufactured by CAM and verified by virtual articulator.

Several commercial solutions have developed a wide range of applications in this field such as 3D scanning, computer-aided design (CAD), engineering, computer-aided engineering (CAE), computer-aided machining (CAM), rapid manufacturing (RM), rapid prototyping (RP), computer-aided inspection (CAI), computer-aided quality (CAQ), etc.

A digital model of the patient's mouth is used for the custom design of the prosthesis using CAD. It is then manufactured using CAM and verified using a virtual articulator that simulates the complex movements of the mandible.

Advantages include improved prosthesis quality, reduced manufacturing time and reduced chair-time, as it is not necessary to make fixes in the patient's mouth. However, the great disadvantage of these systems today is the investment cost they require, so to be economically viable must be achieved through these systems produce a considerable number of quality restorations.

2 State of the Art

2.1 Historical Evolution of the Mechanical Articulator

The first articulators to emerge were adaptive hinge articulators capable only of opening and closing in hinge movements. It had an anterior vertical stop which was a carriage or machine bolt. The instrument was not flexible and lateral movements were reduced to a minimum. Jean Baptiste Garriot, a renowned dentist, in 1805 laid the foundations for the first occluder [1, 2], consisting of a simple hinge, with a screw fixed in the posterior portion against a metal plate that served as a vertical stop.

In 1894 C.B. Bixby invented a connection to mount the models on a hinge articulator, which was the predecessor of the facebow. This tool allows the location of the upper arch with respect to the intercondylar axis to be transferred to the articulator.

In 1899 A.D. Gritman invented and improved the shape of the articulator with fixed condylar guides. In 1902 Carl Christensen introduced the protrusion register as a means of adjusting the articulator. The first articulator with individualized diagonal trajectories, built from a register, was built in 1904 by Parfitt.

In 1950 Bergstróm designed an instrument called "arcón", where the name derives from articulator-condyle. This term is usually used for articulators in which the condyle is located on the lower limb and the condyle guides on the upper limb.

In 1975 the Denar Mark II was designed by the Denar Company with the help of Sumiya Hobo and Frank Celenza, which is considered a semi-adjustable articulator. But in recent years there have been virtually no substantial improvements in dental articulators. However, articulators are becoming more functional, manageable and more precise in their construction and function.

2.2 Description of the Dental Articulator

In dentistry, both for technical and comfort reasons and in order to predict results, some procedures must be tried and performed outside the patient's mouth. Among these procedures, occlusal diagnosis, occlusal balancing and the production of dental prostheses stand out. These procedures make it necessary to have 'extra ore' a mechanical or virtual replica of the human masticatory apparatus that simulates the movements carried out by the Temporomandibular Joints.

A dental articulator is a device equipped with mechanisms, elements equivalent to the anatomy itself, by means of which the mandibular movements can be reproduced, once the patient's position relations and adjustments have been registered and transferred. This transfer is carried out by means of the facebow, and thus the models are placed in the exact position with respect to the intercondylar axis.

Adjustments allowed by a mechanical articulator typically include the condyle inclination, and Bennett angle.

2.3 Digital Systems for Dentistry

There are different CAD solutions for modeling dental prostheses in the market. 3Shape offers its TRIOS intraoral scanner which stands out for its ability to generate meshed models of the patient's mouth very quickly and accurately. Exocad allows working with digital models of the patient's mouth for the design of prostheses and 3D implants from the geometry of the patient's mouth and includes a virtual articulator for design and verification.

The MODJAW system uses an optical scanner to capture jaw movement [3, 4]. It can provide animated three-dimensional models of the patient's occlusion as well as a dynamic map of the dental contacts and calculate the occlusion parameters (Bennett angles, condylar inclination…).

2.4 The Virtual Articulator

No mechanical articulator is capable of imitating all the sophisticated jaw movements, such as the elastic deformations of the jaws, the resilience of the oral mucosa, or the dental mobility allowed by its periodontal ligaments. Virtual articulators based on computer models can provide the solution to this problem.

There are two ways to generate a virtual articulator. The virtual articulator by Gartner and Kordass [5], from the University of Greifswald records the movement of the jaw electronically to simulate the movement of digital arcades.

Szentpétery [6] on the other hand proposes to use mathematical models of the movement of a mechanical articulator. It allows mathematically simulating movements that cannot be simulated mechanically, such as a curved Bennett movement. Theoretically it can simulate the elasticity of the jaws, the oral mucosa and the ligaments [7].

The Exocad and 3Shape virtual articulator module can be cited as a commercial example.

3 Methodology

The objective of this work is to evaluate the incidence of the different parameters of the dental articulator in the simulation of jaw movements.

First, a PSH Panadent Articulator (see Fig. 1) was scanned using the ATOS I 3D scanner, resulting in a point cloud (see Fig. 2). The Reverse Engineering Geomagic program has allowed obtaining the different geometric elements (see Fig. 3). Next, the model has been exported to the generic CAD program Solid Edge.

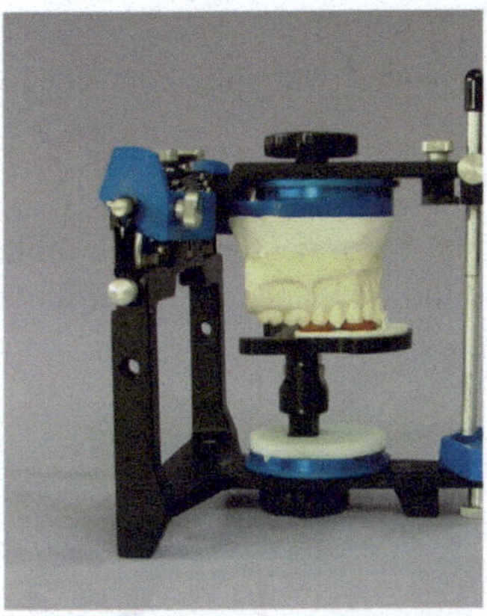

Fig. 1. Articulator Panadent PSH

Fig. 2. Point cloud

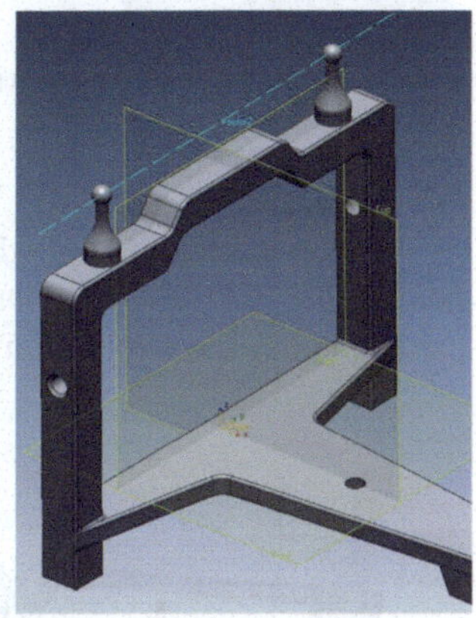

Fig. 3. Basic surfaces

Once the articulator has been imported into Solid Edge, the adjustable parts, such as the condylar fosses (see Fig. 4), are parameterized and assembled (see Figs. 5 and 6). The geometry was then parameterized to simulate the adjustable parameters of the physical articulator, in particular, the inclination of the condylar pathways, the movement of Bennett and the incisal inclination. In addition, two new parameters have been incorporated to modify the intercondylar distance and the radius of condylar pathway, which is not adjustable in the Panadent PSH articulator.

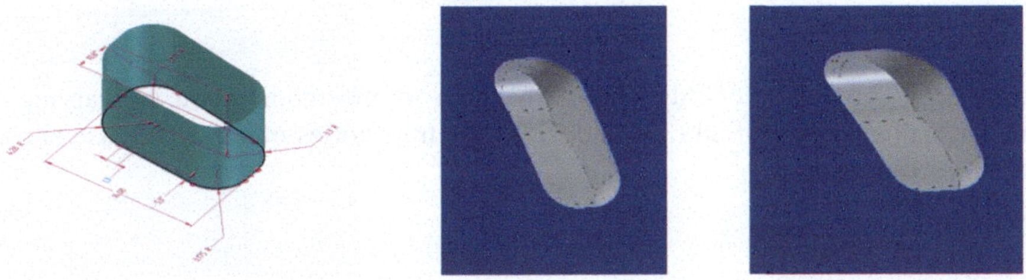

Fig. 4. Parameterization of condylar fossae

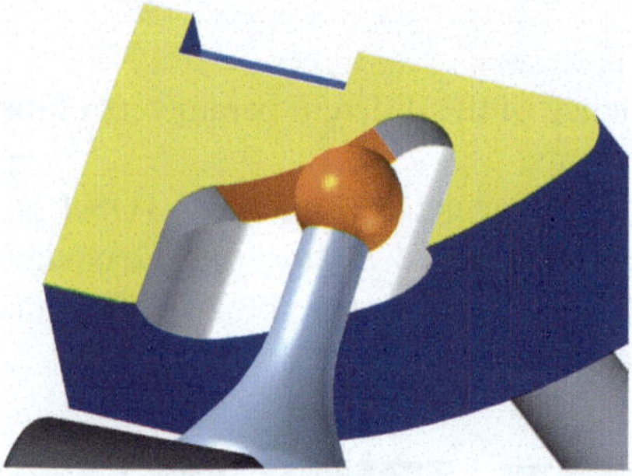

Fig. 5. Application of geometric relations between components

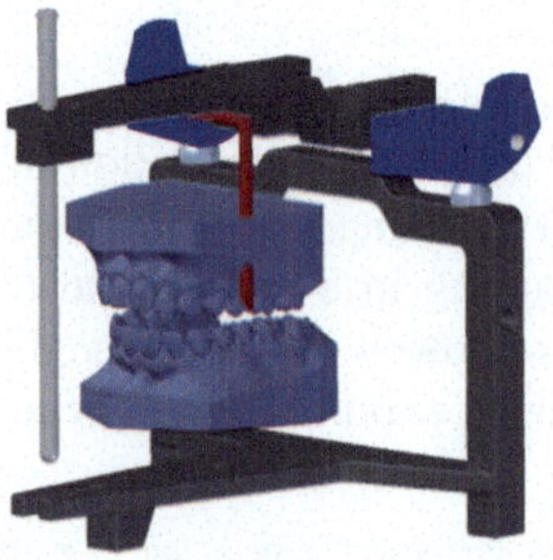

Fig. 6. Model mounted with piece to simulate molar point

Motion was then applied to the assembly to determine the trajectory of the molar point using different parameters. The goal is to evaluate the incidence of the different parameters in the simulation of jaw movements (see Fig. 7).

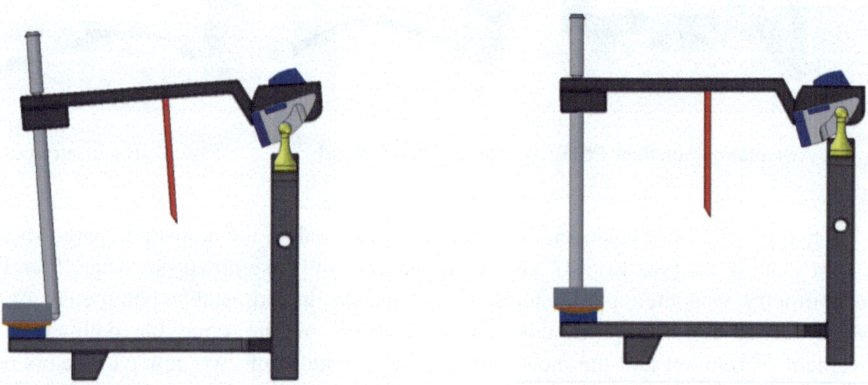

Fig. 7. Simulation of the protrusive movement of the molar point

4 Results

The Table 1 shows the coordinates of the path of the molar point by varying the condylar pathway inclination (32° and 51°). The trajectories obtained are also showed below (see Fig. 8).

Table 1. Coordinates of the path

CI = 32°		CI = 51°	
X (mm)	Z (mm)	X (mm)	Z (mm)
0,25	0,97	0,14	0,99
0,57	1,92	0,29	1,98
0,92	2,86	0,50	2,96
1,34	3,77	0,87	3,90
1,81	4,66	1,30	4,83

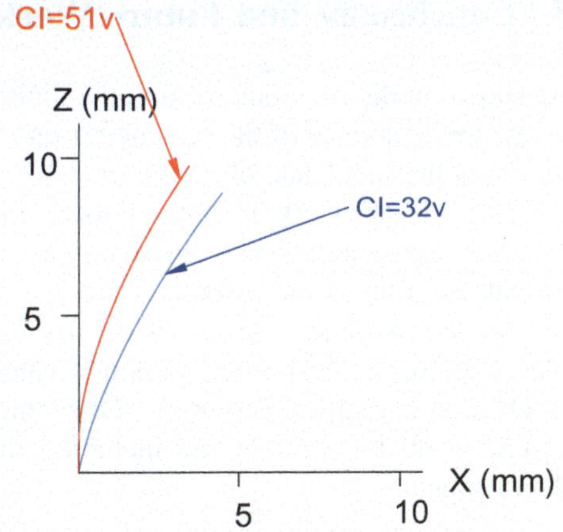

Fig. 8. Trajectories of the molar point using different condylar pathway angles

In order to obtain comparable values, the angles formed by the trajectory with the horizontal plane (AH) at different distances (D) from the origin have been obtained (see Fig. 9).

Table 2. Coordinates of the path

D (mm)	AH(°)				
	CI = 30°	CI = 35°	CI = 40°	CI = 45°	CI = 50°
1,0	62,4	65,6	69,5	73,1	77,1
2,0	61,4	63,9	66,5	69,2	72,1
3,0	60,8	63,0	65,3	67,7	70,3
4,0	60,1	62,2	64,4	66,7	69,1
5,0	59,5	61,5	63,6	65,8	68,1

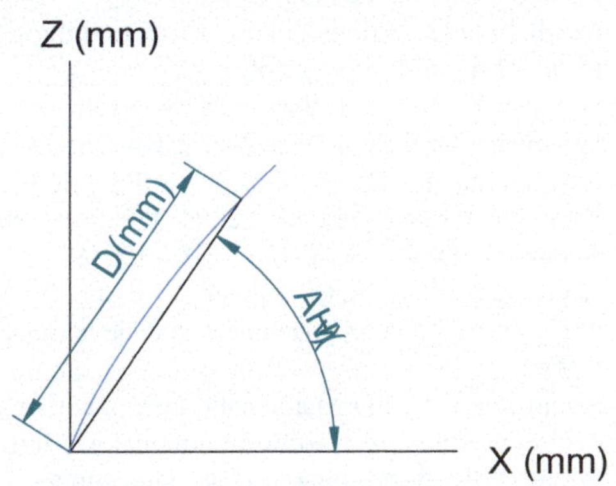

Fig. 9. Trajectories of the molar point using different condylar pathway angles

To assess the degree of dependence of the trajectory as a function of condylar inclination (CI), different AH values have been obtained as a function of CI (Table 2).

The linear correlation between CI and AH for a distance of 3 mm is 0.995, which shows the high degree of dependence of the trajectory on the inclination of the condylar road.

5 Conclusions and Future Work

As shown in the previous section, the digitization of the articulator makes it possible to assess the incidence of the parameters on the trajectories of the molar parts. The aim is to assess the incidence of all parameters.

The present work is limited to a single articulator. It would be interesting to evaluate the simulations carried out by different articulators, both mechanical and virtual, existing in the market.

Another avenue of research involves incorporating other parameters not contemplated by mechanical articulators that value soft tissues, as well as the position of the head. A more realistic geometry of the temporomandibular joint, reflecting the complex geometry of the condyle and including the intercondylar disk, may also be a future development.

Acknowledgments. The authors thank the Faculty of Engineering of Gipuzkoa for locating the DEHI Dental Engineering- Hortz Ingeniaritza (www.ehu.eus/dehi) research group's Laboratory in their facilities and the University of the Basque Country UPV/EHU; and to the Country Council of Gipuzkoa to support this work (grant number: 70/19).

References

1. Mitchell, D.L., Wilkie, N.D.: Articulators through the years. Part I. Up to 1940. J. Prosthet. Dent. **39**(3), 330–338 (1978)
2. Mitchell, D.L., Wilkie, N.D.: Articulators through the years. Part II. From 1940. J. Prosthet. Dent. **39**(4), 451–458 (1978)
3. Gutman, Y., Keller, J.: Jaw tracking device and methods of analysis of patient's specific TMJ kinematics. Acta Sci. Dent. Sci. **2**(10), 73–79 (2018)
4. Solaberrieta, E., Barrenetxea, L., Minguez, R., Iturrate, M., De Prado, I.: Registration of mandibular movement for dental diagnosis, planning and treatment. Int. J. Interact. Des. Manuf. (IJIDeM) **12**(3), 1027–1038 (2018)
5. Kordass, B., Gartner, C., Sohnel, A., Bisler, A., Voss, G., Bockholt, U., Seipel, S.: The virtual articulator in dentistry: concept and development. Dent. Clin. North Am. **46**(3), 493–506 (2002)
6. Szentpétery, A.: 3D Mathematic movement simulation of articulators and its application by the development of a software articulator. Martin-Luther University of Halle (1999)
7. Serrat, E.P.: Bases diagnósticas, terapéuticas y posturales del funcionalismo craneofacial, Ripano (2006)

Medical Play Therapy: Development of the JUNIOR-MRI Role Play

Diego Paderno(✉), Edoardo Alessio Piana, Ileana Bodini, and Valerio Villa

Department of Mechanical and Industrial Engineering, University of Brescia, Via Branze 38, 25123 Brescia, Italy
diego.paderno@unibs.it

Abstract. The purpose of this work is to help the little patients of the Paediatric Onco-Haematology Department of the “Monica and Luca Folonari” bone-marrow Transplant Centre to tackle their anxiety and worries when subjected to Magnetic Resonance Imaging (MRI) clinical examinations, in order to reduce the period of time between the arrival of a patient in the Diagnostics Department and the actual exam execution. To do so, following the principles of the Medical Play Therapy, a user-centred role play called JUNIOR-MRI has been developed. The game is addressed at children between the ages of 5 and 10 and consists of a scaled model of a real MRI machine, with a gurney, a patient doll and a control station. A tablet application has also been developed to simulate the procedure of the clinic exam, reproduce the real sounds of a MRI through a loudspeaker placed on the scale MRI and show some fake images of the investigation carried out on the doll. The child plays the role of the doctor, getting acquainted with the exam procedure and instrumentation and learning not to fear the clinic exam. The experimental phase of the game has begun, and the JUNIOR-MRI is now used in the Pediatric Diagnostics Department of the Spedali Civili di Brescia, following an ad-hoc clinic protocol.

Keywords: Medical Play Therapy · Role play · User-centred design · Design methods

1 Introduction

When children with onco-haematological diseases are long hospitalized, they can feel anxious and stressed, and some of the medical procedures they are subjected to, such as Magnetic Resonance Imaging (MRI), emphasize these feelings. In particular, during MRI, the little patients have to be alone in a little space, without moving for an undefined lapse of time and listening to loud and unpleasant sounds, without understanding of what it is really happening around them. Older children understand the meaning of the diagnostic procedure but for them the examination is exhausting and they are anxious about their recovery.

According to the literature the MRI is unpleasant because of (i) the perception of being closed in a small place; (ii) the exposure to loud and annoying sounds; (iii) the need to remain motionless; (iv) the need to wear receiving coils; (v) the duration of the

F. Cavas-Martínez et al. (Eds.): INGEGRAF 2019, LNME, pp. 219–226, 2020.
https://doi.org/10.1007/978-3-030-41200-5_24

whole procedure [1]. When the little patients are frightened or worried, they are less cooperative, so that the whole examination might be very long, and it might result in low quality images. In some cases, the sedation is required and the children are exposed to the risks of anesthetic, moreover the costs for the hospital stay increase.

To reduce sedations and guarantee the good outcome of the examination it is important to prepare children to the MRI [1–5].

To play takes an important role in the life of a human beings: Platon said that "you can find out more about a person in an hour of play than in a year of conversation", and symbolic play is largely used in psychotherapy and it is also one important step in the development of a child.

In the 40's, the psychologist Virginia Axline defined the term "Play Therapy" as an opportunity for the children to grow in the most favorite possible conditions. In this model the child is able to face, elaborate and overcome disturbing emotions, making sense of pain experiences [6, 7].

The Medical Play Therapy is a specific therapy applied to medical procedures and is based on playing as a therapeutic method to help children to face an actual psychological uneasiness due to both the illness and the treatment [7].

The purpose of this project is to help little patients to tackle their anxiety and worries when subjected to Magnetic Resonance Imaging (MRI) clinical examinations, in order to reduce the period of time between the arrival of a patient in the diagnostics Department and the actual exam execution. To do so, a user-centered role play called JUNIOR-MRI has been developed. The game is addressed at children with an ages 5 to 10 and consists of a scaled model of a real MRI apparatus, with a gurney, a patient doll and a control station. A tablet application has also been developed to simulate the procedure of the clinic exam, reproduce the real sounds of a MRI through a loudspeaker placed on the scaled MRI and to show some fake images of the investigation carried out on the doll. The child plays the role of the doctor, getting acquainted with the exam procedure and instrumentation and learning not to fear the clinic exam.

This work aims to:

1. reduce of the stress of the little patients;
2. induce the children not to need their parents during the medical examination;
3. reduce the duration of the complete medical examination;
4. reduce the number of sedations, without affecting the good quality of the images.

The project develops in three principal phases:

Phase 1 – Design and construction of a Medical Play Room
Phase 2 – Experimentation: medical procedures desensitization paths
Phase 3 – Results analysis and dissemination

Until now the Phase1 is completed, and the Phase 2 has just begun.

This paper describes the work related to the Phase 1 and the method to validate the efficacy of the JUNIOR-MRI role play.

2 Design and Construction of the Medical Play Room

A Medical Play Room has been set up to host the device: a medical clinic has been replicated to simulate a MRI clinical examination, to let the little patients play and learn some typical aspects of this type of investigation. In particular:

- the MRI photographs the insides;
- the patients need to enter the machine, to be scanned and photographed;
- the MRI is loud;
- scanning takes a lot of time;
- the diagnostic procedure is not painful;
- while scanned, the patient must not move, while the doctor looks through a window.

Figure 1 shows the full system, both the render Fig. 1(a) and the real thing installed in the hospital Fig. 1(b): it represents a complete scaled model of a MRI machine, with a gurney a patient doll and a control station. The game is also based on a tablet, which is the user interface, and a system to reproduce the real noises of a MRI.

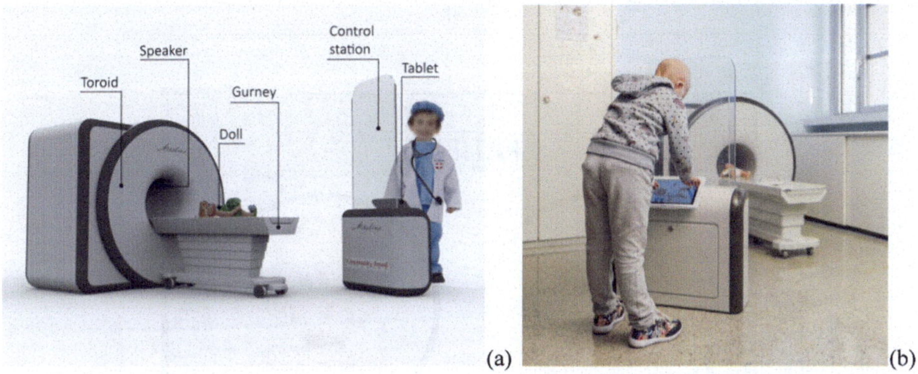

Fig. 1. JUNIOR-MRI (a) render; (b) the role-play installed in the hospital.

When the system is set-up, an area of $2 \times 2.5\ m^2$ is required, and the total weight is about 60 kg. All the components are principally realized in a varnished MDF wood, which can be sanitized. No electronic component is under tension during the game. Both the tablet and the speaker are not powered during game time. There is a Bluetooth® connection between tablet and loudspeaker. The glass of the control station is made of a shatterproof plastic.

Each part has been studied to be intrinsically safe: the weight of each part is sufficiently low not to endanger children and roll-over protection systems have been put in place such as webbing for anti-tip hook and an anti-tip base of the control position.

Both the wooden structures of the game and the application software are user-centered, to help the little patients to learn the MRI procedure steps and not to fear the medical examination, facing it without stress.

In the JUNIOR-MRI role play, the child is the "doctor" which works in the control station where the tablet is located. The application which simulates the real MRI examination is installed on the tablet. A member of the medical staff plays the role of the "assistant", which acts by overseeing operations and acting on both the gurney and the toroid. A doll is the "patient" to be examined through MRI, and it is positioned on the gurney. The latter is brought close to the toroid and partially inserted in the dedicated hole. When the child doctor gives a signal, the assistant slides the moving table of the gurney so that the doll is inside the toroid: during the scanning, the application software controls a little loudspeaker located in the high part of the toroid. This speaker emits sounds similar to the ones generated during a real MRI examination. Meanwhile, the child doctor runs the application and virtually scans the doll, virtually localizing and quantifying the illness, through four different consecutive phases in which the little patient interact with the app, for example choosing the type of generated sounds.

When the doll is finally positioned in the MRI machine, the child can run the application. At first, a screenshot with the start button appears.

After pressing the start button the screenshot in Fig. 2 appears, and the patient can be chosen.

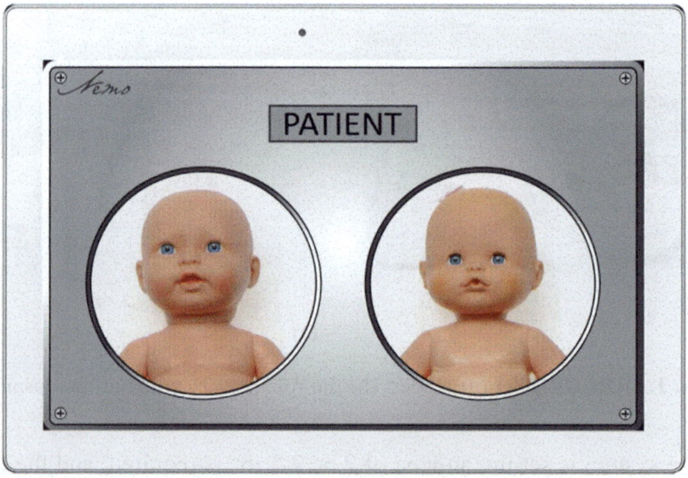

Fig. 2. Patient choice.

Once the patient is chosen, a screenshot with the whole body of the chosen doll is presented and the child doctor can directly click on a part of the body to select the body area where the illness is probably located. There are nine possible body areas: head, neck, shoulder (right and left), torso, and leg (right and left).

The next step is the scan. Figure 3 shows the screenshot in which the child doctor can choose the sound sequence of the MRI. Each colored button corresponds to a different sound sequence, which is interrupted if the button is not pushed.

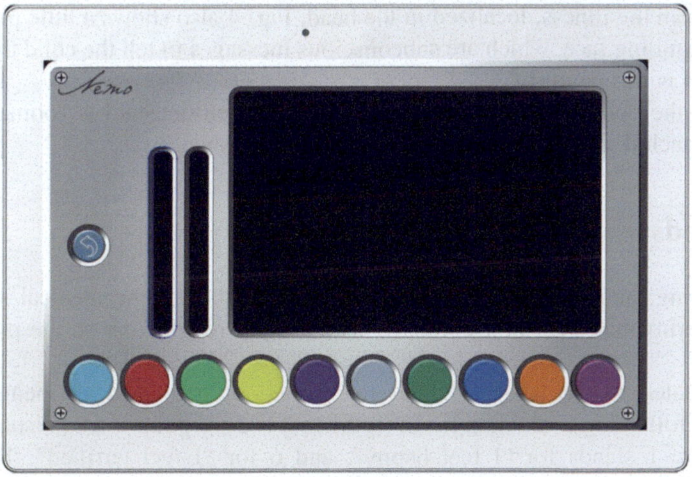

Fig. 3. Screenshot in which the sound sequence can be chosen and reproduced.

The vertical bars in the left side represent the status of the single scan, and the status of the whole medical examination.

There are four scan steps. At first the scanned image is defocused. If the button that generates the sound is continuously pressed, the scanned image becomes more and more well-defined. Then a first localization of the illness is possible, as shown in Fig. 4.

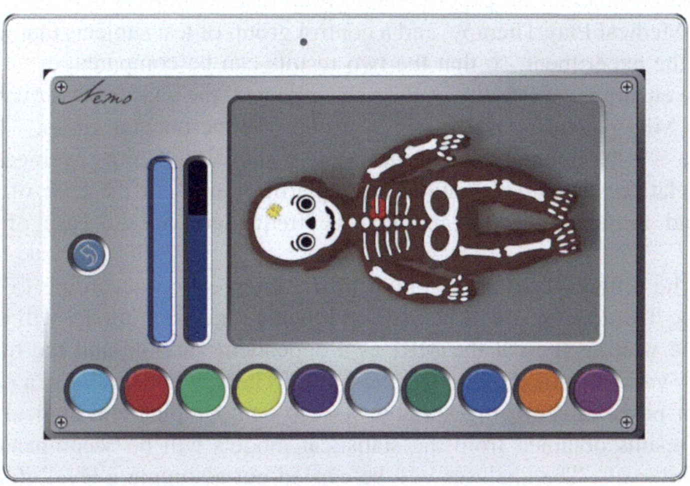

Fig. 4. First localization of the illness.

Apart from the illness, localized in the head, Fig. 4 also shows a little pulsating red heart and a smiling face, which are subconscious messages to tell the child that the MRI examination is not painful.

Finally, the body area in which the illness has been detected is zoomed, the procedure is concluded, and the MRI examination is finished.

3 Methods

Before, during, and after the game session, the member of the medical staff should collect experimental data, in accordance with current regulations on the protection of privacy.

Data related to the level of stress manifested or perceived are being collected through the following evaluation tools: (i) the first is a six-point VAS (Visual Analogue Scale) where 1 stands for "I feel happy", and 6 for "I feel terrified". The scale is correlated with smiley faces representing emotional expressions to facilitate response in younger children; (ii) the second is the aOBSD: Amended Observation Scale for Behavioral Distress [8], an observational tool that allows evaluating the paediatric procedural distress through the analysis of behaviors demonstrated by the child, such as crying, muscles stiffness, verbal resistance. The instrument has not been validated for Italy, but one of its translations has been used in other Italian works [9, 10], so it has been chosen to use the instrument anyway.

To evaluate the time needed for the beginning of the diagnostic examination, the presence or absence of the parent during the examination, the need for sedation and the quality of the image, an observer will be present during the time of preparation of the child for MRI through a special data collection grid.

The research includes an experimental group of test subjects that will be prepared through the Medical Play Therapy, and a control group of test subjects that does not use the path of the experiment, so that the two results can be compared.

To try to ensure the reliability of the data collected, the data collector will not know whether the MRI patient is in the control group or experimental group.

The data will be defined through descriptive statistics such as averages and averages with relative standard deviations and quartile ranges in the case of continuous variables and through absolute and relative frequencies in the case of qualitative variables.

To test the difference in the duration of the intervention according to the type of approach, the T Student test will be used and then a regression model will be built that will have the waiting time at the MRI as a dependent variable and the methodology used (binary variable, standard vs. Medical Play Therapy approach) as a determinant, adjusting for possible confusing variables (socio-demographic and clinical variables).

All the results obtained from the statistical models will be accompanied by 95% confidence intervals. The analyses will be carried out assuming a level of significance equal to 5%.

4 Discussion and Conclusions

As a principal result, the treatment will be considered effective if there is a 40% reduction in the average time elapsed between arrival at the examination site and the actual start of the diagnostic procedure, measured in minutes, between the control group and the experimental group.

Other expected results are: (i) a score of less than 5 on the aOBSD scale in children belonging to the experimental group; (ii) a statistically significant difference in the number of children requiring the presence of the parent in the magnet room between those belonging to the experimental group and those belonging to the control group; (iii) a statistically significant difference in the number of sedations carried out between children in the experimental and control groups.

The experimental phase has started and one year after the beginning of the project, the Medical Play Therapy course has proved to be a valid support tool for children followed by the Department of Pediatric Oncohaematology of the Spedali Civili of Brescia.

Offering to the young patients a space and a path where they can re-elaborate their illness and find standardized procedures has allowed the children to become protagonists of the treatment.

The child develops a self-competence with respect to well-being. The discovery of the meaning and knowledge of all those procedures that are often suffered allows the little patients to contain their emotions through a cognitive control.

Evidence of effectiveness will be evaluated once the case history has been acquired and the data collected has been processed.

Acknowledgments. The research work reported here was made possible by Gilead Italia, through the Community Award Program.

References

1. Carter, A.J., Greer, M.C., Gray, S.E., Ware, R.S.: Mock MRI: reducing the need for anaesthesia in children. Pediatr. Radiol. **40**, 1368–1374 (2010)
2. De Amorim e Silva, C.J., Mackenzie, A., Hallowell, L.M., Stewart, S.E., Ditchfield, M.R.: Practice MRI: reducing the need for sedation and general anaesthesia in children undergoing MRI. Australas. Radiol. **50**(4), 319–323 (2006)
3. De Bie, H.M.A., Boersma, M., Wattjes, M.P., Adriaanse, S., Vermeulen, R.J., Oostrom, K.J., Huisman, J., Veltman, D.J., Delemarre-Van de Waal, H.A.: Preparing children with a mock scanner training protocol results in high quality structural and functional MRI scans. Eur. J. Pediatr. **169**(9), 1079–1085 (2010)
4. Tornqvist, E., Mansson, A., Hallstrom, I.: Children having magnetic resonance imaging: a preparatory storybook and audio/visual media are preferable to anesthesia or deep sedation. J. Child Health Care **19**(3), 359–369 (2015)
5. Edwards, D.A., Arthurs, O.J.: Paediatric MRI under sedation: is it necessary? What is the evidence for the alternatives? Pediatr. Radiol. **41**, 1353–1364 (2011)
6. Axline, M.V.: Play Therapy trad. It. Play Therapy – ed. La Meridiana, Bari (2009)

7. Lawrence, C.: Rubin Handbook of Medical Play Therapy and Child Life: Interventions in Clinical and Medical Settings, 15 December 2017
8. Cromwell, J.A.: An Amended Form of The Observation Scale of Behavioural Distress. A Portfolio of Academic, Clinical and Research Work, Clinical Psychology, University of Surrey (1997)
9. Bana, L., Bruno, I., Geraraduzzi, T., Biasotto, E., Giuseppin, I., Berti, I., Zanazzo, G., Rabusin, M., Vecchi, R., Barbi, E., Sarti, A., Tamaro, P., Ventura, A.: Sedazioni profonde e ripetute in emato-oncologia: vissuto dei bambini affetti da leucemia e delle loro famiglie. Medico e Bambino pagine elettroniche **7**(11) (2004). https://www.medicoebambino.com/?id=RI0411_10.html
10. Kerstulovich, L.: La venipuntura nel bambino: il processo informativo e la sua ricaduta su ansia e dolore. Tesi di laurea, Università degli Studi di Padova, dipartimento di medicina, Corso di laurea in Infermieristica, anno accademico (2014/2015)

Infrared Thermography as a Visualization Technique for Light Muscle Pathologies

Francisco-Javier Soto-Lara[1(✉)] and Manuel Damián Marín Granados[2]

[1] Industrial Technical Engineer, 29130 Alhaurín de la Torre, Málaga, Spain
franciscosoto@intermal.es
[2] University of Málaga, 29017 Málaga, Spain

Abstract. Infrared thermography is a technology that has long been employed in the clinical environment, and its application has been further extended in this sector on account of the technical advances of cameras and the increasing ease of acquiring a computer. With regard to the increased clinical use of this technology, it is very useful to apply this technique to the visualization of mild musculoskeletal pathologies due to its high capacity to produce increasingly sharp and exhaustive images as a result of the infrared radiation emitted (in this case by the patients), whilst modern equipment allows for assigning two pixels of different color at temperatures that differ by 0.01 °C. In this case, the ability to transform the emission of infrared radiation from any body into an image—along with the various graphic patterns that can be observed—allows measurements such as temperature to provide information with very small gradients. A specific example of the utility of this technique will be described in this article, using a case of mild cervical injuries produced by a low-speed traffic accident, which would otherwise be difficult to objectify and evaluate using other techniques.

Keywords: Thermography · Health · Muscular · Pathology · Infrared

1 Introduction

It is well known that heat is transmitted by conduction (within solids), convection (within fluids), and radiation, which is produced from the surface of the body primarily in the visible and infrared wavelengths, which, in addition, are the types of wave primarily capable of transmitting heat.

The infrared spectrum has been studied since the nineteenth century and in a short time Herschel's own son (the first scientist to identify this radiation) was already able to measure temperatures from a distance by creating the first "heat image" [1]. Subsequently, another turning point with respect to the knowledge and use of infrared radiation occurred with the development of the Stefan-Boltzmann Law (1), which is able to relate radiant emissive power with temperature by means of an equation that initially characterized the emission of a black body, but with a small correction known as emissivity (the ability to emit radiation from a surface), which provides us with an

F. Cavas-Martínez et al. (Eds.): INGEGRAF 2019, LNME, pp. 227–233, 2020.
https://doi.org/10.1007/978-3-030-41200-5_25

exact relationship between the radiant emissive power of a real body and its surface temperature:

$$W = \varepsilon \cdot \sigma \cdot Te4 \quad (1)$$

Where ε is emissivity, a value between 0 and 1; σ is Stefan-Boltzmann's constant, and Te is the temperature of the surface of the observed object. With continuous technical and scientific developments, by the 1970's thermographic tests were already being carried out with equipment that could be transported in a van and used for health purposes, with the technique being approved by the FDA [2] in 1982 for the early detection of breast cancer. As can be seen on its website, it later recognized such approval, although its use is restricted to a complementary rather than an independent test (and with good reason, in the opinion of this author) [2].

Along with the development of the applied use of thermography in the health sector, various investigations have allowed for the development of protocols or guides, both for the adjustment of the patient and for environmental conditions, allowing studies to be reproducible and reliable, even defining the characteristics of the equipment to be used in order to obtain results with sufficient clinical value [3]. The cited guides all agree on the need to limit external influences such as air currents or variations in ambient temperature, including the influence of clothing, since working with infrared radiation involves visualizing the emission of the surface of the skin. The guides also recommend the equipment to be used, suggesting a resolution of 320 × 240 pixels. With this resolution the majority of manufacturers of thermal cameras offer a NEDT (noise equivalent differential temperature) of 50 mK at 30 °C room temperature, or similarly, this means that this cameras can offer an image of two pixels with different colors at a temperature difference of 0, 05K, which can decrease even further with higher resolutions. When complying with all the necessary conditions—and thanks to the technological advancements in cameras and improvements in resolution—following the aforementioned guides allows for obtaining images such as the following (Fig. 1):

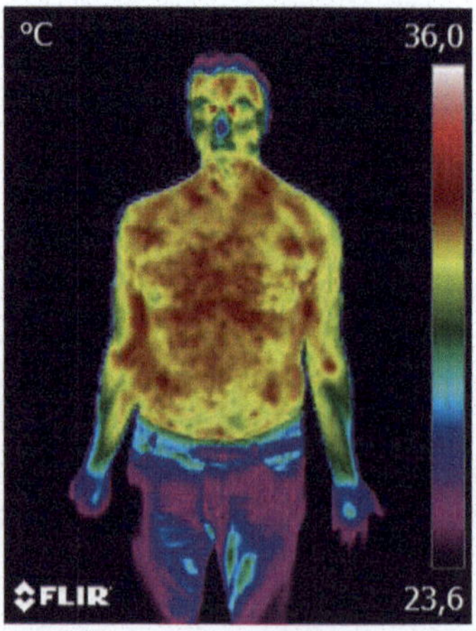

Fig. 1. Thermal image of a patient. Source: Author's own files

This provides the starting point from which the thermographic analysis can begin, this being a matter of both graphic pattern analysis and temperature measurements. Currently, the bases and contour parameters of these analyzes are well defined since it has been shown that a healthy body is thermally symmetric, due to the considerable advances and extensive use of thermography in the prevention of sports injuries. [4] Likewise, the mentioned thermal symmetry means that the graphic patterns adopted by the distribution of heat in the skin are not equal and symmetrical on the opposite side of the patient (always in muscle groups or similar musculoskeletal systems), supposing a region of interest that could be a possible area related to a pathology, given that the standard general patterns are also known and can be compared with any case that is studied such as, for example, the back [5]. This part of the process is known as qualitative thermography (Fig. 2).

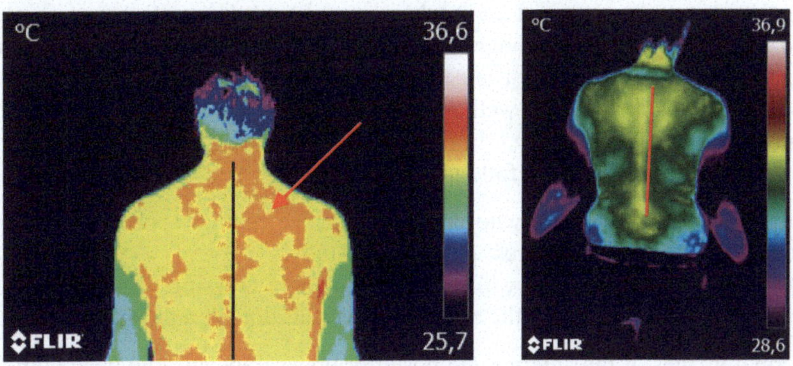

Fig. 2. On the left, thermal asymmetry in the region of the upper trapezius. In the image on the right there are other pathologies in other areas but the upper trapezius is healthy and thermally symmetric. Source: Author's files

In general, the proposed procedure for working with thermography always involves locating thermal anomalies (qualitative) either by observing patterns that differ from those expected or by comparison with symmetrical references without pathologies either in the clinical or industrial environment (including buildings). Assessment of the anomaly requires the measurement of the temperature, comparing the obtained temperature with that expected or with the reference pattern or symmetrical zone without pathology. This is known as quantitative thermography. In this article we will study the case of a patient who had a traffic accident causing damage to different parts of the body for which other, more "invasive" tests could not obtain clear evidence. However, the high sensitivity of thermography allowed for locating and graphically presenting the existence of such pathologies in a way that was compatible with the symptomatology reported by the patient.

2 Materials and Methods

For the case study we used a room without external windows, with the climate controlled by an air conditioning system that established a room temperature of between 21 and 23 °C prior to the study. The camera used was a Flir T-335 of 320 × 240 pixels resolution and with a NEDT of 50 mK at 30°C. For analysis the software used was the Tool + software (Version 6.4.18039.1003) and the ambient temperature and humidity were measured with the Extech MO297 thermo-hygrometer, ensuring that the ambient temperature did not change significantly during the taking of images. According to the American Academy of Thermology [4] the patient had previously been warned to: avoid taking analgesics or anti-inflammatories at least 24 h before the study; if desired, must shower at least 2 h before the test; and the last meal should be eaten a minimum of 4 h before the study. Following the guidelines for these types of studies, the patient is requested to avoid receiving therapies 24 h before the test and to avoid sunbathing or shaving the areas to be observed during that time interval. It is well documented that in order to perform a clinical thermographic study, the patient must be without clothes in the region of interest for at least ten minutes [6], in order to establish a history and obtain data on the symptoms as well as any pathology that could influence the thermal images and hinder the study.

The data obtained were the following:

- Patient: a 48-year-old man who does not practice regular sports or present musculoskeletal diseases of interest prior to the study.
- Case: run over by a vehicle whilst crossing a zebra crossing.
- Symptomatology following the accident: cervical, dorsal, and lumbar pain

Thermal images of the anterior and posterior upper trapezius and even both laterals are then taken in order to compare the resulting graphic patterns. The images were entered into the Tools + software, changing the available color palettes for one that allows for a better appreciation of any possible asymmetries or differences with respect to a healthy pattern. The "RAIN.PAL" was chosen, since it has relatively few colors and offers a sharper contrast to identify (presumably mild) alterations.

Once the palette is selected, a "thermal adjustment" or distribution of the colors of the palette is carried out at temperatures that allow for observing the previously described thermal patterns.

3 Results and Discussion

Following the mentioned adjustments, images are obtained that verify the existence of thermal patterns that differ from the standard patterns in terms of the cervical, dorsal, and lumbar areas, as shown in the following images (Fig. 3):

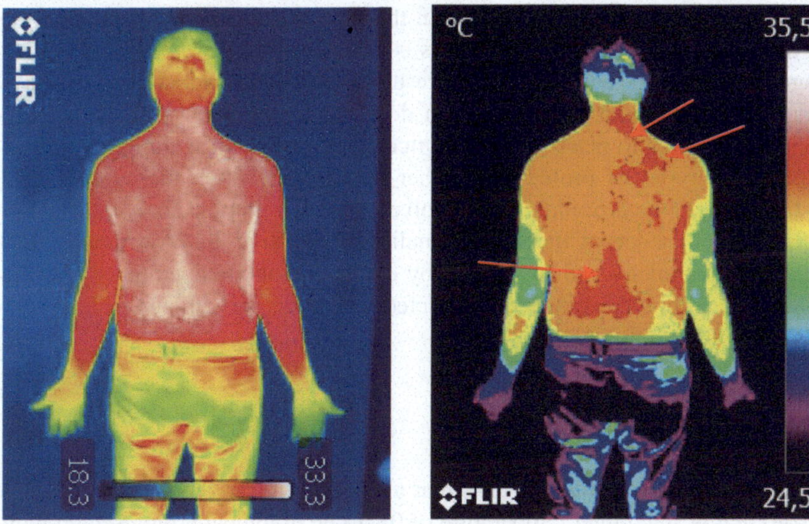

Fig. 3. Image of the back of the patient studied. Left: Image as obtained from the camera. Right: image analyzed with Tools + software with marks indicating the detected thermal abnormalities.

As can be observed, there is an asymmetry with respect to the axis of the vertebral column in both the cervical and dorsal areas. The pattern of the lower back presents a lower lack of symmetry but is still somewhat different from the normal thermal pattern of a healthy back, whose image is shown as follows (Fig. 4):

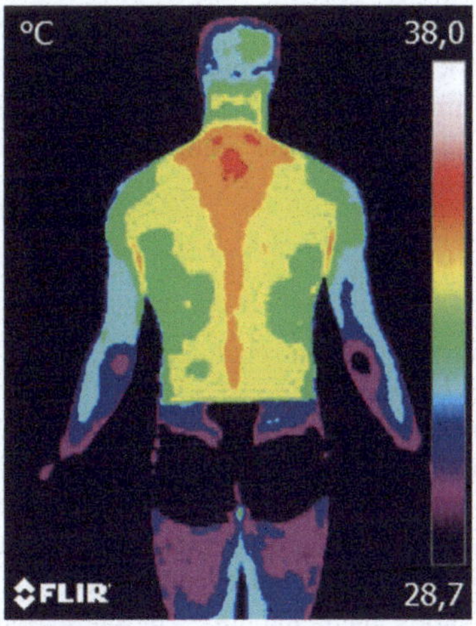

Fig. 4. Image of a healthy back used as a reference in the sports injury risk analysis software "Thermohuman" [7]

The temperature measurement between thermal asymmetries revealed a difference greater than 0.5 °C which verifies that this is a mild pathology. What was observed at this stage is compatible with what was presented by the subject studied and is within the parameters and results that have been demonstrated by research and advances in thermography applied to health. The analytical procedure is also similar to that used to determine other types of problems in other sectors for which thermography is also useful. In any case, the procedure first involves an initial graphic analysis of the images to confirm the existence of thermal anomalies (qualitative analysis) upon which the severity of the pathology is then verified by comparing the temperatures with those of areas that fall within the range of predicted behavior or with the images of other subjects without pathologies.

4 Conclusion

Infrared thermography is a technology that allows the infrared emission of any object (or person) that is at a higher temperature of 0K to be graphically represented with very high quality and precision. This allows for obtaining clear and objective data on pathologies whose manifestations are usually very mild and therefore difficult to objectify using more traditional invasive medical procedures. Infrared thermography is therefore a tool of great interest for objectifying mild muscle pathologies.

Evidence for the utility of thermography in many applications is shown in the work of the Graphic Expression Department of the ETSI at the University of Malaga. This department has added this technology to its instruments for research and teaching, having already conducted several training activities on this technology along with a number of end-of-degree projects allowing many students to gain further skills in addition to those already acquired in their studies.

Acknowledgements. This work has been possible thanks to collaboration with the University of Malaga, particularly the Department of Graphic Expression in Engineering who have provided encouragement and support in the writing of scientific documents.

References

1. F. Systems. https://www.flir.com/globalassets/imported-assets/document/flir-t6xx-series-user-manual.pdf. Accessed 05 Feb 2019
2. U.S. Food and Drug Administration. https://www.fda.gov/ForConsumers/ConsumerUpdates/ConsumerUpdatesEnEspanol/ucm326948.htm. Accessed 05 Mar 2019
3. American Academy of Thermology. https://aathermology.org/organization-2/guidelines/guidelines-for-neuro-musculoskeletal-thermography. Accessed 05 Mar 2019
4. Hildebrandt, C., Zeilberger, K., Ring, E.F.J., Raschner, C.: The application of medical infrared thermography in sports medicine. In: An International Perspective on Topics in Sports Medicine and Sports Injury, 534 p. (2012)

5. Sánchez, E.M.: La termografía por infrarrojo en el traumatismo menor de columna cervical. In: XI Jornadas de Valoración del Daño Corporal: aspectos médico-prácticos, Madrid, Mapfre Foundation, Instituto de Prevención, Salud y Medio Ambiente, D.L. (2015)
6. Bouzas Marins, J.C., Gomes Moreira, D., Piñonosa Cano, S., Sillero Quintana, M., Dias Soares, D., de Andrade Fernandes, A., Sousa da Silva, F., Amaral Costa, C.M., dos Santos Amorim, P.R.: Time required to stabilize thermographic images at rest. Infrared Phys. Technol. **65**, 30–35 (2014)
7. Thermohuman. http://www.thermohuman.com/es/. Accessed 08 Mar 2019

Development of a Methodology of 3D Modeling of Heads for Their Application in the Design of Customized Sports Helmets

David Ranz Angulo(✉), Ramón Miralbes-Buil, and Luis Vidal-Piña

Departamento de Ingeniería de Diseño y Fabricación, EINA, Universidad de Zaragoza, C/María de Luna s/n, 50018 Saragossa, Spain
dranz@unizar.es

Abstract. This works deals with the development of a methodology for 3D modeling of heads, so that, with the digital model that is obtained, the interior of a sports helmet can be customized to improve comfort and protection against impacts during its subsequent use. In order to define the final methodology, according to each one of the business lines (face-to-face and online), different tools have been studied and tested; all intended for the same purpose, which is 3D modeling, but through different techniques. The techniques used have been the one of scanning by means of structured light technology, time-of-flight and photogrammetry with different digital cameras. The three techniques have been tested throughout the study on a dummy bust and have subsequently been applied to real users. This study aimed to determine the influence of certain factors, such as luminosity or the presence of hair among others, in each of the different techniques of modelling. The error between the obtained digital models and the real head has been determined through different control measurements. So establish what conditions are most appropriate for each of the techniques and select the optimal methodology for each of the future lines of business.

Keywords: 3D modelling · Scanning · Photogrammetry · Sport helmets

1 Introduction

Currently, the big problem that exists in the use of sports helmets, is that the measure of them is regulated by some standard sizes [1, 2]; so the helmet does not fit perfectly into the user's head. It means that gaps and possible movements of the head inside the helmet can drive to serious injuries in case of impact.

As background, the application of 3D scanning to sport product customization is rather limited. A case related with the modification of the interior of a Moto GP helmets can be found [3], a Spanish company offers customized shin guards and facial masks [4], likewise customized earphones can be found in the market [5]. Besides, some work about the scanning the head of a baby to make an orthotic protection helmet [6] was done. However, beyond these examples, in terms of customized sports helmets, the application is void.

F. Cavas-Martínez et al. (Eds.): INGEGRAF 2019, LNME, pp. 234–244, 2020.
https://doi.org/10.1007/978-3-030-41200-5_26

Therefore, this works deals with the development of a method by which customized helmets can be designed using 3D scanning technology [7, 8]. The project has been carried out using two 3D scanning equipment: Sense 2 (3D Systems) and Pro S3 Scanner (HP). The first uses time-of-flight technology [9], while the second uses the structured light technique [10]. Both of them provide its own software to process the acquired data. Besides, photogrammetry technique [11] is also tested since it only requires a camera to capture the images and a specific software to process them. In this case, two different cameras were used, Canon EOS1300d and the front camera of an Iphone 6S mobile phone.

In a first stage, the necessary tests are carried out to establish the appropriate scanning conditions to obtain the optimal 3D models by means of each one of the techniques. In addition, the influence of the hair was analyzed when obtaining the 3D models of the head, and a method is proposed to achieve the best results for this specific application, in which a good helmet fit must be achieved. Subsequently, the different 3D models were obtained and the error calculation estimated by comparison with real data in order to determine the reliability of each methodology. Once the methodologies to be followed were well defined, they were applied to real users.

In this way, it is sought to establish the appropriate procedures for two lines of business. First, called on-site, the user goes to the professional's facilities for data acquisition. And a remote mode, the user with assistant support takes the data and upload them to web platform for their treatment and construction of the 3D model.

2 Materials and Methods

2.1 Equipment

The tests throughout this work are carried out using three different 3D scanning technologies. First is the structured light scanning and consists of projecting a geometric pattern on the surface of the object to be scanned in 3D and a camera records the distortion of the image. Depending on how the pattern is deformed, it is possible to calculate the location of each point. In order to apply this technology, the Scanner Pro S3 desktop device is used (Fig. 1, right).

Fig. 1. Scanning equipment: (left) Sense 2; (right) Pro S3 Scanner.

Time-of-flight technique determines the distance between the 3D scanner and an object by measuring the time that the laser or infrared light emitted by the device takes to reflect. For this purpose, the Sense 2 manual scanner is used (Fig. 1, left).

Photogrammetry consists in measuring and determining the position of points in a three-dimensional space using images taken at several slightly different angles. It is done using a Canon EOS1300d camera and an Iphone 6S to compare both devices (Fig. 2). In this case, images processing and model construction, the specific software Agisoft Photoscan Pro is used.

Fig. 2. Photogrammetry equipment: (left) Canon EOS1300d; (right) Iphone 6S.

Most of the tests to be carried out during the project, to set up the optimal conditions for each technique, were done with a dummy's bust, since for practical purposes it provides the necessary information that a head of a person would offer. In addition to this bust was placed 3D scanning targets to facilitate processing when working with software (Fig. 3a).

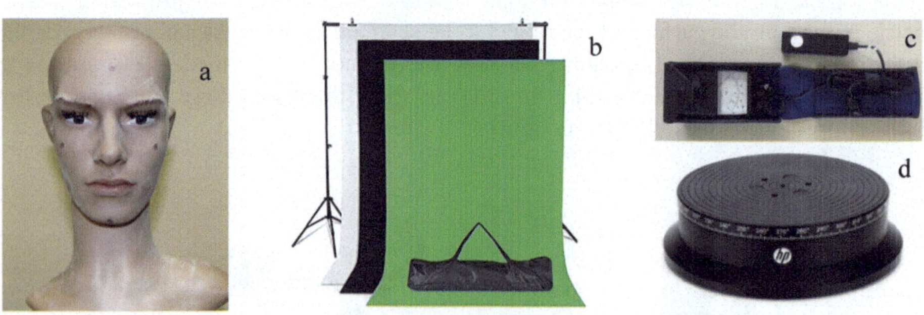

Fig. 3. Ancillary equipment: (a) Dummy's bust, (b) Background, (c) Luxometer, (d) Turntable.

On the other hand, for the desktop scanner and photogrammetry, in order to achieve greater precision, a turntable, that rotates automatically depending on the technique and the scanning equipment, is used. A luxometer was used to determine what lighting conditions make that the scanners provide the optimal models. Since the incident light is determinant, the illuminance was measured over the bust dummy under different conditions. In order to achieve cleaner and more homogeneous scans, a photographic background is used with which several tests can be done by exchanging the different colored fabrics and observing the best results are obtained.

2.2 Tuning

In order to determine the optimal conditions to obtain the suitable 3D models for this application, the tuning of the most appropriate procedure for each technique was developed.

Pro S3 Scanner has a complex software to use, HP 3D Scan Professional, with a slow learning curve. This basically consists of three functions: configuration, scanning and shape fusion. To achieve a high quality of the model, the number of scans needed must be between 14 and 22 shots; with a smaller number of that strip there are not enough images to form a quality model and, with more than 22 it has been observed that there is loss of quality of the model when having to merge so many scans with each other. The lighting conditions are very strict because, thanks to the overexposure indicator screen shown by the software, when you open the window to make way for natural light, scanning becomes impossible. It gives better results an artificial and general light that illuminates the dummy in a homogeneous way, the optimum value of illuminance was set-up at 112 lx (Fig. 4).

Fig. 4. Models obtained with Pro S3 Scanner.

Sense 2 uses an easy-to-use software with a very intuitive user interface. For optimal scanning the scanner-object distance has to be constantly tuned to reach difficult access areas and avoid ‘empty areas’. Although it is possible to install it on a tripod, the results are better when using it manually, because it allows the adjustment of distances, which is not possible with the tripod as it remains static and areas remain undetected. The quality of the resulting model is acceptable for the application for which it is intended. For the Sense 2 hand scanner, the type of light does not have much influence, since it has been proven that there are no notable changes with natural light (intense and focused) or artificial light (soft and general). The calculated illuminance received by the dummy during the tests to obtain the optimal model is: 123 lx (Fig. 5).

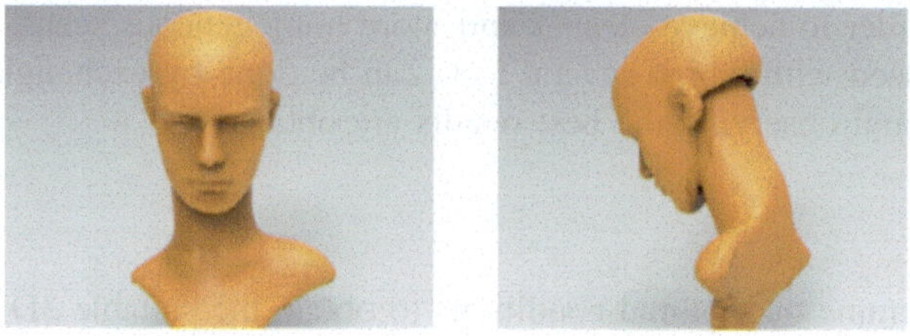

Fig. 5. Models obtained with Sense 2 scanner.

In the photogrammetry technique, Agisoft Photoscan Pro is used. The difficulty of using this software depends on the quality of the photographs that are provided, since the higher the quality, the lower the image processing required. The more photographs are included in the workflow; the better results are obtained. At least 70 photographs taken from different angles are needed to obtain a quality model. Each photograph must have in common at least 50% of the previous one. Scanning by photogrammetry is much more sensitive to brightness than the other two methods seen above. After several tests, it is decided to apply talcum powder to the dummy to reduce the reflections that do not allow obtaining a model of enough quality. The difference in accuracy and quality of the mesh is negligible if we compare the models obtained with the CANON EOS1300d at 12 Megapixeles and the IPHONE 6S at 2 Megapixeles. The quality of the mesh is the lowest of the three techniques tested, however, applying some model repair adjustment, it is perfectly valid for the purpose of the project. The first tests with artificial light only were not valid. It is decided to allow the entrance of natural light which, together with the artificial light, achieves a greater luminance that allows the details to be better appreciated in the photographs. The biggest problem of photogrammetry are the shiny surfaces which we solve by applying the talcum powder, as mentioned above. The calculated illuminance received by the dummy during the test to obtain the optimal model is: 250 lx (Fig. 6).

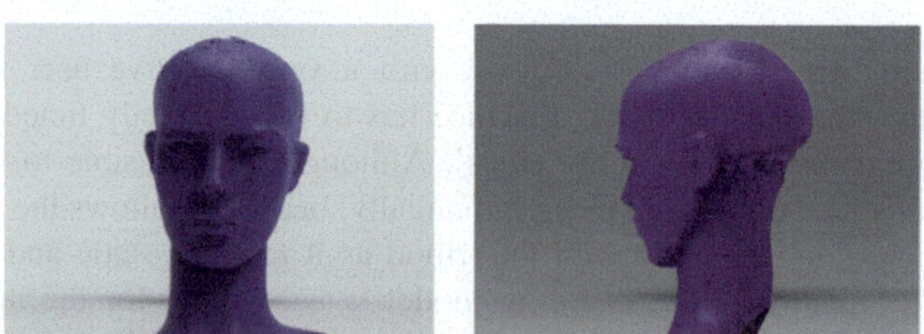

Fig. 6. Models obtained by photogrammetry (Agisoft Photoscan Pro).

2.3 Hair Influence

Once the tests with the dummy have been carried out, it is considered important to put a wig imitating the human hair to see how the scanners respond. As seen in the Fig. 7, the Pro S3 Scanner is not able to detect the hair, because it is a bright surface overall and it has been shown that the brightness causes excess saturation. Therefore, in the entire area where there is hair, the scanner is unable to capture that information. The Sense 2 hand scanner is capable of detecting the wig, however the result of the model is of very low quality. Wearing swimming caps to contain the hair, both in men and women, is proposed to avoid hair glosses and unify the methodology.

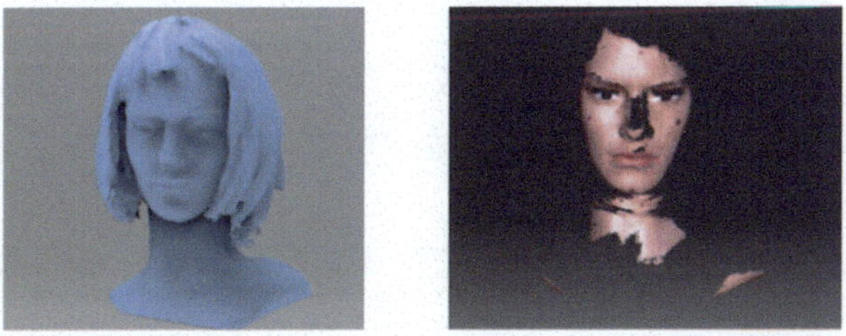

Fig. 7. Models with hair trials: (left) Sense 2; (right) Pro S3 Scanner.

The caps selected for testing are made of lycra; this material is chosen due to the fact that the silicone can reflect the light shinning, while lycra is a fabric that does not makes shine. Tests have been carried out with a totally homogeneous swimming cap without drawings and then with another one that has geometric drawings of different colors.

Fig. 8. Models of swimming caps to contain the hair: (left) No patterns, (right) Patterns.

It has been verified that for the Scanners Sense 2 and Scanner Pro S3 the most suitable cap is the yellow color without patterns (Fig. 8 left), since the one of patterns when having dark colors produces a lot of saturation and geometry located in dark areas can not be captured by the scanner. However, for photogrammetry, the one that works best is that of patterns (Fig. 8 right), it may be due to the fact that since this technology is guided by points, the more it has reference, the easier it is to process the images and the better the model comes out. The obtained models using swimming caps for each method are showed in Fig. 9.

2.4 Benchmarking for Optimal Conditions

After the tuning stage, the theoretical values are established to obtain the optimal 3D models according to the technique used. These parameters are shown in the Fig. 9 and will be used in the error determination and the real user stage to set up the final procedures for each business line.

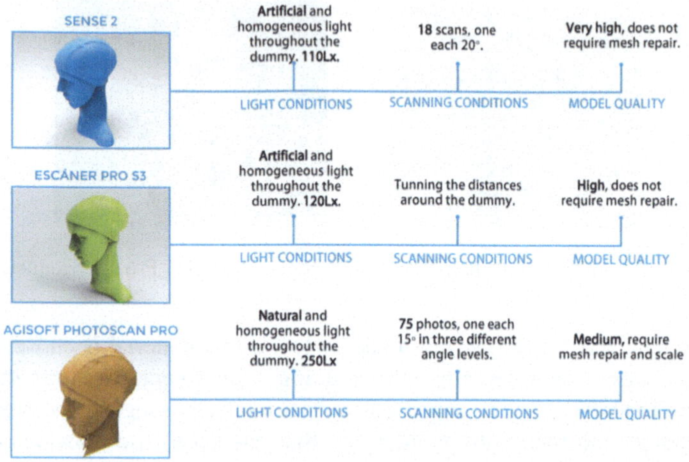

Fig. 9. Scanning and environmental conditions.

3 Results and Discussion

3.1 Results

For each of the techniques, the 3D models were obtained applying the optimal theoretical conditions and the deviations were determined with respect to the real head model. Distances were taken directly between the targets located on the dummy head by direct measurement with a caliper, as can be appreciated in Fig. 10.

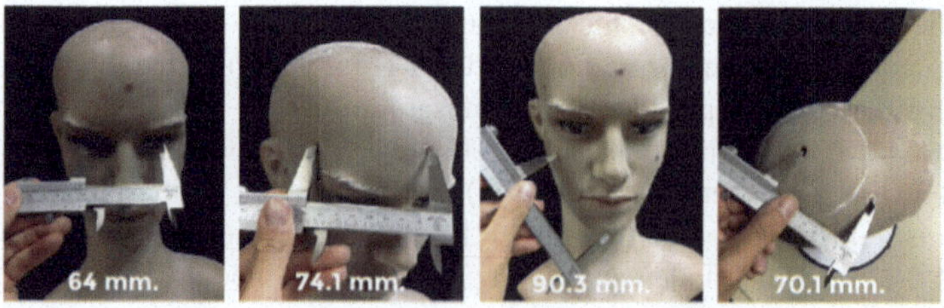

Fig. 10. Measurements: eye-eye; forehead-temple; cheek-chin and center-head.

For each of the techniques, the 3D models were obtained applying the optimal theoretical conditions and the deviations were determined with respect to the real head model. Distances were taken directly between the targets located on the dummy head by direct measurement with a caliper, as can be appreciated in Fig. 10. Subsequently, using the Netfabb software [12], the same distances between the targets were measured in the 3D models and the absolute error and percentage were obtained for each of them, as can be seen in Fig. 11. These distances, which are key to the design of protective helmets, do not offer significant errors for this type of application. Being, the most accurate model obtained by the HP 3D scan with errors not exceeding 0.4%. The least accurate model is obtained by photogrammetry, with maximum errors equal to 2%, being admissible when generating paddings that adjust the helmet to the user's head.

	DISTANCES (mm)											
	EYE - EYE			FOREHEAD - TEMPLE			CHEEK - CHIN			CENTER - HEAD		
DUMMY	64			74.1			93.2			70.1		
3D MODEL	Sense2	HP3Dscan	Agisoft	Sense2	HP3Dscan	Agisoft	Sense2	HP3Dscan	Agisoft	Sense2	HP3Dscan	Agisoft
	63.6	64	65.09	73.2	74.2	74.68	93.6	92.9	91.34	69.4	69.8	69.02
	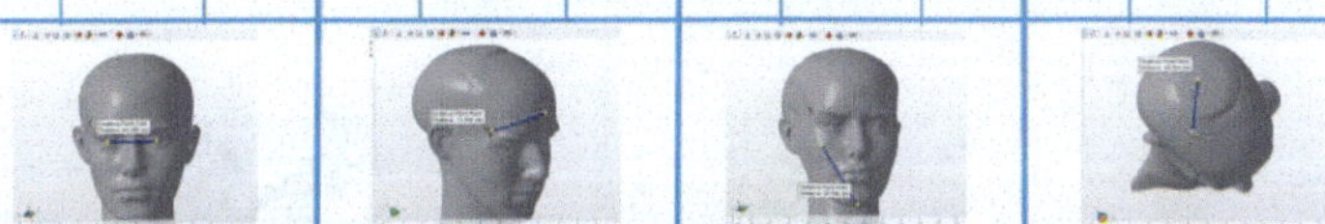											
ABS ERROR	0.4	0	1.09	0.9	0.1	0.58	0.4	0.3	1.86	0.7	0.3	1.08
% ERROR	0.62%	0%	1.70%	1.21%	0.13%	0.78%	0.42%	0.32%	1.99 %	0.99%	0.42%	1.54%

Fig. 11. Error of Sense 2, HP Pro S3 Scan and Agisoft Photoscan Pro measurements.

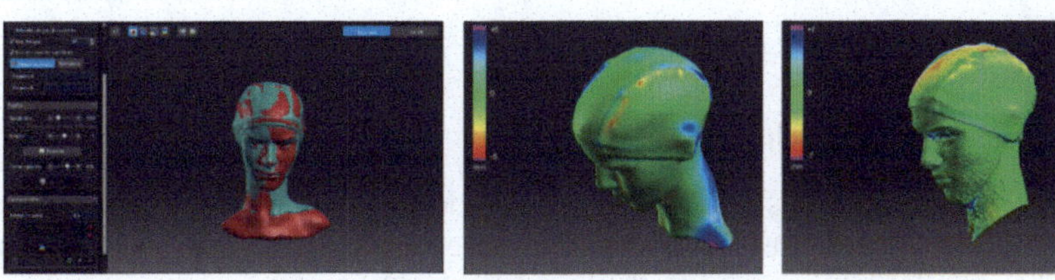

Fig. 12. Error of surface coincidence: (left) Model overlap, (center) Sense 2; (right) Agisoft Photoscan.

Figure 12 shows the deviation between surfaces when both models are overlapped. In both models, greater differences can be seen in the sewing areas of the cap, possibly due to the concentration of hair in those areas. However, in general terms, both models provide very low values of deviation, being able to appreciate a greater deviation, close to 3 mm, in the model obtained by photogrammetry.

3.2 Real User Application

In order to definitively establish a methodology that can be used in a real business model, the optimal theoretical conditions established above on real users are applied, in order to validate and modify them if it would be necessary.

Two previously described scenarios are studied, one in which the user comes to the specialized center in person to build the model (on-site). A second scenario, the user takes the images remotely and uploads these images to the web platform for the subsequent development of the 3D model (remote).

Specifically, for the on-site mode, the methodology is applied using the structured light technique with the Scanner Pro S3 desktop computer, since it has the best results in terms of mesh quality and error. For the remote mode, the use of photogrammetry is selected because of the accessibility for the user to the image acquisition equipment, independently of the device typology. Being this last mode, especially attractive, by enabling the user to scan his head and receive the custom helmet without leaving home. Figure 13 shows the real users and the corresponding 3D models that has been obtained for each of the optimized methodologies.

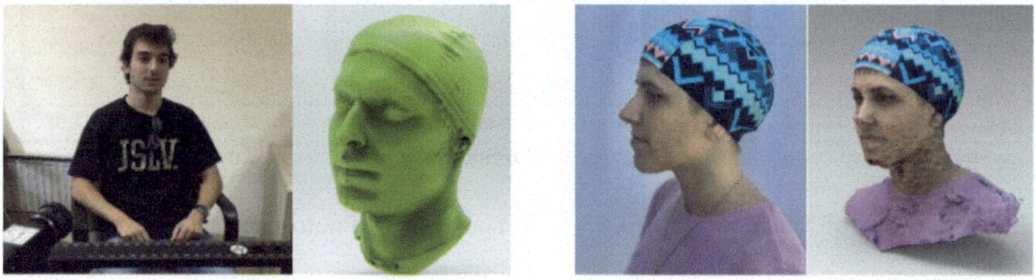

Fig. 13. Real user application: (left) Pro S3 scanner; (right) Photogrammetry-Agisoft Photoscan Pro.

3.3 Proposed Methodologies

After applying the previously established methodologies to the user, some of them are modified and the procedures for the two proposed business lines are established. The proposed methodologies are finally included in the Table 1.

Table 1. Description of the on-site and remote methodologies.

	On-site	Remote
Equipment	Pro S3 scanner	Digital Camera (min. 2 Megapixels)
Personal	Qualified staff	Non-qualified user assistant
Light conditions	Not be very intense. Cover homogeneously the entire head of the user. Luminance of 110 lx	Take the photos in a covered place, with the lights on and let natural light in through the window. Approximated luminance 300 lx
Procedure	Wear the lycra cap and stay in a sitting position for about 10 min while spinning in a swivel chair	Wear the patterned cap, stand in front of a monochromatic background and spin around on your own
Data acquisition	Number of scans: 18. One every 20°	User assistant takes 75 photographs (25 each 15°) in 3 different elevation angles: front, high and low angles
Post-processing	HP 3D SCAN PROFESSIONAL or equivalent software cleans and modifies captures until obtaining the optimal final model	Batch of photos is uploaded to the web platform for further processing with Agisoft Photoscan Pro or equivalent

4 Conclusions

The present paper explores the 3D model construction techniques to be applied to the customization of protection sport helmets. The goal is to provide the optimal conditions to be used with the different techniques to build a 3D model with the enough quality for this application. For this, different influence factors were evaluated, such as, lighting conditions, type of photographic background, position of the capture device, number of captures and hair presence; and its effect on the 3D model quality was analyzed. Besides, the measurement deviations between the 3D models obtained by the different techniques were determined. By this way, it was possible to correctly select the technique and the most suitable values and/or procedures to obtain quality 3D models. This selection is coherent in terms of available resources and preparation time with each line of business, on-site and remote.

References

1. Towner, E., Dowswell, T., Burkes, M., Dickinson, H., Towner, J., Hayes, M.: Bicycle helmets: a review of their effectiveness, a critical review of the literature. Road Safety Research Report, vol. 30 (2002)
2. Chang, C., Ho, C., Chang, S.-Y.: Design of a helmet, ME 499/599 (2003)
3. Metrología, S.: http://www.sariki.es/el-escaner-3d-rexcan-permite-el-diseno-personalizado-y-ajuste-del-casco-de-jorge-lorenzo-para-motogp/. Accessed Mar 2019
4. https://www.espinilleraspersonalizadas.es/. Podoactiva. Accessed Jan 2019
5. https://all3dp.com/normal-earphones-one-size-doesnt-fit/. Hanna Watkin. Accessed Jan 2019

6. https://www.artec3d.com/es/news/escanear-la-cabeza-de-un-bebe-para-hacer-un-casco-tipo-ortesis. Artec3D. Accessed Mar 2019
7. Luximon, Y., Ball, R., Justice, L.: The 3D Chinese head and face modeling. Comput.-Aided Des. **44**, 40–47 (2012)
8. Ellena, T., Skals, S., Subic, A., Mustafa, H., Pang, T.Y.: 3D digital headform models of Australian cyclists. Appl. Ergon. Part A **59**, 11–18 (2017)
9. Diraco, G., Leone, A., Siciliano, P.: Human posture recognition with a time-of-flight 3D sensor for in-home applications. Expert Syst. Appl. **40**(2), 744–751 (2013)
10. Chiu, C.-Y., Pease, D.L., Fawkner, S., Sanders, R.H.: Automated body volume acquisitions from 3D structured-light scanning. Comput. Biol. Med. **101**, 112–119 (2018)
11. Fryer, J., Mitchell, H., Chandler, J.: Applications of 3D Measurement from Images. Whittles Publishing (2007)
12. https://www.autodesk.com/products/netfabb/overview. Netfabb. [En línea]. Accessed Apr 2019

Comparative Analysis of the Adhesion of Metallic Inserts on Dental Implants-Prosthetic Assembly Generated by Polymeric Materials Used for Additive Manufacturing

M. Suffo[1(✉)], J. I. Vilches-Pérez[2], and M. Salido-Peracaula[2]

[1] Department of Mechanical Engineering and Industrial Design, High Engineering School, University of Cádiz, Campus Río San Pedro s/n, 11510 Puerto Real, Cádiz, Spain
miguel.suffo@uca.es

[2] Research Group PATOBIOLOGÍA. Biomedical Research and Innovation Institute of Cádiz (INiBICA) Research Unit, "Puerta del Mar" University Hospital. University of Cádiz, Spain. Faculty of Medicine, SCIBM, Dr Marañon 3, 11002 Cádiz, Spain

Abstract. Materials used in additive manufacturing provide several advantages due to their versatility and availability as inputs but also have a certain degree of uncertainty in their applications to the industry. One of the research fields complementary to this type of materials relates to the adhesives used to attach different components of an assembly whose dimensions are defined by a machine. Clinics and laboratories in the dental sector are trying to adopt CAD-CAM techniques, including 3D printers and numerical control machines, as milling devices. Usually, dentists must face long waiting periods until the laboratory provides suitable and implantable dental pieces, which would be significantly shortened if the additive manufacturing is applied, as it will allow the professional to provide customized solutions and reduce the wait time. In this work, a case-control comparison on adhesives was performed, considering both those commonly used in medicine and others more oriented to the non-medical industry and using samples as classic implant-prosthetic sets. The dental crowns used in restorative dentistry that are part of the assemblies were digitized with hand scanners, modeled and edited on 3D meshes and, finally, printed on a medium-sized 3D printer. The results obtained in tensile tests showed the suitability of alternative commercial adhesive products in biocompatible materials that are commonly used in additive manufacturing, such as ULTEM™ 1010 or Polycarbonate.

Keywords: Dental adhesives · Additive manufacturing · Implant-prosthetic dental sets · Adhesion strength

1 Introduction

Implant-Prosthetic Assemblies (IPAs) are generally used in dental restoration. An IPA is composed by a series of simple elements whose optimal fit, connection and ultimate adherence determine the final performance of the assembly and maintain the concept of

F. Cavas-Martínez et al. (Eds.): INGEGRAF 2019, LNME, pp. 245–253, 2020.
https://doi.org/10.1007/978-3-030-41200-5_27

minimally invasive restoration [1, 2]. The dentist must decide the type of restoration required by each patient depending on their situation by choosing a series of alternatives or solutions that ensure a minimum of success but lead to the reduction or extension of the assembly´s life cycle [3, 4].

A typical case of dental restoration are the implantable crowns, which are fixed to the jaw by a series of mechanical parts designed to connect each other and form a perfect functional set, as those shown in Fig. 1.

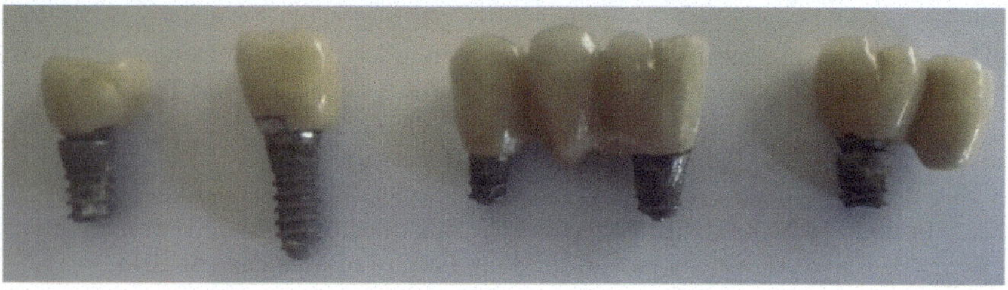

Fig. 1. Different examples of failed implants-prosthetic assemblies (IPA).

Figure 2 outlines the configuration of the elementary pieces that form a complete IPA.

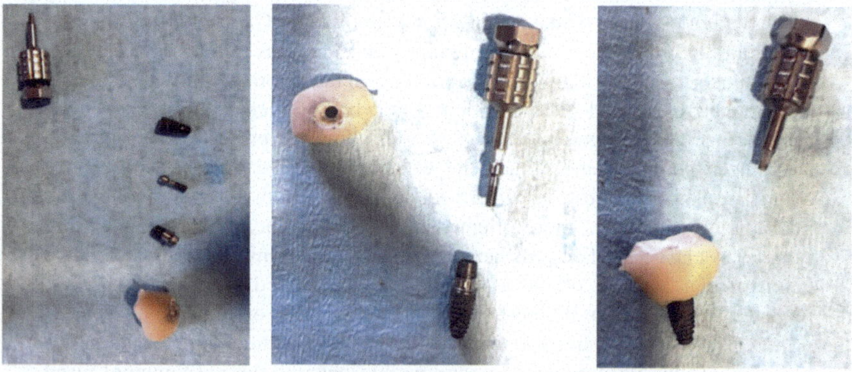

Fig. 2. Real assembly of the implant-dental prosthesis assembly; (a) Simple elements; (b) Implant-interface connection by a prosthetic screw; (c) Crown-Interface sticked by adhesive and screwed to the implant at 30 Nm

It is evident that the metallic pieces are screwed and tightened to each other, but the metallic assembly must be also bonded to the crown, which is the component replacing the dental piece. They are made by non-metallic material in a dental laboratory by independent prosthetic personnel through manufacturing processes, generally by molding.

Materials used in the manufacture of the crowns are usually feldspathic porcelain, metal-porcelain, high-resistance porcelain (zirconium, disilicates, …) or resins. The effectiveness of the dental unit depends ultimately on the correct adherence between the crown material and the metallic elements [2].

The adhesive systems that are currently used to connect crowns to metallic elements such as pillars or interfaces, are based on resins that react to the photopolymerization, chemical polymerization or dual (photo/chemical) polymerization and are cemented until their complete union [5]. Methacrylate-based resins are not effective in PEEK (Polyether-ether-ketone) materials whereas adhesion lasts longer with other types of materials.

There are several brands of dental adhesives for clinical use, although their differences are relative, as they are mainly composed by the same products that only vary in proportion. However, choosing from one to another according to their composition is not trivial, as a detailed description of such composition is normally not provided [4].

Dental cements or adhesives in the context of implant prostheses usually consist of two phases. The liquid phase or matrix that provides adhesive properties and the solid phase or filler that determines mechanical and optical properties. Moreover, chemical, photopolymerizable or dual adhesives can be also found, and hence, variable combinations of monomers, solvents, photo-initiators and filler particles are available [6].

Currently, the adhesives and cements on the interface pieces ensure the union to metals and porcelains by the incorporation of Silanes or specific molecules for metal binding, such as the 4-Meta. Different works discuss the correct treatment for metals and porcelains either by blasting or by acid etching - hydro-fluoric acid [7–9]. Bacteria, body fluids (including those from the tooth), hygiene and food, directly influence the duration of the adhesives. The length period is also affected by the presence or absence of water, which is very difficult to control by the clinician.

The protocol for the adhesion of the interfaces to prosthetic rehabilitation varies depending on the material used and the preferences of the dentist and/or the dental laboratory. Thus, interfaces made of Titanium grade 5 or CrCo are usually drilled with Al_2O_3 particles or acids, such as hydrofluoric (HF), in order to achieve a greater roughness and promote adhesion. The nature of the adhesive is directly related to its ability to maintain the bond, with resin-based cements having a lower loss of retention with respect to those made of glass ionomer or zinc-eugenol oxides [10].

The rise of CAD/CAM technologies in implant-supported restorations has considerably increased the use of interfaces, especially through the incorporation of milling systems and 3D printers in the dental clinics, which allow the dentists to avoid dependence from external laboratories. This has also facilitated the use of new materials, such as hybrid resins (composite resins with a high percentage of porcelain particles) and 3D printing resins [11–13].

The current interest in dentistry lies on simplifying the adherence procedures by reducing the time of clinical application, also increasing retention capacity and long-term stability. At present, additive manufacturing provides alternatives for the design and manufacture of prototypes of crowns made by different biocompatible materials that provide new restorative and biomechanical solutions. Furthermore, restoration times are shortened, which leads to cost savings and ultimately results in tangible benefits for the patient [14].

This work presents a comparative analysis based on the case-control methodology, between a control, which is feldspar porcelain crown, that is attached through an adhesive commonly used in dentistry and 4 particular cases considering different adhesives: two of them use crowns made in the laboratory by regular techniques with PMMA (polymethyl methacrylate) materials and reinforced resins of commercial brands and two others using techniques of additive manufacturing with ULTEM™ 1010 materials and polycarbonate (PC) [15].

2 Materials and Methods

In order to carry out the analysis, a series of tests were applied to the samples considering as control cases. Materials used as samples and the tests applied are described below.

2.1 Materials

Testing materials are shown in Fig. 3.

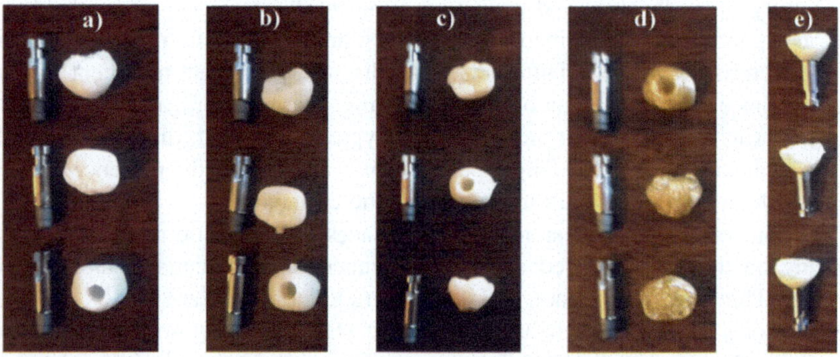

Fig. 3. Particular cases to be tested over the control sample. Inserts were always made of Titanium grade 5; (a) Polycarbonate case printed in 3D (PC sample); (b) PMMA case manufactured by molding (PMMA sample); (c) CRIOS COLTENE™ reinforced resin case manufactured by molding (CC Sample); (d) ULTEM™ 1010 case printed in 3D (Sample ULTEM); (e) Brilliant CRIOS COLTENE™ reinforced resin control Case manufactured by molding and sticked by a dental adhesive (Sample CC + A).

Sample shown in Fig. 3(e) and used as a control (CCRR+A) was composed by a hybrid resin of the trademark Brilliant CRIOS COLTENE™ and manufactured by molding (milled with Cerec Compact™) that was sticked to a Vulkan™ Titanium grade 5 IPA. PC sample (Fig. 3(a) was obtained by FDM additive manufacturing techniques with a Stratasys Fortus 450mc printer equipped with a T16 nozzle (0.2540 mm). RRCC sample in Fig. 3b was made of Telio CAD™ PMMA material manufactured by molding in a prosthetic laboratory. RRCC sample in Fig. 3c was made of hybrid material that was manufactured by molding in a similar way as the control case but with no adhesive. ULTEM™ 1010 sample in Fig. 3(d) was obtained by FDM additive manufacturing techniques in a similar way as the PC case, but using Stratasys biocompatible ULTEM™ 1010 material with a T14 nozzle (0.2540 mm); identical procedure was followed for the supporting material but a T16 nozzle (0.2540 mm) was used.

The prosthetic crown was obtained by scanning the titanium insert and processed with the Cerec3™ system and InLab 3.88™. Once the piece was obtained, it was 3D scanned with a CS3600 Cares-tream™ scanner. The digitized 3D model was obtained through photographs that progressively overlap and assembly together until generation was completed. The model was formed by a 20 μ resolution mesh. The samples made by additive manufacturing (PC and ULTEM™ 1010), were generated from the digitized 3D model from a previous scan of the RRCC sample. As shown in Fig. 4a, the model contained some defects that had be corrected with a CAD platform [16].

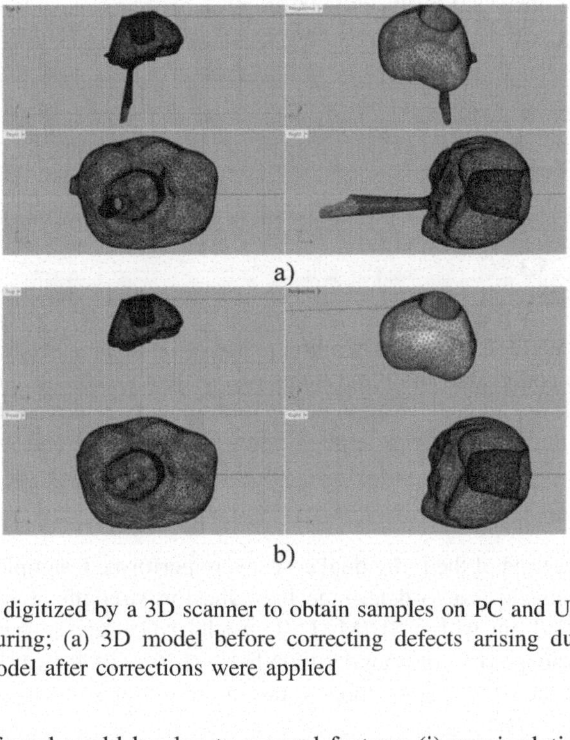

a)

b)

Fig. 4. 3D model digitized by a 3D scanner to obtain samples on PC and ULTEM™ 1010 by additive manufacturing; (a) 3D model before correcting defects arising during the scanning process; (b) 3D model after corrections were applied

The defects found could be due to several factors: (i) manipulation as the crowns were manually scanned and fastened by a light-curing resin mechanism, which could cause waste and excess; (ii) presence of brightness in the material; (iii) presence of waste or dirt in the vicinity; particles, moisture, or even, dust residues arising during scanning. Finally, the presence of illuminated or dark areas on the piece could also alter the process. Figure 4b illustrates the 3D model corrected with the tools of mesh editing.

The IPA was then added to all cases through a series of three different technologies of medical line adhesives that had not been tested previously on this type of dental products, which introduced a certain degree of uncertainty. The three adhesion technologies were property of Henkel Ibérica and are listed in Table 1. The control type samples were sticked with a 3M Spain S.L. RelyX Unicem commercial cement, following the instructions given by the manufacturer.

Table 1. Adhesives and equipment used in the tests

Product	Batch #	Expiry date	
LOCTITE AA 326	71438413	04/19	Acrylic adhesive
LOCTITE SF 7649	17 CM 273	03/19	Activator
LOCTITE HY 4090	82140579	09/19	Cyanoacrylate-epoxy hybrid adhesive
LOCTITE M31 CL	LN 8JAA 1296	09/20	Medical epoxigrade adhesive
LOCTITE SF 7063	18KM 1137	10/21	Cleaner

2.2 Adhesion Test

Samples were sticked under the following environmental conditions:

Temperature: 25.3 °C
Relative humidity: 23.5%

The adhesion method consisted in a first phase of IPAs cleaning. In a second phase, the adhesive was added into the blind holes of the dental crowns and the metal parts were sticked, with the adhesive excess being subsequently removed. In the case of the AA326 product, the activator was applied before the adhesive and let evaporate.

2.3 Tensile Test

Once the adhesion tests of the individual cases were performed, samples were subjected to a mechanical tensile test to determine the adhesion strength or friction resistance. The test was carried out with a SHIMADZU model AGS-X at a speed of 2 mm/min. The method was adapted to the standard UNE-EN 1465: 2009 (ISO 4587) but, given the complexity of the IPA, specific tools to fasten the samples had to be built, as shown in Fig. 5.

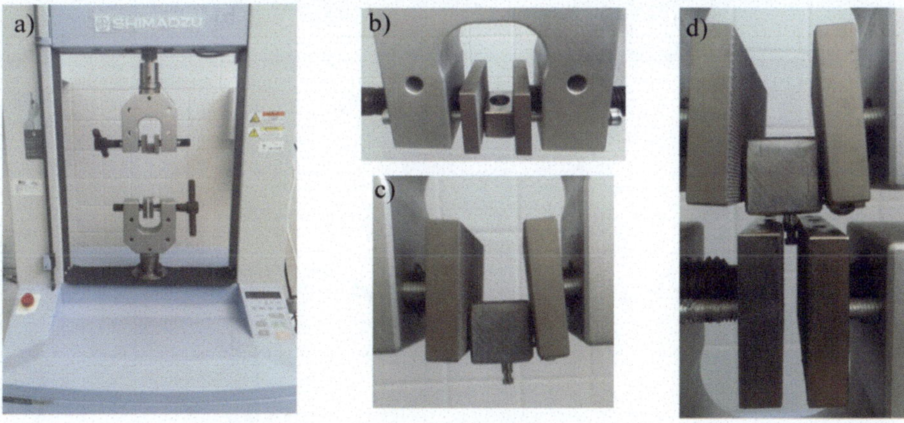

Fig. 5. Equipment used in the tensile test. (a) SHIMADZU AGS-X; (b) Adaptation tool for dental samples; (c) Sample introduced in the tool; (d) Jaws before the test.

3 Results and Discussion

The results obtained during the tensile test are presented and classified on the basis of the materials tested in the study. This facilitates to establish comparisons between crowns made of the same material but sticked to the inserts by different adhesives. The graph of the control type sample sticked with the 3M adhesive is also shown for the sake of clarity. Figure 6a depicts the results of the tensile test applied to the CC material.

The two control type samples ("control 2" and "control 3") fell apart at the beginning of the test, and hence, only the graph corresponding to the sample type control 1 could be obtained (yellow line).

It is evident that the commonly used 3 M adhesive used in the control type samples exhibited the worst response to the tensile test, with the adhesive behaving as an aesthetic filler rather than an elastic-plastic retainer. The same CC material bonded with the three types of Henkel adhesives presented the best results. It is worthy to emphasize that the best response was obtained when this material was sticked with the medical grade epoxy adhesive (gray line, M31), reaching a maximum adhesion strength of 5442.45N. This value exceeded the responses of all tests performed to all the materials sticked with their respective adhesives. This result was expected since it is the most widely used material in dentistry. However, acrylic and hybrid adhesives also achieved high levels of resistance, but friction occurred before and remained longer with respect to others. Figure 6c showed that the PMMA respond well to two of the adhesives: hybrid (orange line, 4090) and epoxy (M31); although, not at the same high level as the CC case. Both adhesives had the lowest strength values.

Finally, results obtained by the two materials used in additive manufacturing can be considered as novel data and the most remarkable information drawn from our study. Both the ULTEM^{TM} 1010 and the PC, (Figs. 6b and 6d, respectively) exhibited similar results, being comparable to the CC material commonly used in dentistry. The two

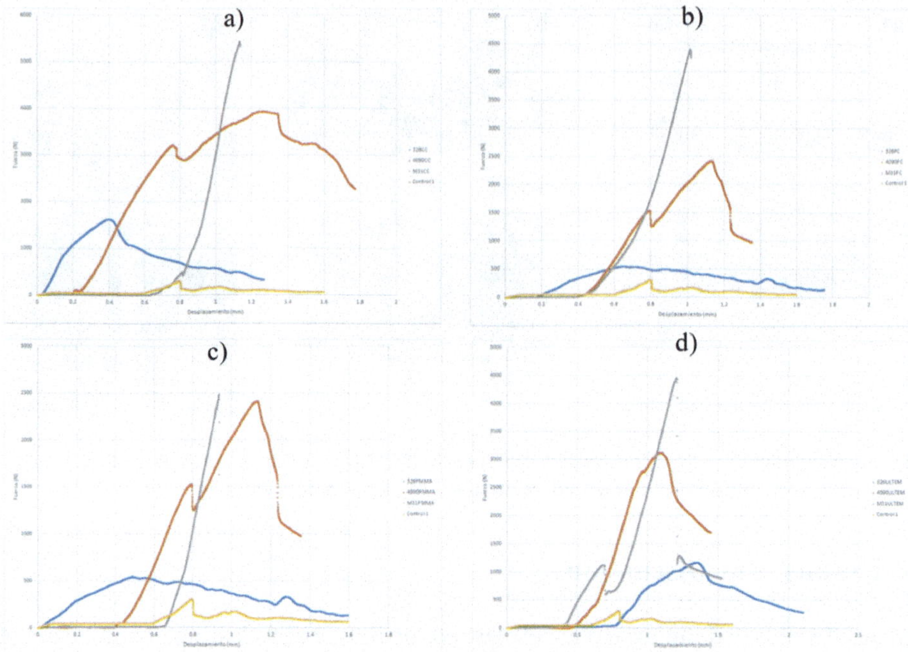

Fig. 6. Results of the tensile test on the Crios Cortene material.

materials are biocompatible although the ULTEM^{TM} 1010 has better mechanical characteristics than the PC, but the most important aspect from an innovative point of view was to find out a medical grade epoxy adhesive that responds well to the adhesion of these polymers used in manufacturing additive.

4 Conclusions

Even though our study has some limitations, adhesive products of two different manufacturers of recognized international reputation in the sector have been explored in order to assess their application to biomaterials commonly used in additive manufacturing and oriented to dental implantology. The comparative analysis performed between different biomaterials and adhesives have allowed to identify the response of these biomaterials subjected to a tensile test. The epoxi medical grade adhesive had turned out to be the most suitable according to its resistance or strength friction. The crowns obtained by additive manufacturing techniques using ULTEM^{TM} 1010 and PC also showed very promising results for subsequent medical use. These results open up an innovative research field in the manufacture of larger parts exceeding the size of small and medium 3D printers. In addition, this work demonstrates the suitability of biomaterials obtained by 3D printing for the dental sector. Finally, our results allow to continue the research oriented to the application of the medical grade epoxy adhesive on implantable elements as protective coating of surfaces that easily oxidize, such as Titanium.

Acknowledgements

- This work was partially funded by the project PI-0013-2017 (Call ITI Cádiz 2017 - R & D Projects). Progress and Health Foundation. Consejería de Salud y Familias. Junta de Andalucía.
- Authors wish to thank the Henkel group for products supply.

References

1. DeGrange, M., Francois-Roulet, J.: Minimally Invasive Restorations with Bonding, Quintessen. Quintessence Publishing, Chicago (1997)
2. Sakr, O.M.: Microshear bond strength of bioactive restorative materials to dentin, vol. 6, no. 4, pp. 90–94 (2018)
3. De Munck, M., Van Landuyt, K., Peumans, M., Poitevin, A., Lambrechts, P., Braem, M., Van Beerbeek, B.: A critical review of the durability of adhesion to tooth tissue: methods and results. J. Dent. Res. **84**(2), 118–132 (2005)
4. Cadenaro, M., Maravic, T., Comba, A., Mazzoni, A., Fanfoni, L., Hilton, T., Ferracane, J., Breschi, L.: The role of polymerization in adhesive dentistry. Dent. Mater. **35**(1), e1–e22 (2018)
5. Salou, L., Hoornaert, A., Louarn, G., Layrolle, P.: Enhanced osseointegration of titanium implants with nanostructured surfaces: an experimental study in rabbits. Acta Biomater. **11**, 494–502 (2015)
6. Van Landuyt, K.L., Snauwaert, J., De Munck, J., Peumans, M., Coutinho, E., Suzuki, K., Yoshida, Y., Lambrechts, P., Van Meerbeek, B.: Systematic review of the chemical composition of contemporary dental adhesives, vol. 28, pp. 3757–3785 (2007)
7. Cadenaro, M., Antoniolli, F., Sauro, S., Tay, F.R., Di Lenarda, R., Prati, C., Biasotto, M., Contardo, L., Breschi, L.: Degree of conversion and permeability of dental adhesives. Eur. J. Oral Sci. **113**(6), 525–530 (2005)
8. Yoshida, Y., Inoue, S.: Chemical analyses in dental adhesive technology. JDSR **48**(2), 141–152 (2012)
9. Dentária, M., Sezinando, A.: Looking for the ideal adhesive – a review. Rev. Port. Estomatol. Med. Dentária e Cir. Maxilofac. **55**(4), 194–206 (2014)
10. Pinés-Hueso, J., Alvarez-Arenal, A., Gonzalez-Gonzalez, I.: The effect of compressive cyclic loading on the retention of cast single crowns cemented to implant abutments. Int. J. Prosthodont. **29**(1), 80–82 (2016)
11. Lumpe, T.S., Mueller, J., Shea, K.: Tensile properties of multi-material interfaces in 3D printed parts. Mater. Des. **162**, 1–9 (2019)
12. Yang, Y., Li, L., Zhao, J.: Mechanical property modeling of photosensitive liquid resin in stereolithography additive manufacturing: bridging degree of cure with tensile strength and hardness. Mater. Des. **162**, 418–428 (2019)
13. Clark, E.A., Alexander, M.R., Irvine, D.J., Roberts, C.J., Wallace, M.J., Sharpe, S., Yoo, J., Hague, R.J.M., Tuck, C.J., Wildman, R.D.: 3D printing of tablets using inkjet with UV photoinitiation. Int. J. Pharm. **529**(1–2), 523–530 (2017)
14. Yilmaz, N.: Luting agent effectiveness on implant crown retention, vol. 2, no. April, pp. 7–11 (2011)
15. Wu, H., Sulkis, M., Driver, J., Saade-castillo, A., Thompson, A., Koo, J.H.: Multi-functional ULTEM TM 1010 composite filaments for additive manufacturing using Fused Filament Fabrication (FFF). Addit. Manuf. **24**(August), 298–306 (2018)
16. Roschli, A., Gaul, K.T., Boulger, A.M., Post, B.K., Chesser, P.C., Love, L.J., Blue, F., Borish, M.: Designing for big area additive manufacturing. Addit. Manuf. **25**(November 2018), 275–285 (2019)

3D-Printed Canine Tibia Model from Clinical Computed Tomography Data

Fátima Somovilla-Gómez[1(✉)], Saúl Iñiguez-Macedo[1], Eduardo Jiménez-Ruiz[2], Laura Muro-Fraguas[1], Gonzalo Gañán-Catalina[1], Álvaro Leciñana-Soldevilla[1], Marina Corral-Bobadilla[1], Carmen Díaz-Bertrana-Sánchez[3], and Rubén Lostado-Lorza[1]

[1] University of La Rioja, 26004 Logroño, La Rioja, Spain
fatima.somovilla@unirioja.es
[2] Public University of Navarre, 31006 Pamplona, Navarra, Spain
[3] Autonomous University of Barcelona, 08193 Barcelona, Spain

Abstract. The goal of this study is to show the process for obtaining a 3D model of a canine tibia using free software from clinical quality computed tomography (CT) images. In this case, the obtained model was used for generating a computer simulation with the Finite Element Method (FEM) and for generating a 3D printed canine tibia. First, a real canine tibia was scanned in order to generate Dicom (Digital Imaging and Communication on Medicine) CT images. Using the open source 3D Slicer, the Dicom CT images, the cortical and trabecular bones were segmented to construct to build a first 3D tibia model. The defects of this first 3D model were examined, and its surfaces were smoothed using open source MeshMixer software. Subsequently, the smoothed 3D tibia model was exported to STL file (Standard Triangle Language), and then imported to Mentat-Marc FEM software in which uniform tetrahedral elements meshes for the trabecular and cortical bones were generated. These 3D meshes were used to develop a Finite Element Analysis for studying the mechanical stiffness of the tibia. Also, the stl file obtained was imported to the Wanhao cura v.18.04 software, and the G-code file were created. Finally, a 3D canine tibia model was printed in a Wanhao Duplicator 6 3D printer with ivory-colored PLA. This 3D printed model was used for later for educational training.

Keywords: Tibia · Computerized axial tomography · Canine · Cortical · Trabecular · FEM · 3D printed model

1 Introduction

The fibula and tibia are the two bones that form the lower rear leg of dog. The tibia is a long bone located in the lower thigh that is the part of the hind leg beneath the knee to the hock [1]. The tibia is one of the major weight bearing bones of the hind limb. Fractures of the fibula and tibia are commonly encountered in the dogs. The fracture of the tibia in dogs occurs when any bone region exceeds the ultimate stress of the material (in this case the bone). Usually, the fibula and tibia fracture at the same time

F. Cavas-Martínez et al. (Eds.): INGEGRAF 2019, LNME, pp. 254–262, 2020.
https://doi.org/10.1007/978-3-030-41200-5_28

due to conformation of the lower rear leg. Depending on the fracture type and configuration, they may be treated the different manner [2, 3]. Current bio-imaging devices such as Computed Axial Tomography (CAT) scan can represent and reconstruct three-dimensional anatomical details to facilitate quality diagnoses and thus improve the patient's clinical treatment planning [4]. Also, to perform an analysis related to dog bones, some aspects should be taken into account, such as cost, ethical concern and limited availability of bone samples. For this reason, 3D model bones is commonly used for various experimental and computer simulation works in multidiscipline research including orthopedic surgery, biomechanics engineering and medical implants. Furthermore, CT images are also used to study different biomechanical systems with several purposes: to analyze pathological bone conditions, to describe and simulate the mechanical behavior of biological systems through FEM, to support and improve the surgical planning with a 3D solid model printer reducing the cost of the surgical operation [5, 6]. The quickly prototyping process, known as 3D printing, is expanding widely in clinical application. This method also provides the possibility of modeling and simulating anatomical structures [7]. Moreover, it can also be useful for educational training for medical students [8, 9]. In this study, a special image processing tools for image segmentation and 3D reconstruction was used to generate the stl files. This file obtained was used for generating a computer simulation with the Finite Element Method (FEM) and for generating a 3D printed canine tibia.

2 Materials and Methodology

2.1 Case Description. Canine Anatomy

The work is concentrated on a middle-aged canine tibia that belongs to a corpse, with 30 kg of weight. The tibia bone structure is formed by two parts: cortical bone and the trabecular bone (See Fig. 1).

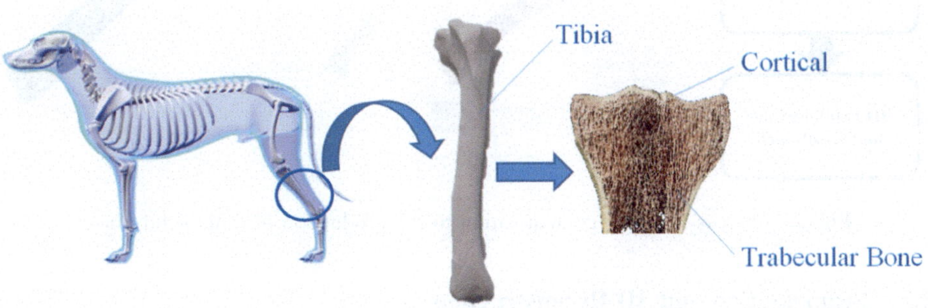

Fig. 1. Canine Tibia to be studied (left). Cortical and trabecular bone structure (right).

2.2 CT Medical Imaging Acquisition (CT Scan)

Medical imaging technique is the process of creating a visible representation of the interior of the body, and it is usually developed with Computed Axial Tomography (CAT) scan. This device applies a helical tomography technique, which generates 2D image of a thin section of the studied body. In CT, a beam of X-rays spins around the studied body acquiring a serial of 2D images by sensitive radiation detectors [10]. This two-dimensional (2D) scan shows a “slice” of the inside of the studied body, and the computer stacks these scans, or serial of slices, one on top of the other to create a detailed image of the studied body such as organs, bones, or blood vessels. Firstly, to acquire corpse canine tibia CT images, the CT conditions were configured: 512 × 512 matrix size, 168 mm field of view, pixel spacing 0.32 mm, slice thickness 0.6 mm, 120 kVp and 80 mA. Dicom files were generated by the CT medical imaging equipment, and consisting of a series of cross-sectional image slices across the interest region, in this case, the tibia of a canine cadaver. Figure 2 shows the process from image acquisition to the production of a 3D solid object.

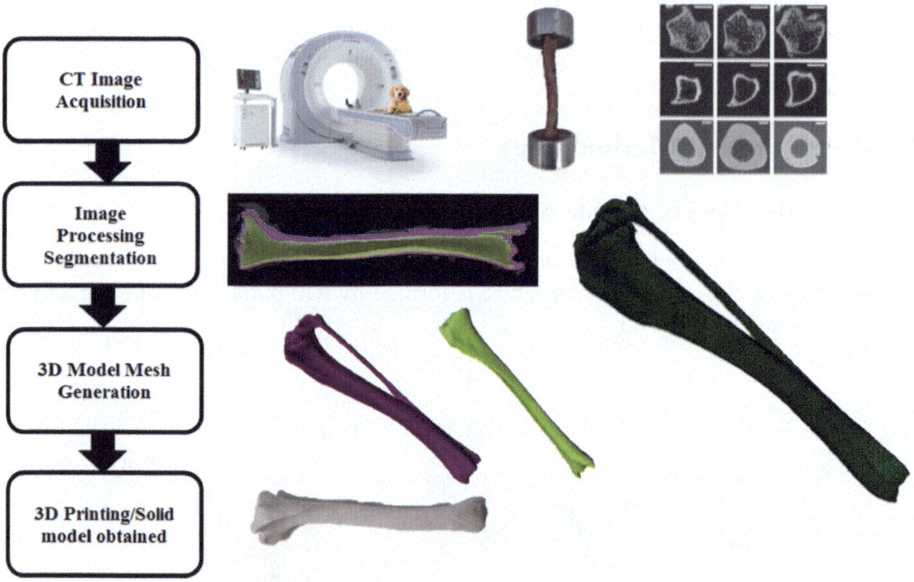

Fig. 2. Process from image acquisition to the production of a 3D solid object

2.3 Segmentation and 3D Reconstruction

Tibia Segmentation with 3D Slicer Software

Image segmentation was applied to the acquired CT DICOM dataset of a canine cadaver tibia, using 3DSlicer software Slicer v4.10.1 [11]. Software 3D Slicer is an open source software package available for download on Windows, Linux and Mac Os X. DICOM CT dataset, which are obtained in the CT scan, were imported into 3D Slicer.

After the CT Dicom images were imported, three different views (axial, sagittal, and coronal) of the CT scan were obtained. The three 2D anatomical viewers in 3D Slicer are defined by Red, Yellow and Green windows. In addition, 3D Slicer automatically created a volume of the entire tibia (See Fig. 3). In order to separate the cortical bone and the trabecular bone of the tibia, a segmentation process based on the threshold of Hounsfield units (HU) values were applicated. Then, two different masks for the cortical and trabecular bone were created. In the case for the cortical bone mask, a threshold with a range of values from 700 Hu to 3070 Hu was applied. For trabecular bone, extraction a thresholding using value range from 150 HU to 700 HU was performed. Then, a separate masks of cortical and trabecular bones were obtained. The two different masks for cortical (purple) and trabecular (yellow) bones are shown in Fig. 4.

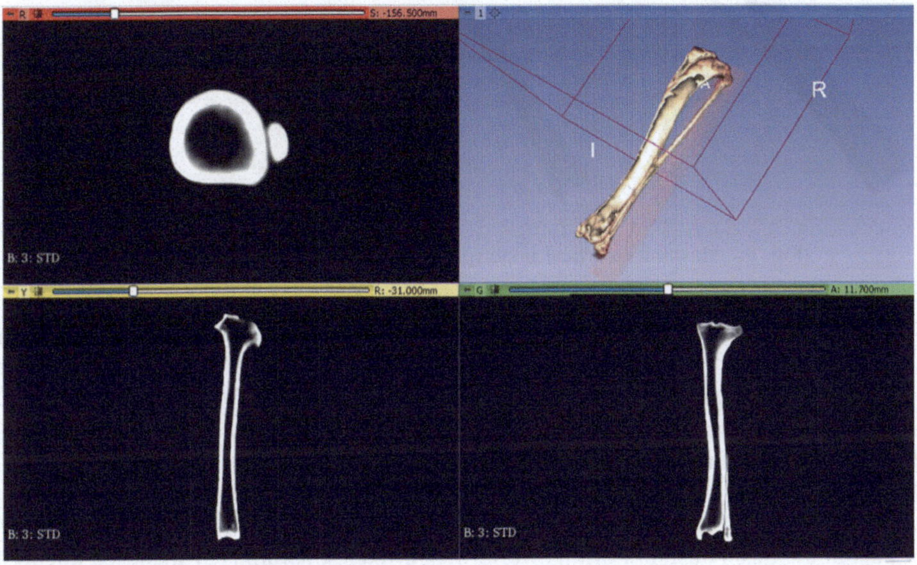

Fig. 3. Three 2D anatomical views and volume rendered Tibia image

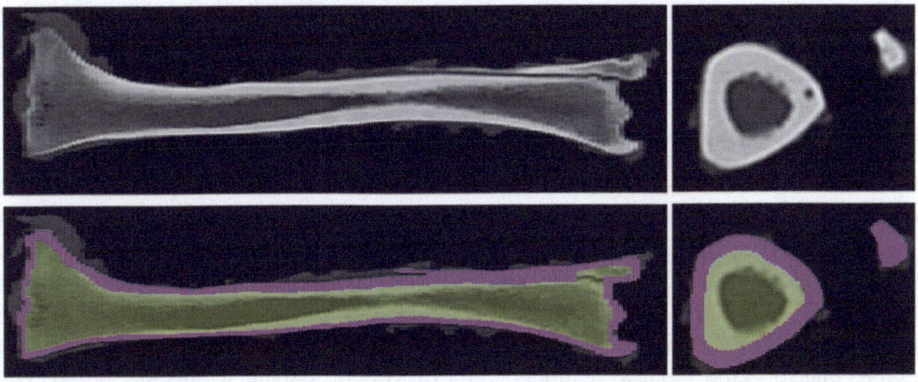

Fig. 4. Masks for cortical bone (purple) and trabecular bone (yellow)

The edition and correction of the mask was required prior to obtain solid 3D model. These corrections were manually developed by adding or removing pixels. Through a manual segmentation, 3D Slicer can create a volume of the specific structures such as cortical and trabecular bone, generating a 3D model of both. Figure 5 shows the completed volume generation for both cortical and trabecular bone.

Fig. 5. Masks for cortical bone (purple) and trabecular bone (yellow)

Image 3D Reconstruction. Once the definitive volumes have been created 3D Slicer software is capable of generating and export STL files by itself. STL (Standard Triangle Language) is a computer-aided design (CAD) computer file format that defines 3D object geometry, excluding information such as colour, textures or physical properties that do include other CAD formats. It was created by the company 3D Systems, designed for use in the rapid prototyping industry and computer-aided manufacturing systems. Before printing the 3D model, it was check that there were no unwanted holes or cuts in the mesh design. For this work, MeshMixer software [12] was used due to it is free open source software from AutoDesk. It is one of the most outstanding programs to edit and prepare mesh type files like.stl and.obj. MeshMixer offers a great variety of different tools to prepare and improve 3D objects for 3D printing. The.stl files generated in the Slicer are imported into the MeshMixer software to be analyzed. First, it was checked if there were irregularities in the model through a complete analysis with the Inspector tool function. After the inspection several holes were detected and repaired automatically (Fig. 6).

3D modelling Mesh Generation. The final 3D Tibia model was imported to Mentat Marc Finite Element Method (FEM) software [13]. In this work, a Finite Element (FE) model of the tibia was prepared for future studies of mechanical behavior. FEM has been widely applied to various medical imaging applications over the past two decades [14–16]. In this case, a 3D FE model that consisted of a combination of 4-node tetrahedral solid FE elements was proposed to model the canine Tibia. The two distinct regions that form the Tibia were modelled by FEM (Cortical and Trabecular bone). The

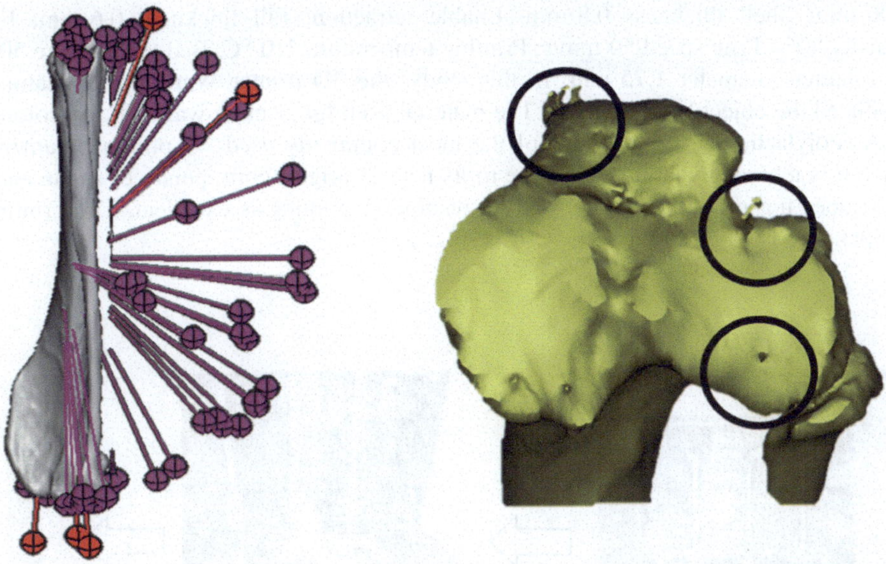

Fig. 6. MeshMixer software Analysis with the Inspector tool (left); Holes and irregularities of the mesh (right)

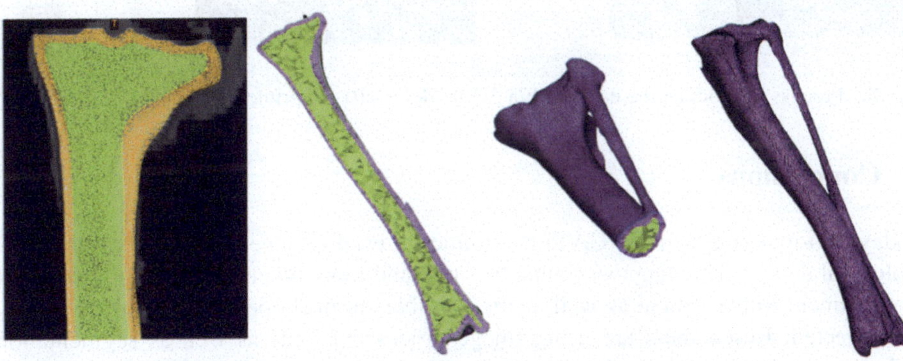

Fig. 7. Trabecular (left) and Cortical (right) mesh generation with Mentat Marc Software

mechanical properties of the trabecular and cortical bone, Young Modulus (E) and Poisson modulus (u) were, respectively: 12000 MPa and 0.3, 100 MPa and 0.2 (Fig. 7).

3D Printing. The.stl format file was imported to the 3D printer. Wanhao Cura software [17] uses the 3D model (stl), and determines the 3D printer toolpath for create the solid tibia according to the selected options. A G-code file is generated and exported out of the program. The file contains a list of the complete X, Y and Z axis coordinates, temperatures, thicknesses, speeds and materials, which are used for print the 3D tibia model. The printing parameters for the 3D model were as follows: Layer height

0.06 mm; Shell thickness 0.8 mm, Enable retraction; Fill thickness 0.6 mm; Fill Density 80%; Print speed 50 mm/s; Printing temperature 210 °C; Bed temperature 50 °C; Filament diameter 1.75 mm. In this study, the 3D printer Wanhao Duplicator 6 produced the object layer by layer. The material used for printing was an ivory-colored PLA (Polylactic acid) that is one of the most commonly used 3D printing filaments since it is a biodegradable plastic due to its natural origin (corn, potato or sugarcane). Also does not produce warping and its resolution printing is excellent. The printing temperature is from 200 to 220 °C (See Fig. 8).

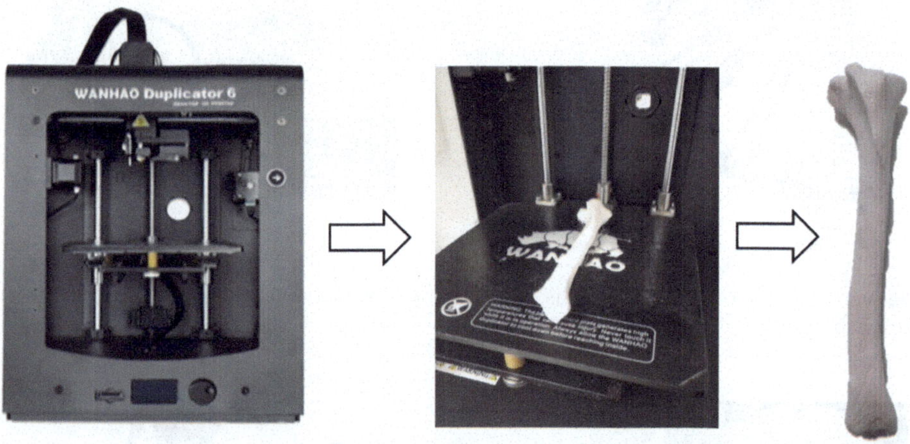

Fig. 8. Process to produce the canine tibia 3D model. (left) 3D printer; (Right) 3D solid model.

3 Conclusions

Today, bio-imaging devices (CAT) is commonly used to generate and reconstruct 3D biological details. The diagnoses must be more and more precise for the application of the treatment to the patient as well as the possible surgical operations to which it may be subjected. Researchers are using image processing tools as well as segmentation techniques to analyze, among other things, bone structures that ultimately generate solid models that can be printed in 3D inexpensively and quickly. Radiographic images are generally used to study biomechanical systems, since they allow: describe and simulate the biomechanical behavior of biological systems, analyze the pathological conditions of biological systems, study the structure and properties of tissues and create a solid model, using 3D printers, which could be used as a reference for surgical planning, which reduces the cost and time of surgery. In this work, the process of obtaining a 3D model printed for educational and research use is shown from a tibia of the corpse of a dog. Two open source software were used in this case (3D Slicer and MeshMixer) with which a virtual model was built (stl files). These.stl files served as the basis for the creation of the mesh for a FE model as well as for the generation of a G-code file for subsequent 3D printing.

Acknowledgments. The authors wish to thank the University of La Rioja, Insorvet veterinary orthopedics (https://www.insorvet.com/en_US/), Albeitar Veterinary Hospital (https://hospitalalbeitar.com/) and to the Autonomous University of Barcelona (AUB).

References

1. Evans, H.E., De Lahunta, A.: Miller's Anatomy of the Dog-E-Book. Elsevier Health Sciences, St. Louis (2013)
2. Gemmill, T.J., Cave, T.A., Clements, D.N., Clarke, S.P., Bennett, D., Carmichael, S.: Treatment of canine and feline diaphyseal radial and tibial fractures with low-stiffness external skeletal fixation. J. Small Anim. Pract. **45**(2), 85–91 (2004)
3. Johnson, A.L., Kneller, S.K., Weigel, R.M.: Radial and tibial fracture repair with external skeletal fixation: effects of fracture type, reduction, and complications on healing. Vet. Surg. **18**(5), 367–372 (1989)
4. Esses, S.J., Berman, P., Bloom, A.I., Sosna, J.: Clinical applications of physical 3D models derived from MDCT data and created by rapid prototyping. Am. J. Roentgenol. **196**(6), W683–W688 (2011)
5. Rengier, F., Mehndiratta, A., Von Tengg-Kobligk, H., Zechmann, C.M., Unterhinninghofen, R., Kauczor, H.U., Giesel, F.L.: 3D printing based on imaging data: review of medical applications. Int. J. Comput. Assist. Radiol. Surg. **5**(4), 335–341 (2010)
6. Chua, C.K., Chou, S.M., Lin, S.C., Eu, K.H., Lew, K.F.: Rapid prototyping assisted surgery planning. Int. J. Adv. Manuf. Technol. **14**(9), 624–630 (1998)
7. Mertz, L.: Dream it, design it, print it in 3-D: what can 3-D printing do for you? IEEE Pulse **4**(6), 15–21 (2013)
8. Lee, M.Y., Chang, C.C., Ku, Y.C.: New layer-based imaging and rapid prototyping techniques for computer-aided design and manufacture of custom dental restoration. J. Med. Eng. Technol. **32**(1), 83–90 (2008)
9. Harrysson, O.L., Hosni, Y.A., Nayfeh, J.F.: Custom-designed orthopedic implants evaluated using finite element analysis of patient-specific computed tomography data: femoral-component case study. BMC Musculoskelet. Disord. **8**(1), 91 (2007)
10. Goerne, H., Rajiah, P.: Computed tomography. In: Right Heart Pathology, pp. 601–612. Springer, Cham (2018)
11. Fedorov, A., Beichel, R., Kalpathy-Cramer, J., Finet, J., Fillion-Robin, J.C., Pujol, S., Bauer, C., Jennings, D., Fennessy, F., Sonka, M., Buatti, J., Aylward, S., Miller, J.V., Pieper, S., Kikinis, R.: 3D Slicer as an image computing platform for the Quantitative Imaging Network. Magn. Reson. Imaging **30**(9), 1323–1341 (2012)
12. Autodesk Meshmixer: http://www.meshmixer.com. Accessed 19 Sept 2019
13. Öchsner, A., Öchsner, M.: The Finite Element Analysis Program MSC Marc/Mentat. Springer, Singapore (2016)
14. Somovilla-Gómez, F., Lostado-Lorza, R., Corral-Bobadilla, M., Escribano García, R.: Improving the process of adjusting the parameters of finite element models of healthy human intervertebral discs by the multi-response surface method. Materials **10**(10), 1116 (2017)
15. Somovilla-Gómez, F., Lostado-Lorza, R., Corral-Bobadilla, M., Escribano-García, R.: Improvement in determining the risk of damage to the human lumbar functional spinal unit considering age, height, weight and sex using a combination of FEM and RSM. Biomech. Model. Mechanobiol. **19**(1), 351–387 (2019)

16. McCartney, W., MacDonald, B., Ober, C.A., Lostado-Lorza, R., Somovilla-Gómez, F.: Pelvic modelling and the comparison between plate position for double pelvic osteotomy using artificial cancellous bone and finite element analysis. BMC Vet. Res. **14**(1), 100 (2018)
17. Wanhao 3D printer: http://wanhao3dprinter.com/Index.html. Accessed 19 Sept 2019

Efficacy of Morpho-Geometrical Analysis of the Corneal Surfaces in Keratoconus Disease According to Moderate Visual Limitation

J. S. Velázquez-Blázquez[1], D. G. Fernández-Pacheco[1], J. Alió del Barrio[2,3], J. L. Alió[2,3], and F. Cavas-Martínez[1](✉)

[1] Department of Structures, Construction and Graphical Expression, Technical University of Cartagena, 30202 Cartagena, Spain
francisco.cavas@upct.es
[2] Keratoconus Unit of Vissum Corporation Alicante, 03016 Alicante, Spain
[3] Department of Ophthalmology, Miguel Hernández University of Elche, 03202 Alicante, Spain

Abstract. The cornea is a complex hemispheric structure, made of collagen fibres that provide it a homogenous and stable geometry. During keratoconus disease, a loss of tenacity takes place in the collagen fibres that form the corneal structure, producing an alteration of its geometry, this is, a change of its curvature, and therefore, a loss of visual quality of patients. The geometric characterization of the hemispheric structure by means of biometric parameters is a very solid technique of diagnosis, based in a virtual 3D model, which has already been validated for several degrees of severity of keratoconus pathology. In this prospective comparative study, 93 corneas (50 healthy subjects and 43 patients with keratoconus with moderate visual limitation) were geometrically modelled. The results obtained in this work suggest that the best predictive biometric parameters are anterior corneal surface area and posterior apex deviation, and that the strongest correlation is produced between sagittal plane apex area in minimum thickness point and sagittal plane apex area. The studied biometric parameters have shown significant differences between groups. Therefore, the analysis of the biometric parameters that register the geometric decompensation that locally appear in a corneal region, as a response to the asymmetry produced during the development of keratoconus disease with a moderate visual impairment, is a new approach that may lead to a better understanding of the disease with this degree of optical limitation.

Keywords: Computer-Aided Geometric Design (CAGD) · Optical Aberrometry · Scheimpflug · 3D modelling

1 Introduction

The cornea is an avascular structure with a hemispheric shape that is part of the anterior segment of the human eye [1]. It is a highly differenced tissue with five layers which have different functions, being the most important the one of the first ocular refractive element, due to its properties of transparency and curvature along all the structure [2].

F. Cavas-Martínez et al. (Eds.): INGEGRAF 2019, LNME, pp. 263–272, 2020.
https://doi.org/10.1007/978-3-030-41200-5_29

One of the main pathologies of the cornea is keratoconus, which is a clinical term that describes a corneal condition characterized by its progressive thinning, which causes a cone-shaped corneal protrusion [3]. This degenerative geometrical deformation leads to a decrease of cornea's optical quality.

The knowledge, from a clinical point of view, of the geometrical characteristics of the cornea is important in clinical practice [1]. There are several studies in scientific literature which have validated indexes or topographic descriptors of the cornea [3, 4] in order to define the cut off values that allow the discrimination between normal and keratoconus corneas. From this indexes, several keratoconus classifications have been proposed, such as a morphological classification [1], keratometric [5], of keratoconus severity [6], or Amsler-Krumeich [1]. However, none of these classifications measures the degree of severity of keratoconus depending on the visual limitation level.

In addition, computer-aided design is a widespread tool for biological structure modelling [7–9], and its later use for disease diagnosis [10, 11]. Our research group have developed some new virtual geometric models of the human cornea that have been used to characterize corneal ectasia disease basing on Amsler-Krumeich classification [12–14], and for other classification, named RETICS [15], which allows characterizing keratoconus' pathology considering visual performance of patients in its initial phases [16, 17].

This investigation work proposes a way to characterize cornea's morphogeometry in keratoconus pathology with a moderate degree of visual performance, and also to quantify the existence of correlations between the geometries of corneal surfaces for this degree of severity of the disease.

2 Materials and Methods

2.1 Participants

This observational case series study included 93 corneas of 93 patients (selected at random to avoid interference) split into two groups: a "normal" group of healthy individuals, that included 50 subjects which presented no ocular pathology (36.80 $\pm$ 15.67 years); and a "moderate KC" group, composed of 43 individuals diagnosed with moderate KC (46.67 $\pm$ 24.99 years). The classification procedure to discriminate between normal and moderate KC cases was run according to well-known and widespread clinical and topographic evaluations [18].

All patients were selected according to the RETICS grading. The criteria followed for patient inclusion were: patients diagnosed as Grade III KC eyes (moderate visual limitation, 6/15 < CDVA $\leq$ 6/9.5 in Snellen scale, or 0.4 < CDVA $\leq$ 0.6 in decimal), corneal topography showing a localized steepening, and/or a pattern of asymmetric bowtie shape, with or without skewed radial axis. The exclusion criteria were the following: wearing contact lenses during a month before the first visit, moderate to severe dry eye, ocular surface irritation or other active ocular comorbidity, or any previous ocular surgery.

These evaluations were made at Vissum Instituto Oftalmologico Alicante, Spain (Vissum) whose patients were adequately informed about the clinical study and voluntarily signed their consent to participate. The study was authorised by the clinic's Institutional Review Board, according to the tenets of the Declaration of Helsinki (7th revision, October 2013, Fortaleza, Brazil).

2.2 Examination Protocol

Every subject selected for this study was examined by means of a Sirius System® tomographer (CSO, Florence, Italy), under the directives of a validated protocol previously created by our research group. This protocol has been profusely explained in previous studies [12–14], and it has proved its efficiency when it comes to diagnose and characterize KC in asymptomatic (pre-clinical) and mild visually-impaired eyes [16, 17].

After applying this protocol, the outcome obtained is a patient-specific 3D virtual model of the cornea, which is then analysed to stablish several biometric parameters (Fig. 1). These biometric parameters studied herein, along with their characteristics, have been previously described in [19], and are summarized in Table 1, but this is the first time that they have been used to study KC eyes with moderate visual impairment. In this study, the Rhinoceros' surface was deformed to minimise the nominal distance between the spatial points and the surface. This deviance was later estimated by the software, giving a mean value of the distance error for the solution surface of $5.560 \times 10^{-17} \pm 5.81 \times 10^{-17}$ mm (mean ± standard deviation).

Table 1. Biometric parameters analysed in the study [19].

Biometric parameter	Description
Total corneal volume (Vol_{tot}) [mm^3]	Volume limited by front, back and peripheral surfaces of the solid model generated
Anterior/posterior corneal surface area ($Corarea_{ant}$/$Corarea_{pos}$) [mm^2]	Area of the front/exterior and rear/interior surfaces
Total corneal surface area ($Corarea_{tot}$) [mm^2]	Sum of anterior, posterior and perimeter corneal surface areas of the solid model generated
Sagittal plane apex area ($Splarea_{papex}$) [mm^2]	Area of the cornea within the sagittal plane passing through the optical axis and the highest point (apex) of the posterior corneal surface
Anterior and posterior apex deviation (Dev_{aapex}/Dev_{papex}) [mm]	Average distance from the optical axis to the highest point (apex) of the anterior/posterior corneal surfaces
Sagittal Plane Area in minimum thickness point ($Splarea_{minthk}$) [mm^2]	Area of the cornea within the sagittal plane passing through the optical axis and the minimum thickness point (maximum curvature) of the posterior corneal surface
Anterior and posterior minimum thickness point deviation ($Dev_{aminthk}$/$Dev_{pminthk}$) [mm]	Average distance in the XY plane from the optical axis to the minimum thickness points (maximum curvature) of the anterior/posterior corneal surfaces
Centre of mass X, Y, Z (COM_X, COM_Y, COM_Z) [mm]	Centre of mass coordinates X, Y, Z of the solid

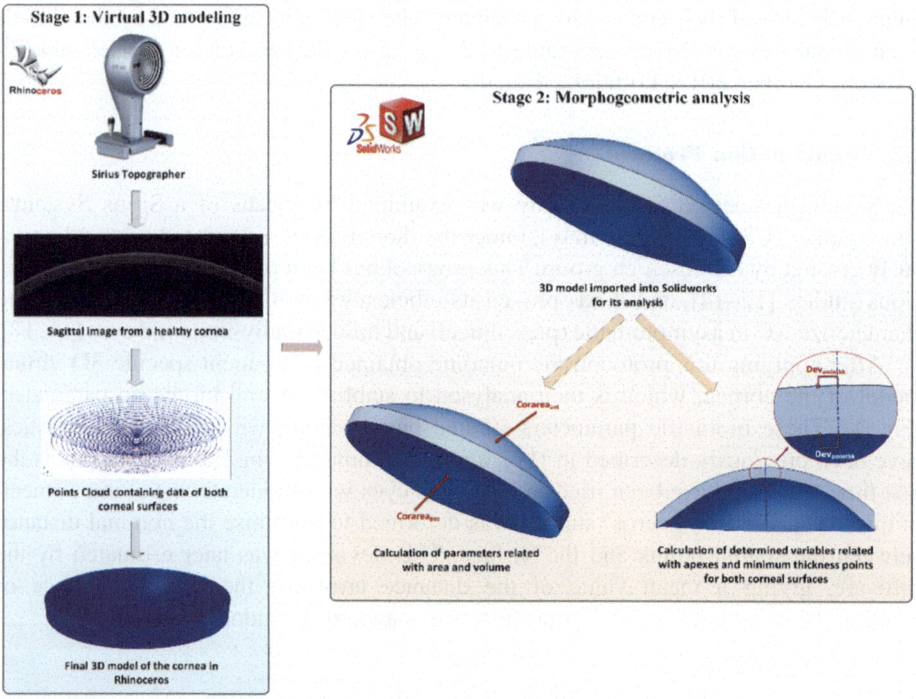

Fig. 1. Scheme of the procedure for the generation of a virtual model and later analysis of the corneal structure.

2.3 Statistical Analysis

Data normality was checked by means of Kolmogorov-Smirnov and Shapiro-Wilks tests. Depending on the results of these tests, Student's T-test or U-Mann Whitney Wilcoxon tests were used, when appropriate. Pearson coefficients were used to assess the correlation between data with normal distributions, while Spearman coefficients were used for not normally distributed ones. The level of significance for p-values fixed for all statistical tests was 0.05. Receiver-operating characteristics (ROC) curves were the method chosen to find out which parameters better characterized diseased corneas, fixing optimal cut-offs by means of Youden's J index, considering sensitivity and specificity values [20, 21]. The software used to perform all analyses were Graphpad Prism V6 (GraphPad Software, La Jolla, USA) and IBM SPSS V 23.0 software (SPSS, Chicago, USA).

3 Results

The majority of analyzed parameters presented statistically significant differences when comparing healthy eyes with moderate KC ones, as shown in Table 2 below.

Table 2. Descriptive values and differences in the modelled biometric parameters among the normal and moderate KC groups. SD: standard deviation. P: statistical test, Z: z-score.

Biometric parameters	Normal group (n = 50)				Moderate KC group (n = 43)				z	P
	Mean	SD	Min	Max	Mean	SD	Min	Max		
Vol_{tot} (mm^3)	25.71	1.53	23.23	29.07	23.97	1.52	21.06	28.74	−5.16	0.000
$Corarea_{ant}$ (mm^2)	43.08	0.14	42.77	43.33	43.67	0.47	42.79	45.09	−7.37	0.000
$Corarea_{pos}$ (mm^2)	44.24	0.26	43.53	44.71	45.22	0.80	44.17	47.53	−7.24	0.000
$Corarea_{tot}$ (mm^2)	103.87	1.12	100.73	105.66	104.91	1.70	101.91	109.95	−3.47	0.001
$Splarea_{papex}$ (mm^2)	4.32	0.26	3.93	4.87	3.93	0.29	3.19	4.83	6.72	0.000
$Splarea_{minthk}$ (mm^2)	4.31	0.26	3.92	4.86	3.92	0.29	3.19	4.82	6.61	0.000
Dev_{aapex} (mm)	0.00	0.00	0.00	0.01	0.02	0.00	0.00	0.07	−7.57	0.000
Dev_{papex} (mm)	0.07	0.02	0.04	0.09	0.21	0.08	0.03	0.34	−7.37	0.000
COM_X (mm)	0.04	0.02	0.01	0.09	0.02	0.05	−0.15	0.20	3.39	0.001
COM_Y (mm)	0.03	0.02	0.00	0.08	0.00	0.05	−0.14	0.16	−5.14	0.000
COM_Z (mm)	0.77	0.02	0.71	0.81	0.83	0.06	0.72	1.00	−6.26	0.000
$Dev_{aminthk}$ (mm)	0.83	0.21	0.44	1.27	0.81	0.09	0.35	1.81	0.37	0.713
$Dev_{pminthk}$ (mm)	0.76	0.20	0.38	1.24	0.74	0.26	0.30	1.31	0.57	0.570

3.1 Roc Analysis

A ROC analysis was used to ascertain the predictive value of the modelled parameters (Fig. 2). Five biometric parameters showed an area under the ROC (AUROC) above 0.7 (Table 3).

Table 3. The area under the ROC results.

Biometric parameters	AUROC	Sensitivity	Specificity	Cut off value
$Corarea_{ant}$	0.945	86.0	96.0	≥ 43.2595 mm^2
$Corarea_{pos}$	0.937	83.7	96.0	≥ 44.6205 mm^2
Dev_{aapex}	0.910	83.7	94.0	≥ 0.0010 mm
Dev_{papex}	0.945	93.0	99.9	≥ 0.0965 mm
COM_Z	0.878	69.8	98.0	≥ 0.0809 mm

Table 4 shows a summary of all correlations between the moderate KC group's modelled biometric parameters found significant. Mutual relations between normal group's parameters were addressed in a previous study [16], so they have not been included in this one.

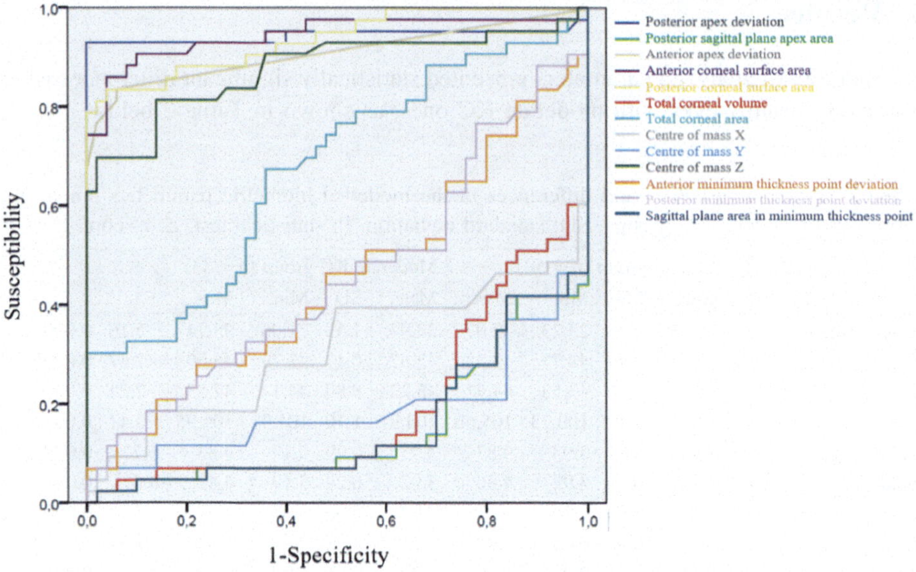

Fig. 2. Curves for modelled parameters detecting moderate KC.

Table 4. The significant correlation coefficient values for the modelled variables in the moderate KC group.

Measurement correlation	Moderate KC group (n = 43)	
	Correlation coefficient	P value
$Splarea_{papex}/Vol_{tot}$	0.965	0.000
$Corarea_{ant}/Dev_{papex}$	0.313	0.041
$Corarea_{pos}/Dev_{papex}$	0.361	0.017
$Corarea_{ant}/Corarea_{pos}$	0.941	0.000
$Corarea_{ant}/Corarea_{tot}$	0.778	0.000
$Corarea_{pos}/Corarea_{tot}$	0.864	0.000
$Vol_{tot}/Corarea_{tot}$	0.354	0.020
$COM_X/Corarea_{ant}$	−0.340	0.025
$COM_X/Corarea_{pos}$	−0.317	0.038
$COM_X/Corarea_{tot}$	−0.361	0.017
COM_X/COM_Z	−0.369	0.015
$COM_Z/Corarea_{ant}$	0.938	0.000
$COM_Z/Corarea_{pos}$	0.882	0.000
$COM_Z/Corarea_{tot}$	0.844	0.000
$Dev_{aminthk}/Dev_{papex}$	0.440	0.003
$Dev_{pminthk}/Dev_{papex}$	0.532	0.000
$Dev_{aminthk}/Dev_{pminthk}$	0.979	0.000
$Splarea_{minthk}/Splarea_{papex}$	0.997	0.000
$Vol_{tot}/Splarea_{minthk}$	0.963	0.000
Dev_{aapex}/Dev_{papex}	0.403	0.007

4 Discussion

The cornea is a complex hemispheric structure, made of collagen fibres that provide it a homogenous and stable geometry. In a pathological scenario, a loss of tenacity takes place in the collagen fibres that form the corneal structure, producing an alteration of its geometry, this is, a change of its curvature, and therefore, a loss of visual quality of patients [22].

There are in scientific literature several works that have studied the geometry of corneas with a moderate degree of optical aberration [13, 23, 24]. Several studies [23] observed that corneal thickness is not constant in this pathology, as a thinning is produced in the region in which the loss of structural tenacity appears. These results are consistent with the ones obtained by other researchers [24] that analysed the thickness of corneas with an advanced degree of optical aberration. Other recent study has demonstrated that some morphological singularities as well as corneal thickness are related with disease progression [13]. Therefore, the cornea in this pathology shows a region of geometrical decompensation as a non-symmetric response to the disease progress. However, the parametrization and study of the correlation between the biometric parameters that register this decompensation, which allow to evaluate the progression of keratoconus with moderate visual loss, have not been analysed for each patient from a virtual model of their own eye.

In this study, all measures made have resulted consistent, and almost all mean differences have been statistically significant between healthy corneas and corneas with keratoconus showing a moderate degree of visual impair, except for the anterior/posterior minimum thickness point deviation. These results are in accordance with the ones reported by previous studies that evaluated some of these anatomic parameters with equipment based in Scheimpflug technology [25]. In our study, we have used a technology based in a double rotating camera based in the Scheimpflug projection system. This system has shown, in a previous study [26], being a robust and repeatable system for spatial data acquisition of the hemispheric structure, from which basing the generation of the virtual model with CAD software [17].

Regarding the AUROC, four parameters from all the ones analysed in our study showed very high discriminant values (AUROC > 0.9). Two parameters, which are the anterior and posterior corneal surface areas showed AUROC values of 0.937 y 0.945 respectively, due to the fact that in the pathological scenario a biomechanical weakening takes place, caused by the loss of tenacity of the hemispheric structure and consequently, an increase of the corneal surface curvature is produced. This tendency is in line with some previous works that made ratio studies between anterior and posterior areas for different degrees of severity of the disease [25]. The other two parameters with high discrimination values were anterior apex deviation (AUROC $= 0.910$) and posterior apex deviation (AUROC $= 0.945$), this is because the corneal apex is the point of maximum curvature in corneal surfaces, therefore, according with other authors [27], the apex deviation is a parameter that could affect the visual performance of the patients.

In this study, we have analysed the correlation between biometric parameters that register the structural alteration which manifests during the disease development. So, the parameters related with the areas show a strong positive correlation, specifically between anterior and posterior surface areas (R^2 = 0.941, p = 0.000), anterior surface area and total corneal surface area (R^2 = 0.778, p = 0.000) and posterior and total corneal surface areas (R^2 = 0.864, p = 0.000), due to the fact that corneal surfaces tend to behave proportionally during the evolution of the disease. These results are consistent with the ones obtained for corneal curvature in eyes with keratoconus, with the biomechanical weakening of the cornea, and with the possible effect of intraocular pressure in the weakened corneal structure [28].

Regarding anterior and posterior deviations, both between corneal apex points (R^2 = 0.832, p = 0.000) and minimum thickness points (R^2 = 0.997, p = 0.000), there is a strong positive correlation, this is in line with the results reported by other authors [25] that demonstrated a robust correlation between apex deviation and the minimum thickness points and their relation with the pachymetric progression of corneal thickness.

5 Conclusion

The analysis of the biometric parameters that register the geometric decompensation that locally appear in a corneal region, as a response to the asymmetry produced during the development of keratoconus disease with a moderate degree of optical aberration, is a procedure that provides the ophthalmologist a new tool for the diagnosis of this pathology with this degree of development. In this work it has been ascertained that the most predictive biometric parameters are the anterior corneal surface area and the deviation of the posterior apex, detecting that the strongest correlation is the one between the areas of the sagittal planes defined through the minimum thickness points and the posterior corneal apex. The studied biometric parameters showed significant differences between groups. Therefore, this new approach based in the personalised diagnosis from a computational model may lead to a better understanding of the disease with this degree of optical limitation.

Funding. This publication has been carried out in the framework of the Thematic Network for Co-Operative Research in Health (RETICS) reference number RD16/0008/0012 financed by the Carlos III Health Institute-General Subdirection of Networks and Cooperative Investigation Centres (R&D&I National Plan 2013–2016) and the European Regional Development Fund (FEDER).

Conflict of Interest. The authors have no conflict of interest to declare.

Financial Disclosure. Neither author has a financial or proprietary interest in any material or method mentioned.

References

1. Albertazzi, R.: Queratocono: pautas para su diagnóstico y tratamiento. Buenos Aires: Ediciones Científicas Argentinas **6**, 123–135 (2010)
2. Buey Salas, M., Peris, M.: Biomecánica y arquitectura corneal. Elsevier, Spain (2014)
3. Cavas-Martínez, F., De la Cruz Sánchez, E., Martínez, J.N., Cañavate, F.F., Fernández-Pacheco, D.: Corneal topography in keratoconus: state of the art. Eye Vis. **3**(1), 5 (2016)
4. Piñero, D.P., Alió, J.L., Alesón, A., Vergara, M.E., Miranda, M.: Corneal volume, pachymetry, and correlation of anterior and posterior corneal shape in subclinical and different stages of clinical keratoconus. J. Cataract Refract. Surg. **36**(5), 814–825 (2010)
5. Zadnik, K., Barr, J.T., Edrington, T.B., Everett, D.F., Jameson, M., McMahon, T.T., Shin, J. A., Sterling, J.L., Wagner, H., Gordon, M.O.: Baseline findings in the Collaborative Longitudinal Evaluation of Keratoconus (CLEK) study. Invest. Ophthalmol. Vis. Sci. **39** (13), 2537–2546 (1998)
6. McMahon, T.T., Szczotka-Flynn, L., Barr, J.T., Anderson, R.J., Slaughter, M.E., Lass, J.H., Iyengar, S.K., Group CS: A new method for grading the severity of keratoconus: the Keratoconus Severity Score (KSS). Cornea **25**(7), 794–800 (2006)
7. Ukwatta, E., Arevalo, H., Rajchl, M., White, J., Pashakhanloo, F., Prakosa, A., Herzka, D. A., McVeigh, E., Lardo, A.C., Trayanova, N.A., Vadakkumpadan, F.: Image-based reconstruction of three-dimensional myocardial infarct geometry for patient-specific modeling of cardiac electrophysiology. Med. Phys. **42**(8), 4579–4590 (2015). https://doi.org/10.1118/1.4926428
8. Henckel, J., Holme, T.J., Radford, W., Skinner, J.A., Hart, A.J.: 3D-printed patient-specific guides for hip arthroplasty. J. Am. Acad. Orthop. Surg. **26**(16), e342–e348 (2018). https://doi.org/10.5435/JAAOS-D-16-00719
9. VanKoevering, K.K., Zopf, D.A., Hollister, S.J.: Tissue engineering and 3-dimensional modeling for facial reconstruction. Facial Plast. Surg. Clin. North Am. **27**(1), 151–161 (2019). https://doi.org/10.1016/j.fsc.2018.08.012
10. Liew, Y.M., Khalid, A., Tan, L.K., Lim, E., Chan, B.T., Md Sari, N.A.B., Chee, K.H., Abdul Aziz, Y.F.: Spatial cardiac dysfunction assessment via personalized modelling from MRI. In: 2018 IEEE EMBS Conference on Biomedical Engineering and Sciences, IECBES 2018 - Proceedings, pp. 69–74 (2019). https://doi.org/10.1109/iecbes.2018.8626668
11. Naser, M.A., Sayed, A.M., Wahba, A.A., Eldosoky, M.A.A.: Modeling procedures for breast cancer diagnosis based on clinical elastography images. In: Proceedings - 2018 13th International Conference on Computer Engineering and Systems, ICCES 2018, pp. 671–677 (2019). https://doi.org/10.1109/icces.2018.8639338
12. Cavas-Martínez, F., Bataille, L., Fernández-Pacheco, D.G., Cañavate, F.J.F., Alio, J.L.: Keratoconus detection based on a new corneal volumetric analysis. Sci. Rep. **7**(1) (2017). https://doi.org/10.1038/s41598-017-16145-3
13. Cavas-Martínez, F., Bataille, L., Fernández-Pacheco, D.G., Cañavate, F.J.F., Alió, J.L.: A new approach to keratoconus detection based on corneal morphogeometric analysis. PLoS One **12**(9) (2017). https://doi.org/10.1371/journal.pone.0184569
14. Cavas-Martínez, F., Fernández-Pacheco, D.G., De La Cruz-Sánchez, E., Nieto Martínez, J., Fernández Cañavate, F.J., Vega-Estrada, A., Plaza-Puche, A.B., Alió, J.L.: Geometrical custom modeling of human cornea in vivo and its use for the diagnosis of corneal ectasia. PLoS One **9**(10) (2014). https://doi.org/10.1371/journal.pone.0110249

15. Alio, J.L., Pinero, D.P., Aleson, A., Teus, M.A., Barraquer, R.I., Murta, J., Maldonado, M.J., Castro de Luna, G., Gutierrez, R., Villa, C., Uceda-Montanes, A.: Keratoconus-integrated characterization considering anterior corneal aberrations, internal astigmatism, and corneal biomechanics. J. Cataract Refract. Surg. **37**(3), 552–568 (2011). https://doi.org/10.1016/j.jcrs.2010.10.046
16. Cavas-Martínez, F., Fernández-Pacheco, D.G., Cañavate, F.J.F., Velázquez-Blázquez, J.S., Bolarín, J.M., Alió, J.L.: Study of morpho-geometric variables to improve the diagnosis in Keratoconus with mild visual limitation. Symmetry **10**(8) (2018). https://doi.org/10.3390/sym10080306
17. Cavas Martinez, F., Garcia Fernandez-Pacheco, D., Fernandez Cañavate, F.J., Velázquez Blázquez, J.S., Bolarin, J.M., Tiveron, M., Alio, J.L.: Detección De Queratocono Temprano Mediante Modelado 3D Personalizado Y Análisis De Sus Parámetros Geométricos. Dyna Ingenieria E Industria **94**(1), 175–181 (2019). https://doi.org/10.6036/8895
18. Huseynli, S., Salgado-Borges, J., Alio, J.L.: Comparative evaluation of Scheimpflug tomography parameters between thin non-keratoconic, subclinical keratoconic, and mild keratoconic corneas. Eur. J. Ophthalmol. **28**(5), 521–534 (2018). https://doi.org/10.1177/1120672118760146
19. Cavas-Martínez, F., Fernández-Pacheco, D.G., Parras, D., Cañavate, F.J.F., Bataille, L., Alió, J.: Study and characterization of morphogeometric parameters to assist diagnosis of keratoconus. BioMed. Eng. Online **17** (2018). https://doi.org/10.1186/s12938-018-0564-7
20. Lasko, T.A., Bhagwat, J.G., Zou, K.H., Ohno-Machado, L.: The use of receiver operating characteristic curves in biomedical informatics. J. Biomed. Inform. **38**(5), 404–415 (2005). https://doi.org/10.1016/j.jbi.2005.02.008
21. Pepe, M.S.: The Statistical Evaluation of Medical Tests for Classification and Prediction. Medicine (2003)
22. Piñero, D.P.: Technologies for anatomical and geometric characterization of the corneal structure and anterior segment: a review. Semin. Ophthalmol. **30** (2015). https://doi.org/10.3109/08820538.2013.835844
23. Anderson, K., El-Sheikh, A., Newson, T.: Application of structural analysis to the mechanical behaviour of the cornea. J. R. Soc. Interface **1**(1), 3–15 (2004). https://doi.org/10.1098/rsif.2004.0002
24. Gefen, A., Shalom, R., Elad, D., Mandel, Y.: Biomechanical analysis of the keratoconic cornea. J. Mech. Behav. Biomed. Mater. **2**(3), 224–236 (2009). https://doi.org/10.1016/j.jmbbm.2008.07.002
25. Montalbán, R., Alio, J.L., Javaloy, J., Piñero, D.P.: Correlation of anterior and posterior corneal shape in keratoconus. Cornea **32** (2013). https://doi.org/10.1097/ico.0b013e3182904950
26. Cavas-Martínez, F., Fernández-Pacheco, D.G., Cruz-Sánchez, E., Nieto Martínez, J., Fernández Cañavate, F.J., Vega-Estrada, A.: Geometrical custom modeling of human cornea in vivo and its use for the diagnosis of corneal ectasia. PLoS One **9** (2014). https://doi.org/10.1371/journal.pone.0110249
27. Tomidokoro, A., Oshika, T., Amano, S., Higaki, S., Maeda, N., Miyata, K.: Changes in anterior and posterior corneal curvatures in keratoconus. Ophthalmology **107**(7), 1328–1332 (2000)
28. Piñero, D.P., Alió, J.L., Aleson, A., Escaf Vergara, M., Miranda, M.: Corneal volume, pachymetry, and correlation of anterior and posterior corneal shape in subclinical and different stages of clinical keratoconus. J. Cataract Refract. Surg. **36** (2010). https://doi.org/10.1016/j.jcrs.2009.11.012

Innovation in Design

Experimental and Analytical Approaches for Determining the Axial Behavior of Cylindrical Extension Springs

Manuel Paredes[1(✉)], Thomas Stephan[2], and Hervé Orcière[2]

[1] ICA, Université de Toulouse, UPS, INSA, ISAE-SUPAERO, MINES-ALBI, CNRS, 3 rue Caroline Aigle, 31400 Toulouse, France
manuel.paredes@insa-toulouse.fr
[2] CGR International, 1 rue Frederic Joliot Curie, 93274 Sevran, France

Abstract. Cylindrical springs have been commonly exploited in mechanical systems for years. While the behavior of compression springs can be considered as well identified, the understanding of the behavior of extension springs still needs improvement as it appears that the influence of the loops on the global stiffness is not yet taken into account properly. Moreover, it would be of key interest to analyze how initial tension in extension springs influences the beginning of the load-length curve. The paper investigates these topics using analytical and experimental approaches in order to help engineers design extension springs with greater accuracy. Several loop geometries are tested and Castigliano's theorem is exploited to propose an analytical formula for the stiffness of each. Experimental tests have been performed on a set of extension springs with non-basic loops in order to evaluate the accuracy of the formulae.

Keywords: Extension springs · Spring design · Springs · Initial tension

1 Introduction

Cylindrical extension springs are commonly exploited to store and return energy in mechanical devices. Most of the help available to designers, European standards (EN 13906-2) or industrial software, such as Advanced Spring Design or Spring Calculator Professional, exploit common analytical formulae from the reference book on spring design written several years ago by Wahl [1]. Since that time, a few works on fatigue resistance [2], optimal design [3] and simulation using FEA [4] have improved our knowledge of extension springs. Hopefully, recent works related to compression springs may also be of key interest for our study. Paredes [5] and Dym [6] have performed experimental and analytical studies that may be successfully extended to cylindrical extension springs. Pretension springs are commonly designed with an initial tension (i.e. the coils of the body are manufactured with an axial pitch smaller than the wire diameter). The associated load-deflection curve and the main parameters are described in Fig. 1. We can see that, at the very beginning of deflection, a smooth curve can appear. This is mainly due to the behavior of the loops. We propose to

F. Cavas-Martínez et al. (Eds.): INGEGRAF 2019, LNME, pp. 275–283, 2020.
https://doi.org/10.1007/978-3-030-41200-5_30

develop both experimental and analytical approaches in order to increase our understanding of extension springs at the beginning of deflection.

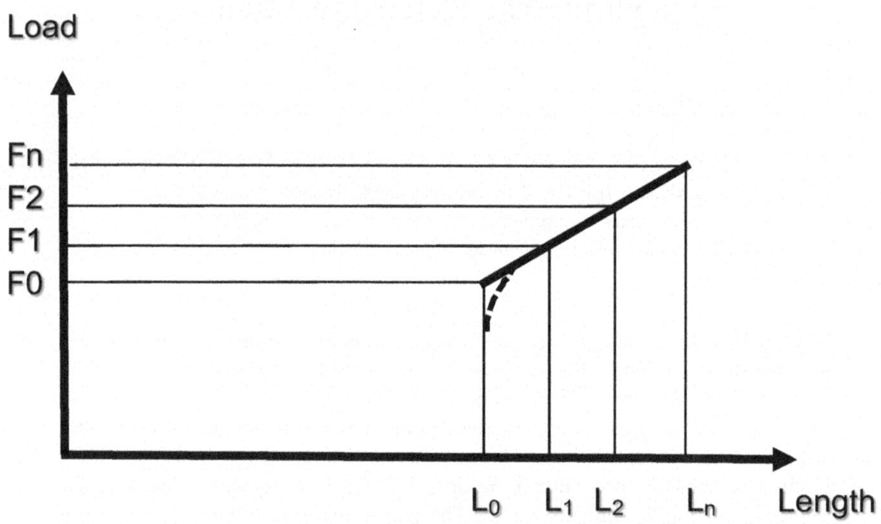

Fig. 1. Load-length relation for extension springs, from European standards.

2 Study of an Extension Spring with Extended Loops

2.1 Details of the Extension Spring

In order to highlight the influence of loops on the global load-length curve, an extension spring with extended loops is studied (see Fig. 2). This spring is used in the automotive industry for the wiping system.

The left loop is a common extended machine loop while the right loop is more extended with an intermediate curvature.

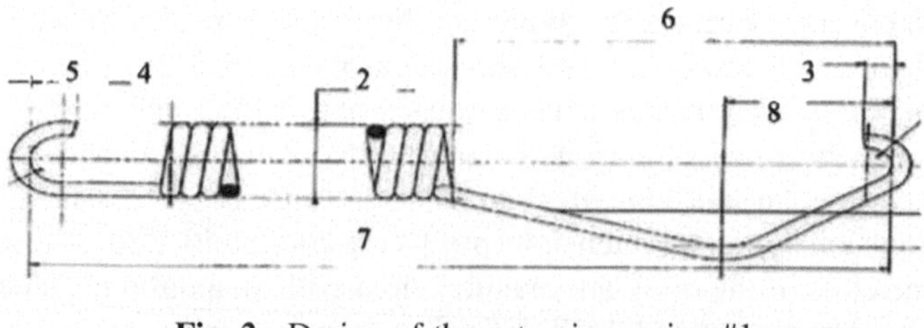

Fig. 2. Design of the extension spring #1

The spring is made of steel (G = 81500 MPa, E = 206000 MPa). The wire diameter, d, is 2.10 mm, the external diameter, Do, is 12.67 mm, the body of the spring includes n = 12 coils and the initial tension, P0, is 15 N. Consequently, the standard formula for extension springs induces a rate of k = 14.3 N/mm.

2.2 Experimental Test Bench

A test bench from Andilog (Spring Test 1, see Fig. 3) was exploited to test four specimens of the studied spring. The force gauge had 500 N capacity, with 0.1% accuracy and 0.04 N resolution. The displacement transducer had a resolution of 0.01 mm.

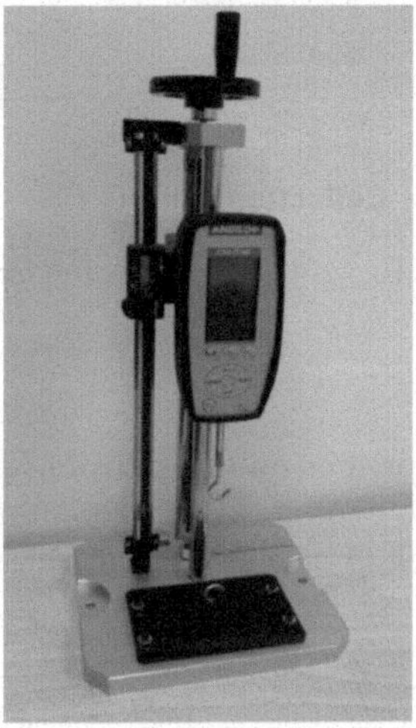

Fig. 3. Andilog Spring test 1.

2.3 Experimental Results

The results obtained are presented in Fig. 4. In this case, it appears that the model proposed by standards differs markedly from the experimental results. A first linear part is highlighted, where the loops deflect while the coils of the body are still in contact due to the initial tension. Moreover, the global rate of the spring (average 10.4 N/mm) is significantly lower (−27%) than expected in the standards.

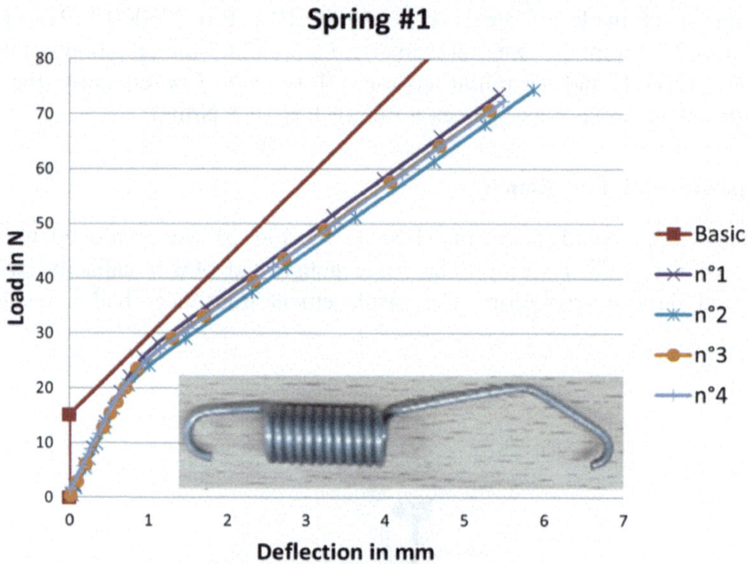

Fig. 4. Experimental results for spring #1.

2.4 Analytical Study

Shimoseki [4] has shown that pretension of extension springs can hardly be simulated in 3D FEA. We thus propose to exploit Castigliano's theorem, as done by Dym [6], in an advanced analytical approach that determines the load-deflection curve of extension springs.

$$\frac{dU^*}{ds} = \frac{1}{2}\left[\frac{N^2}{ES} + \frac{T^2}{GS} + \frac{M^2}{EI} + \frac{To^2}{GJ}\right] \text{ and } F_L = \frac{1}{F} \times \frac{dU^*}{dF}$$

To determine the flexibility, FL, the normal load, N, transversal load, T, bending moment, M, and torque. To, have to be expressed as functions of the external axial load, F, all along the geometry that is considered. In the case under study, each loop is modeled in four parts. Figure 5 describes the parameters used to design the loop (center of the wire).

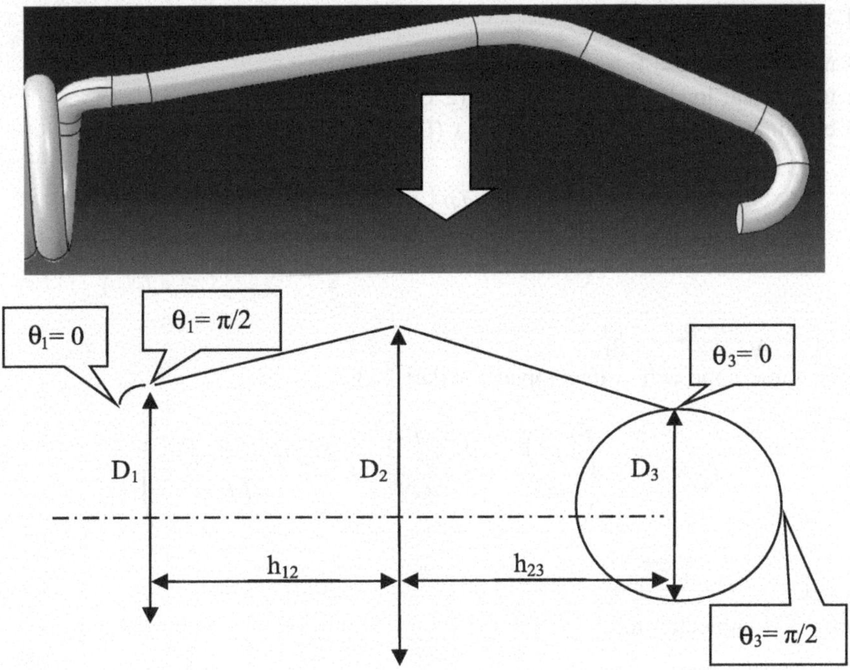

Fig. 5. Model of the loop

In part 1, the local radius R1 is assumed to be small, so the geometry can be considered as contained in a plane parallel to the axis of the spring. Thus in Part 1, an axial load, F, on the spring induces:

a normal load $N = F \sin(\theta_1)$
a transversal load $T = F \cos(\theta_1)$
a bending moment $M = F R_1 (1 - \sin(\theta_1))$
a torque $T_o = \frac{1}{2} F D_1$

such that

$$F_{L1} = R_1 \int_0^{\pi/2} \left(\frac{\sin^2\theta_1}{ES} + \frac{\cos^2\theta_1}{GS} + \frac{R_1^2 \left(1 + \sin^2\theta_1 - 2\sin\theta_1\right)_1}{EI} + \frac{D_1^2}{4GJ} \right) d\theta_1$$

with $J = \frac{\pi d^4}{64}$; $I = \frac{\pi d^4}{32}$; $S = \frac{\pi d^2}{4}$
Thus

$$F_{L1} = R_1 \left(\frac{\pi}{4ES} + \frac{\pi}{4GS} + \frac{R_1^2(3\pi - 2)}{EI} + \frac{\pi D_1^2}{8GJ} \right) \tag{1}$$

Part 1–2 is subjected to:

a normal load N = F h_{12}/L_{12}
a transversal load T = F (D_2 − D_1)/(2 L_{12})
a bending moment M = ½ F (D_1 + x (D_2 − D_1)/L_{12} with $0 \leq x \leq L_{12}$

$$F_{L12} = L_{12}\left(\frac{\left(\frac{h_{12}}{L_{12}}\right)^2}{ES} + \frac{\left(\frac{D_2-D_1}{h_{12}}\right)^2}{4GS} + \frac{D_1D_2 + \frac{(D_2-D_1)^2}{3}}{4EI}\right) \tag{2}$$

with $L_{12} = \sqrt{\frac{(D_2-D_1)^2}{4} + h_{12}^2}$

The same approach is developed for Part 2–3

$$F_{L23} = L_{23}\left(\frac{\left(\frac{h_{23}}{L_{23}}\right)^2}{ES} + \frac{\left(\frac{D_3-D_2}{h_{23}}\right)^2}{4GS} + \frac{D_2D_3 + \frac{(D_3-D_2)^2}{3}}{4EI}\right) \tag{3}$$

with $L_{23} = \sqrt{\frac{(D_3-D_2)^2}{4} + h_{23}^2}$

Part 3 is subjected to:

a normal load N = F cos(θ_2)
a transversal load T = F sin(θ_2)
a bending moment M = ½ F D_1 cos(θ_2)

such that

$$F_{L3} = \frac{\pi D_1}{8}\left(\frac{1}{ES} + \frac{1}{GS} + \frac{D_1^2}{4EI}\right) \tag{4}$$

The four local flexibilities can be added to obtain the global flexibility of the loop, FL:

$$F_L = F_{L1} + F_{L12} + F_{L23} + F_{L3}$$

The flexibility of each loop and the flexibility of the body can be added to obtain the global rate of the spring. In this case, the number of coils of the body is decreased by R1/(πD) to take the rapid transition of each loop into account as this transition is already included in the calculation of the flexibility of the loop (part1).

This approach also enables the spring behavior to be estimated at the very beginning of the deflection ki, when the coils of the body are still in contact due to the pretension. Half a coil of the body is added to consider the initial flexibility of the body (without pretension) and the flexibilities of the loops are referenced as FLi and FLj.

$$k_i = \frac{1}{\frac{4D^3}{Gd^4} + F_{Li} + F_{Lj}} \tag{5}$$

2.5 Results and Discussion

The spring loops can be modeled with the following parameter values (Table 1):

Table 1. Parameters of the loops (in mm).

Loop	R_1	D_1	h_{12}	D_2	h_{23}	D_3
#1	3	10.56	4.5	10.56	4.5	10.56
#2	3	10.56	26	24.23	14.6	7.11

According to the proposed analytical approach, the spring can thus be modeled with bilinear behavior and the results are presented in Fig. 6.

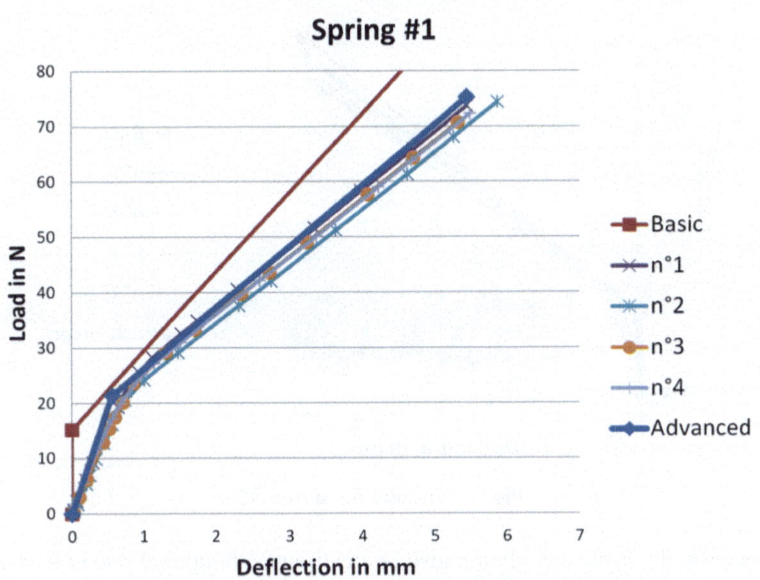

Fig. 6. Comparison with the analytical study

The proposed approach exhibits a global rate of 11.0 N/mm (+6% error with respect to experimental data) that is clearly more accurate than the one commonly proposed by standards and can be a useful time-saver for designers. Moreover, the analytical initial rate of 38.3 N/mm is quite a good approximation of the experimental one of 29.7 N/mm (+30%). For this particular spring, the loop flexibility is approximately equivalent to having 2.8 more coils in the body of the spring. Estimating the loop flexibility is thus a key issue to improve the accuracy of analytical calculations.

We can also see that the tested spring exhibits a rapid transition between the two linear behaviors. The next section highlights the influence of this transition on the global behavior.

3 Study of an Extension Spring with a Smooth Decrease of Pretension in the Body

The second spring studied is used in the automotive industry for the wiping system and presents a smooth transition between the two linear behaviors. The spring is made of steel (G = 81500 MPa, E = 206000 MPa). The wire diameter is d = 3.20 mm, the external diameter is Do = 13.36 mm, the body of the spring includes n = 24.25 coils and the initial tension is P0 = 93 N. Consequently, the standard formula for extension springs leads to a rate of k = 39.6 N/mm. The proposed analytical approach suggests an initial rate ki = 545 N/mm and a global rate k = 37 N/mm.

Once again, four specimens were tested and the results are presented in Fig. 7.

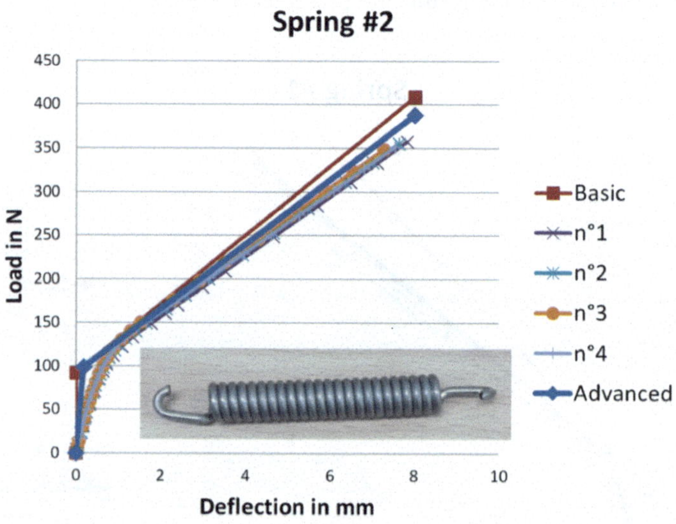

Fig. 7. Results for spring #2

In this case, the basic calculation also overestimates the global rate (+14% from the mean experimental results) and the proposed approach is more accurate (+7%).

Nevertheless, at the beginning of deflection, the spring exhibits a smooth curve rather than linear behavior. According to the spring manufacturers, this is due to the manufacturing process. In the coiling machine, a transition has to be managed between the body (with pretension, which means an axial pitch smaller than the wire diameter) and the loops that are free to deflect (with no pretension). This transition greatly influences the behavior of the spring.

4 Conclusions

This study highlights the influence of the loops on the load-deflection behavior of extension springs. As far as we are aware, FEM has difficulty simulating the pretension of extension springs, so this study exploits both experimental and analytical approaches.

An analytical approach has been developed in order to be able to evaluate the global rate of the springs (when all the coils deflect) and also that at the very beginning of deflection, when the coils of the body are still in contact due to the initial tension.

Until now, the proposed analytical approach can be considered as accurate for a spring with a rapid transition (i.e. with bilinear behavior).

The effects of the transition of the global load-deflection curve are illustrated in Fig. 8. For a given design (i.e. same external geometry, material and pretension F_T) several load-deflection curves can be obtained.

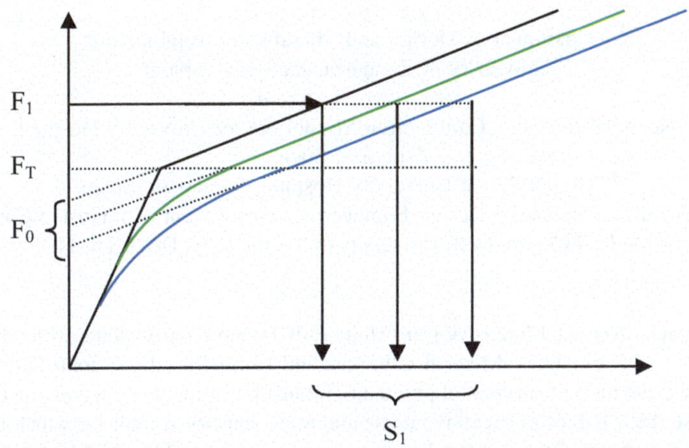

Fig. 8. Influence of the coiling process on the global behavior

For a given external axial length, F_1, the smoother the transition, the greater the associated deflection S_1. This is another key issue for manufacturers and further investigations will be necessary to fully cover the range of industrial applications.

It would be interesting to develop accurate FEA simulations that would take account of both preload in the body and transitions between the body and the loops in order to build new analytical models for extension springs, which could be used by designers and manufacturers.

References

1. Wahl, A.M.: Mechanical Springs. McGraw-Hill Book Company, New York (1963)
2. Gluck, J.V.: Fatigue of small helical extension springs subjected to static torsion and alternating lateral deflection. J. Eng. Mater. Technol. **104**(3), 180–185 (1982)
3. Paredes, M., Rodriguez, E., Sartor, M.: Optimisation methods for compression, extension and torsion spring design. In: International Conference IDMME, Bath (2004)
4. Shimoseki, M., Hamano, T., Imaizumi, T.: FEM for Springs. Springer, Heidelberg (2003)
5. Paredes, M.: Analytical and experimental study of conical telescoping springs with nonconstant pitch. ASME J. Mech. Des. **135**, 094502 (2013)
6. Dym, C.: Consistent derivations of spring rates for helical springs. ASME J. Mech. Des. **131** (2009)

Design Guidelines for Light and Noise Management in the Neonatal Intensive Care Unit

Rosana Sanz-Segura[1(✉)], Eduardo Manchado-Perez[1], Maria Pilar Ferrer-Duce[2], Delia Gonzalez de la Cuesta[3], and Elif Özcan[4]

[1] Department of Design and Manufacture Engineering, University of Zaragoza, Zaragoza, Spain
rsanz@unizar.es
[2] Neonatal Intensive Care Unit at Miguel Servet University Hospital, Zaragoza, Spain
[3] Miguel Servet University Hospital, Saragossa, Spain
[4] Faculty of Industrial Design Engineering, Department of Intensive Care Erasmus Medical Centre, Delft University of Technology, Delft, The Netherlands

Abstract. Neonatal Intensive Care Units (NICU) are environments with a high level of sensory stress. Medical advances and technology have contributed to increase the ratio of survival of premature infants but some devices and practices expose these babies to excessive noise and toxic sensory stimuli for which they are not prepared. This is related to an increase of neonatal morbidities, that are considered as minor sequelae, but that can greatly alter the life of the child and the family. Those responsible for hospital management and caregivers who want to take a step forward, need standards to guarantee the benefit of neonatal health and a proper physical and cognitive development of these babies. Design activity, from a Human-Centered Design approach (HCD), together with Developmental Centered Care (DC) contribute to identify and reduce adverse environmental conditions for newborns and premature infants. The purpose of this paper is to establish a method to provide design recommendations and good practice guidelines from evidence and especially from in-situ observations carried out in neonatal units by a multidisciplinary team (i.e., nurses, NICU supervisors and designers). Thus, we identify proposals to reduce stress situations and obtain potential benefits in the development of the hospitalized infant through adaptation of the NICU macroenvironment (i.e. the reduction of light and noise).

Keywords: NICU · Design guidelines · Alarm fatigue · Human-Centered Design

1 Introduction

Medical advances and technology have contributed to increase the ratio of survival of premature infants. This increasing of technological devices nevertheless exposes these babies to toxic sensory stimuli for which they are not prepared. This is related to a rise

F. Cavas-Martínez et al. (Eds.): INGEGRAF 2019, LNME, pp. 284–293, 2020.
https://doi.org/10.1007/978-3-030-41200-5_31

in neonatal morbidities, considered as minor sequelae, but that can greatly alter the life of the child and the family. A premature baby passes directly from the intrauterine environment to an incubator in the neonatal unit. The mother's womb is a dark and warm environment with soft auditory stimuli where the predominant sound is the voice of the mother; the effect of gravity does not exist and the movement of the baby is limited to his/her own gestures and the mother's movement. Conversely, a neonatal intensive care unit (NICU) is a cold and noisy environment with inappropriate sound stimuli coming from several sources, many of them artificial. The exposure to excessive noise and light stimuli affects the physiological neonate and the neurobehavioral development of an immature brain [1].

The current actions implemented to reduce these unwanted effects are based on the so-called Developmental Centered Care [2, 3]. These evidence-based practices favor the sensorineural and emotional development of newborns and the reduction of stress. Within this approach, the adequacy of the environment (i.e. the reduction of light and noise) is one of the aspects to address as it is directly related to neurobiological development in an ordered chain: touch at 8 weeks, vestibular system at 12 weeks, taste at 16, smell at 22 weeks, hearing at 20 weeks and sight at 30 weeks. Design activity, from a Human-Centered Design approach (HCD) [4], together with Developmental Centered Care (DC) can also help to take a step further and reduce adverse environmental conditions for newborns and premature infants, improving at the same time the wellbeing of the rest of inhabitants in critical care units.

Human Centered Design consist of a nearly anthropological approach that puts the user at the very core of the design process. Its principles are based on obtaining a close empathy with users to better understand their needs. The product itself is thus conceived as a system of services whose functions are devoted to fulfil the user expectations (beyond an artefact constructed of mechanisms and components), providing the better possible user experience. Human Centered Design relates closely to open innovation practices as it aims users to actively participate in the development process of the product, contributing to define design specifications, co-create solutions and evaluate prototypes [4–7]. Consequently, our contribution exposes the observational study carried out in a NICU in order to provide design recommendations and good practice guidelines obtained from a multidisciplinary collaboration of designers and clinicians (nurses and clinical supervisors) [8]. The objective is to get a better environmental management of the NICU (i.e., light and noise control) to attend in a proper way the needs of its inhabitants (patients, relatives and clinicians).

2 Effects of Noise and Benefits of Noise Reduction in Newborns

2.1 Negative Effects of Noise

The negative effects of noise detected in this environment stress and alter the state of the newborn, but also of the rest of the inhabitants of the unit (clinicians, family).

- Affectations to the newborn: alterations of the autonomic nervous system (tachycardia, desaturations, apneas…); sleep disruptions; changes in brain activity; an energy expenditure increase; behavioral changes; alterations in the development of speech and language; difficulties in listening to the mother's voice; and hearing loss. Bilateral hearing loss is ten times more frequent in premature babies, being the environmental noise one of the main factors to which they are subjected in critical units [9].
- In the clinician: cardiovascular alterations; increase in stress hormones and blood glucose; increase in muscle tension and in gastrointestinal motility; sleep disturbances; changes in behaviour and mood (fatigue, irritability, anger …) and alarm desensitization; interference at work (increased cognitive workload affects efficiency and safety and consequently, a more risk of errors) [10].
- In the family: the noise generates anxiety and worry; it produces interferences in the communication with practitioners and their child; difficult family interactions; and the perception of a chaotic and disorganized environment, with emotional implications [11].

2.2 Benefits of Noise Reduction

The benefits observed in the newborn are:

- an improvement in the control of the autonomic system: reduction of heart rate and blood pressure, as well as the number of apneas.
- reduction of ventilation days, improves digestive tolerance, which reduces the length of hospital stay.
- improves the sleep cycle: lengthens the peaceful sleep.
- avoid increases in intracranial pressure and episodes of hypoxemia.
- normal development of language, attention and perception is favored.

Regarding health professionals, a better quality of care is observed: paying more attention and reducing errors.

3 Effects of Light Stimulus and Benefits of Light Reduction in Newborns

3.1 Negative Effects of Light Stimulus Increase

The development of the visual system in a premature should continue without luminous stimulus, mainly in those babies under 30 weeks of gestational age (GA) who do not have defensive capacity for the regulation of the entry of light into the retina (thin eyelids, or inability to constrict pupil). Excess light can affect an adequate sleep pattern. Sleep and mainly the REM phase must be as long as possible for an adequate visual development.

3.2 Benefits of Light Reduction

AAP (American Academy of Pediatrics) recommends that the luminosity be less than 60 lx, the WHO (World Health Organization) less than 45 lx. In the case of preterm infants less than 30 weeks old, levels greater than 20 lx are not recommended. Nevertheless, there are demanding tasks in which it will be necessary to provide higher levels of light (taking an intravenous catheter, preparation of medication…). In these cases, the high light should not increase the intensity more than 10% in the rest of children.

The benefits of light reduction to newborns are:

- Improves psychomotor development;
- Increases the stability of the autonomic system and as a consequence lower heart rate, lower blood pressure and apneas;
- Reduces stress and greater weight gain is obtained;
- It favors the reduction of noise and manipulations, therefore, periods of sleep and rest increase;
- More periods of ocular opening, decreasing the motor activity;
- In greater than 32 weeks of GA exposure to light in short periods, establishing a sleep-wake cycle, favors the establishment of the circadian rhythm and increases weight gain.

Likewise, a series of detrimental effects of light reduction have been observed such as a negative influence on the mood and work performance for parents and professionals, so there should be resting areas with light. The scarcity of light hinders proper monitoring and early detection of possible complications arising from the maintenance of venous lines, nasogastric tubes, etc. [12].

4 Objectives and Procedure

The main objective is to raise an environment of light and noise as low as possible to promote the proper neurodevelopment of newborns, especially preterm infants. The procedure is based on a multidisciplinary approach that will include Human Centered Design principles together with the existent Developmental Centered Care basis. Although health education is carried out through awareness-raising routines with individual and global actions, environmental changes should involve a detailed study of medical devices as well as the understanding and interaction of the inhabitants (i.e., nurses) with advanced technology and graphic interfaces [8, 13, 14]. Human Centered Design offers a systematic, structured methodology of work that might be adapted to different situations and helps to control the process of analysis of problems, generation of ideas and evaluation, making its further implementation easier. Thus, graphical engineering and product design and development might play an important role in successfully designing medical equipment and alarm interfaces compatible with noise and light management.

To this end, all the inhabitants of the neonatal intensive care unit (clinicians, babies, families and caregivers) have been taken as the focus of our observational study, paying special attention to premature babies under 32 weeks and/or under 1500 g due to the risks involved in being subjected to an excessively noisy and luminous environment. As a first step to noise control and management, measurements were made during the month of March 2019 in a public hospital, using a sonometer model KOBAN KS1351 with a measurement range between 35 and 130 dB. Table 1 shows the true peak sound pressure levels generated by different sources in a NICU. Newborns are exposed to noises that exceed 88 dBA during their stay in the neonatal unit. The sources of noise that contribute the most to this aggressive environment are the staff voices and the alarms from medical devices such as respirators, monitors, pumps and other life support devices. The American Academy of Pediatrics (AAP) recommends that back-ground noise level in neonatal units should be less than 45 dBA with transient sound to 65–70 dBA; The World Health Organization (WHO) recommends levels below 30–35 dBA [15, 16]. Once the design guidelines and recommendations have been implemented, the new results and (noise and light) measurements will be compared with the results compiled in Table 1.

Table 1. Frequent sources of noise and measured number of decibels.

Sources	Sound pressure level in dBC (in the incubator)	Sound pressure level in dBC (out of the incubator)
Close door of small incubator	89	84
Close drawer	–	76
CPAP/respirator	78	65
Incubator motor	70	50
Sensor medics	76	76
Place analytical tubes	72	–
Hit with a finger	88	84
Pulse oximeter alarms	81	86
Infusion pump alarm	70	78
Respiratory machine alarm	76	82
Monitor alarms		85
CPAP alarms		82
Close the door	–	85
Move chair	–	80

4.1 A Developmental Centered Care and Human-Centered Design Multidisciplinary Approach

The proposed procedure for obtaining a series of design guidelines for noise and light reduction in NICUs is shown in Table 2. The contribution consists in the development of some tasks to be carried out separately by clinicians from a Developmental Centered Care approach and designers with a Human Centered Design focus; and some complementary activities to be carried out by a multidisciplinary team comprising participants from both disciplines.

Table 2. Tasks to be developed separately, and by a multidisciplinary team.

Phase	Clinicians	Designers	Multidisciplinary team
Initial research	Previous investigation: -Analysis of the general context in NICU -Gathering of existing literature on the relevance of the environment in possible sequelae on neonates -Support of scientific data on the influence of sensory stimulation on neonates in sleep and wakefulness -Support of scientific data on the development of the auditory and visual system of the newborn	Previous investigation: -Study of alarms in diverse critical environments, but especially in hospital settings (more general, not just NICU) -Centre in the different perspective of the clinicians, babies and relatives. -Support of scientific information from other analogous areas	Focus the investigation on: - Influence of the alarms in a NICU unit on the presence of environmental noise and its impact on the development of the newborn - Approach from a broad context that starts from the study of alarms in general critical contexts, then sanitary, then NICU
Specific research	Empirical observation: -Description of noise and light sources in NICU -Measurement of environmental noise and light -Observation of the effects of noise and light in newborns, personnel and family.	Focus the study on alarms in healthcare environments: -Description of types of alarms -Multimodal perspective (sound alarms vs other types of alarm) -Perspective psychology of perception and communication theory	Detailed description of the NICU ecosystem in relation to the presence of noise originating from alarms and its influence on newborns, staff and family members Detailed reference of the theoretical and scientific framework that contextualizes the empirical observations
Analysis	General comparison of the characteristics of intrauterine vs external environment	Categorization of the types of sounds present in the external environment, produced by machines and equipment (Causal/Intentional)	Comparison of the sonorous characteristics of the intrauterine versus external environment, describing the latter in a more systematic, extended and detailed way respecting to machines and devices

(*continued*)

Table 2. (*continued*)

Phase	Clinicians	Designers	Multidisciplinary team
Needs identification	Identification of opportunity: Improvement of care (Description of requirements)	Identification of opportunity: Improvement of care (Description of requirements)	Identification of intervention points and opportunities for improvement
Ideation	List of proposals for NOISE reduction and adaptation of the LIGHT	Adaptation of the model Event/Alarm/Reaction. This relates and allows positioning: -The event to be measured by the machine or installation -The type of alarm used to provide staff with a type of information and trigger their response -The type of response, its risk of success/error, its promptness	Mapping of the particular NICU case on the model of the Event/Alarm/Reaction of: 1. Situation previous to the list of proposals of the clinical team 2. Situation expected after implementation of proposals
Proposal prototyping	Drafting of a decalogue/compliance/protocol of action	Validation of the adapted-to-case Event/Alarm/Reaction model	Review of the decalogue Drafting of conclusions Justification of the adequacy of the measures proposed in the decalogue Implementation and evaluation of the obtained results

5 Design Guidelines for Noise and Light Reduction

As a result of the multidisciplinary iteration, the proposals listed below were considered as the most relevant design recommendations for noise and light reduction (see Tables 3 and 4). Work is currently under progress and the benefits are still to be evaluated to obtain quantitative data after a reasonable implementation time. Data compiled in Table 1 will be used as a reference to stablish comparison and rank future achievements.

Table 3. Measures for noise reduction.

Measures to be adopted for noise reduction
Staff sensitization: a brochure will be delivered to the parents and staff
Avoid talking loud, and when close to incubators speak in whispers
Avoid the accumulation of medical personnel: talk in restricted areas or offices. Avoid accumulation in the NICU entrance and especially in changes of guard
Reduce slamming doors in main entrance, vending machines and lockers
Switch or mute mobile phones
When intermittent withdrawals of CPAP or high flow are made, withdraw them from inside the incubator and turn them off, closing the flow
When the BiPAP Infant flow is withdrawn intermittently, remove it from the incubator and place an adhesive in the areas of the cannulas to avoid the alarm jumping due to low flow
Control the alarm of the minute volume of the respirator
Control periodically the accumulation of water in the breather pipes and remove it
Careful opening and closing of incubator
Do not touch the incubator with fingers or place objects on it or write on them
Reduce alarms volume, transform into light alarms if possible and respond asap
Telephone: reduce the volume and move it away from the incubator areas
Install and maintain a sound level meter or decibel meter

Table 4. Measures for light reduction.

Measures to be adopted for light reduction
Light in the unit to be tenuous, being almost obscure where the most premature newborns are. To the extent possible, use individual and controlled intensity light bulbs for each child, covering the eyes to avoid direct light to the eyes
Light sleep-wake transition: keep dark environment in the night hours and dim light during the day. When the children are manipulated, try to perform when they are awake or wake them up in a soft and progressive way, not abruptly
Provide use of covers in younger children on incubators
Close monitoring of children who carry nasogastric tube or via, controlling it every hour through the use of flashlights
Use natural light whenever possible. Blinds will be kept low both in the NICU and in neonates, closing them completely when direct light comes in

6 Conclusions

Available knowledge from product design expertise of aspects related to the identification of user needs can be greatly appreciated in areas closely related to human experience, such as healthcare [13, 17]. In addition to this, design activity contributes significantly to the improvement of the proposals arising from professionals from other fields of expertise. Multidisciplinary team working and cocreation provide a capacity for problem-solving, consistent with open-innovation strategies. Besides, a multidisciplinary approach takes advantage of the human centred design approach, the knowledge of technology implied in the development of advanced devices (such as those present in NICUs) and the possibilities of the state-of-the-art graphic interface design, in collaboration with the expertise contributed by the clinicians. All in an organized and objective focused manner. Future measurements of noise and light are still to be taken in order to quantify the obtained benefits, and compared to those compiled in Table 1. Also, further observation must be carried out at different moments of activity (i.e. day vs night, or week day vs weekends, or while different nurse teams are at work).

The multidisciplinary iteration can provide very useful results, although sometimes team working can be difficult due to the different technical formation, knowledge and professional culture of the participants. Thus, a future line of investigation is to provide tools and a general framework to ease the incorporation of design procedures to the ambit of healthcare.

References

1. Cardoso, S.M.S., Kozlowski, L.D.C., Lacerda, A.B.M.D., Marques, J.M., Ribas, A.: Newborn physiological responses to noise in the neonatal unit. Braz. J. Otorhinolaryngol. **81** (6), 583–588 (2015). https://doi.org/10.1016/j.bjorl.2014.11.008
2. Als, H.: The newborn communicates. J. Commun. **27**(2), 66–73 (1977). https://doi.org/10.1111/j.1460-2466.1977.tb01828.x
3. Als, H., Lawhon, G., Duffy, F.H., McAnulty, G.B., Gibes-Grossman, R., Blickman, J.G.: Individualized developmental care for the very low-birth-weight preterm infant: medical and neurofunctional effects. JAMA **272**(11), 853–858 (1994). https://doi.org/10.1001/jama.1994.03520110033025
4. Giacomin, J.: What is human centred design? Des. J. **17**(4), 606–623 (2014)
5. Maguire, M.: Methods to support human-centred design. Int. J. Hum.-Comput. Stud. **55**(4), 587–634 (2001)
6. Wilkinson, C.R., De Angeli, A.: Applying user centred and participatory design approaches to commercial product development. Des. Stud. **35**(6), 614–631 (2014)
7. Rejón, J.V.: Showing the value of ethnography in business. In: Ethnographic Praxis in Industry Conference Proceedings, vol. 2009, no. 1, pp. 162–169. Wiley/Blackwell, August 2009. https://doi.org/10.1111/j.1559-8918.2009.tb00136.x
8. Özcan, E., Birdja, D., Edworthy, J.R.: A holistic and collaborative approach to audible alarm design. Biomed. Instrum. Technol. **52**(6), 422–432 (2018). https://doi.org/10.2345/0899-8205-52.6.422
9. Romeu, J., Cotrina, L., Perapoch, J., Linés, M.: Assessment of environmental noise and its effect on neonates in a Neonatal Intensive Care Unit. Appl. Acoust. **111**, 161–169 (2016). https://doi.org/10.1016/j.apacoust.2016.04.014
10. Honan, L., Funk, M., Maynard, M., Fahs, D., Clark, J.T., David, Y.: Nurses' perspectives on clinical alarms. Am. J. Crit. Care **24**(5), 387–395 (2015). https://doi.org/10.4037/ajcc2015552

11. Benzies, K.M., Shah, V., Aziz, K., Lodha, A., Misfeldt, R.: The health care system is making 'too much noise' to provide family-centered care in neonatal intensive care units: Perspectives of health care providers and hospital administrators. Intensive Crit. Care Nurs. **50**, 44–53 (2019). https://doi.org/10.1016/j.iccn.2018.05.001
12. Lasky, R., Williams, A.: Noise and light exposures for extremely low birth weight newborns during their stay in the neonatal intensive care unit. Pediatrics **123**(2), 540–546 (2009)
13. Sanz-Segura, R., Romero-Piqueras, C., Manchado-Pérez, E., Özcan, E.: Service design and sound: a chance for exploration in oncological treatment rooms. In: Advances on Mechanics, Design Engineering and Manufacturing II, pp. 639–648. Springer, Cham (2019). https://doi.org/10.1007/978-3-030-12346-8_62
14. Sanz Segura, R., Manchado Pérez, E.: Product sound design as a valuable tool in the product development process. Ergon. Des. **26**(4), 20–24 (2018). https://doi.org/10.1177/1064804618772997
15. Carvalhais, C., da Silva, M.V., Xavier, A., Santos, J.: Good practices to reduce noise levels in the neonatal intensive care unit. In: Occupational and Environmental Safety and Health, pp. 297–302. Springer, Cham (2019). https://doi.org/10.1007/978-3-030-14730-3_32
16. Weber, A., Harrison, T.M.: Reducing toxic stress in the neonatal intensive care unit to improve infant outcomes. Nurs. Outlook (2018). https://doi.org/10.1016/j.outlook.2018.11.002
17. Carmel-Gilfilen, C., Portillo, M.: Designing with empathy: humanizing narratives for inspired healthcare experiences. HERD: Health Environ. Res. Des. J. **9**(2), 130–146 (2016)

Inspiration in Elements of Nature for the Promotion of Creativity in the Conceptual Design of Products

D. Parras-Burgos(✉), D. G. Fernández-Pacheco, F. J. F. Cañavate, F. Cavas-Martínez, F. L. Sáez-Gutiérrez, and J. Nieto

Technical University of Cartagena, C/Doctor Fleming s/n, 30202 Cartagena, Murcia, Spain
dolores.parras@upct.es

Abstract. Creativity concept has been widely studied in the literature and is considered an innate factor in human being that is vital to grow as a society. In the design field, creativity is considered a fundamental factor for obtaining innovative products, and inspiration is regarded as a process that can be enhanced. Nature can be an inexhaustible source of resources for obtaining a great diversity of ideas that can be applied in product design. In addition to functionality, the visual appearance of the objects is also decisive to secure a favourable response from consumers and obtain greater success in the market. This paper presents, through a design example, a tool that favours the cognitive process from analogies while looking within different domains that support the creation of novel and original ideas. Through a selection process, the user relates different types of geometry to obtain images that help generate those new ideas. Therefore, this tool combines observation and the use of visual stimuli to promote creativity, using shapes and forms present in nature as a source of inspiration.

Keywords: Creativity · Product design · Bioinspiration · Visual stimuli · Geometry

1 Introduction

Creativity is considered an innate capacity of the individual and a phenomenon linked to human being since its origin [1]. Nowadays, it is considered that creativity has a great scope and is important both individually and socially: at the individual level, it can be used to solve problems at work and in everyday life; at the level of society, creativity can lead to new scientific discoveries, emerging movements, novel inventions and recent social programs. Therefore, individuals, organizations and societies must adapt existing resources to changing demands in order to remain competitive [2]. Moreover, creativity starts innovations, helps in solving problems and allows a company to capture a larger market share [3–6].

Inspiration has been associated for a long time with creative disciplines, such as music, art, science, engineering, architecture and industrial design [7–9]. The latter, industrial design, is closely related to the process of shaping the visual appearance of

F. Cavas-Martínez et al. (Eds.): INGEGRAF 2019, LNME, pp. 294–302, 2020.
https://doi.org/10.1007/978-3-030-41200-5_32

products. This procedure plays a key role in the perception and evaluation of the properties of the product performed by the consumer [10, 11]. The user's response to a product is often stimulated by visual information, since it provides a large amount of information in a very short time [12]. The visual appearance of a product is, therefore, a key determinant of the consumer response and the success of the product [13, 14].

2 The Inspiration in Nature and the Use of Analogies in the Design Process

Nature can inspire solutions in any field related to design and engineering. These solutions can be efficient, practical, sustainable and with the necessary potential to discover new technologies, materials and processes [15]. In this sense, two ways of being inspired by nature can be found: (i) from a "reductionist/superficial vision", which consists of an imitation of natural forms or processes in the domain of engineering/design; and (ii) from a "holistic/deep vision", which is understood as the way to achieve ecologically sustainable products, that is, products that do not harm the environment during their manufacture, use or disposal [16].

Observation can be considered as one of the first tools to enhance creativity, as well as analogy is a methodological strategy that can serve to innovate and solve different project problems [17]. The design from analogies has been recognized by researchers as an important method for the generation of novel and creative ideas [18, 19]. Biologically inspired design implies cross-domain analogies, and has led to the achievement of many innovative designs in practice. Therefore, the design developed from a biological inspiration can be considered an ideal domain for the study of analogies that can be creative [20]. The Italian designer Carmelo di Bartolo [17] introduces the "bionics" term in the world of design with its etymological meaning "the study of life forms". In this sense, nature invites us to put thought into practice: be creative is to look beyond what others see.

Tools are needed to support the ability of people to use analogies. A deeper understanding of the cognitive mechanisms that underly design and analogy is crucial for the development of these tools. In this regard, cognitive studies that evaluate the impact of biological examples in the process of generating ideas during the conceptual design can be found. The results suggest that exposure to biological examples in the generation of ideas may increase their novelty, in contrast to the exhibition of human engineering examples [21].

In this context, there exist some tools and resources in which designers and engineers can easy access to biological information very useful to solve their design problems. Table 1 shows the most relevant and representative tools and resources that use nature as a database or as a reference.

Table 1. Main tools and resources based on bioinspiration.

Authors	Name	Description	Functional aspects	Shape aspects	Aesthetics aspects
[22, 23]	Ask Nature (Web-Online Database)	2406 functional characteristics obtained from elements of nature. When specifying a functional problem, Asknature will show a number of relevant biological phenomena that solve similar functional problems	Yes	No	No
[24, 25]	Triz/BioTriz (Database)	Solutions are related to similar principles through the use of a matrix. BioTriz shows how problems are solved by natural systems based on the analysis of about 500 biological phenomena	Yes	No	No
[16, 26]	Biomimetic cards	Four categories of biomimetics are defined in relation to design: materials, mechanical/dynamic (general engineering and locomotion), structures (structural engineering and architecture) and shapes (architecture and art)	Yes	Yes	No
[27–29]	Idea-Inspire (modelo SAPPhiRE, Database)	Database with more than 700 entries of natural and artificial phenomena that can be searched using 7 attributes. These attributes are used to describe functionalities, behaviours and structures that help designers with automated analogy searching	Yes	No	No
[30]	DANE (Design by analogy to nature engine, CAD System)	DANE is a CAD system that provides access to functional models of biological systems. Four characteristics are considered: cognitive, collaborative, conceptual and creative	Yes	No	No
[31]	Method to unite systematically the disparate domains of biology and engineering	Method to unite systematically the disparate domains of biology and engineering using natural language analysis. This method is based on WordNet, a lexicon database organized according to theories about how people use and remember the language, and not alphabetically as dictionaries	Yes	No	No

(*continued*)

Table 1. *(continued)*

Authors	Name	Description	Functional aspects	Shape aspects	Aesthetics aspects
[32]	The BioM Innovation Database	The global biomimetic activity is shown since 1960. It is the first tool to include data related to the process, focus, geography and vocabulary. The database currently contains more than 350 cases of products or concepts that have been identified as biomimetics	Yes	No	No
[33, 34]	The engineering-to-biology Thesaurus	Dictionary that relates terms of engineering and biology as an aid in the teaching of biomimicry in the in-engineering. This dictionary has been designed to allow engineers, without a prior knowledge in advanced biology, to take advantage of the ingenuity of nature for their purposes	Yes	No	No
[35, 36]	Method based on the use of descriptions of biological phenomena for the generation of ideas	Use of descriptions of biological phenomena to develop concepts to solve a simple problem	Yes	No	No

However, all the tools described in Table 1 are only focused on finding ideas from a functional point of view. The study of elements of nature considering shape or aesthetics aspects has not yet been used, nor is there a resource that offers geometric information during the creative process in the conceptual design phase.

For all the above, this communication presents a novel tool that can be used to promote creative stimulation and generation of ideas taking into consideration geometry and elements of nature as sources of inspiration.

3 Materials and Methods

Creativity in design is considered an iterative process, with comings and goings, in search of the best idea. Wang and Xiao [37] call this process "chaotic mechanism" of the designer's thinking. These authors consider the design process as a nonlinear dynamic system, defining four phases (design-analysis-evaluation-redesign) within which the designer moves iteratively. In another study, Noguchi and Nagai [38] propose a model that represents the trajectory of thought during the creative design process, arguing the importance of searching as a fundamental strategy in design. Basing on these two premises, this communication proposes a tool in which the designer will look for the shape that best suits the product to be designed by interacting with different geometries along several consecutive stages. This process does not necessarily follows a linear trajectory, as iterations are allowed as many times as needed, being able to return to other phases and change the selected geometries in order to obtain different results [39]. From a geometrical study, different forms have been related to images of elements of nature, which favours the cognitive process of the search by analogies (Fig. 1).

The Eclipse programming environment and the JAVA programming language have been chosen for the implementation of the proposed tool. The tool is based on a decision system that consists of several screens were different geometric options are offered depending on the selections made by the user. Finally, a final stage where the images of those elements of nature that contain or are related to the selected geometries is shown. The designer can use one of these images as an inspiration resource and capture the design concept by performing a sketch.

The next step is the virtual modeling of the attained sketch, where the shape, dimensions and characteristics of the new product are specified with more detail. In this work, the computer-aided design program SolidWorks 2016 was used to model the final designs.

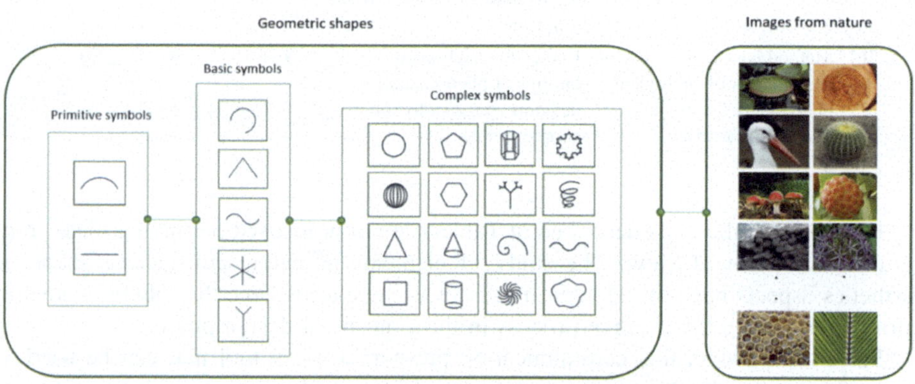

Fig. 1. Diagram of relationships between geometric shapes and images of nature.

4 Results

An example of the design process of a specific case is shown below to better understand the operation of the proposed tool. In this example, the designer is entrusted with the task of designing a perfume container with shapes inspired by nature. Among all the possible solutions, two examples are shown below: a male version and a female version of the perfume container. The following figures show captures of the different stages of the tool according to the selections made by the user (for both examples) until reaching the final image that will be used as reference for the sketch:

(a) **Male version perfume container:** in stage 1, two primitive symbols are offered, and the designer selects the *Curve* shape (Fig. 2a). When selecting this symbol, in stage 2, two basic symbols are shown, selecting in this case, *the Compound curve* symbol (Fig. 2b). In stage 3, five complex symbols appear, and *Wave* is selected (Fig. 2c). In stage 4, the gallery window shows photographs of elements of nature that contain this type of wavy shapes (Fig. 2d). To filter the obtained results, the *Filtered by groups* windows is accessed, showing four categories (flora, fauna, rocks and fluids). For this example, the *Fluids* filter was applied (Fig. 2e). This way, only the results that fall into this category are displayed in the gallery window. Among all the images, the photograph shown in Fig. 2f finally serves the designer as inspiration for the concept he was looking for. The tool also offers the possibility of drawing sketches and saving them in an image file format (Fig. 2g).

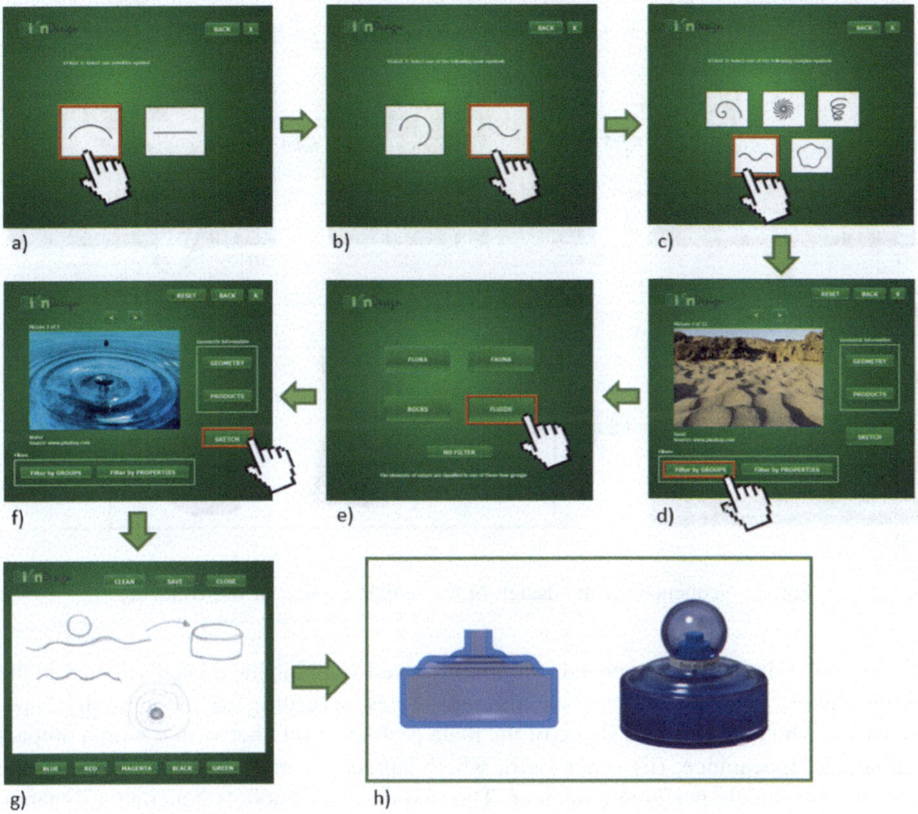

Fig. 2. Sequence for the design of the male version of the container.

(b) **Female version perfume container:** in stage 1 and 2 the same curved symbols as in the previous example are selected. On this occasion the *Swirl* complex symbol is selected in stage 3 (Fig. 3c). In the gallery window, new photographs related to this type of forms are now shown and are later filtered using the *Fluids* group (Fig. 3d). Among the results displayed in the gallery, the photograph shown in Fig. 3f inspires the designer to generate the concept of this second model. Then, new sketches are performed to define the design of this new container (Fig. 3g).

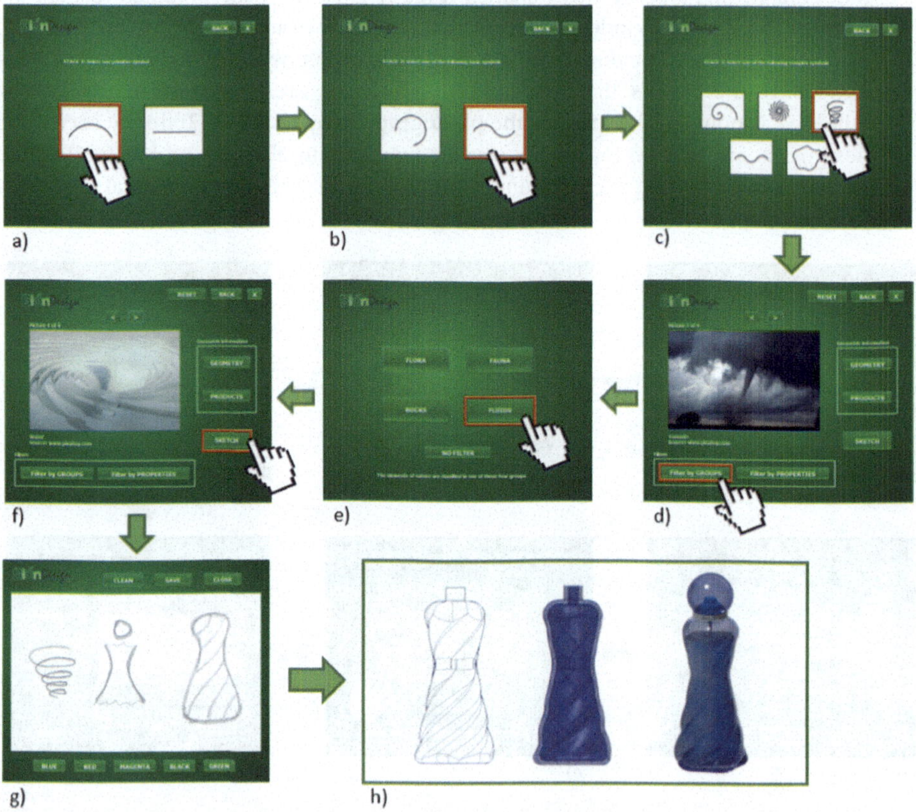

Fig. 3. Sequence for the design of the female version of the container.

As shown in the two proposed examples, curved symbols have been selected in the initial stages, finally defining two different styles according to the chosen shape: (i) waves, which suggest the shape of the male perfume container with a more compact and simple appearance; (ii) water swirl, which suggests a more stylized and feminine form for the female perfume container. The two designed models generate a dynamic and moving sensation. The idea of a plug, like a drop of water without gravity, is unified in the two proposed designs (Figs. 2h and 3h).

5 Conclusions

In this communication, a new application approach based on decision systems is presented to enhance the creativity of the designer. Users can obtain novel and original ideas based on forms and aesthetic aspects of elements of nature. To do this, the designers begin their creative process by observing and selecting basic geometrical symbols/concepts and increasing the level of complexity according to the responses received at each stage. The proposed tool allows users to obtain an adaptable solution from analogies based on the observation of forms of nature, which can serve as inspiration in the conceptual design process.

The application has been developed as a scalable resource, what permits to gradually increase the number of figures and information in the gallery for future versions of the tool.

References

1. De la Torre, S.: Educar en la creatividad, p. 187. Editorial Narcea, Madrid (1987)
2. Sternberg, R.J., Lubart, T.I.: The concept of creativity: prospects and paradigms. In: Handbook of Creativity, vol. 1, pp. 3–15 (1999)
3. Chakrabarti, A., Morgenstern, S., Knaab, H.: Identification and application of requirements and their impact on the design process: a protocol study. Res. Eng. Des. **15**(1), 22–39 (2004)
4. Eisentraut, R., Badke-Schaub, P.: Creativity: a personality trait or an illusion. In: Proceedings of the International Workshop Engineering Design and Creativity, Pilsen, Czech Republic (1995)
5. Sarkar, P., Chakrabarti, A.: Assessing design creativity. Des. Stud. **32**(4), 348–383 (2011)
6. Westwood, A.R., Sekine, Y.: Fostering creativity and innovation in an industrial R&D laboratory. Res.-Technol. Manag. **31**(4), 16–20 (1988)
7. Eckert, C., Stacey, M.: Sources of inspiration: a language of design. Des. Stud. **21**(5), 523–538 (2000)
8. Goldschmidt, G., Sever, A.L.: Inspiring design ideas with texts. Des. Stud. **32**(2), 139–155 (2011)
9. Herring, S.R., et al.: Getting inspired!: understanding how and why examples are used in creative design practice. In: Proceedings of the SIGCHI Conference on Human Factors in Computing Systems (2009)
10. Crilly, N., Moultrie, J., Clarkson, P.J.: Shaping things: intended consumer response and the other determinants of product form. Des. Stud. **30**(3), 224–254 (2009)
11. Veryzer, R.W.: Key factors affecting customer evaluation of discontinuous new products. J. Prod. Innov. Manag. **15**(2), 136–150 (1998)
12. Ulrich, K.T.: Design: Creation of artifacts in society. University of Pennsylvania (2011)
13. Bloch, P.H.: Seeking the ideal form: product design and consumer response. J. Mark. **59**(3), 16–29 (1995)
14. Coates, D.: Watches Tell More Than Time: Product Design, Information, and The Quest for Elegance, p. 296. McGraw-Hill, London (2003)
15. Armendáriz, E.N., et al.: Ingeniería Bioinspirada. OmniaScience, p. 130 (2014)
16. Volstad, N.L., Boks, C.: Biomimicry–a useful tool for the industrial designer? In: DS 50: Proceedings of NordDesign 2008 Conference, Tallinn, Estonia (2008)
17. Coronado, R.: Biónica, metodología e innovación. in II Encuentro Latinoamericano de Diseño "Diseño en Palermo", Buenos Aires, Argentina (2007)

18. Goel, A.K.: Design, analogy and creativity. IEEE Expert: Intell. Syst. Appl. **12**(3), 62–70 (1997)
19. Keshwani, S., Chakrabarti, A.: Towards automatic classification of description of analogies into SAPPhIRE constructs. In: International Conference on Research into Design (2017)
20. Vattam, S.S., Helms, M.E., Goel, A.K.: Nature of creative analogies in biologically inspired innovative design. In: Proceedings of the Seventh ACM Conference on Creativity and Cognition (2009)
21. Wilson, J.O., et al.: The effects of biological examples in idea generation. Des. Stud. **31**(2), 169–186 (2010)
22. Deldin, J.-M., Schuknecht, M.: The AskNature database: enabling solutions in biomimetic design. In: Biologically Inspired Design, pp. 17–27. Springer (2014)
23. Winters, J.: Learning from nature. Mech. Eng. **131**(7), 64–65 (2009)
24. Glier, M.W., et al.: Evaluating methods for bioinspired concept generation. In: Design Computing and Cognition, vol. 12, pp. 41–57. Springer, Netherlands (2014)
25. Vincent, J., et al.: Biomimetics: its practice and theory. J. R. Soc. Interface **3**(9), 471–482 (2006)
26. Lenau, T.A., et al.: Biocards and level of abstraction. In: 20th International Conference on Engineering Design (ICED15), Milan Italy (2015)
27. Chakrabarti, A., et al.: A functional representation for aiding biomimetic and artificial inspiration of new ideas. AIE EDAM **19**(02), 113–132 (2005)
28. Kolle, A.: Analogical problem-solving in design, evaluating the feasibility and benefits of the methodical use of biological analogies in solving design problems. Produktdesign 9 Fordypning, Artikkelsamling Høst (2005)
29. Sarkar, P., Chakrabarti, A.: The effect of representation of triggers on design outcomes. Artif. Intell. Eng. Des. Anal. Manuf. **22**(02), 101–116 (2008)
30. Goel, A.K., et al.: Cognitive, collaborative, conceptual and creative—four characteristics of the next generation of knowledge-based CAD systems: a study in biologically inspired design. Comput. Aided Des. **44**(10), 879–900 (2012)
31. Chiu, I., Shu, L.: Biomimetic design through natural language analysis to facilitate cross-domain information retrieval. AI EDAM: Artif. Intell. Eng. Des. Anal. Manuf. **21**(01), 45–59 (2007)
32. Jacobs, S.R., Nichol, E.C., Helms, M.E.: "Where are we now and where are we going?" The BioM innovation database. J. Mech. Des. Trans. ASME **136**(11), 111101 (2014)
33. Nagel, J.K.S., Nagel, R.L., Eggermont, M.: Teaching biomimicry with an engineering-to-biology thesaurus. In: ASME 2013 International Design Engineering Technical Conferences and Computers and Information in Engineering Conference, IDETC/CIE 2013. American Society of Mechanical Engineers, Portland (2013)
34. Nagel, R.L., et al.: Exploring the use of functional models in biomimetic conceptual design. J. Mech. Des. **130**(12), 121102 (2008)
35. Mak, T., Shu, L.: Abstraction of biological analogies for design. CIRP Ann. Manuf. Technol. **53**(1), 117–120 (2004)
36. Mak, T., Shu, L.: Using descriptions of biological phenomena for idea generation. Res. Eng. Des. **19**(1), 21–28 (2008)
37. Wang, X., Xiao, R.: Research on the thinking of creative design and the measure of design creativity. In: Computer-Aided Production Engineering, pp. 67–70 (2001)
38. Noguchi, H., Nagai, Y.: A study on the thinking path model of creative design process. In: DS 30: Proceedings of DESIGN 2002, The 7th International Design Conference, Dubrovnik (2002)
39. Parras-Burgos, D., et al.: An agent-based approach for the application of nature's forms to product conceptual design. PLoS ONE **13**(12), e0208930 (2018)

Teaching – Learning in Graphic Engineering

Objective Evaluation of Geometrical Product Specification/Geometrical Dimensioning and Tolerancing Basic Skills

Gabriele Baronio, Ileana Bodini(✉), Diego Paderno, Stefano Uberti, and Valerio Villa

Department of Mechanical and Industrial Engineering, University of Brescia, Via Branze 38, 25123 Brescia, Italy
ileana.bodini@unibs.it

Abstract. The aim of this work is to study methods and criteria to objectively evaluate Geometrical Product Specification (GPS) and Geometrical Dimensioning and tolerances (GD&T) basic skills of engineering students. To do so, it is important to define who is the examiner, what is the topic of the exam, how to examine, when and why. In particular, for what concerns the topics, knowledge, competences, skills, concepts and abilities should be assessed. Basic knowledge is the easier topic to evaluate in an objective and reliable way, for example using closed-ended questions, but skills like creativity are difficult to measure. Following the principles of Bloom's taxonomy, a technical drawing evaluation grid has been developed and used by the authors. The grid has eight learning levels, and target knowledge, competences, and skills have been defined for each learning level, then specific tests to verify the target for each level have been studied and developed. In this paper some examples of application of the grid are described and some preliminary results and considerations are reported. In particular, using an objective closed-ended question test, which rigorously assesses the basic levels of the grid, "residual" students decrease in number, and the students increase in number, the test is cost-effective.

Keywords: GPS/GD&T · Engineering students evaluation · Engineering education · Bloom's taxonomy · Technical drawing

1 Introduction

The use of objective and rigorous evaluation methods and techniques makes it possible to compare classifications, so as to ensure better selection and greater school justice [1–3]. But what exactly is meant by objective evaluation? In the Italian and French areas "docimology" gives some answers, while in the Anglo-Saxon areas "educational measurement" studies objective evaluation. The two things are very close but not perfectly overlapping, in fact docimology is a branch of pedagogy while educational measurement is in the area of psychometrics.

In particular, docimology is one of the experimental techniques that deal with the study of evaluation systems of verification tests; its purpose is to give indications on how to find objective parameters for the attribution of scores in tests; it aims to find

F. Cavas-Martínez et al. (Eds.): INGEGRAF 2019, LNME, pp. 305–312, 2020.
https://doi.org/10.1007/978-3-030-41200-5_33

objective evaluation methods, through various types of tests, such as: tests, structured tests, unstructured tests. Somehow, docimology is a metrology of knowledge.

To do a good job of objective evaluation we must first define the points that correspond to the classic journalistic sextet [4, 5]: who evaluates and/or who has to be evaluated; what has to be evaluated, how can we evaluate, where the evaluation takes place.

Concerning "who" evaluates or is evaluated, when the judgment on learning is expressed by a teacher or by an examiner (internal or external, as in the case of commissions of final examinations) or when students examine other students we speak of hetero-evaluation; when instead the student judges his own learning we speak of self-evaluation. Self-assessment is important: it serves to strengthen personal awareness, to increase motivation in learning and can integrate hetero-assessment in the formulation of synthetic judgments. In general, the distorting effect of the examiner coinciding with the teacher is greatly underestimated: a coach should not also be a referee [6].

The most substantial aspect, however, concerns what has to be evaluated: knowledge, skills, capabilities but also notions and abilities. In particular, the American pedagogue Benjamin Samuel Bloom defines the taxonomy of educational objectives, shown in Fig. 1, which describes the stages of learning from the simplest to the most complex one [7–9].

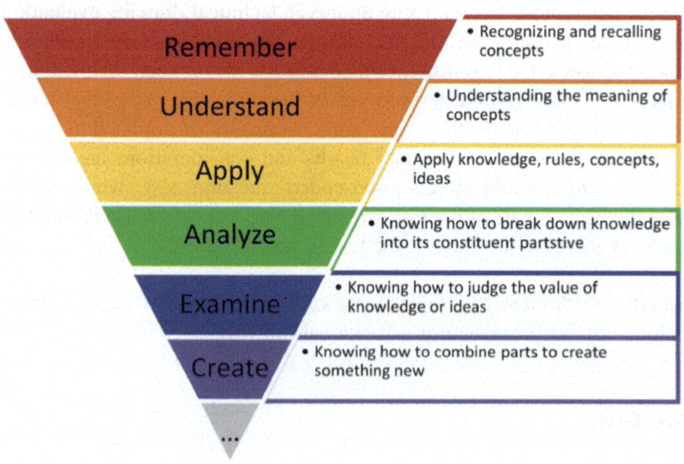

Fig. 1. Bloom's taxonomy.

From the point of view of evaluation, the simplest competences are also the easiest to evaluate objectively and reliably, while more complex competences such as creativity are difficult to be objectively evaluated.

In other words: (i) to under understand a concept you must first know it; (ii) to apply a concept you must first understand it; (iii) to analyze a concept you must first be able to apply it; (iv) to judge the value of knowledge you must first have analyzed it; (v) to create something new you must first have judged the values at stake.

With regard to "how" evaluate, the subjective evaluation is not to be discarded altogether. The holistic approach to evaluation based on impression, which avoids the use of shared analytical criteria, is widespread at all levels. Sometimes the first impression is actually confirmed by more rigorous subsequent evaluations, but this does not in any way legitimize the use of this form of judgment.

In order to solve some subjective aspects, a series of methodologies can be adopted, which lead to an increase in the degree of validity and reliability of the evaluations: (i) specify the "content" of the evaluations using shared reference frameworks; (ii) assign judgments collectively, after having agreed on criteria and methods of assignment; in this case more than a referee is required; (iii) use standardized tests to evaluate competences not explicitly expressed; (iv) use clear and shared indicators and survey tools for direct observation.

"Where" the evaluation takes place is partly related to "who" evaluated and "who" has to be evaluated. In the case of self-assessment there are no problems: everyone can do it where they want. In the case of assessments carried out by examiners the location of the test, its structure and the distribution of the candidates are very important. In the language and computer skills certification tests (ECDL type), there are precise rules on the minimum distance between candidates, and, in some cases, a map with their position during the test is required.

"Why" evaluate? Which is to say: "why measure"? "Why" is linked to "when" evaluate: (i) diagnostic or prognostic or initial assessment is made at the beginning of a program of study, to determine at what level of teaching a student should be, or it is made as an entrance exam, to seek guidance on the opportunity to undertake a particular curriculum, or course; (ii) formative or "in itinere" evaluation accompanies the didactic process in its development and serves to give or have indications on how to better direct the formative activities; (iii) overall or final evaluation is carried out at the end of a teaching phase, to certify the achievement of certain levels of competence.

As an obstacle to objective evaluation, there is also a fundamental incompatibility between the psychological nature of the teaching activity and the mathematical character of the proposed measurement procedures (problems well known to those who deal with measurements in the field of social sciences) [10].

Another questionable argument in defense of subjective evaluation is that every good teacher correctly judges his students. In this regard, to resume the comparison with sport: even every coach knows his players well, but to have the most valid references you need to play with other teams, with shared rules and an impartial judge.

2 Limits of the Traditional Mode for Design Examinations and Evaluation Grid

Although each university has slightly different traditions, the traditional way for a basic Industrial Technical Design examination is the "extraction of a part from an assembly". Figure 2 shows an exam carried out by a student and then corrected.

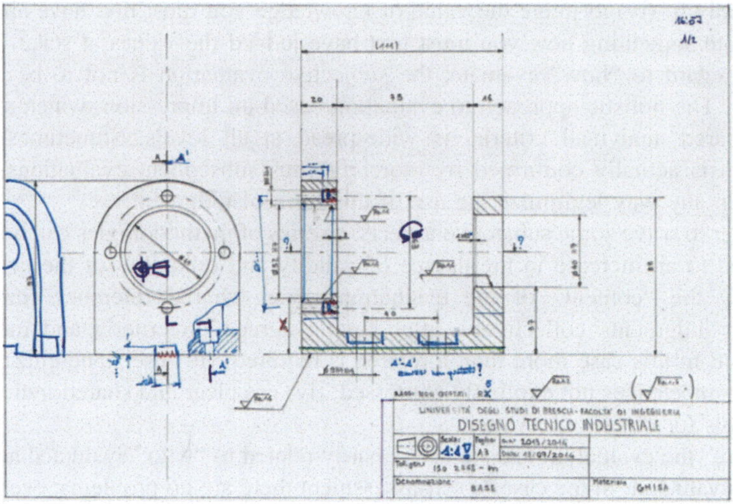

Fig. 2. Exam carried out and then corrected.

In this type of test the student has to: (i) understand morphology; (ii) represent morphology with the minimum number of views and sections; (iii) enter all the necessary dimensions without redundancies; (iv) indicate the dimensional tolerances, if necessary, after having understood the functionality of the couplings and having solved any compromises due to the chains of tolerances; (v) indicate the roughness of the surfaces, if necessary.

The limit of this mode is that only those students who achieve a very high evaluation are correctly evaluated, because it is difficult to give intermediate scores.

In fact, taking up what was said in the introduction regarding the taxonomy of Bloom: (i) it is possible to correctly represent the morphology of the part with the minimum number of views and sections only when the morphology has been understood in a perfect way; (ii) you can only add correctly the dimensions in the drawing after you have represented views and sections in a perfect way; (iii) you can only correctly indicate tolerances after perfect dimensioning.

To overcome these limits we decided to restructure the examination mode so that we could evaluate the individual skills independently.

Before doing this, however, we needed a grid of multi-level qualifications to ensure that the assessments are carried out in a uniform and repeatable manner.

A Technical Drawing Evaluation Grid (TDEG) has been developed [11]. The type of grid used is the one adopted in the European Qualifications Framework (EQF) [12]. Regarding "what" has to be evaluated, the EQF considers three aspects: knowledge, skills, and competences. In elaborating the grid we thought about the double nature of drawing, tool and language, as it happens also for writing. As suggested by the EQF, eight levels have been recognized, and for each of these levels two descriptions were found: one that looks at the drawing as a communication language, the other that looks at the drawing as a working tool.

3 Applying the Evaluation Grid

To apply the grid to the evaluations of our students we developed two systems of questions: the former is composed of closed-answer questions, and it is called TDT (Technical Drawing Test) [13], the latter is composed of open-answer questions, and it is called QRA.

A question with a closed answer foresees one and only one correct answer: it can be a numerical value or a simple statement.

An open-ended question provides for the independent production of a correct but not univocal solution. Maybe, a better solution is possible, but the alternatives are not completely wrong.

Closed-ended questions are suitable to evaluate the knowledge and understanding of concepts while open-ended questions are suitable to evaluate the ability to apply elaborated knowledge to solve circumscribed cases.

For higher level competences, our experience is that the evaluation is carried out in a less structured and more time-distributed way: elaborated with the help of reviews, internships, degree theses [14].

Closed-ended questions have been implemented in "Moodle", which is the most widespread IT environment for university teaching, at the moment. Figure 3 shows a screenshot of the test.

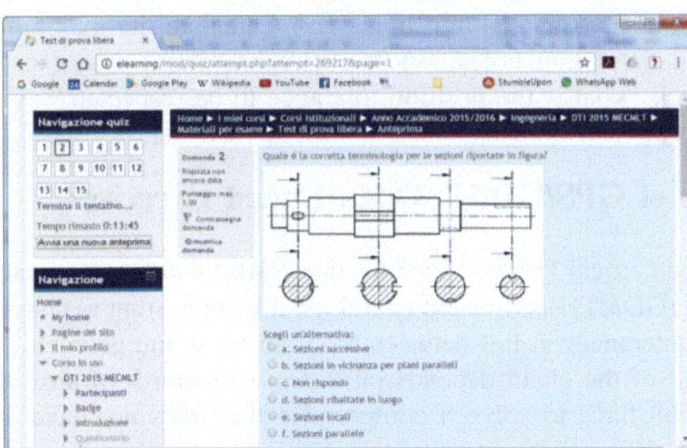

Fig. 3. Screenshot of a closed-ended test.

The open-ended questions were implemented with exercises to be filled, printed on four or five sides of A4 paper.

4 Examples of GPS/GD&T Closed-Ended Exercises

Figure 4 shows a closed-ended exercise, found in [15].

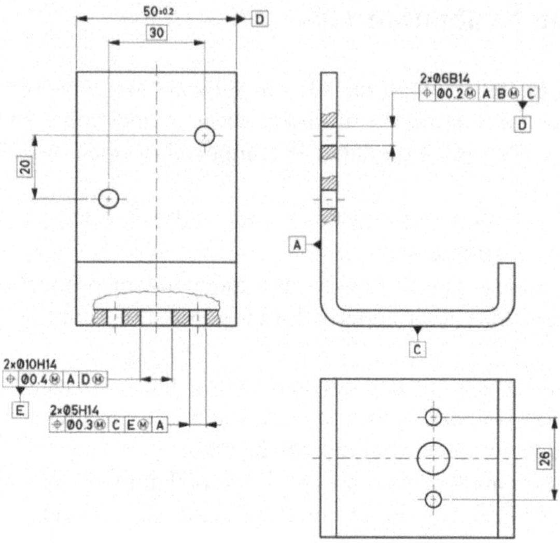

Fig. 4. GPS/GD&T closed-ended exercise [11].

The following closed-ended questions can be asked with regard to the drawing in Fig. 4: (i) What are datum B and D? (ii) How is the datum D simulated? (iii) What is the maximum Bonus for the 10 mm hole? (iv) Calculate the maximum shift for the 10 mm hole and the 5 mm holes; (v) A 10.15 mm diameter is measured from the 10 mm hole check: what is the position tolerance to be respected?

5 Examples of GPS/GD&T Open-Ended Exercises

To introduce Geometrical Product Specification (GPS) and Geometrical Dimensioning and Tolerancing (GD&T) open-ended exercises, it is interesting to consider an exercise on the chain of tolerances: it lies between the traditional and geometric dimensioning. In fact, the result of the chain depends on a sum of features of size, therefore errors propagate and you have to solve a compromise that does not have a single correct answer. Figure 5 shows an example of this type of exercise.

Which criteria do we use to evaluate the exercises carried out? First of all, it is necessary that the examined student has correctly identified the chain. Otherwise, the exercise is null and the correction is interrupted. The next steps are as follows: (i) the sum of the nominal dimensions must produce the nominal value of the mechanical backlash; (ii) the sum of the tolerances on the individual dimensions must produce the tolerance of the mechanical backlash; (iii) the tolerances must produce the minimum and maximum clearance values.

Other types of exercises concern: the application of the various definitions of geometric tolerances; the insertion of a Datum Reference Frame; the application of required geometric tolerances; the insertion of partial references; the dimensioning of partial references.

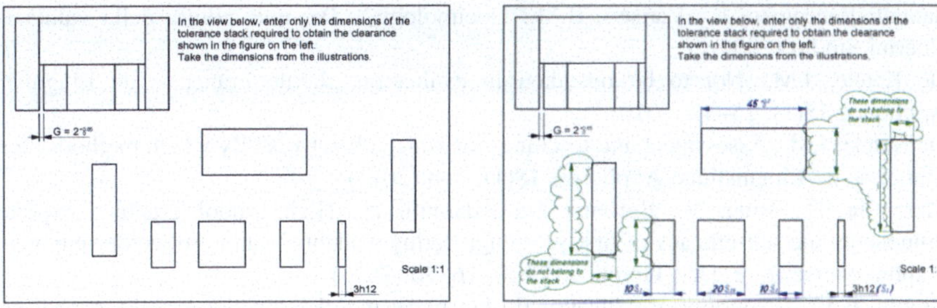

Fig. 5. Example of a chain of tolerances.

6 Conclusions

In general, all evaluation systems are criticized.

The first criticism is that the examination is often seen as an end in itself, and not to learn the subjects of the exam. It should also be added that many studies indicate that at fifteen months of the examination only 20% of the notions in the memory of the examined person. This depends on how the tests were structured: well-structured tests are representative not only of the notions but also of the skills required. In these cases, the long-term storage rate is better, and the proposed tests are structured to be representative of both notions and skills.

An interesting aspect that is often overlooked is that the performance of examinations in the traditional way is in diseconomy of scale: taking the double examination of candidates costs more than twice as much as the time. We simply need to coordinate, replicate structures, repeat controls, etc. When we started to carry out the closed-ended tests extensively, we observed that the procedure was in economies of scale: taking the double examination of candidates costs much less than twice as much as the time.

Moreover, thinking about our examination, we realize that we remember more clearly the things that were asked of us to pass those exams. Lowering the cost of evaluation would allow for more "memorable events". By properly designing the modalities and contents of the tests, a deep learning could be obtained, not only simple notions but also skills and abilities.

Acknowledgments. In the last ten years, many colleagues have contributed to the contents of this article: Riccardo Adamini; Danilo Cambiaghi; Emilio Chirone; Alessandro Copeta; Luca Dassa; Riccardo Metraglia.

References

1. De Landsheere, G.: Évaluation continue et examens. Edition Labor, Précis de docimologie (1992)
2. Benvenuto, G.: Mettere i voti a scuola - Introduzione alla docimologia, Carocci editore (2003)

3. Santelli Beccegato, L., Varisco, B.M.: Docimologia - Per una cultura della valutazione, Guerini Studio (2004)
4. De Ketele, J.M.: Not to be mistaken in evaluation. Revue Francaise de Linguistique Appliquee **15**(1), 25–37 (2010)
5. De Ketele, J.M.: Assessment and teaching: for which objectives? By which methods? Revue Francaise de Linguistique Appliquee **18**(1), 3–8 (2013)
6. Metraglia, R., Villa, V., Baronio, G., Adamini, R.: High school graphics experience influencing the self-efficacy of first-year engineering students in an introductory engineering graphic course. Eng. Des. Graph. J. **79**(3), 16–30 (2015)
7. Bloom, B.S.: Taxonomy of educational objectives: the classification of educational goals. D. McKay, New York (1956)
8. Bloom, B.S.: Human Characteristics and School Learning. McGraw-Hill, New York (1976)
9. Anderson, L.W., Krathwohl, D.R., Bloom, B.S.: A Taxonomy for Learning, Teaching, and Assessing. Longman, Harlow (2001)
10. Scouller, K.: The influence of assessment method on students' learning approaches: multiple choice question examination versus assignment essay. High. Educ. **35**(4), 453–472 (1998)
11. Metraglia, R., Baronio, G., Villa, V.: Learning levels in technical drawing education: proposal for an assessment grid based on the european qualifications framework (EQF). In: ICED 11 - 18th International Conference on Engineering Design - Impacting Society Through Engineering Design, vol. 8, pp. 161–172 (2011)
12. Council recommendation of 22 May 2017 on the European Qualifications Framework for lifelong learning and repealing the recommendation of the European Parliament and of the Council of 23 April 2008 on the establishment of the European Qualifications Framework for lifelong learning (2017/C 189/03)
13. Villa, V., Motyl, B., Paderno, D., Baronio, G.: TDEG based framework and tools for innovation in teaching technical drawing: the example of LaMoo project. Comput. Appl. Eng. Educ. **26**(5), 1293–1305 (2018)
14. Metraglia, R., Baronio, G., Villa, V.: Issues in learning engineering graphics fundamentals: shall we blame cad? In: Proceedings of the International Conference on Engineering Design ICED, vol. 10 (DS 80-10), pp. 31–40 (2015)
15. Chirone, E., Tornincasa, S.: Disegno Tecnico Industriale, vol. II. Il Capitello, Torino (2015)

How to Assess the Impact of ICT Contribution in the Graphic Expression Subjects for Engineering Degrees?

Triviño-Tarradas Paula(✉), Carranza-Cañadas Pilar, Burgos-Ladrón de Guevara Enrique, and Hidalgo-Fernández Rafael Enrique

Departamento de Ingeniería Gráfica y Geomática, Universidad de Córdoba, Campus de Rabanales, Ctra. Nacional IV, km. 396, 14014 Córdoba, Spain
ptrivino@uco.es

Abstract. Traditional lectures are still widely implemented in university teaching. Nevertheless, new approaches based on digital resources and on new technologies are of a high interest for teaching engineering subjects, such as the 'engineering drawing' or 'graphic expression'. The application of Information and Communication Technologies (ICT) is providing new strategies in university teaching, thanks to the digital transformation and to the technological advances occurring in the last decades.

The aim of this research is to provide with a methodology, based on motivational concepts validated, to assess the potential of the ICT implementation for teaching subjects related to graphic engineering in the current university teaching system for engineering degrees. Analysis of both teaching systems, by the determination of the effectiveness of each instructional approach through a benchmarking on traditional lectures and new ICT for 'engineering drawing' subject teaching will be performed in first year students belonging to four different degrees: Industrial Electronic Engineering, Electrical Engineering, Agricultural Engineering and Forestry Engineering.

Keywords: ARCS · Motivation · Engineering drawing

1 Introduction

The constant endeavor to find a variety of teaching methods in education is important to lead to a teaching-learning process healthier and more sustainable [1]. According to some scientific literature, motivation is a corner stone in the teaching-learning process [2]. Therefore, any kind of fact or element encouraging and improving students' motivation would be of high interest to achieve their goals.

Theoretical lectures, which have been traditionally regarded as indispensable when teaching 'systems of representation' or 'engineering drawing' are now been partially replaced by an increasing use of practical sessions in computer-aided design [3]. According to statements of Villa et al. [4], graphical communication skills are fundamental for engineering graduates, and future engineering professionals. In this

F. Cavas-Martínez et al. (Eds.): INGEGRAF 2019, LNME, pp. 313–319, 2020.
https://doi.org/10.1007/978-3-030-41200-5_34

context, digital skills implementation, via the greater use of ICT tools in the teaching-learning process, makes digital competences, acting as communicative competences in general and graphical in particular, would be better understood.

The inclusion of ICT within the Spanish University system was crucial to fully committed to the European Higher Education Area (EHEA) since 2003. Nevertheless, there is not much literature on how the introduction of ICT influences the teaching-learning process in graphical engineering subjects. However, some authors highlight the relevance of ICT technologies have brought to the university teaching, e.g. using Computer-Aided-Design (CAD), increasing the academic results and the student satisfaction level with the subject [3], and the increase of internet resources [5]. Ayala-Alvarez et al. [6] taught the earthworks using other type of ICT tools such as the 3D models. The important role that 3D models use have in improving the better understanding of the teaching-learning process was highlighted. Certainly, any education field should adapt and introduce the needed innovation techniques that help students increasingly face complex problems in the current society [1, 4]. This research aims to provide more insight and knowledge about the utilization of these innovative systems, since they can help improving students' motivation as one of the main factors related to success. Therefore, monitoring motivation will enable having more insight and perception about any influence a (positive or negative) that the introduction of ICT techniques has in the teaching-learning process. In this context, the ARCS model [7] can make the difference, since it has been successfully validated in many research studies in different countries and cultures [8].

Improvements in academic results and in the students' satisfaction level are expected to increase due to an increased utilisation of these innovative and user-friendly techniques [3].

Digital tools provide professors with some different means of the traditional ones to support their work. In addition, these new teaching models based on virtual tools enable students profit on the alternative to physical attendance in the classroom, such as the purely virtual or blended teaching models [7].

ICT utilization enables the online or/and the blended learning. These new technologies offer one of the most relevant improvements to the teaching-learning process, especially in the normal and standard instructions in the Graphic Expression subjects in Engineering courses [2]. Despite the many benefits the e-learning approach can deliver, facilitating students the access of the information without the need of their presence (handouts and presentation material), the online learning has its drawbacks [9].

There is a broad consensus that educational approaches are not a 'one size fits all', since each of them has pros and cons [6, 10].

2 Methodology

2.1 Framework of Study

Description of the subject ('Systems of representation' and 'Engineering drawing'). Both, are divided into two types of educational credits: Theory sessions (36 h) and practical sessions (24 h). The theory is mainly guided on traditional lecturing supported by additional information given through a Moodle online platform. In this context, questionnaires of each part of the theoretical contents are conducted online to all the students. And the practice is mainly based on exercises of the subject. The timing for theory is two sessions of 1.5 h per week for the Industrial Electronic Engineering degree and the Electrical Engineering one, whereas the timing allocated for the Agricultural Engineering and Forestry Engineering is of three 1-h sessions per week. However, the practical timing is the same for all degrees; 2-h sessions long.

There will always be a selected group performing the traditional learning and other one utilizing the innovative one. In order not to unbalance the whole and entire learning of the subject, these groups will shift from every 3 weeks.

2.2 Learning Systems

Traditional system

The traditional system is based on theoretical classroom lectures counting with the professor in the class. This system tends to utilize the conventional material available in place; chalk, blackboard, beamer, transparencies projector.

Innovative system

Whereas, in the innovative system, this prototype of learning is also accompanied with the implementation of ICT techniques, such as online questionnaires online, 3D model software, CAD drawings, Sketches, or any new technology that could be used to improve the students understanding on the subject. Students following the class based on ICT tools, should have an average level on computing tools usage. This requirement should be referred to the students at the time of doing the enrollment.

Theory contents, exercises (data and solutions) could be uploaded to the Moodle platform. This online platform encourages the autonomous learning of students [11]. This online learning approach enables the existence of room for forums, exercises, texts, scoring marks, warnings related to the subject, and even messages to any the participants.

2.3 ARCS Model

According to Keller [8], students' motivation in learning can be tracked and determined measuring four components:

- *Attention (A): material must be able to capture the curiosity, enthusiasm, and interest of students.*
- *Relevance (R): mainly refers to the students' thoughts towards the link between the new element used and introduced into the process and their former experiences.*
- *Confidence (C): this component relies on the expectation of success that students think they will reach at the end of the learning process.*
- *Satisfaction (S): to the positivity and mood with which students face the learning experiences.*

Students would be more motivated to learn when implementation the ICT tools, since the already mentioned factors would certainly enhance [12]. These four factors should be measured by the means of a survey called 'Instructional Materials Motivation Survey' (IMMS), which originally has 35 questions on a five-point Likert scale.

2.4 Assessment of Satisfaction, Motivation and Effectiveness

Students belonging to all the cited Engineering degrees mentioned in point 2.1., will be surveyed about their overall satisfaction and motivation with the system ARCS model modified. Their perceptions of the subject effectiveness in enhancing students' engineering degree skills will be assessed.

After the implementation of the ARCS model to all students, an interview to 25% of the students' sample, randomly selected, will be conducted. The purpose of this interview is to obtain data about the effectiveness of the ICT tools in graphic engineering teaching.

Questions such as:

- How did you find the course using the ICT tools?
- According to you, what were the positive and negative aspects of using ICT tools in learning the subject of graphic engineering?
- Did this method's implementation contribute to your achievement in positive aspect?

Despite conducting the IMMS questionnaire to each student to assess their motivation, their attendance to practical lessons should be also monitored along with the students delivering the required exercises. Additionally, quantitative data, such as the number of students attending the final exam and the students' final marc were collected.

2.5 Data Analysis

After conducting the questionnaires and collecting them, some potential target students will be invited to be interviewed to better understand their views and gain a clear understanding from them. A sensible number of students will be selected for an interview. Data from students' open-ended question will be analysed using content analysis to help identify and summarize relevant themes that arise.

3 Results

The benchmarking among these two different systems will be performed (traditional learning system and blended learning). For making the comparative a survey will be conducted and some interviews. The survey to be conducted should contain questions to be asked to students about these items:

(A) Background
(B) Spatial visualization ability
(C) Learning results
(D) Advantages and disadvantages of ICT tools
(E) Usability of ICT tools (in teaching and in learning)
(F) Acceptance
(G) Students' marks

Table 1 will show the results of the questionnaires conducted to students of the 4 different degrees.

Table 1. Results of the four categories of ARCS.

Items	
Attention	
1	There was something interesting at the beginning of the lesson that got my attention
2	These new materials are eye-catching
3	The quality of the online materials helped to hold my attention
4	This theoretical lesson was so abstract that it was hard to keep my attention on it
5	The theoretical content of this lesson looked unappealing
6	The way the information is arranged on Moodle platform helped keep my attention
7	This lesson has contents that stimulated my curiosity
8	The amount of repetition in this lesson caused me to get bored sometimes
9	I learnt contents that were surprising or unexpected
10	The variety of type of contents (draws, texts, exercises, illustrations) helped keep my attention on the lesson
11	There are so many texts (writing) on the page that it is irritating
Relevant	
12	It is clear to me how the content of this material is related to contents I already know
13	There was material on this lesson that showed me how this content could be important to some people
14	Completing this lesson successfully was important to me
15	The content of this material was relevant to my interest
16	There are explanations of examples of how people use the knowledge of this lesson
17	The content uploaded to the Moodle platform convey the impression that this material is worth knowing for engineering degrees
18	This lesson was not relevant to my needs because I already knew most of it
19	I could relate the content of this lesson to things I have seen, done, or thought about the professional I want to become
20	The content of this lesson will be useful to me
Confident	
21	When I first looked at this lesson, I had the impression that it would be easy for me
22	This material was most difficult to understand that I would like for it to be
23	After reading the introductory information, I felt confident that I knew what I was supposed to learn from this lesson
24	Many of the theoretical content had so many information that it was hard to pick out and remember the important concepts
25	As I worked on this lesson, I was confident that I could learn the content
26	The exercises in this lesson were too difficult
27	After working on this lesson for a while, I was confident that I would be able to pass a test on it
28	I could not really understand quite a bit of the material in this lesson
29	The good organization of the content helped me be confident that I would learn the material

(*continued*)

Table 1. (*continued*)

Items	
Satisfaction	
30	Completing the exercises in this lesson gave me a satisfying feeling of accomplishment
31	I enjoyed this lesson so much that I would like to know more about this topic
32	I really enjoyed studying this lesson
33	The wording of feedback after the exercises, or of other comments in this lesson, helped me feel rewarded for my effort
34	I felt good to successfully complete this lesson
35	It was a pleasure to work on such a well-designed lesson

Some questions to be asked to deepen in the feasibility of the innovative learning are:

- The handout material uploaded on line the University Moodle platform was enjoyable?
- Did it reinforce your interest?
- Did you find it appropriate?
- Did it increase your understanding?

The main goal is determining the feasibility of using each system, with and without innovative techniques, and identify the best learning practices of each approach. In terms of their effectiveness in enhancing student learning of structural engineering subjects.

The autonomous learning should be studied for both systems. In this context, autonomous learning practices in the engineering drawing subject, belonging to most of the engineering degrees is widely utilised. This task is widely helped and supported by means of these groundbreaking materials that can be uploaded to the net, making them available for students.

The results will deepen on the dichotomy between autonomous learning and instructed learning. Each of the studied system could provide with advantages for its particular learning.

4 Conclusions

ICT communicative competence, not understood as a digital competence, but a communicative and dissemination one, is currently integrated to improve the quality of high-level education at the Universities. The use of these technologies should not imply or lead to give up the traditional ones, especially for certain subjects such as Graphic Engineering (Descriptive geometric, spatial skills).

ICT techniques encourage student participation and motivation, and as a result, help improve academic results. Certainly, innovative technologies, assessed and monitored through a validated methodology, bring about an increase in the academic results and student satisfaction with the subject. Therefore, all elements aiming to improve the online learning environment would enhance students' motivation and their involvement and success within the academic sector.

Acknowledgments. The research work reported here was made possible by the thoughtfulness and the experience of all professors belonging to the graphic expression area of the Graphic Engineering and Geomatic Department involved in the learning process for the first-year students of the grades Industrial Electronic Engineering, Electrical Engineering, Agricultural Engineering and Forestry Engineering at the University of Cordoba.

References

1. Akkuzu, N., Akçay, H.: The design of a learning environment based on the theory of multiple intelligence and the study its effectiveness on the achievements, attitudes and retention of students. Procedia Comput. Sci. **3**, 1003–1008 (2011)
2. Ayala-Alvarez, F.J., Blazquez-Parra, E.B., Montes-Tubio, F.: Incorporation of 3D ICT elements into class. Comput. Appl. Eng. Educ. **25**(3), 542–549 (2017)
3. Olmedo-Torre, N., Farrerons-Vidal, O., Lapaz-Castillo, J., Bermúdez-Rodriguez, F.: The influence of ICT on learning in graphic engineering. Procedia – Soc. Behav. Sci. **237**, 737–744 (2017)
4. Villa, V., Motyl, B., Padermo, D., Baronio, G.: TDEG based framework and tools for innovation in teaching technical drawing: the example of LaMoo project. Comput. Educ. Eng. Educ. **26**, 1293–1305 (2018)
5. Pusca, D., Chiliban, M.: Reforming the engineering and design course for the first year engineering students: with or without CAD. In: Proceedings Paper. Balkan Regional Conference on Engineering and Business Education & ICEBE, vol. I and II, 369 p. (2009)
6. Ayala-Alvarez, F.J., Blazquez-Parra, E.B., Montes-Tubio, F.: Improving graphic expression training with 3D models. Visual. Soc. Jpn. **20**, 889–904 (2017)
7. Keller, J.: Development and use of the ARCS model of instructional design. J. Instr. Dev. **10** (3), 2–10 (1987)
8. Keller, J.: Motivational Design for Learning and Performance. The ARCS Model Approach. Springer, New York (2010). https://books.google.es/books?hl=en&lr=&id=HRCQlZzMwhsC&oi=fnd&pg=PR5&dq=motivation+design+for+learning+and+performance+keller&ots=RaFTS2a52G&sig=NJoG2DVrQMPfxfbAT_Eg-DruAYk#v=onepage&q=motivation%20design%20for%20learning%20and%20performance%20keller&f=false. Accessed 28 Jan 2019
9. Heinze, A., Procter, C.: Reflection on the use of blended learning. Education in a changing environment, pp. 1–12, September 2004. http://usir.salford.ac.uk/1658/1/4247745025H__CP_-_paper9_5.pdf. Accessed 23 Jan 2019
10. Ruiz de Miras, J., Expósito, D., Rocha, D., Robles, M.D.: A web application to support teaching of computer graphics to engineering students. Int. J. Eng. Educ. **34**(1) 34–44 (2018)
11. Garcia-Rubio, A., Diaz-Morilla, A., Vereda-Alonso, C., Gomez-Lahoz, C., Garcia-Herruzo, F., Rodriguez- Maroto, J.M.: Comparative analysis of the teaching enhancements produced by the use of the information and communication technologies (ICT) in two subjects of different science degrees: academic results. In: Proceedings of ICERI2012 Conference, 19th–21st November 2012, Madrid, Spain (2012)
12. Bolliger, D.U., Supanalorn, S., Boggs, C.: Impact of podcasting on student motivation in the online learning environment. Comput. Educ. **55**, 714–722 (2010)

Industrial Design as a Tool for Enhancing Entrepreneurship

Cristina Martin-Doñate[1](✉), Fermin Lucena-Muñoz[2], Lina Garcia-Cabrera[3], Elisabeth Estevez[4], and Jorge Manuel Mercado-Colmenero[1]

[1] Department of Engineering Graphics, Design and Projects, University of Jaen, Campus Las Lagunillas, s/n, 23071 Jaen, Spain
cdonate@ujaen.es
[2] Department of Management, Marketing and Sociology, University of Jaen, Jaen, Spain
[3] Department of Computer Sciences, University of Jaen, Jaen, Spain
[4] Department of Electronics and Automatic Engineering, University of Jaen, Jaen, Spain

Abstract. The entrepreneurial spirit is currently considered as the engine of a more dynamic industrial society. Creativity, as well as the capacity to take new initiatives, are fundamental values for students, with these capacities being increasingly in demand by companies. However, despite the importance of developing these skills, subjects in the technical field do not include activities that contextualize theoretical knowledge in the entrepreneurship field. This paper presents the results of a set of activities related to the development of entrepreneurial capacity in students of mechanical engineering by means of new activities related to the business field in the development of new mechanical designs. The designs developed have been planned as technical solutions to problems of some industrial companies. The design results have been presented in the form of prototypes. In a parallel manner, the students have complemented the product development project with a product business model. The results obtained show an increase in the capacity of transferring industrial knowledge from the classrooms to the industrial environment as well as a growth in motivation on the part of the students who have carried out the innovative activities related to entrepreneurship in design.

Keywords: Mechanical design · Entrepreneurship · CAD · Product development

1 Introduction

Entrepreneurship has emerged as an important topic in economic development. Defined as the act of organizing resources to initiate commercial activity, entrepreneurship has been studied extensively from a variety of perspectives [1]. An entrepreneur in design must have the skills of both an industrial designer and a business manager. However, although in recent years the field of entrepreneurship has taken on great importance as a

F. Cavas-Martínez et al. (Eds.): INGEGRAF 2019, LNME, pp. 320–329, 2020.
https://doi.org/10.1007/978-3-030-41200-5_35

means of self-employment in the field of design, academic education still focuses on the teaching of CAD and manufacturing subjects. The four main components of NPD (New Product Development) are classified into: Research, Design (including engineering and industrial design), Marketing and Manufacturing [2]. These tasks are usually carried out by specialists in each of the areas indicated, especially in large organizations with large-scale projects. Teaching in design has focused on preparing professionals according to this technical profile. However, every day the demand for small-scale projects increases, there being a great need for technical professionals with an adaptable profile and flexible skills. The business professional in design demands a different model of training. This trend presents a very different professional landscape compared to the past industrial structure that created the field of product design. At the present time there are many more options available for industrial design graduates, in addition to working for large manufacturers or design consultancies [3]. Students can produce and market their own designs or work with colleagues to create a start-up company.

The current environment demands a new type of design engineer characterized by thinking widely in different disciplines such as creativity, sustainability [4, 5] and human requirements [6]. Engineers have many of the skills needed to be successful entrepreneurs. The 21st century engineer must have common sense and a broad technical background, and be able to make quick information searches using this information to develop a project. The purpose of business engineering is to design products and value-added processes that generate demand through innovation [7].

The development of the entrepreneurial capacity in engineering requires a significant change in the training processes integrating educational and research systems in order to generate and experience sustainable creative processes and practices [8, 9]. However, although academic institutions are the most important part of this training process they must be complemented with actions and initiatives undertaken by other institutions outside the educational system. Entrepreneurship requires ecosystems in which multiple stakeholders play key roles. The entrepreneurial engineer can understand basic engineering concepts and quickly assess what needs to be done, and can acquire the necessary tools and use these tools competently. This engineer must know how to work in a team and in countries other than his own and to have the necessary communication skills and understanding of current global problems to work effectively with other people. Additionally, the engineer must have the ability to conceptualize and turn the concept into reality. An engineer with these characteristics requires a business spirit, broad imagination and management skills to identify needs and find new solutions.

This study presents the results of a set of activities related to the development of entrepreneurial capacity in students of mechanical engineering by means of new activities related to the business field in the development of new mechanical designs.

The designs developed have been planned as technical solutions to problems of some industrial companies. The design results have been presented in the form of prototypes. In a parallel manner, the students have complemented the product development project with a product business model.

2 Materials and Methods

With the aim of developing entrepreneurial skills in students of mechanical engineering, a set of activities was carried out during the academic year 2018/2019. These activities have been framed in an entrepreneurship project funded by Jaen University. The vast majority of students who study mechanical engineering have planned to develop their work in large companies. However, it is not always possible for students to find a qualified job in this environment. Self-employment is considered as an alternative to depopulation in less industrialized environments. However, engineering students do not have the tools to propose an initial business plan, mainly because they finish their studies with a solely technical profile.

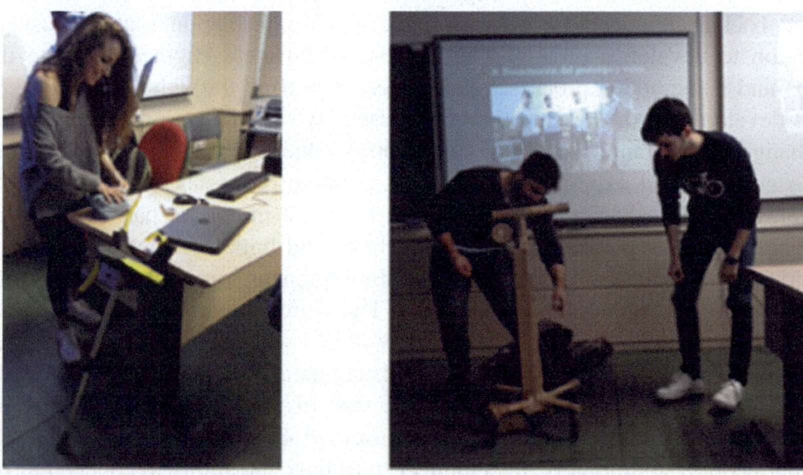

Fig. 1. Motivation of the students with the activities proposed in the project

With this objective, a set of activities consisting of developing new products and proposing a business plan based on product design were carried out. The purpose of these activities has been to motivate and train mechanical engineering students in the use of industrial design as a tool for self-employment (see Fig. 1).

Fig. 2. Presentation of the problem to be solved

An industrial company from the plastics sector has collaborated in the entrepreneurship project. The company presented to the mechanical engineering students a design problem to solve. This design problem consisted of developing orthopedic aid devices for customers with a lack of physical mobility (Fig. 2).

Fig. 3. Seminars for business training and self-employment

Additionally, in order to complement the student's training in entrepreneurship issues several seminars related to business model methodologies were given. These seminars were taught by professors of the area of business organization and professional experts in entrepreneurship (Fig. 3). The goal was to help the mechanical engineering student to understand that obtaining the design concept is only a small part of the process of creating a business. The tool used to propose the business model based on product was the Canvas methodology. The Canvas diagram is a practical and simple tool that allows to perform a global analysis of all the important aspects that make up the business model, working on problems as a team. This method is based on an analysis of the environment in which the company is going to operate: identifying the target customer segment; how to reach customers, the type of relationship with customers and finally the means of payment. After analyzing the company's environment, the "value proposal" had to be prepared through alliances with partners, studying in detail the cost structure of the product. In this way it is possible to familiarize engineering students with technical training with business language and terminology, which will improve communication with future clients.

Fig. 4. Visit to the business incubator and training with entrepreneurs

The activities were complemented with several talks given by staff specialized in entrepreneurship, and visits to business incubators (Fig. 4) where students were able to talk and discuss their doubts with entrepreneurs, who told them their experience in creating a startup (Figs. 5, 6 and 7).

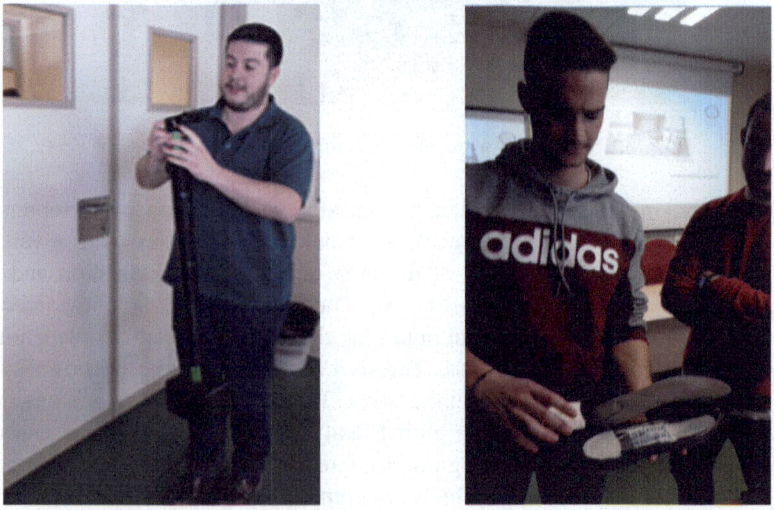

Fig. 5. The students present their designs to the representatives of the company

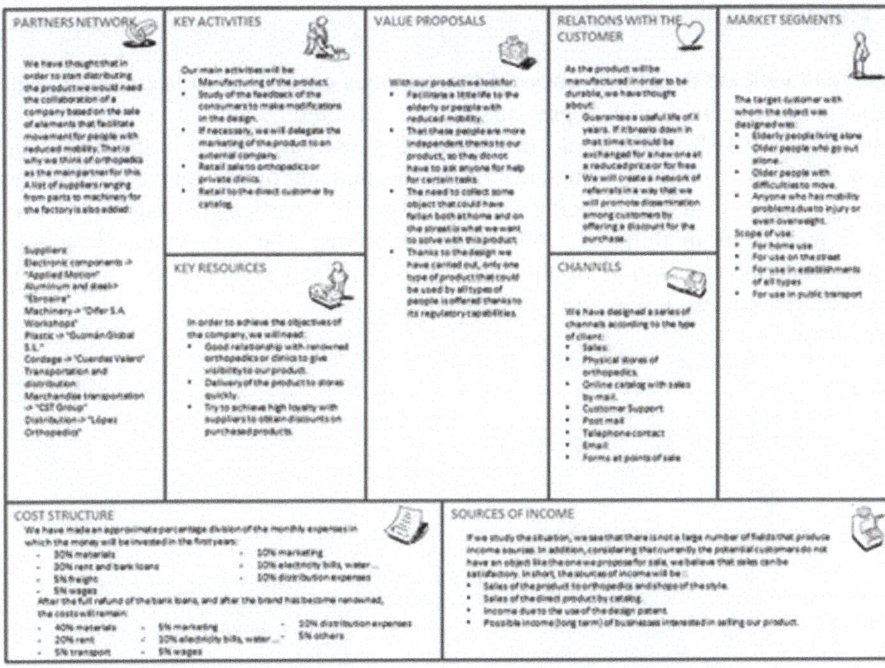

Fig. 6. Canvas model based on product completed by the students

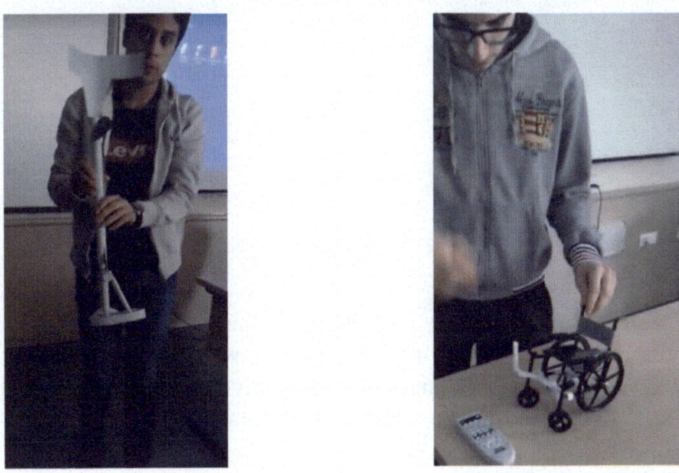

Fig. 7. Selected minimum viable products

3 Results

In order to validate the effectiveness of the activities carried out concerning students´ motivation in the field of entrepreneurship and self-employment, a survey was conducted on the ILIAS platform of the students that had participated in these activities. The survey was completed voluntarily by all the students of the course graphic engineering in mechanical degree, being a total number of 51 participant students (41 males and 10 female) from different working design teams. The survey aimed to assess whether the implementation of this activity had helped students to improve their knowledge about business models in industrial design.

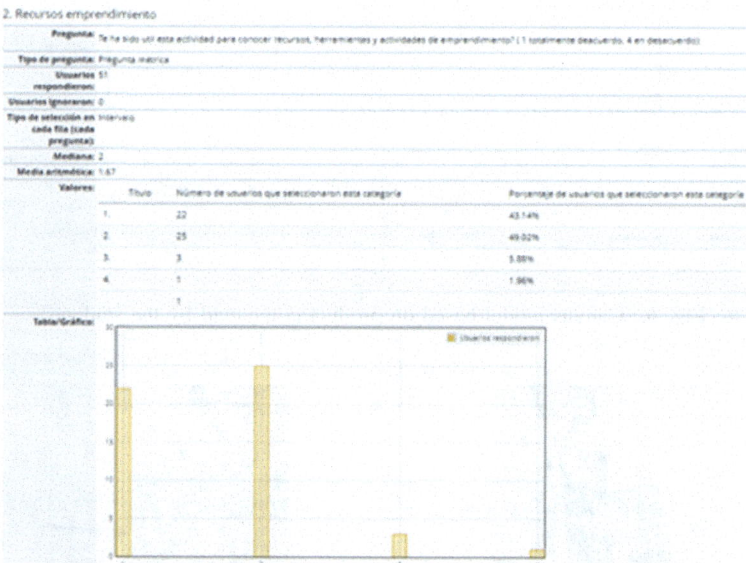

Fig. 8. Survey results

Students were asked about the usefulness of the activities carried out for their training in entrepreneurship. Following a criterion of assessment between [1 fully agree and 4 totally disagree] Students answered affirmatively about the usefulness of the activities in 92.17% of the total answers. 43.14% totally in agreement and 49.02% in agreement (Fig. 8).

Fig. 9. Survey results

Students were surveyed about the consideration of entrepreneurship as a future employment option, to which 94.13% answered affirmatively, with 56% totally agreeing and 37.25% agreeing. The results of the surveys have been indicated in Fig. 9.

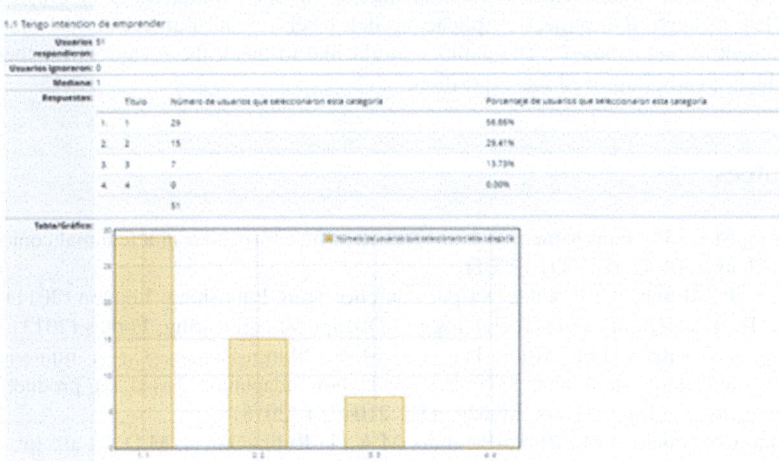

Fig. 10. Survey results

One of the main objectives of the project has been to encourage the motivation for entrepreneurship and self-employment in engineering students, for which the students were asked about their intention to create a start up at the end of their studies, to which

86,27% answered affirmatively, with 56,86% totally agreeing and 29,41% agreeing. The results of the surveys have been indicated in Fig. 10.

4 Conclusions

The development of the entrepreneurial capacity in engineering requires a significant change in the training processes, integrating educational and research systems to generate and experience sustainable creative processes and practices. Following this line, in the course 2018/19 in the degree of mechanical engineering at the University of Jaen a set of activities related to developing the entrepreneurial capacity in students of mechanical engineering have been realized. Third-year students in groups created several designs according to the requirements imposed by an environmental company, solving real problems. Additionally, students proposed a product-based business model for each of their designs using a Canvas diagram. The training activities in CAD and design methodology have been complemented with seminars and practical classes related to the creation of a startup. Designs were presented to members of the company who reviewed and complemented the proposals of the students from a technical and commercial point of view, improving students learning process. The results obtained from the surveys carried out indicate a great interest in entrepreneurship, with 86,27% of the number of participating students responding affirmatively to the option of self-employment as a way of working at the end of their studies.

Acknowledgments. This research work was supported by the University of Jaen as part of the Plan de Apoyo a la Transferencia del Conocimiento, el Emprendimiento y la Empleabilidad (2018-2019) through the project "Aplicación del diseño industrial como herramienta de emprendimiento y autoempleo". The authors would like to thank the reviewers for their comments, which greatly improved the exposition.

References

1. Feldman, M.P.: The entrepreneurial event revisited: firm formation in a regional context. Ind. Corp. Change **10**(4), 861–891 (2001)
2. Rodgers, P., Milton, A.: Product Design. Laurence King Publishing, London (2011)
3. Upbin, B.: How Quirky Uses Technology to Disrupt Manufacturing. Forbes (2013)
4. Mercado-Colmenero, J.M., Rubio-Paramio, M.A., Martin-Doñate, C.: A numerical and experimental study of a new Savonius wind rotor adaptation based on product design requirements. Energy Convers. Manag. **158**, 210–234 (2018)
5. Mercado-Colmenero, J.M., Rubio-Paramio, M.A., la Rubia-Garcia, M.D., et al.: Int. J. Adv. Manuf. Technol. (2019). https://doi.org/10.1007/s00170-019-03626-0
6. Habash, R.: Green Engineering: Innovation, Entrepreneurship and Design. CRC Press, Boco Raton (2017)
7. Kriewall, T.J., Mekemson, K.: Instilling the entrepreneurial mindset into engineering undergraduates. J. Eng. Entrep. **1**(1), 5–19 (2010)

8. Martin-Doñate, C., Rubio-Paramio, M.A., Montalvo-Gil, J.M., Mercado-Colmenero, J.M.: Application of creativity and competition analysis in teaching two-dimensional representation of mechanical assemblies. In: EDULEARN16 Proceedings, pp. 7215–7223 (2016)
9. Martin-Doñate, C., Lucena-Muñoz, F., Gallego-Alvarez, J.: Integration of marketing activities in the mechanical design process. In: Eynard, B., Nigrelli, V., Oliveri, S., Peris-Fajarnes, G., Rizzuti, S. (eds) Advances on Mechanics, Design Engineering and Manufacturing. Lecture Notes in Mechanical Engineering. Springer, Cham (2017)

On the Internationalization of CAD Learning Through an English Glossary

Raquel Plumed[1(✉)], Victoria Pérez-Belis[1], Anna Agustín-Claramonte[2], Néstor Jarque-Bou[1], Carmen González-Lluch[1], and Ana Piquer[1]

[1] Department of Mechanical Engineering and Construction, Universitat Jaume I, Av. Sos Baynat s/n, 12071 Castellon, Spain
plumed@uji.es
[2] IES Matilde Salvador, Av. de Casalduch 120, 12005 Castellon, Spain

Abstract. The internationalization of higher education is an essential factor to improve the quality and efficiency of Spanish universities, providing students with the main skills, and knowledge to interact effectively in an international and multicultural work context as professionals. The internationalization of universities must be a transversal process, not exclusive of its territorial dimension, aimed at advancing towards a society and a knowledge economy that propitiate a solid and stable model of development and growth. To this end, professors in the area of Graphic Expression for Engineering at the Universitat Jaume I (UJI) have developed an online glossary of specific terms in English related to the 3D modelling CAD tools used in Graphic Engineering subjects. This new online tool seeks to train students to increase their technical vocabulary in English and improve their learning and communication skills to face possible collaborations in future European projects. The glossary is introduced weekly to the students during the course. Subsequently, a survey is conducted to the students to verify the effectiveness of the training. This work collects the results and conclusions of this analysis.

Keywords: Online glossary · 3D modelling · Engineering graphics

1 Introduction

Nowadays employers prefer to take employees who have soft skills or a professional training in transversal competencies. The word soft skill is a sociological term referring EQ-the Emotional intelligence quotient of a person [1, 2]. Soft skills are the abilities that help to become a complete professional. On the other hand, hard skills are the technical skills achieved by the students mainly in the traditional academic courses. During the last years, university programs have increased in emphasis on soft skills and particularly communication skills.

Globalization demands a more competitive workforce and opens the labor market to a broader area than a local region. Besides, globalization forces companies to work in an international environment. Different languages should not be an obstacle in the globalized market. According to the number of native speakers, the three most spoken

F. Cavas-Martínez et al. (Eds.): INGEGRAF 2019, LNME, pp. 330–338, 2020.
https://doi.org/10.1007/978-3-030-41200-5_36

languages in the world are in order: Chinese, Spanish and English. But according to the total number of people who speak a language, whether or not the language is their mother tongue, the ranking changes. The order in this case is: English, Chinese and Hindi. Spanish is in the fourth place according to Ethnologue's study [3]. UJI university is located in Castellon (Spain), then Spanish is a mother tongue for most of the students. Nevertheless, demands of English skills by potential employers and professional bodies reflect the great importance of including English language communication skills in the university curricula.

Internationalization is also part of the university aims. Students can take a semester or a course in other university all around the world with a similar technical curriculum to the original grade at their own one.

The authors are lecturers and professors in the area of Graphic Expression for Engineering at the UJI university. The present work is focused on the subject "Technical Graphics Communication". This subject is taught in the second semester of the first course in Engineering: Industrial Engineering, Chemical Engineering, Electrical Engineering and Mechanical Engineering. With the aim of introducing students to English language communication skills, and considering that virtual learning allows the interactivity and promotes the motivation, efficiency and the improvement of the knowledge in a flexible environment [4], they have prepared an English glossary including terms related with the subject.

This paper is structured in the next sections: Sect. 2 describes the methodology employed to design and create the educational website and the study conducted on the usability of the website by the users, Sect. 3 presents the analysis of the results and Sect. 4 sums up the main conclusions obtained.

2 Methodology

Three main stages were followed in order to complete the experience: the selection and definition of main terms, the design and use of the educational website and finally, the definition of a survey for students to evaluate the glossary and the usability of the educational website.

2.1 Selection and Definition of Specific Terms of 3D Modelling CAD Tools

After a review of different material related to 3D modelling, such as books [5, 6], website pages [7], and support provided by different CAD software's [8], a list of 40 terms was defined. These were divided between two main groups: those related to the sketch stage (previous step required to define the geometry of the figure) and some others related to modelling (providing a solid feature from a previous sketch). The list of several terms was reviewed by teachers, who were responsible for the selection of the 21 final terms. Table 1 shows the terms selected and defined to be used by students.

Table 1. Terms selected by teachers.

Content	Term	Content	Term
Sketch terms	Main plains	Model terms	Extrusion
	Origin		Revolution
	Circle		Loft
	Line		Swept
	Arc		Fillet
	Polygon		Feature tree
	Spline		
	Axis		
	Fix constraint		
	Dimensional constraints		
	Geometrical constraints		
	Over-constrained sketch		
	Fully constrained sketch		
	Under-constrained sketch		

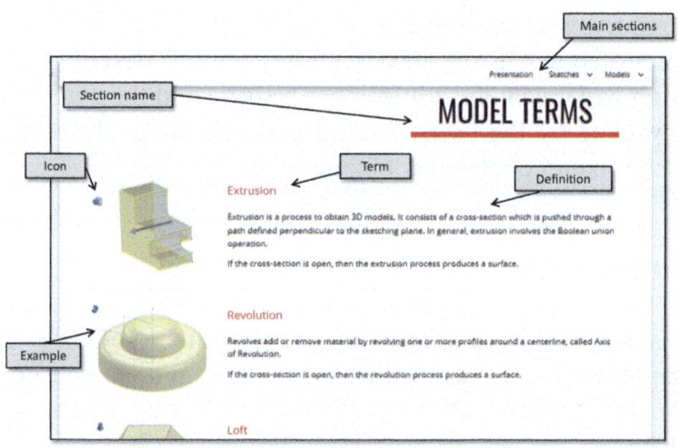

Fig. 1. Website caption.

2.2 Design and Use of the Educational Website

Once the terms were selected, an educational website page was developed in order to host the glossary for students [9]. The website was developed and designed using Google Sites, as it takes advantage of working in the cloud (allowing collaboration of all teachers) and no hosting fees.

The aim of the website design was to provide an easy and intuitive tool for students to be used. Figure 1 shows and example of the design of the website.

2.3 Definition of a Survey for Students

After students were using the website during some sessions, a survey to evaluate their usefulness was developed in order to know their opinion about that.

The survey consists of three main blocks. The first one helps to characterize the basic traits of the people who take part in the study (gender, bachelor degree, and level of English).

The second block extracts information on the general opinion that the proposed resource deserves the participants. Some of the questions formulated in this block are open-ended to receive comments and ideas for improvement proposed by the participants.

The third block is related to a usability study. To obtain a consistent result we have applied a standard analysis known as System Usability Scale (SUS). The SUS was invented by John Brooke in 1996 [10, 11] and has proven to be a reliable method to evaluate the usability of the systems compared to industry standards. It consists of a 10-item questionnaire with a Likert scale. Participants rank each question from “strongly agree” to “strongly disagree”. These responses are subsequently translated into a numerical score. Applying the SUS calculation method, a usability score is obtained in the range of 0–100. The results do not represent a percentage, but a score to measure the level of usability when users interact with the website. The average score of the SUS is 68. With a score of 80.3 or higher, users express their full satisfaction with the use of the website. If the score is less than 51, the website must be redesigned to improve user satisfaction. Questions of the survey are shown in Table 2.

Table 2. Questions of the students’ survey.

Aim	Question	Possible answer
To know the profile of students	Please, indicate your bachelor degree	Industrial engineering
		Chemical engineering
		Electrical engineering
		Mechanical engineering
	Gender	Male
		Female
		Other
	English level	Lower intermediate B1 (PET)
		Intermediate B2 (FIRST)
		Upper intermediate C1 (ADVANCED)
		Advanced C2 (PROFICIENCY)
		Pre-intermediate A2
		Without English studies
To know the need of the glossary	Do you think it is necessary to introduce English in this type of subjects? Why?	Open answer
	Do you consider that its domain in this area is necessary for your professional future? Why?	Open answer

(*continued*)

Table 2. (*continued*)

Aim	Question	Possible answer
To know the usefulness of the glossary	Have you found the terms shown on the website useful?	Yes/No/Other
	Which terms have you found most useful?	1. Strongly disagree/2. Disagree/3. Neither agree nor disagree/4. Agree/5. Strongly agree
	Which other terms would you add?	Open answer
	Which other sections do you think may be helpful?	Evaluation questionnaires
		Games to learn the vocabulary
		Spelling exercises of the basic terms
		Tutorials in English
SUS study	I think I would use this website frequently	1. Strongly disagree/2. Disagree/3. Neither agree nor disagree/4. Agree/5. Strongly agree
	I find this website unnecessarily complex	1. Strongly disagree/2. Disagree/3. Neither agree nor disagree/4. Agree/5. Strongly agree
	I think the website is easy to use	1. Strongly disagree/2. Disagree/3. Neither agree nor disagree/4. Agree/5. Strongly agree
	I think I would need help from a person with technical knowledge to use this website	1. Strongly disagree/2. Disagree/3. Neither agree nor disagree/4. Agree/5. Strongly agree
	The functions of this website are well integrated	1. Strongly disagree/2. Disagree/3. Neither agree nor disagree/4. Agree/5. Strongly agree
	I think the website is very inconsistent	1. Strongly disagree/2. Disagree/3. Neither agree nor disagree/4. Agree/5. Strongly agree
	I imagine that most people would learn to use this website very quickly	1. Strongly disagree/2. Disagree/3. Neither agree nor disagree/4. Agree/5. Strongly agree
	I find that the website is very difficult to use	1. Strongly disagree/2. Disagree/3. Neither agree nor disagree/4. Agree/5. Strongly agree
	I feel confident in using this website	1. Strongly disagree/2. Disagree/3. Neither agree nor disagree/4. Agree/5. Strongly agree
	You need to learn many things before being able to use this website	1. Strongly disagree/2. Disagree/3. Neither agree nor disagree/4. Agree/5. Strongly agree

3 Results

A total of 50 students answered the online questionnaire, all of them enrolled in the first-year subject of Technical Drawing Communication. Students belonged to different Engineering degrees, specifically the sample consists of 7 students of chemical engineering, 7 of electrical engineering, 18 of industrial engineering and 19 of mechanical engineering. Of the total participants, 15 were women and 35 were men.

The participating students signed electronically a data protection agreement. This document showed the questionnaire's purpose and explained that teachers assure confidentiality and protect the personal data handled.

As Fig. 2 shows, more than half of the participants has an intermediate level of English (level B1 or B2).

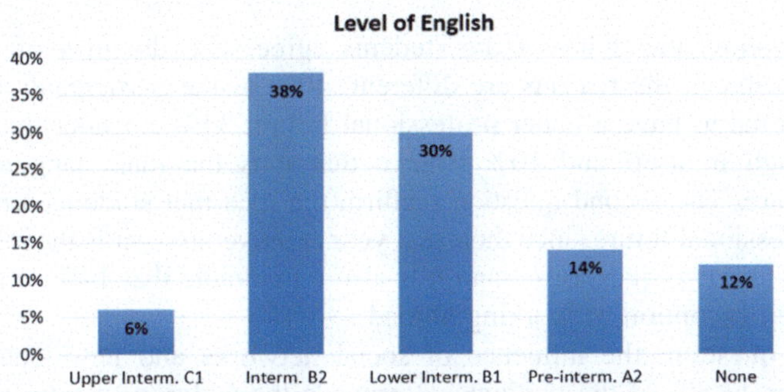

Fig. 2. Level of English of the participants.

Table 3. Questions and relative frequencies in answers of the second block.

Questions	Answers	Rel. Freq.
Do you think it is necessary to introduce English in this type of subjects? Why?	Yes, to have a better professional future	40%
	Yes, because English is important	34%
	Yes, because many software works in English	10%
	No, it is not necessary	10%
	No, that would only complicate the subject	6%
Do you consider that its domain in this area is necessary for your professional future? Why?	Yes, to work in international companies	8%
	Yes, to work abroad	14%
	Yes, to have a better professional future	34%
	Yes, because English is important	29%
	Not much in this area.	6%
	DK/NA	10%

(continued)

Table 3. (*continued*)

Questions	Answers	Rel. Freq.
Which other terms would you add?	I would not add anything else	26%
	More modelling terms	18%
	More detailed explanations	8%
	Tutorials from the teachers	8%
	More exercises/examples	6%
	DK/NA	34%

Table 3 presents a summary of the open questions (first column) and the answers obtained (second column). Answers are described by their relative frequency (right column).

The results show that 84% of the students agree with the idea of introducing English in the subject, the reasons are different: 40% of the participants believe that English is essential to have a better professional future; 34% consider that English is important enough in itself and 10% believe that it is the main language used in technical software. The second question reaffirm the idea that students are concerned about their professional future since their first year at university—it is the most frequent response in the first two questions—and it is also remarkable that 14% of the students already consider the option of working abroad.

In the last question, the influence of social networks and new technologies is evident when up to 8% of the participants propose the addition of tutorials from the teachers.

In general, the participants consider that the terms defined in the website glossary are useful—reflected by 88% of the polled people (Fig. 3 left)—and there are not many differences in the preference between the terms referred to sketches or models according to Fig. 3 right: 74% of the participants consider the terms related to sketches useful or very useful and 78% for the terms referred to the models.

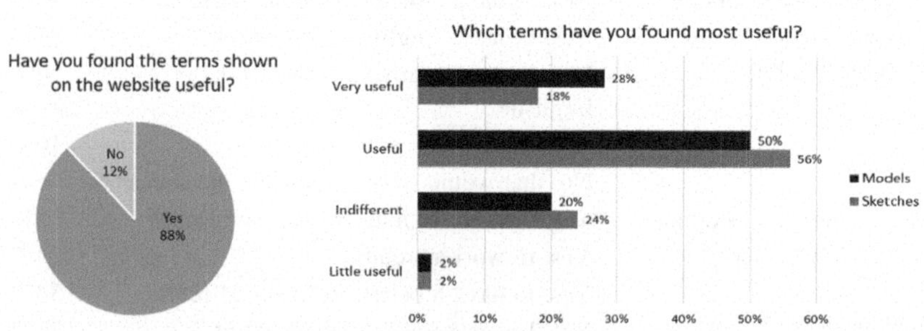

Fig. 3. Equation formatting.

Finally, after collecting the scores given by the participants in the SUS analysis to measure the usability of the website, the correspondent calculation is applied. The average score obtained is 68.5. This means that users define the Glossary website as a useful resource; however, it can still be improved.

As future work, the teachers involved in the development of this tool will have to continue working to make it more usable. A first step should be to analyse the ideas and proposals launched by the students in order to adapt the glossary to their tastes and needs.

4 Conclusions

The knowledge of several languages in general and of English in particular has become a challenge and a necessity for students of engineering degrees.

It is important that students do a gradual immersion in English from the first year of their studies. For this purpose, a website glossary of terms in English related to the subject of Technical Graphic Communication has been developed.

The paper also describes the subsequent questionnaire addressed to engineering students to know their opinion and evaluate the usability of the tool.

Results show that students are aware of the importance of a knowledge of English, both personally and professionally.

The SUS test reveals that students value the usability of the developed website resource in a positive way. At the same time, students have launched some ideas and proposals to improve the glossary website. Then, the glossary website has turned out to be an effective resource to work on immersing students in English. It also represents an element of continuous improvement in teaching.

Acknowledgments. The research work reported here was made possible by the Innovative Education Project: "Aprofundiment en conceptes de modelatge 3D parametric" USE - UJI.

References

1. Choudary, V.: The importance of training engineering students in soft-skills. Int. Mon. Ref. J. Res. Manag. Technol. **III**, 61–64 (2014)
2. Schulz, B.: The importance of soft skills: Education beyond academic knowledge. J. Lang. Commun. **2**, 146–154 (2008)
3. Summary by language size. Ethnologue. https://www.ethnologue.com/. Accessed 03 Apr 2019
4. López de la Madrid, M.: Uso de las TIC en la educación superior de México. Un estudio de caso. Apertura **7**, 63–81 (2007)
5. Bertoline, G.R., Wiebe, E.N., Hartman, N., Ross, W.: Fundamentals of Graphics Communication. McGraw-Hill, New York (2010)
6. Lieu, D.K., Sorby, S.: Visualization, Modeling and Graphics for Engineering Design. Delmar, Cengage Learning, Boston (2009)
7. Technical graphics communication glossary. http://highered.mheducation.com/sites/0073655988/student_view0/glossary27/index.html. Accessed 30 Mar 2019

8. SolidWorks. http://help.solidworks.com/2018/English/SolidWorks/sldworks/c_solidworks_web_site.htm. Accessed 30 Mar 2019
9. CAD 3D English glossary. https://sites.google.com/uji.es/cad-3d-english-glossary/presentation. Accessed 01 Apr 2019
10. Brooke, J.: SUS: a "quick and dirty" usability scale. In: Jordan, W., Thomas, B., Weerdmeester, B.A., McClelland, A.L. (eds.) Usability Evaluation in Industry. Taylor and Francis, London (1996)
11. Brooke, J.: SUS: a retrospective. J. Usability Stud. **8**(2), 29–40 (2013)

Analysis of Spatial Comprehension Through a Retrospective Study of Its Effects Among First Year Electrical Engineering Students

Manuel-Damián Marín-Granados(✉),
Fernando Gómez-Hermosa,
María-del-Carmen Ladrón-de-Guevara-López,
Laia Miravet-Garret, and Francisco-Javier Soto-Lara

University of Málaga, 29017 Málaga, Spain
mdmarin@uma.es

Abstract. To achieve the competences required by engineering students, spatial reasoning skills—understood as the ability to mentally visualise, position, and manipulate an object—are of critical importance. Whilst some authors regard these skills to be innate, other studies show that these can be trained with interactive exercises. The objective of the present study is to identify the impact of these skills in our students, regardless of the metric with which these competences are measured. In the School of Industrial Engineering at the University of Malaga, students took part in a practical exercise in spatial reasoning by reconstructing 3D objects from 2D images. In the present study, the results of more than 400 engineering students were analysed during the periods of the 2012/13 and 2018/19 courses. In addition to describing the time course of the results obtained, we also compared the improvements in performance over time.

Keywords: Spatial reasoning · Spatial ability · Spatial vision · Graphic expression

1 Introduction

Spatial skills are linked to success in technical professions [1]. Among our students, these skills can either be measured directly through specific diagnostic tests, or indirectly through spatial control exercises within the assessment tests used in graphic expression subjects. There are a number of studies that diagnose these skills directly and that also achieve, through a corresponding intervention [2], an improvement in such skills. For most of the studies the diagnosis of these skills is largely based on the use of visuospatial tests [3–5] both before and after the intervention. There are also specific tasks or exercises that can be developed for identifying levels of spatial comprehension without employing any direct measurement of these skills.

For various authors [1, 6–9] the spatial capacity of an individual can be divided into two components; spatial orientation—understood as the ability to imagine an object from another perspective—and spatial visualization, which is the capacity to mentally manipulate objects. These skills are put into practice by obtaining views from a physical model and the inverse process, reconstructing a model from its views.

F. Cavas-Martínez et al. (Eds.): INGEGRAF 2019, LNME, pp. 339–346, 2020.
https://doi.org/10.1007/978-3-030-41200-5_37

In this study we analysed the results of spatial comprehension exercises from the Degree in Electrical Engineering at the University of Malaga. The sample consisted of 438 students, evaluated between February 2013 and February 2019, and enrolled in the first year course of Graphic Expression in Engineering.

It is also of interest to note the distinction that is made between the different forms of learning and their influence on the subjects of Graphic Expression. In particular, visual learning, "learns by seeing the specific information presented", whilst the kinaesthetic learning style involves "learning through experience" which is easily distinguished from the auditory style, in which learning involves listening to "instructions and verbal explanations" [10].

Polycube models were used in an experimental test to reinforce spatial visualization. This experiment was conducted during the course 2018/19, and we assessed the impact of this experience on the levels of performance achieved in the spatial comprehension exercises.

Future research will analyse the correlation between the results obtained on the spatial comprehension exercises with those of the visuospatial tests.

1.1 Previous Works

In the analysis of Beltrán [2], the study focused on obtaining the views through different techniques such as traditional teaching or the use of models and non-computer multimedia support. In their studies, Casati Calzada et al. [11] and Melían Díaz et al. [12] focus more specifically on the creation of models as the main means to achieve an improvement in spatial awareness. Melián Díaz et al. [12] report that when students build topographic models from digital 3D models this helps to improve the interpretation of relief and other concepts associated with the three-dimensional interpretation of terrain forms. Casati et al. [11] confirmed that this improvement is evidenced in the reduced dropout rates, as well as through the results of a survey conducted with the students.

Some investigations have also explored the potential for different didactic techniques to improve spatial skills. For example, Beltrán Polaina [2] makes a distinction between the following three educational techniques: the traditional method through slate tracings, the 'shadow mode', and the 'view mode', which consist of projecting shadows by means of a light source and a wall, or the projection of images by replacing the light source with a video camera. In spite of confirming that the latter two educational techniques are the most effective, this study does not rule out the possibility that the traditional method could be more effective in the later stages of student learning, once the student has reached certain levels of perception and spatial comprehension.

More recently, with the arrival of virtual realiy (VR) and augmented reality (AR) technologies, the possible contribution of these tools to the improvement of spatial reasoning has also been investigated. Martín Gutiérrez et al. [13] conclude that the use of RA has a positive effect on learning and improvement of spatial abilities, while García Domínguez et al. [14] and Ha Fang [15] have combined the use of physical objects with computer graphics to achieve the same purpose. Other authors

[16] investigate the possible improvement in spatial visualization derived from the frequent use of video games.

Unlike the experiences that use VR and AR, the previously described studies all share in common the use of 3D elements, tangible, and perceptible in physical space to obtain 2D elements as a result, establishing a link between what is drawn in 2D and the object/volumetric space associated with said drawing. The work presented here aims to strengthen this link through the use of physical polycube parts, as explained below.

2 Materials and Methods

During a semester, for students in the first year of the Degree in Electrical Engineering, through the subject of "Graphic Expression in Engineering", the Representation Systems serve as a reversible geometric process between the plane and the space. Until the evaluation test, students mentally reconstruct the pieces that the teacher describes in the exercises using 2D images. However, due to multiple factors (e.g., group size, lack of resources, and classroom capacity), the student does not have access to 3D pieces according to their level of spatial comprehension.

During the course of 2017/18, polycubes of a 12-piece SOMA cube [17] were produced by the 3D Printing Department at the Department of Graphic Expression, Design, and Projects of the University of Malaga. These impressions were made in ABS in sufficient quantity to support 150 students per course.

Following a presentation, during which the objectives of the subject were explained, questionnaires were firstly administered to the students to collect their personal data, followed by the Mental Rotation Test (MRT) tests that were used to assess their skills.

The MRT test is considered by experts in the Engineering Design Graphics Division (EDGD) to be one of the top three tests of 24 models [18] and the literature shows that it is one of the best tools to measure spatial visualization. Its high reliability and internal consistency has also been confirmed [19].

Three polycube pieces (Fig. 1) were then distributed to each student. At this point, each student was required to assemble at least two pieces to generate a new 'object' that would be later sketched with normalized views. In pairs, they exchanged the normalized views they had made between them. Finally, the exercises were corrected in pairs and the improvement obtained following this activity was evaluated by means of the same tests that were administered in the beginning.

After a week of peer-to-peer comparison, the spatial skills post-tests were passed.

3 Type of Exercises for the Evaluation of Spatial Comprehension

The exercise used to evaluate spatial reasoning abilities consists of representing axonometric views of a piece. All the evaluations correspond to those used in the first year Graphic Expression course in the Degree in Electrical Engineering.

Fig. 1. Construction of figures through polycube pieces.

On all the occasions this was divided into three views (ground, elevation, and profile in the European diedric system for the representation of the pieces. However, the proposed representation system has varied between the isometric, trimetric, and cavalier perspective. Students are always given an hour and a half to solve the exercise, and it has always been composed of, at least, the following sections:

i. Represent the requested perspective at the indicated scale.
ii. Find a secant plane to the figure, given 3 points
iii. Find the section produced by said plane on the piece
iv. In the case of representation in the cavalier perspective, the student is traditionally requested to reduce the section produced in the piece.

The correct execution of the exercise accounts for between 15% and 25% of the total grade given for the subject. To homogenize the grades obtained in these exercises, we analysed only the score obtained on the first section, as a representative value of the student's abilities to visualize the piece from the images. This score does not take into account a poor application of the scale of representation and/or the reduction

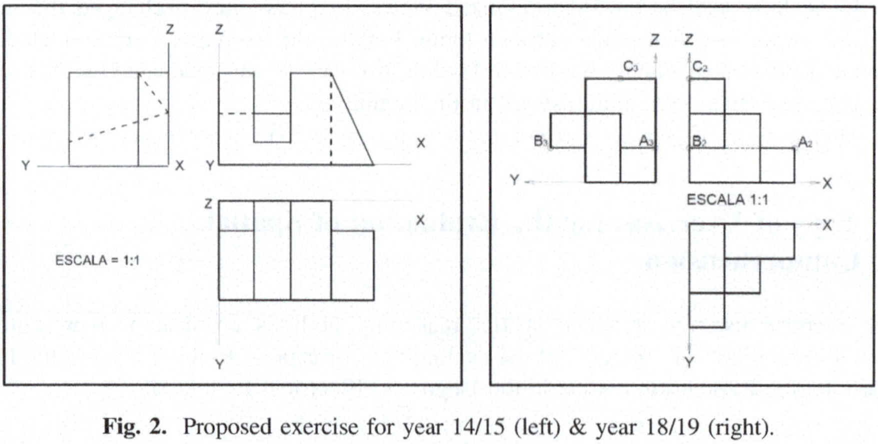

Fig. 2. Proposed exercise for year 14/15 (left) & year 18/19 (right).

coefficients. Likewise, this rating was weighted on 5 points, standardising the evaluation criteria of the teachers who proposed and corrected these exercises.

Regarding the degree of difficulty of the pieces to be represented, we can assume that these were highly similar on all occasions, although in the last course we considered the pieces to have the highest degree of difficulty (Fig. 2 right) despite having no oblique face. This is because all the views in the figure are composed of combinations of square modules, thus lacking a representative view.

4 Results and Discussion

The evolution of the performance on the spatial reasoning test is displayed in Fig. 3. It appears that the dispersion of the scores is reduced in the experimental course (course 18/19), in addition to being the highest with respect to the rest of the courses.

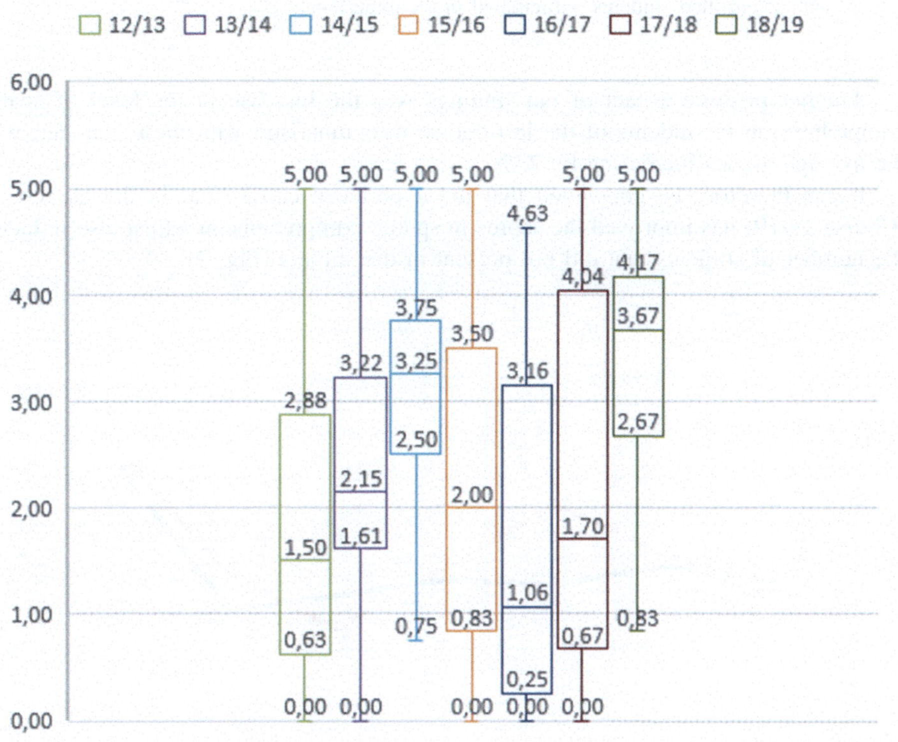

Fig. 3. Evolution of the score of the spatial reasoning test from the 2012/13 to the 2018/19 academic year.

The majority of students participated in the experience with polycubes during the 18/19 academic year (around 67% of the students enrolled participated). The fact that this experience was carried out at the beginning of the semester allowed for this high rate of participation. Of those who participated, 49% of students (Table 1) presented at the first ordinary examination of the subject (at the end of the semester). However, this represents 18% more participation than the average recorded in the previous years, or similarly, a reduced proportion of those who did not present.

Table 1. Key data in the courses 2012–2019.

Year	12–13	13–14	14–15	15–16	16–17	17–18	18–19
N° presented*	37	55	67	75	71	53	80
N° enrolled	118	163	217	237	243	196	164
Performance level (%)	31%	34%	31%	32%	29%	27%	49%
Average grade (up to 5 pts)	1,86	2,39	2,87	2,24	1,62	2,19	3,42
Use of the polycubes	No	No	No	No	No	No	Sí **

*Students that presented in the first ordinary evaluation
**68% of enrolled students participated in the experience

Another positive aspect of our findings was the increase in the level of spatial comprehension in students of the last course in comparison with the remainder, with the average grades increasing by 23%.

It can therefore be concluded that the experience carried out in the last course (Course 18/19) has improved the scores in spatial comprehension whilst also reducing the number of students that did not present in the subject (Fig. 4).

Fig. 4. Comparison of level of reasoning with performance level.

Currently, the correlations between spatial comprehension test scores and improvements in spatial abilities are being investigated according to visuospatial tests.

5 Limitations of the Research Study

The difficulty of the exercises in the different years is subjective, which may be a bias in the final interpretation of the results of the average grade, but not the rate of return.

Acknowledgments. This study has been made possible by the support of the Educational Innovation Project PIE011-1719 of the University of Malaga.

References

1. McGee, M.G.: Human spatial abilities: psychometric studies and environmental, genetic, hormonal, and neurological influences. Psychol. Bull. **89**, 889–918 (1979)
2. Beltrán Polaina, J.M.: Sistema diédrico. Técnicas educativas con ayudas 3D en el espacio real, y su simulación en el espacio virtual. Rev. Medios y Educ. **36**, 151–170 (2010)
3. Cohen, C.A., Hegarty, M.: Inferring cross sections of 3D objects: a new spatial thinking test. Learn. Individ. Differ. **22**(6), 868–874 (2012)
4. David, L.T.: Training of spatial abilities through computer games – results on the relation between game's task and psychological measures that are used. Procedia - Soc. Behav. Sci. **33**, 323–327 (2012)
5. Hauptman, H.: Computers & education enhancement of spatial thinking with virtual spaces 1, 0. Comput. Educ. **54**(1), 123–135 (2010)
6. Bishop, A.: Space and geometry. Acquis. Math. Concepts Process. 176–204 (1983)
7. Clements, M.A.: The question of how spatial ability is defined and its relevance to math education. **15**, 8–20 (1983)
8. Lohman, D.F.: Spatial ability: review and re-analysis of the correlational literature. Stanford University (1989)
9. Tartre, L.A.: Spatial orientation skill and mathematical problem solving. J. Res. Math. Educ. **21**(3), 216–229 (1990)
10. Hauptman, H., Cohen, A.: The synergetic effect of learning styles on the interaction between virtual environments and the enhancement of spatial thinking. Comput. Educ. **57**(3), 2106–2117 (2011)
11. Casati Calzada, M.J., Rúa Armesto, J.J., Pérez Benedito, J.L., Aliaga Maraver, J.J., Pérez Álvarez, J.: Metodologías activas como mejora de la visión espacial. VI Jornadas Innovación Universitaria, pp. 1–12. Universidad de Málaga (2009)
12. Melián Díaz, D., Carbonell Carrera, C., Saorín Pérez, J.L., de la Torre, Cantero J.L., Martín Dorta, N.: Fabricación digital de maquetas para la mejora de la interpretación cartográfica y el fomento de la competencia creativa. ABE **1**, 11–26 (2017)
13. Martín-Gutierrez, J., Trujillo, R., Acosta-gonzalez, M.M.: Augmented reality application assistant for spatial ability training. HMD vs computer screen use study. Procedia - Soc. Behav. Sci. **93**, 49–53 (2013)
14. García Domínguez, M., Martín Gutiérrez, J., Roca González, C., Mato Corredeaguas, M.C.: Methodologies and tools to improve spatial ability. **51**, 736–744 (2012)
15. Ha, O., Fang, N.. Development of interactive 3D tangible models as teaching aids to improve students. In: Spatial Ability in STEM Education, pp. 13–15 (2013)

16. Appelbaum, L., Cain, M., Darling, E., Mitroff, S.: La reproducción de videojuegos de acción está asociada a una mejor sensibilidad visual, pero no a alateraciones en la memoria sensorial visual. Atención, Percepción y Psicofísica **75**(6), 1161–1167 (2013)
17. Brunvoll, J., Cyvin, B., Cyvin, E., Cyvin, S., Paus, A., Stolevik, M., Stolevik, R.: The computerized SOMA cube. Comput. Math. Appl. **128**, 113–121 (1986)
18. Kelly, W.F.: Measurement of Spatial Ability in an Introductory Graphic Communications Course (2013)
19. Vandenberg, S.G., Kuse, A.R.: Mental rotations, a group test of three-dimensional spatial visualization. Percept. Motor Skills **47**, 599–601 (1978)

Self-assessment in Industrial Engineering. Pilot Test in Graphic Engineering at the Polytechnic University of Catalunya-BarcelonaTech

Anna Pujol-Ferran and Oscar Farrerons-Vidal

Escola d'Enginyeria de Barcelona Est-EEBE, Polythecnic University of Catalunya-Barcelona Tech, C/Eduard Maristany 10-14, 08019 Barcelona, Spain
{anna.pujol.ferran, oscar.farrerons}@upc.edu

Abstract. We present a study on Self-evaluation in the subject of Graphic Engineering in industrial engineering degrees. It has been implemented at the Polytechnic University of Catalunya in its Terrassa campus (Barcelona). This pilot project started in the 2017–18 academic year and it has continued in 2018–19 school year. The research begins with a survey completed by the students after they have developed and presented the final project of Graphic Engineering. Students must respond about the contents and competences they acquire throughout the course.

Although the survey also asks for teamwork on the whole subject, we will focus on data corresponding to the project and self-evaluate it to its final result.

Students must assess their learning from different graphic skills such as: pieces representation, dimensioning, sections and the knowledge of 3D geometry.

The self-assessment is not yet a binding evaluation, but it helps to obtain more consensual assessment, taking into account, the teacher evaluation first, without underestimating what the student does.

Keywords: Self-evaluation · Graphic engineering · Project · Skills

1 Introduction. Self-evaluation and Its Context

This pilot test has been developed at Higher School of Industrial, Aerospace and Audiovisual Engineering of Terrassa (ESEIAAT) belonging to UPC (Polytechnic University of Catalunya). It is an ingrained engineering school with a close regional scope of great industrial importance. Graphic Engineering subject is taught to 260 students by 10 professors.

Like all university education, the field of engineering has also evolved in recent years, with new methodologies to improve in all teaching areas. One of these areas, the evaluation is being experimented new formulas, denying the classical evaluation-qualification by the teacher. New forms of evaluation are emerging with all naturalness such as co-evaluation, shared evaluation, democratic evaluation, dialogue qualification and self-evaluation.

Self-evaluation is the assessment that students make about themselves or about a process and/or personal result. The E.C.T.S (European Credit Transfer) system has

F. Cavas-Martínez et al. (Eds.): INGEGRAF 2019, LNME, pp. 347–355, 2020.
https://doi.org/10.1007/978-3-030-41200-5_38

modified the relationship between teacher-student-learning process, with methodological and organizational changes.

The main changes in evaluation are:

- continuous and formative evaluation, more than final and summative
- evaluation of learning process (not only the apparent final grade)
- evaluation of different types of learning and skills
- evaluation to improve, not only as a control [1].

Evaluation has a positive effect on student learning when it relates to authentic tasks, it represents reasonable demands. This encourages students to use knowledge in a realistic context, which motivates the development of a wide variety of skills and is perceived as beneficial to long-term [2].

Specific objectives are set out from the use of new methodologies in teaching-learning and evaluation towards [3]:

- Strengthen the autonomy of students, creative thinking, reflective and critical, and the ability to self and co-evaluation.
- Interact socially and professionally with their environment through the study of cases, proposing solutions to problems and working in a collaborative way.
- Checking if the work that incorporates self-reflection reinforces students' learning.
- Analysing critically and providing proposals for improvement in the development of individual and collaborative work and put them into practice.

Likewise, these valuations become individual correction factors that, applied to the common qualification of the project, allow to obtain specific punctuations for each team's student [4].

2 Self-assessment in Graphic Engineering

A popular method for projects evaluation is used by engineering teachers with Evaluation Instruments [5], based on objective criteria, agreed between different parts of the subject. In addition, these criteria are previously communicated, so students know them very well. This practice avoids note claims, because the criteria are well established, although it is common for students to ask for corrections.

Group feedback significantly increases in group projects [6], where the students inform in writing the common errors committed, concepts that have detected when they are not clear, and other doubts.

This study aims to compare the self-assessment survey that began in the course 2017–2018 in which the results in Engineering Graphic of 1st year, 135 self-evaluated students were evaluated [7]. In this 2018–19 academic year, 220 students have been self-evaluated.

Both in the past and this year in which the self-assessment survey was carried out, results have not been binding in the final grade of the subject.

It should be noted that in the subject of Graphic Engineering taught in the first semester on the first year of undergraduate studies, the students are still adapting to university, its operation, its infrastructures and equipment, the dynamics of the classes and teachers. These students are 18–19 years old on average.

2.1 Dynamics of the Graphic Engineering Subject

The general aim of Graphic Engineering in the first year is the apprenticeship about the representation of plans, both individual or pieces as planes of objects together, assemblies, explosions, etc.

The focus of the subject is practical. Students develop a group project of 3 people between 3–6 weeks. During the classes, an exhaustive follow-up is carried out.

The achievement contains the following parts:

- idea-approach
- sketch-measurement
- representation of plans
- memory development
- oral-visual presentation before the class

It should be noted that there is a follow-up of the project and there are tutorials by the teacher, both in person and online, since students can send parts of the project by e-mail and they are reviewed by the teacher.

Consequently, not only is there a resolution of doubts, but in most cases, they already represent a first project correction.

A potential application is found when the project is defended in oral presentations scheduled during last class-session of the subject. In an oral-visual presentation, each group of 3 students has 10–12 min to explain their project: objectives, approach, development and conclusions. After the presentation, the classmates must make contributions and oral evaluations of each project.

Teacher presents an assessment at the end of each presentation where he comments on strengths and achievements of each project but also weak or unpalatable points.

At this moment, each student, individually, can make an assessment of their work, after listening to the contributions of both, peers and teacher, and has more arguments for self-evaluation.

2.2 Self-assessment Surveys

Self-evaluation is intended for a person to make an assessment of himself or on a process and/or personal result [8].

A fundamental issue about self-assessment survey includes 3 aspects to be assessed: project, group work, and subject.

The project evaluation and subject evaluation are personal valuations, about oneself. On the other hand, assessment of group work is a group co-evaluation, an evaluation among students from the same group, referring to their contribution [9].

Each of the 3 parts include: 8 questions about the project, 5 questions about the group work and 5 about the subject. In total there are 18 questions. Students are assessed with the possibility of 5 ratings notes (from 1 to 5), with 1 being the lowest and 5 being the highest.

The survey asks about some specific aspects of the projects, such as:

- The representation of pieces
- The dimensioning of the views
- The representation of sections
- 3D geometry knowledge

In addition, they are asked in a more general way: the time allocated to the project, the functioning and cohesion of the work team and if the objectives of the subject have been achieved.

2.3 Survey Processing

From the self-assessment surveys, a project note has been collected, which is given by the student himself. The numbering of self-evaluation has the following equivalence:

- 5 - Excellent (9–10)
- 4 - Notable (7–8,99)
- 3 - Approved (5–6,99)
- 2 - Suspense (3–4,99)
- 1 - Deficient (0–2,99)

Ratings notes are ones the student himself puts on, referent to the project he/she has made. Therefore, we present the results of two studies:

- PROJ_EG. The self-assessing notes of Graphic Engineering Project-
- DES_PROJ_EG. The lag of notes comparing self-assessment with the real score corrected by teacher of the Graphic Engineering Project.

For the phase shift tables, the following 5 values will be taken into account:

- Same note: When student self-evaluates with the same grade that teacher has evaluated, the grade coincides.
- Bottom note −1: When real grade obtained by the student is a lower level than grade which he has self-evaluated.

- Bottom note −2: When real grade obtained by the student is 2 levels lower than grade which he has self-evaluated. In this case, student has a very optimistic perception how he is doing, because afterwards he produces very inferior grades.
- Top note +1: When real grade obtained by the student is a higher level than grade which he has self-evaluated. Therefore, in this case, the student thinks that he is doing more worse, than he really does afterwards.
- Top grade +2: When real grade obtained by the student is 2 levels higher than grade which he has self-evaluated. We confirm we have not found any case.

3 Self-evaluation Results

The most outstanding results of the course 17–18 are, taking into account the 135 processed surveys of the participating students:

3.1 PROJ_EG 17–18. The Notes that Self-evaluate the Project in Course 2017–18

Some considerable numbers: 92% of students value their project with a high score, which 54% do with a note of noteworthy. Only one student among the 135 respondents believes that they have suspended the project (Fig. 1).

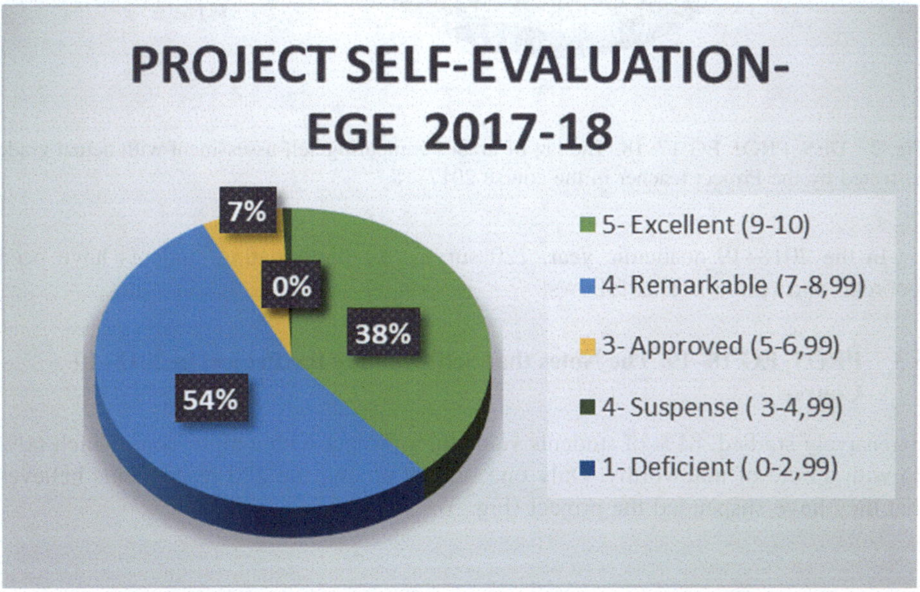

Fig. 1. PROJ_EG 17–18. Ratings Project notes in the course 2017–18

3.2 DES_PROJ_EG 17–18. The Lag of Grades Comparing Self-assessment with Actual Grade Corrected by Project Teacher in the Course 2017–18

A remarkable variety of results: 38% of the students agree with the same grade teacher has given them, while 58% put on a higher grade than they actually get (Fig. 2).

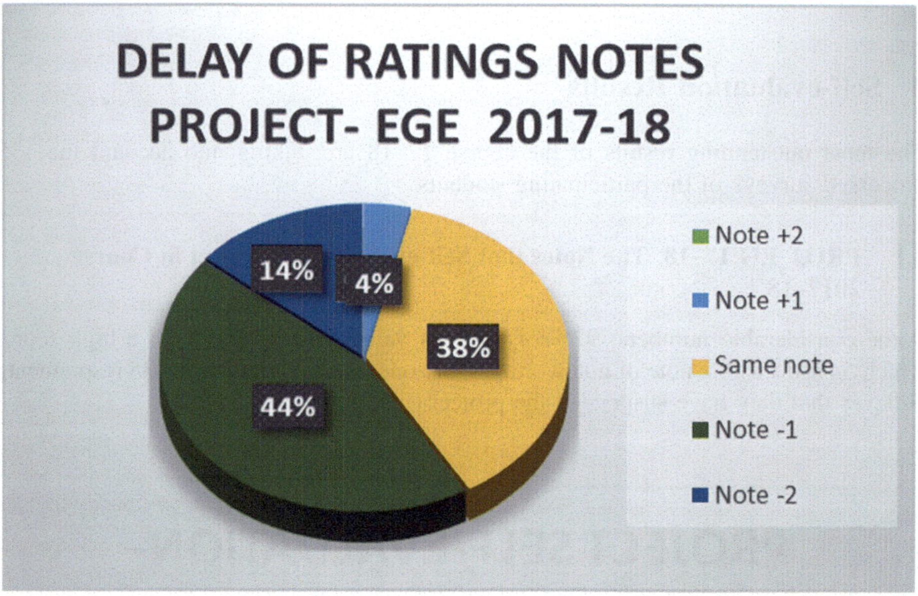

Fig. 2. DES_PROJ_EG 17–18. The lag of grades comparing self-assessment with actual grade corrected by the Project teacher in the course 2017–18

In the 2018–19 academic year, 220 surveys of participating students have been recorded and results are as follows:

3.3 PROJ_EG 18–19. The Notes that Self-evaluate the Project in 2018–19 Course

Extensively studied, 81% of students value their project with a high score, which 60% do with a note of noteworthy. Only one student among the 220 respondents believes that they have suspended the project (Fig. 3).

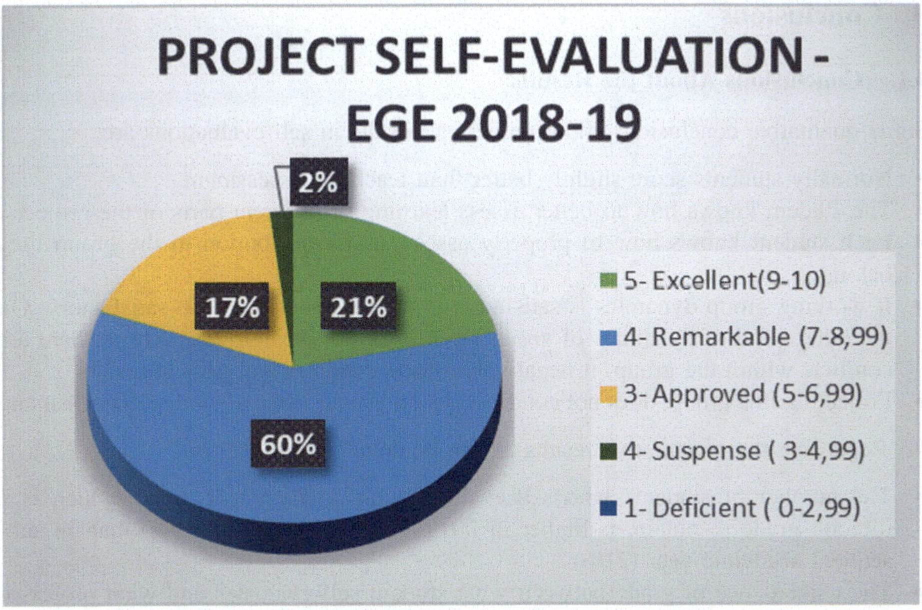

Fig. 3. PROJ_EG 18–19. Ratings Project notes in 2018–19 course

3.4 DES_PROJ_EG 18–19. The Lag of Notes Comparing Self-assessment with Actual Score Corrected by the Project Teacher in 2018–19 Course

Well-documented, 47% of students agree with the same grade given by teacher, while 41% put a higher grade they actually get (Fig. 4).

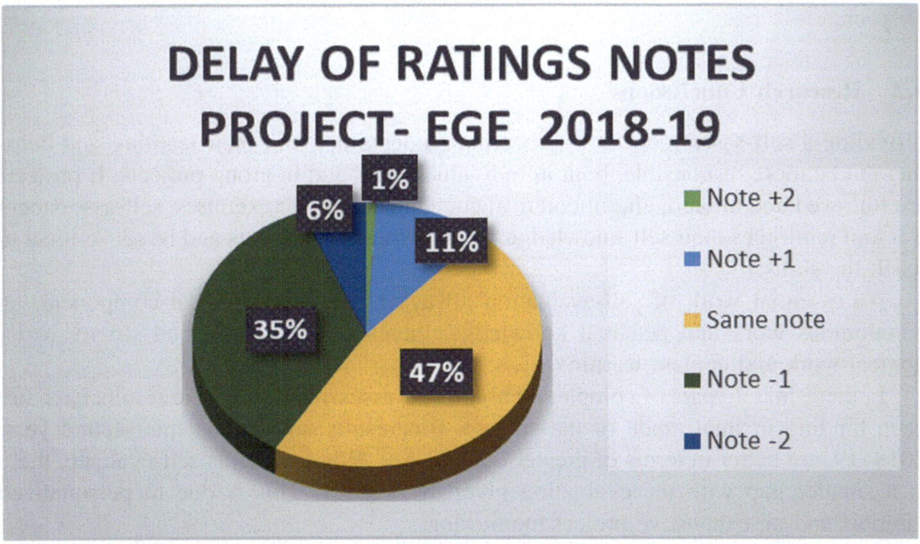

Fig. 4. DES_PROJ_EG 18–19. The lag of notes comparing self-assessment with actual grade corrected by the Project teacher in 2018–19 course

4 Conclusions

4.1 Conclusions About the Results

Some qualitative conclusions about numerical results in self-evaluations are:

- Normally students score slightly better than teacher's assessment.
- The student knows how to better assess learning of different parts of the project.
- Each student knows how to properly assess their contribution to the group they belong.
- If working group dynamics is satisfactory, project results are also satisfactory. On the other hand, if dynamics of group work does not develop properly or there are conflicts within the group, it negatively affects progress and project results.
- Projecting in a group, does not condition its criteria to make a good self-assessment.

Regarding the quantitative results, it can be concluded that:

- The number of students who self-evaluate with an Excellent grade, which is a difficult grade to obtain, is higher in 2017–18 academic year (38%) than in subsequent academic year (21%).
- The coincidence of grade between what student self-evaluates and what professor issues is higher in 2018–19 academic year (47%), than in first year of research (38%). It is positively valued that in this second year we are reaching almost 50% of students who coincide in the grade.
- The gap of grade to one or two higher levels, that is, the student scores slightly better than the actual grade is lower in 2018–19 course (41%) while in previous year it was 58%, having much more lag of note.
- Considerable numerical results and well-known conclusions are obtained for self-evaluation, results in the 2018–19 course have been better than 2017–18 academic year.

4.2 Research Conclusions

Providing a self-assessment to Engineering students helps in their learning and helps them to be more responsible, both in individual work and in group projects. If projects are followed and tutored, after theoretical content and similar exercises, self-assessment is a tool reinforces their self-knowledge, to criticize other projects and be self-critical to reach the skills.

An essential work of self-evaluation always has a psychological component, an autonomous work and personal knowledge, therefore, it goes beyond strictly pedagogical work and content learning.

In these first 2 years of completion about self-assessment form, the results have not been binding in final grade of the subject. But results obtained in this second year, 2018–19, are better in terms of greater coincidence. When students self-evaluate, there is a smaller gap with the evaluation given by teacher. This is due to personalized tutorials and an exhaustive project monitoring.

References

1. Pastor, V.M.: La participación del alumnado en la evaluación: La autoevaluación, la coevaluación y la evaluación compartida La participación del alumnado en la evaluación: la autoevaluación, la coevaluación y la evaluación compartida (2005)
2. Sambell, K., McDowell, L., Brown, S.: "But is it fair?": an exploratory study of student perceptions of the consequential validity of assessment. Stud. Educ. Eval. (1997). https://doi.org/10.1016/S0191-491X(97)86215-3
3. Palomares, A.: El modelo docente universitario y el uso de nuevas metodologías en la enseñanza, aprendizaje y evaluación. Revista de Educacion **355**, 591–604 (2011). https://doi.org/10.4438/1988-592X-RE-2011-355-038
4. Jimenez Valverde, G., Llitjos Viza, A.: Methodological optimization of cooperative online environments (BSCW & Synergeia) as Chemistry teaching aids in hypermedia authoring. Enseñanza de las Ciencias (2008)
5. Cadenato, A., Martinez, M., Gallego, I., Amante, B., Jordana, J., Sanchez, R.F., Farrerons, O., Isalgue, A., Fabregat, J.: Criterios para prácticas de evaluación de calidad. Una propuesta de Grapa-Rima. In VII Congrés Internacional Docència Universitària i Innovació: La universitat: una institució de la societat. La universidad: una institución de la sociedad. The University: an institution of Society (2012)
6. Ibarra Sáiz, M., Rodriguez Gómez, G.: Los procedimientos de evaluación como elementos de desarrollo de la función orientadora en la universidad. REOP - Revista Española de Orientación y Psicopedagogía (2010). https://doi.org/10.5944/reop.vol.21.num.2.2010.11558
7. Pujol-Ferran, A., Farrerons-Vidal, O.: L'autoavaluació en enginyeria. L'aplicació de l'autoavaluació en projectes del Grau d'Enginyeria Industrial com a eina d'avaluació i autoaprenentatge. Congrés Internacional Docència Universitària i Innovació (2018)
8. García Sanz, M.P.: La evaluación de competencias en Educación Superior mediante rúbricas: un caso práctico. Revista Electrónica Interuniversitaria de Formación del Profesorado (2014). https://doi.org/10.6018/reifop.17.1.198861
9. Crebert, G., Bates, M., Bell, B., Patrick, C.-J., Cargnolini, V.: Developing generic skills at university, during work placement and in employment: graduates' perceptions. High. Educ. Res. Dev. (2007). https://doi.org/10.1080/0729436042000206636

About BIM in Civil Engineering

José Ángel Aranda Domingo(✉) and Miguel Ángel Gil Saurí

Department of Graphic Engineering, Universitat Politècnica de València, Valencia, Spain
jaranda@dig.upv.es

Abstract. On 26 February 2014, it was approved the European Directive of the ***European Parliament and of the Council** 2014/24/EU* that established the mandatory use of BIM in public infrastructure tenders for all Member States of the European Union as of 26 July 2019.

BIM (Building Information Modelling) is an n-dimensional model that includes all phases of the life cycle of an infrastructure, that is, it goes beyond the 3D methodology. It covers design, construction and operation/maintenance.

BIM is based on parametric drawing from the simplest elements to the NURBS modelling. For its teaching, it is sufficient a first year of graphic mathematics at the university level.

The communication here presented deals with the experiences gained from the implementation of this methodology in both the Degree and the Master's Degree in Civil Engineering at the *ETS de Ingenieros de Caminos* of *Universitat Politècnica de València.*

Keywords: European directive · BIM · Civil engineering · Parametric drawing · SDOL · SCE · UPV

1 What is BIM?

BIM, Building Information Modelling (Integrated Information Modelling for Construction) is a collaborative digital work methodology that attempts to compile a database from all available project information, integrating all disciplines that are involved in it [1].

As such, it can be said that it is an n-dimensional model that can bring all phases of the construction together (from its initial design to its demolition).

By way of a summary, the dimensions of the model are indicated next: 3D - design and parameterisation of the model, 4D - time integration, 5D - cost estimate (budget), 6D - maintenance and operation, and 7D - incorporation of additional information.

A BIM model has its origin in 3D. It is here where the first difference with 2D modelling takes place. It can be said that working in 2D (set of straight lines and/or curves) is no longer required but that a virtual model is built instead. There is therefore a transformation in such a way that the lines become virtual constructive elements (real). As a result, model elements can be visualised from any point of view and all kind of cross-sections can be obtained.

Once the 3D model has been defined, a great range of factors such as the technical feasibility of the project or the potential construction-related problems can be analysed

F. Cavas-Martínez et al. (Eds.): INGEGRAF 2019, LNME, pp. 356–364, 2020.
https://doi.org/10.1007/978-3-030-41200-5_39

and evaluated a priori, that is, before its construction and even before the award of the contract. Thus, allowing to identify and solve the initial problems in the design phase, when decisions can have greater impact on cost, time, and sustainability [2].

One of the great advantages of these models lies in the parameterisation of the model elements, meaning that any modification applied to them results in the automatic regeneration of the complete model [3].

For all the above reasons, the interest in its implementation in the field of architecture, engineering and construction (AEC) is increasingly growing [4–6].

2 BIM in Civil Engineering

2.1 Present Situation

Although BIM has already come a long way in the field of architecture, this is not the case when it comes to civil engineering. Everyone knows the different types of civil engineering works, among which it can be highlighted linear works (road design, railways, hydraulic works, tunnels, etc.), maritime works or large hydraulic infrastructures (dams).

It is well established that the typology of works listed above currently follow a design methodology based on the simplification of three-dimensional reality. That is to say, it is still possible to appreciate a differentiation between the plan, the elevation and a set of detailed plans that make it possible to create detailed 2D elements (plans) that allow the understanding of the three-dimensional reality they represent.

The transition from the current situation to the parametric 3D model represents a major change. This is the reason why grammar is currently being developed that allows for the correct understanding between the initial 3D model and the rest of the dimensions that have been previously commented. It is here where all attention is being focused now, particularly in the elaboration of files with a determined structure capable of being legible by all the parties involved in the construction phase. These are known as IFC (Industry Foundation Classes) files, which are a data format that allows the exchange of an information model without losing or distorting the data or the information it contains. It is characterised by having an open and neutral code, not controlled by the software producers. It was born to facilitate interoperability between the different parties and designed to elaborate all the necessary information of the construction model throughout its life cycle: from the preliminary design until its execution and its later maintenance, going through the different phases of design and planning.

This methodology is making its way in the sector of transport infrastructure, roads, railways, bridges, tunnels, ports, airports and even in their refurbishment works [7–13].

2.2 Design Software Implementation

The different software for the design of civil works are currently being updated. These must carry out a certification process that ensures the accuracy of the import and export of IFC data according to the appropriate standards. This task is competence of the buildingSMART (formerly the International Alliance for Interoperability), which is embodied in an international organisation whose objective is to improve the exchange of information between the software applications used in the construction industry.

The buildingSMART Spanish Chapter Association has developed a series of BIM guides called euBIM that aim to facilitate the implementation of the methodology in the Spanish industry.

All certified IFC software are capable of reading, writing and exchanging information with other programs. According to the data provided by buildingSMART, the IFC standards are supported by more than 140 software platforms.

3 Description of Subject Contents

The subject is called Building Information Modelling. It has been taught during the first four months of the academic year 2018–2019 in the fourth year of both the Degree of Public Works Engineering (GIOP) and Civil Engineering (GIC) at the *Escuela Técnica Superior de Ingenieros de Caminos* of the *Universitat Politècnica de València (UPV)*, with a total teaching load of 4.5 credits. This subject is divided into two parts: a theoretical and a practical part, both consisting of two hours per week over a period of 12 weeks. The theory is taught in a single classroom while the practical sessions take place in a computer room equipped with 25 computers. The number of enrolled students (37) forced to split the class into two groups.

The program of the subject can be summarised as follows:

1. An introductory session to explain the passage from CAD to BIM.
2. Cartography import. Georeferencing.
3. Linear works.
4. Intersections.
5. Structure modelling (bridge).
6. Federation of BIM models.
7. Planning, time management. 4D model.
8. Cost management. 5D model.
9. Realistic presentation.
10. Building model.

The teaching staff assigned to the subject is interdepartmental. Specifically, the departments involved in teaching this subject are the following:

- Graphic engineering.
- Ground engineering.
- Engineering and transport infrastructure.
- Construction engineering and civil engineering projects.

For the development of the subject, it is necessary the use of different software:

With regards to the cartography import, design of linear works, intersections and design of structures, the ISTRAM software is used.
For the calculation of previously designed structures it is used the CIVILESTUDIO software.

For the federation of models, the NAVISWORKS software (Autodesk) and a free BIM viewer, BIM VISION, are needed.
The development of the 4D planning is carried out with the tool Microsoft Project. Regarding the cost management, Microsoft Excel is used.
INFRAWORKS (Autodesk) is used for the development of a realistic presentation. Finally, the building model is carried out using the software REVIT (Autodesk).

4 Case Study

The case study presented here consists of the parameterisation of a reinforced and prestressed concrete structure. It is the same case study that is presented in one of the lessons in the GIOP and GIC Degrees at *Escuela de Ingenieros de Caminos* of *Valencia.*

In this chapter, the application of a 3D BIM model to a section of road in which it is necessary to define a reinforced and prestressed concrete structure to avoid a small watercourse is shown.

Particularly, this part of the 3D modelling is taught by the graphic engineering department to which the authors of this publication belong.

The process commences importing the topographic surface, which is necessary to start with the design. There are currently three types of formats available for this: vector (DWG or native formats of the different software), digital terrain models (DTM with different resolutions) and dense point clouds or LIDAR models (LAS or LAZ formats).

For the development of this research, a vector format detailed topography at a 1: 500 scale was used. Data collection was carried out with a total station and georeferenced to the projected coordinate system UTM ETRS 89 H30.

Following this, it was carried out a triangulation and the subsequent creation of a topographic surface, the output of which was in IFC format.

Figure 1 shows the triangulation (left) and the terrain model in IFC format (right) that is represented using a BIM viewer.

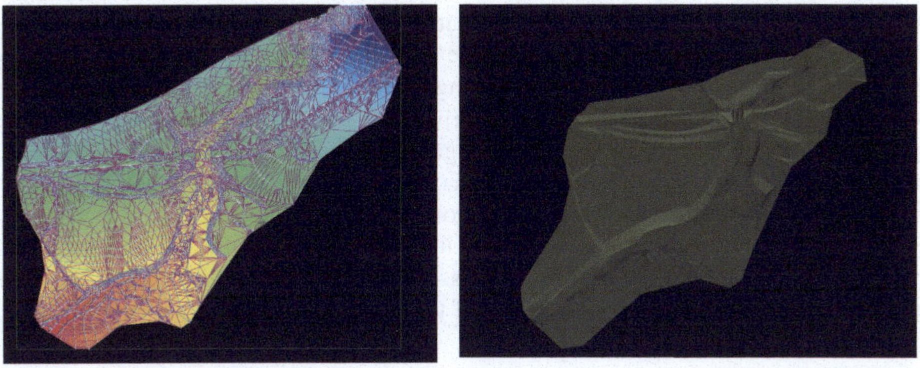

Fig. 1. Triangulation (left) and modelled surface in IFC format (right).

On the georeferenced topographic surface, a 241 m long section of road belonging to the local red network of Teruel region is parameterised. This section of road presents the singularity that it intersects with the Ortigal watercourse, a hydrographic network attached to the Ebro River Basin Authority and, therefore, it is necessary to build a structure to avoid it.

The first step is the design of the linear work (road) and the implicit structural elements, in this case, a precast beam board structure.

Initially, a linear work design software (SDOL) is used. Here, both the road and the structure are parameterised, see Figs. 2 and 3. The data that define the structure are initially introduced based on the experience of the designer. Obviously, these parameters will be verified later by the Structure Calculation Software (SCE), which will determine how accurate they were. As it will be seen later in this document, the process enters in a loop between both programs (i.e. SDOL and SCE) in such a way that the fluency of parameter exchange is key.

Fig. 2. Linear work – topographic surface integration

Once the structure is defined, the geometric data that allow its location in the space and its characteristics are extracted from the SDOL. These elements are: the axes of the bearings, the axes of the n-beams and the outline of the deck (Fig. 4). With regards to the other elements that complete the bridge structure (bearings, abutments and wing walls), these are conditioned by the arrangement of the above. For example, the location of the elastomeric or neoprene bearings corresponds to the intersection of the bearings axes with the beam axes. The arrangement in height, coordinate Z of the deck, is conditioned by the coordinates of the axis of the structure, 3D polygonal. If the deck is not horizontal, this polygonal has the cross gradient value attached to each point. Regarding the abutments and wing walls, it should be noted that their definition is conditioned by the location of the bearings plus a series of parameters that must be entered manually, such as the dimensions of the foundation or the abutment back wall, Fig. 3.

Fig. 3. Structure parameterisation. Precast concrete beams bridge.

Obviously, in the design of a structure not only the parameters indicated above intervene. It is true, however, that the estimation of these are the ones that suppose greater efforts. In addition, any change that is introduced in the layout of a linear work implies that the design has to be started again. The remaining elements that complete the definition of the structure are related to concrete and steel properties, along with the loads that the structure has to cope with (actions). These data generally remain constant.

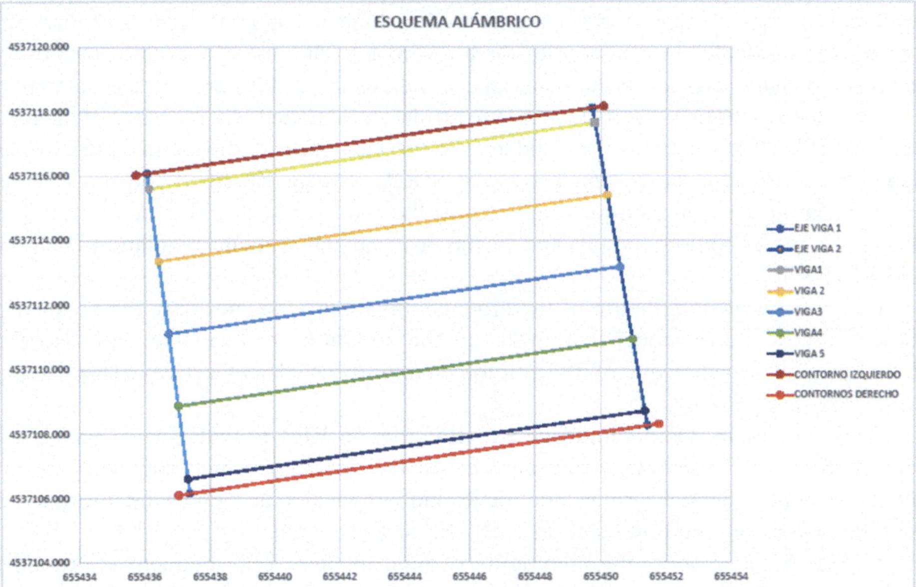

Fig. 4. Coordinate exchange overview of the most relevant elements for the parameterisation of a structure.

Continuing on methodology, the geometric data extracted from SDOL are introduced into the structure calculation software (SCE). This process consists of extracting the coordinates of the most significant elements; bearings axes, beams axis and deck contour.

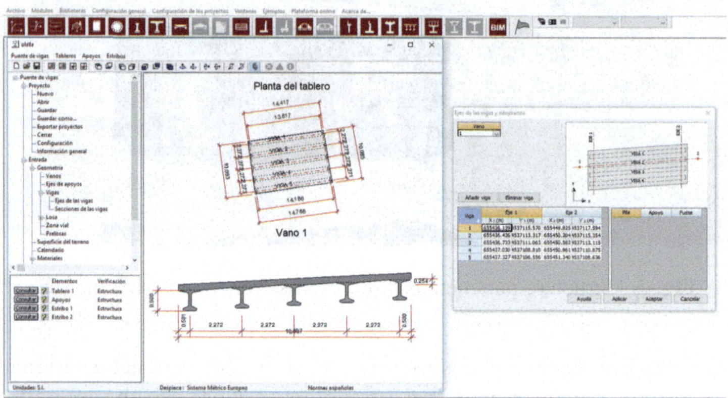

Fig. 5. Structure parameterisation using CIVILESTUDIO from the data obtained with the software for the design of linear works ISTRAM.

Figure 5 shows a connection window between the design software (ISTRAM) and the structure calculation software (CIVILESTUDIO).

Previous data is exported by SDOL and directly entered into the SCE. This process represents a considerable time saving, as well as higher data confidence and pinpoint accuracy (coordinates in meters with three decimals). The rest of the data concerning the bearing characteristics of the materials, as well as the actions are entered manually.

Once the structure is calculated and the results are obtained, the accuracy of the data entered at the beginning can be validated. If the validation is successful, the process concludes, otherwise the parameters in the SDOL must be modified and the process enters again in the loop mentioned before. At the end of this, it is obtained a structure that meets the conditions established in the government structure regulation (IAP 11 and EHE 08).

In this way, on completion of the aforementioned process, not only the geometric dimensions of the elements that make up the structure are available, but also the definition and arrangement of the reinforcements (passive and active). That is, the structure is fully defined.

At this moment, there should be two IFC files in BIM format, one of them corresponding to the linear work and another for the structure, which contains its geometric definition (deck, bearings, abutments and wing walls) as well as the information with regards to the reinforcement in each one of these elements.

Figures 6 and 7 show the integration of both models. The abutment is provided with some transparency for the visualisation of the reinforcements within.

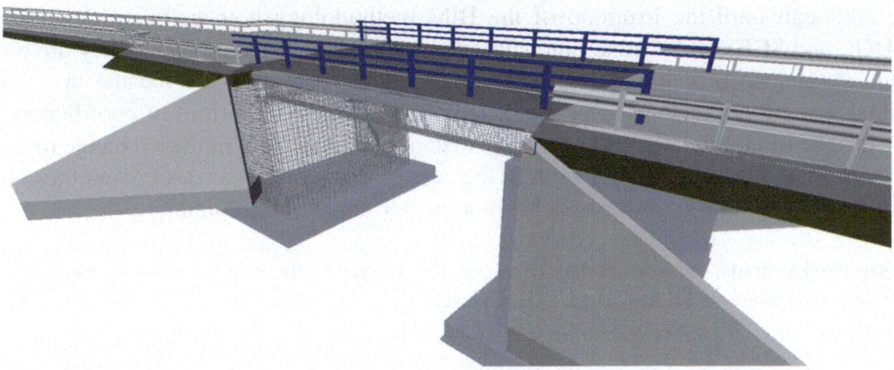

Fig. 6. Integration of the reinforcement obtained through the structure calculation software and the linear work model.

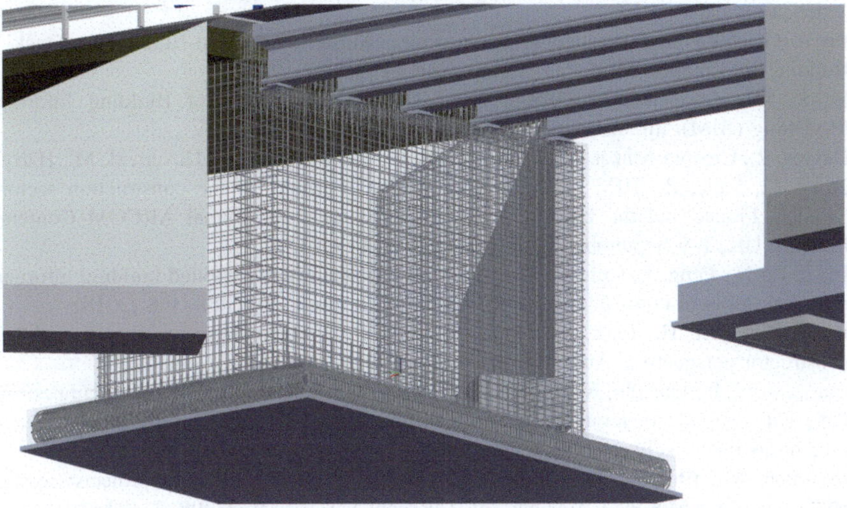

Fig. 7. Visualisation of the reinforcement in one of the abutment.

5 Conclusion

This document presents a case study that is framed within the didactic research. It is the first time the BIM subject is taught at the *Escuela Técnica Superior de Ingenieros de Caminos de Valéncia.*

It consists of a direct application of parametric drawing that while it has already been consolidated in other disciplines, it is not the case in civil engineering. Thus, it has to be highlighted the importance of the parameterisation of an integrated structure in a linear work. Evidently, this represents a breakthrough in terms of its geometric definition, since it allows to assume intrinsic modifications to its spatial location with a reasonable investment of time.

Although until the irruption of the BIM methodology both software developers (SDOL and SCE) of different nationalities were pursuing separate paths, they are now both working in this line. The calculated structures were spatially located in a simplified way by referencing some characteristic element in the origin of coordinates.

In spite of all this, it is still necessary to continue advancing in the exchange of data between the different software, both in the case of those for linear design and those for structure calculation in order to achieve a greater degree of automation.

Acknowledgements. This work has been possible thanks to the collaboration of the Spanish software companies ISTRAM and CIVILESTUDIO.

References

1. es.BIM a. Sobre BIM. https://www.esbim.es. Consulta: 20 de marzo de 2019
2. Ramos Rojas, J.M.: Obtención e implementación de datos 3D en un sistema de gestión BIM. Valencia, España (2017)
3. Elvin, G.: Integrated Practice in Architecture: Mastering Design-Build, Fast-Track, and Building Information Modeling. Wiley, Hoboken (2007)
4. Bryde, D., Broquetas, M., Volm, J.M.: The project benefits of Building Information Modelling (BIM). Int. J. Project Manag. **31**, 971–980 (2013)
5. Davies, R., Crespin-Mazet, F., Linne, A., Pardo, C., Ingemannson Havenvid, M., Harty, C., Ivory, C., Salle, R.: BIM in Europe: innovation networks in the construction sectors of Sweden, France and the UK. In: Proceedings of the 31st Annual ARCOM Conference, Lincoln, UK, 7–9 September 2015, pp. 1135–1144 (2015)
6. Li, J., Li, N., Peng, J., Cui, H., Wu, Z.: A review of currently applied building information modeling tools of constructions in China. J. Clean. Prod. **201**, 358–368 (2018)
7. Bradley, A., Li, H., Lark, R., Dunn, S.: BIM for infrastructure: an overall review and constructor perspective. Autom. Constr. **71**, 139–152 (2016)
8. Campestrini, F., Garrido, M., Scheer, S., Freitas, M.: Entendendo BIM. In: Campestrini, F. (ed.) Universidade Federal do Paraná: Curitiba, Brazil (2015). http://www.gpsustentavel.ufba.br/documentos/livro_entendendo_bim.pdf. Accessed 26 Dec 2018
9. Bensalah, M., Elouadi, A., Mharzi, H.: Integrating BIM in railway projects: review & perspectives for Morocco & Mena. Int. J. Recent Sci. Res. **9**, 23398–23403 (2018)
10. Bensalah, M., Elouadi, A., Mharzi, H.: BIM integration to railway—literature & experiences critical review. In: Proceedings of the 11th International Colloquium of Logistics and Supply Chain Management LOGISTIQUA 2018, Tangier, Morocco, 26–27 April 2018, vol. 11 (2018)
11. Costin, A., Adibfar, A., Hu, H., Chen, S.S.: Building Information Modeling (BIM) for transportation infrastructure—literature review, applications, challenges, and recommendations. Autom. Constr. **94**, 257–281 (2018)
12. Zak, J., Macadam, H.: Utilization of building information modeling in infrastructure's design and construction. In: IOP Conference Series: Materials Science and Engineering, vol. 236, p. 012108 (2017)
13. Neves, J., Sampaio, Z., Vilela, M.: A case study of BIM implementation in rail track rehabilitation. Infrastructures **4**, 8 (2019)

Where is Descriptive Geometry Heading?

Rafael Ortiz-Marín, Gloria Del Río-Cidoncha(✉), and Juan Martínez-Palacios

Dpto. Ingeniería Gráfica, Universidad de Sevilla, Camino de los Descubrimientos s/n, 41092 Seville, Spain
cidoncha@us.es

Abstract. Has descriptive geometry died? Although the question can be impressive for Graphic Engineering, it is thought that the answer would be yes, with some nuances. Descriptive geometry has allowed solving all kinds of three-dimensional problems, reducing them to two-dimensional problems. This paper aims to demonstrate that currently with the new tools available in geometry such as CAD/CAM applications; you can not only model, assemble and reproduce sets of parts, but also solve all kinds of geometry spatial problems. The possibility of solving these three-dimensional problems using CAD applications, using very simple orders for their implementation and subsequent determination, opens a new path in geometry. It will be explained, throughout the article, how the different methods that the descriptive has to show the distances and the angles in their true magnitudes, are not necessary if the tools of the CAD/CAM applications are used, in particular Solid Edge.

Keywords: Descriptive geometry · Spatial geometry · CAD/CAM · Teaching-learning

1 Introduction

There are programs that allow imitating the dihedral system with a vertical projection and a horizontal one. Amongst them are:

PROCECAD [1], computerized drawing assistant, based on the sequential execution of procedures through the use and processing of graphical functions managed by the user. The program reads and analyzes each line to execute it and thus creates the drawing.

This software of descriptive geometry for Windows has been designed as an alternative drawing tool and not as a three-dimensional element processor, however, it is possible to create and use procedures that represent 3D elements, since it has an assistant to enter functions and facilitate the creation of macros, which are the grouping of several graphical functions into one, so that it can be called at any time from a line. Today PROCECAD has stopped updating.

F. Cavas-Martínez et al. (Eds.): INGEGRAF 2019, LNME, pp. 365–373, 2020.
https://doi.org/10.1007/978-3-030-41200-5_40

DESCRIPTIVE GEOMETRY [2], software was created to easily solve the tasks of descriptive geometry. Its operation is based mainly on the use of algorithms for the realization of drawings. The programming language used derives from C. Programming consists in the creation of functions for the treatment of different objects. Its author does not provide service anymore.

ARCHIMEDES GEO3D [3], dynamic geometry software in three dimensions that allows the construction of geometric elements and their intersections. It is characterized by the existence of free objects that can move, the rest of the construction suffer a change according to that movement.

The created objects tree always shows the coordinates and characteristic equations of the same, the existence of a command line allows generating objects through commands and the program has the ability to work with vectors and perform vector calculations.

Therefore, the possibility of solving three-dimensional problems using CAD applications, with very simple orders for their implementation and post-determination, opens a new path in geometry.

2 Planning and Development

It will be explained, solving an exercise, how to avoid the methods that descriptive geometry presents to solve spatial problems.

The problem that arises is the following, given a plane defined by three points ABC. Point A is the vertex of an equilateral triangle whose other vertices have more reference than this; in addition, one of the sides of the triangle is a parallel segment to a horizontal plane. This equilateral triangle is the face of an octahedron. You are asked to draw the octahedron, among all the possible solutions you will choose the one in which the other vertices have a higher dimension than those of the previous face [4].

The data will be implemented in the Solid Edge 2019 CAD/CAM program, (SE) [5].

This program from the synchronous versions has the possibility of entering the coordinates of a point. The axes used (SE) for the coordinates are those shown in Fig. 1, the X coordinate is identified with the reference, the Y with the distance and Z with the elevation of the point.

Also the (SE) has 3 reference planes to which it calls upright (XZ), plant (XY) and refers to the right profile as right (YZ). For the points to remain in the first octant, the Y coordinate is assigned a negative value.

The points that define the plane of the example are the following: A (30, −50, 20); B (20, −5, 60); C (60, −30, 40). The equilateral triangle is 100 mm on the side and point A will be one of its vertices.

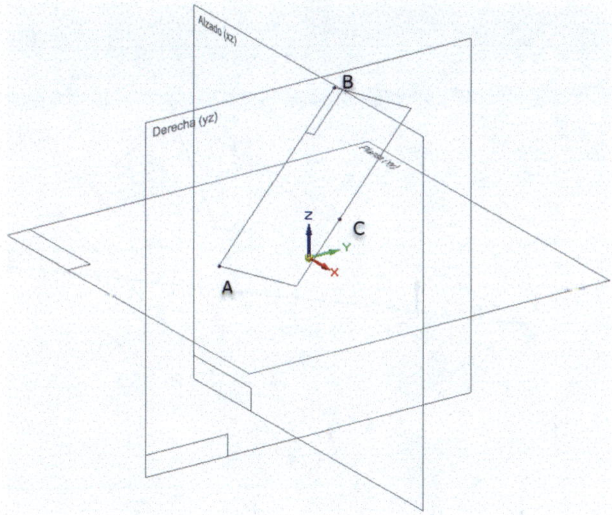

Fig. 1. Coordinate axes, planes of projection and plane ABC

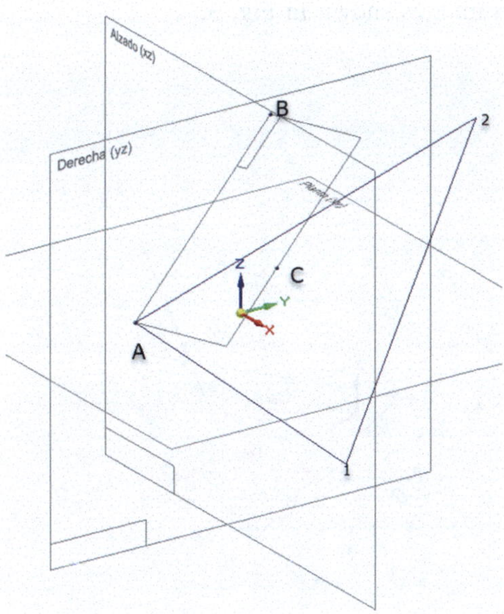

Fig. 2. Equilateral triangle with one side parallel to the XY plane (horizontal).

The parallel side to the horizontal plane (XY) is the segment A1 like it can be appreciated in Fig. 2.

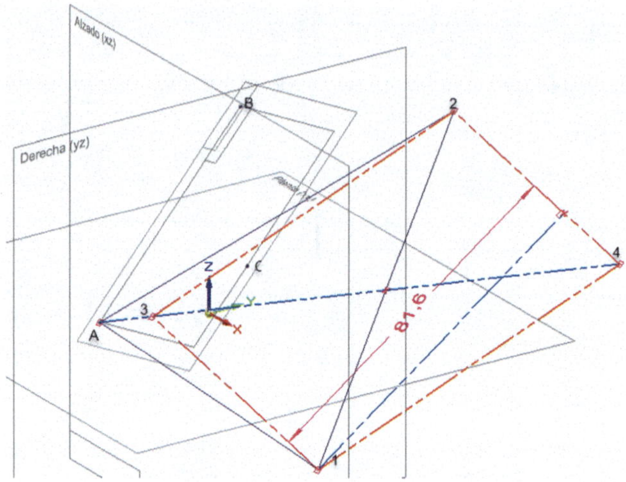

Fig. 3. Principal section of the octahedron.

In the same ABC plane, the main section of the octahedron is built, which will be the diamond 1423 in red. On the main section, the distance between parallel faces of the octahedron is 81.6 mm, as shown in Fig. 3.

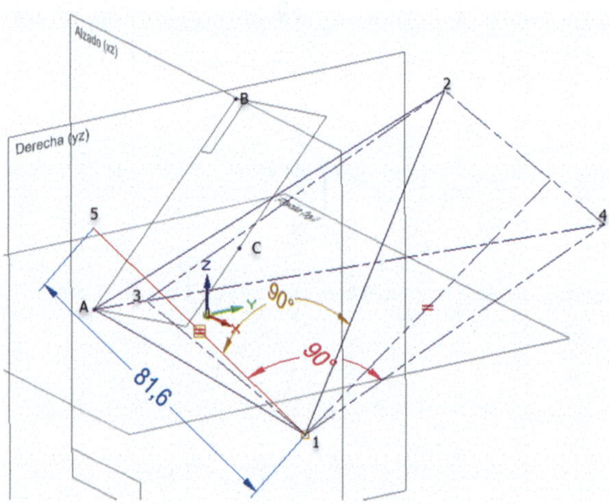

Fig. 4. Perpendicular segment to plane ABC o A12 o 1423.

From point 1, a straight line is drawn perpendicular to the plane ABC; the program has the possibility of plotting lines in 3D and then adding a restriction of perpendicularity to a plane. Then, the distance between faces (81.6 mm) with the equal restriction is carried on that perpendicular, which means that in a 3D sketch the perpendicular is equal to another element. Observe in Fig. 4 the restrictions of perpendicularity and equality.

By point 5 of the line perpendicular to the plane ABC, a plane parallel to ABC is drawn, (see Fig. 5).

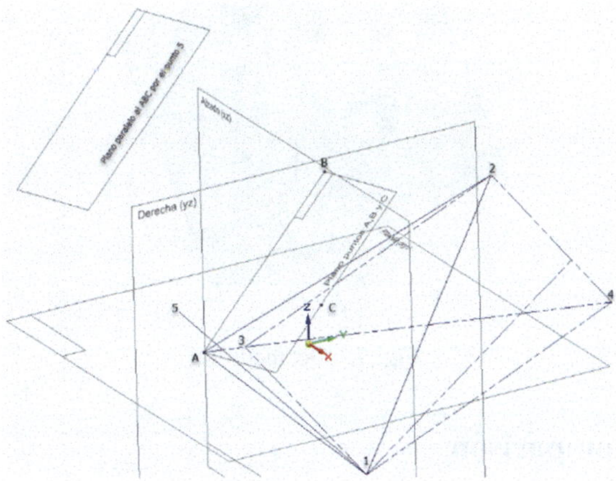

Fig. 5. Parallel plane to ABC from point 5

In the plane parallel to ABC, using the command "project in sketch", which allows to copy a sketch, an equilateral triangle is drawn equal to that in the plane ABC. This equilateral triangle is rotated 180º with respect to its center and the two opposite sides of the octahedron are shown, as it can be seen in Fig. 6.

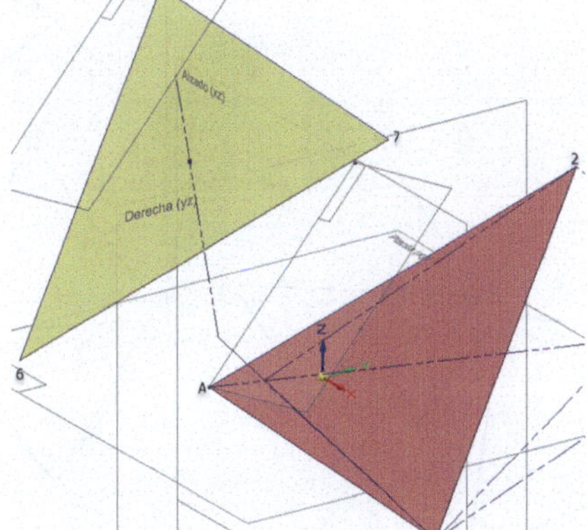

Fig. 6. Parallel triangle 678 to A12.

It only remains to join the vertices of the two faces in an orderly fashion. The vertex 6 with the A and with the 1 plus the vertex 7 with 1 and 2 and finally the vertex 8 with 2 and A, (see Fig. 7).

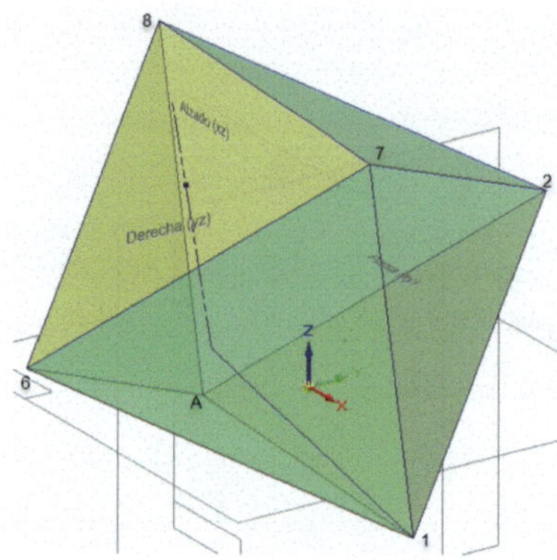

Fig. 7. Octahedron

3 Methods Comparison

We will briefly compare the previous method with the method that descriptive geometry would use.

To solve the problem with the tools of descriptive geometry we would have to do what is shown in Fig. 8, that is, two plane changes to solve the problem, and to undo the plane changes again.

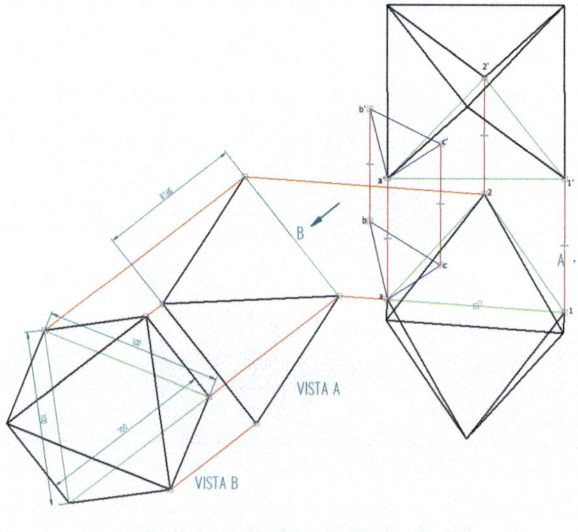

Fig. 8. Problem resolution applying descriptive geometry method.

4 Results

This teaching-learning methodology has been used in the teaching of the polyhedra from the subject Graphic Expression with the purpose of extending its application to other areas of the subject in the next course, according to the degree of success achieved.

In order to asses it, a brief questionnaire of seven items measured on a Likert scale with 5 response levels was proposed to the students at the end of the experience. (1 = Strongly Disagree, 2 = Disagree, 3 = Neither agree nor disagree, 4 = Agree, 5 = Strongly Agree).

The items designed questionnaire were:

1. This experience increases interest in the subject.
2. This experience allows to develop better the spatial vision.
3. The objectives in the explained subject are clear.
4. In each lecture the content explained has been adequate.
5. This experience should be incorporated to the rest of the subject.
6. Class attendance is essential for the follow-up of the subject with this method.
7. The teaching/learning used methodology is adapted to the needs of the students.

The study sample consisted of 23 students, in which 21 answered the survey, obtaining the results shown in Tables 1 and 2.

Table 1. Mean.

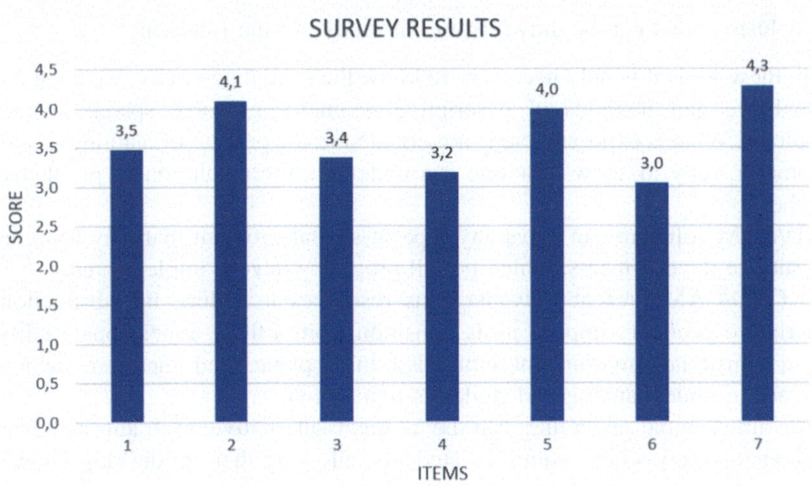

In both graphics, it is evidence, that the answers to items 2 and 5 have obtained the highest values, only exceeded by the answers to item 7. This may be due to the fact that, with the methodology used for the resolution of problems, the spatial vision is at all times present, not being necessary a great effort to spatially visualize the problem, as it happens in the descriptive geometry. The great results achieved in item 5 make it clear that students are committed to apply the methodology to the rest of the subject.

Table 2. Median and Mode

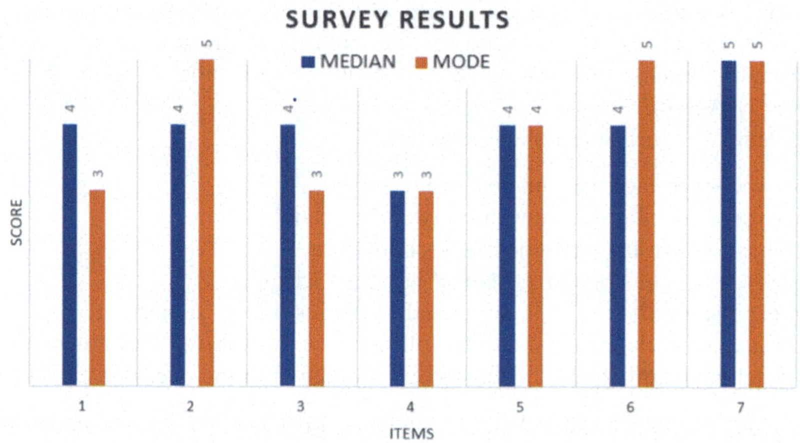

5 Conclusions

The conclusions that can be drawn from this study are the following:

1. With these tools it is only necessary to know the spatial geometry, we must not add knowledge and methods of descriptive geometry to solve spatial geometrical problems. What is exposed here is not oriented to the process of learning descriptive Geometry, only to show how one can undertake the resolution of problems with another tool.
2. CAD/CAM software can solve any type of spatial problem that may arise, except for surface developments, which not all programs have it implemented.
3. The CAD/CAM presents advantageous resources to reduce the abstraction that descriptive geometry implies in its transition from a three-dimensional reality to a two-dimensional environment embodied in a plane, and therefore achieve an increase in understanding and student's motivation.
4. Three-dimensional modeling can be a great alternative to traditional teaching methods and, especially useful for students, allowing them to develop knowledge, skills and abilities applicable during their professional life.

References

1. PROCECAD. https://www.youtube.com/watch?v=zPd5Lf_srHU. Accessed 01 Mar 2019
2. DESCRIPTIVE GEOMETRY. https://en.freedownloadmanager.org/Windows-PC/Descriptive-Geometry.html. Accessed 06 Mar 2019
3. ARCHIMEDES GEO3D. http://archimedesgeo3d.weebly.com/. Accessed 06 Mar 2019
4. Izquierdo-Asensi, F.: Geometría Descriptiva. 24ª ed. Paraninfo, Madrid (2000). ISBN: 84-922109-5-8
5. Online Instructor: Solid Edge 2019 Basics and Beyond: Part Modeling, Assembly Design, Drawings, Sheet Metal, and Surface Design. Independently Published (2018). ISBN-13: 978-1790600007

Active Learning, Improvement of Spatial Abilities and Understanding of Technical Drawing in the Graphic Expression Course

Laura Diago Ferrer(✉), Verónica Zubiaurre Eizaguirre, and Enrique Tardio Monreal

Department of Design and Manufacturing Engineering, Maria Luna 3, 50018 Saragossa, Spain
lauradf@unizar.es

Abstract. Over years of experience in the teaching of Graphic Expression in Engineering, a number of difficulties have been observed. These difficulties pertain to the homogenous participation of students in practical classes, and also to difficulties encountered in achieving a three-dimensional understanding of exercises proposed in a two-dimensional format. The utilization of 3D-PDF methodologies, special awareness, and three-dimensional applications has obtained satisfactory results in a variety of cases. For this reason, the following text presents examples of teaching experience in the utilization of an active learning methodology which was applied in the 2018–2019 academic year. This methodology was based upon the execution of supervised practical work which was supported by the use of three-dimensional tools and Information Computer Technology (ICT). The practical work in question was part of a Graphic Expression course within the degree of Electronics and Automatics at the University of Zaragoza. Importantly, the combination of three-dimensional tools and ICT served as a vertebral axis for the project. The following text describes the teaching methods used and the results obtained in one of the tests from the 2018–19 academic year. These results are then compared with previous years in order to examine the various roles played out within the process and ultimately to open up a debate through which to improve it.

Keywords: Active learning · 3D models · Teaching · Engineering

1 Introduction

Currently, the subject of Graphic Expression in the degree of Electronic and Automatic at the University of Zaragoza faces the problem that between 20–25% of students have never previously encountered materials pertaining to technical drawing or descriptive geometry. Furthermore, there is a general perception amongst alumni that the subject is abstract and complex. As a result, levels of demotivation and frustration are high, since students lack the necessary spatial reasoning skills in order to visualize 3D figures from which to obtain 2D views.

In addition, difficulties have also been observed in the way students manage their study time. This is particularly evident off campus, where preparation is often reduced

F. Cavas-Martínez et al. (Eds.): INGEGRAF 2019, LNME, pp. 374–385, 2020.
https://doi.org/10.1007/978-3-030-41200-5_41

to as little as a week prior to each test or exam. Consequently, the possibilities of students attaining the necessary spatial awareness in order to pass the course are further reduced.

Bearing in mind the difficulties students currently face in obtaining sufficient spatial awareness for the course, and also the obstacles they face in managing their study time, we have prepared a Teaching Innovation Project (PIIDUZ) for the 2018–19 academic year. This project aims to reduce the identified difficulties by shifting focus towards a new perspective. Essentially, the objective of the project is threefold: first, to modify teaching methodologies and to facilitate active, independent and structured learning; second, to improve both individual and group work in order to promote student participation; third, to enhance student learning, performance and motivation towards the contents of the course.

Howard Gardner's theory of multiple intelligences [1] poses a plural vision of intelligence which acknowledges eight different classes of cognitive capacity. Amongst these is the visual-spatial capacity necessary for successful completion of the Graphic Expression course. Arrieta defines spatial capacity as a component of human intelligence which allows us to form, recognize and manipulate images, figures and objects mentally [2]. Various authors have verified that students' acquisition of spatial vision in engineering correlates directly with their potential for success in the professional field [3–5]. This attests to the importance of the Graphic Expression course in achieving spatial vision, and thus to the need for vehicular tools which adapt to modern classrooms in order for students to achieve it.

Traditional teaching techniques in this area have generally been of a passive nature, based upon 2D representations which students must glean from the explications and demonstrations given by the teacher on the board or in PowerPoint. Importantly, these techniques have been observed to limit real interaction between the three-dimensional element and the student, thus reducing any likelihood that the necessary skills will be acquired independently. In the same way, development of spatial capacity is relegated almost entirely to a student's natural capacity and previous experience.

Theses such as those offered by Gutierrez, Villa or Melgosa suggest that the best way to obtain an efficient development of spatial ability in the classroom is through hand-eye exercises [6–8]. Here, the student must perform rotations, visualizations from various view-points, mental modifications of geometry etc. In this sense, there are various types of academic test which allow students to hone and measure these skills, or at least some of their principal components. Saolín et al. compile and summarize the different types of test used for this purpose [9].

Alternatively, there are studies which have recently shown the benefits of 3D-PDF files as a tool for hand-eye development and improvement of spatial ability [10].

Therefore, the methodology which we propose for the 2018–19 academic year combines those tools which have proved most effective in other areas. The following were used as a base both at the beginning and end of the course in order to measure students' capacity for spatial reasoning: the Purdue Spatial Visualization Test (PSVT-R); the Differential Aptitude Test – Spatial Relation Subset (DAT-SR); The Mental Rotation Test (MRT); the Mental Cutting Test (MCT). 3D-PDF were also used for resolving tasks in the supervised practical work.

Finally, in order to address the issues of demotivation and poor time-management, it was decided that additional and immediate motivation would be applied to parts of the course to be able to complete a long-term task. This was done by way of questionnaires, e-evaluation, and e-feedback [11, 12]. Throughout the course, we have proposed weekly tasks with auto evaluations, in which students are presented not only with real-time feedback, answers and scores, but also with their ranking with regard to their peers on the course. It is hoped that this will provide additional motivation in students' learning.

2 Graphic Expression in Electronic and Automatic Engineering

The Graphic Expression course aims to train students to design and represent graphically geometric figures, industrial parts and objects through a universal language that allows to be understood by third parties for manufacturing process later.

That's why the subject is taught over a semester and is divided into three parts with the following teaching hours and percentages of the total of the subject:

- **Theory and problems (45 h):** In this block, students are taught about standardization and also about the dihedral system as a main system of geometric representation. This part of the course is distributed into three teaching hours per week, of which two correspond to the teaching of theory (2 h $\times$ 15 weeks = 30 h), and one to the resolution of problems. These problems pertain both to standardization (during the first five weeks), and geometry (during the remaining ten weeks). The practical classes are always linked to information given in the theory lectures (1 h $\times$ 15 weeks = 15 h). The block constitutes 60% of the global score and is assessed via an end-of-term exam.
- **Standardization (20–30 h):** In this section, continuous assessment is used in order to evaluate a student's progress throughout the course. Tests similar to those carried out in class consist of the completion of plans through the necessary dihedral views, cut views and sections. These must be precisely dimensioned following specific guidelines for the correct interpretation of the objects represented. Students have access to between 20 and 30 h of tutorials outside of class where necessary. The tests represent 30% of the global grade.
- **Computer aided design (15 h):** Distributed into five 3-h sessions, the classes introduce students to the handling of a graphic design program in both 2D and 3D. Student's work is evaluated via continuous assessment, which represents 10% of the final grade.

The results obtained are intended to encourage students to value the possibilities of technical drawing as an investigation tool, and also to appreciate standardization as an ideal convention for the simplification of both production and communication. In this way, it is hoped that students will recognize the universal nature of the target language in the transmission and understanding of communication.

The subject and its results respond to the following proposition and general objectives, i.e. to acquire the following competences: knowledge of the profession;

capacity to learn; capacity to generate new ideas; capacity to solve problems; capacity to put knowledge into practice; capacity of written and oral communication; responsibility at work; motivation for the task at hand; capacity to work independently; interpersonal skills; concern for achieving excellence and a willingness to improve.

In particular, on completion of the course, students should be competent in; precise, clear and objective expression of graphic solutions; visualization of objects from different spatial positions; use of ICT for the graphic representation of project plans.

2.1 Teaching Experience in the Course Prior to the 2018–19 Academic Year

With the implementation of the Bologna Plan in 2009–10, a change of teaching methodology was put into practice. This consisted of evaluating grades of learning and use of standardization via a system of supervised practical work. Previously, this part of the course utilized the same methodology as those of theory and problems, and could only be passed via a final exam, rather than continuous assessment.

Thus, students were assessed via standardized and individualized monitoring throughout the course, a process which consisted of 6 exercises in A4 format. These were taken from a book of design problems for the subject, whose rubrics were presented to the students at the beginning of the course and which students carried out over a series of 2 months. These exercises comprised 6 industrial pieces presented from axonometric or dihedral views which students were to represent from different views, i.e. through the use of cut views, dimensioning and correct management of scale. The main objective was to improve students' spatial capacity.

The 6 exercises were to be completed 3 at a time, with a margin of one month between the two submission dates. Within two weeks of handing in both sets of exercises, students were assigned a meeting with their tutor – to be attended in groups of three – where the work could be corrected.

In these tutorials, attention was placed on the correct representation of the views, cut views and elevations, whilst adhering to UNE regulations, and also to the correct delineation of the complete plan. Student errors were corrected on an individual basis. Conversely, group corrections allowed students to learn from their peers' mistakes, thus fostering a critical approach. After this tutoring process, students were given the remaining two weeks in order to correct their work before the final submission date.

Although this methodology, based on the Problem-Based Learning approach (PBL), improved academic results in comparison with those prior to the Bologna plan, it was noted that students were now copying each others' work. Thus, whilst the evolutionary nature of the methodology meant that emphasis was no longer placed of a single final assessment, it was now impossible to guarantee a homogenous learning experience for all. This became particularly evident when students were presented with two additional regulated tests. The results revealed a number of cases in which a student had passed the six tutored exercises with ease, yet failed to achieve even the minimum grade of 4 necessary to pass the additional tests. In fact, only 20% of students achieved a proportionate grade between both parts of the assessment. The remaining 80% were achieving between grades 7 and 9 in the tutored exercises, whist obtaining a 4 in approximately 55% of cases, and even lower amongst 25% of students.

Importantly, although practical work compensated for poor results in the additional tests, and students were therefore able to pass the subject, it was felt that they were not acquiring a sufficient level of skill. Furthermore, many of the students failed to attend the scheduled meetings with their professors, thus limiting achievement in terms of the objectives set out in the Graphic Expression Course.

On reviewing results between the 2014–15 and 2016–17 academic courses, in which the six supervised tests were combined with the two additional ones with a weighting of 50% of the total module grade, it was decided that the tutor-supervised tests were no longer serving their intended purpose. Finally, in the 2016–17 academic year, evaluation was reduced to standardization only, with just two obligatory practical tests.

3 Teaching Innovation Project (PIIDUZ). Active Learning

The PIIDUZ was proposed at the beginning of the 2018–19 academic year, in order to change teaching methodologies in the theory-practical parts of the Graphic Expression Course. The aim was to achieve greater student independence as well as to improve spatial learning, academic results, motivation, and time management.

3.1 Materials and Methods

Participants. All students following the course – approximately 70% of the total number of students enrolled – participated in the project. 140 students were divided into a morning group (G-311), and an afternoon one (G-312).

Instruments. Two questionnaires of auto evaluation and feedback were developed. Two of these corresponded to the fundamentals and applications of standardization, whilst the other two concerned the application of dimensioning. Each questionnaire comprised between 30 and 40 questions – either of multiple choice, box ticking, or short texts. The Google Forms tool was used to create the questionnaires, since it allowed them to be programmed as tests and therefore gave students immediate feedback.

A further questionnaire was created on the Moodle platform, with a total of 40 questions. These were distributed into 11 DAT-SR spatial tests, 9 MCT, 10 MRT, and 10 PSVT-R. Questions 1–29 were multiple choice (3–4 possible answers), whilst the final 10 questions (30–40) required students to choose 2 correct answers from a possible 4. The final questions pertained to the Purdue Spatial Visualization Test (PSVT-R).

Students completing the questionnaire were given 20 min, in order to homogenize results. No prior knowledge was required.

Finally, the industrial pieces which were later to be proposed as tasks were modelled using the Solid Works program. These would later be converted into 3D-PDF so that students could visualize them during the work process.

Six weekly tasks, with their corresponding exercises, were created on Moodle for the standardization part of the course.

Design and Project Procedure. On the first day of class, students carried out a general visualization test, based on Vandemberg's Mental Rotation Test (MRT) - Mental Cutting Test (MCD); Differential Aptitude Test (DAT-SR), and also the Purdue Spatial Visualization Test – Visualization of Rotations (PSVT:R). The objective was to evaluate students' spatial reasoning before commencing the course. These tests would be repeated at the end of the course.

During the standardization phase, the part which was developed at the time of publication and whose results are represented below, 6 tasks were proposed. These were organized in ascending order of difficulty in order to allow students to prepare for the two obligatory tests progressively. Thus, by the end of the fourth task, they had completed the first test of standardization, and by the end of the sixth task, they had completed the second.

The first task consisted of resolving all possible views of 12 different industrial pieces, presented to the students from an isometric view. This was to be done following the European system of representation. Furthermore, students had to respond to questions, such as which views had the same elevation, the same plan etc. In the second task, work continued on the views of plan, elevation and profiling, adding the simple auxiliary view. The third task involved executing a double auxiliary view and a cut view, as well as an auto evaluation test on standardization. In the fourth task, students were presented with two industrial parts in order to resolve the indicated cut views and also a second auto evaluation questionnaire on standardization. The fifth task presented students with an isometric view with which to formulate the necessary cut views, along with their correct interpretations and dimensioning. Finally, in the sixth task, students were given an industrial part, viewed from different cut views, which had to be resolved and dimensioned. This task was associated with the final auto evaluation on dimensioning.

All tasks were activated on a Monday and deactivated at midnight on Sunday, so that students had immediate access to the next task on completion of the previous one.

Each week, class exercises corresponded to those of the evaluation task, analyzing it in detail through 3D-PDF resources and slides of the main mistakes made in previous tasks. The students completed the practice work and the auto evaluation test and uploaded them onto the course website. These were also handed in to the teacher on paper at a later date. After the deadline, the solutions were uploaded onto the Moodle course, so that students could autocorrect and identify their own mistakes.

A further two auto evaluation tests on standardization were linked to the Moodle platform (one from task 3 and the other from task 4). Meanwhile, an additional 2 auto evaluations on dimensioning were also uploaded (one from task 5 and the other from task 6). These auto evaluations served to monitor students' knowledge, the advantage being that they centered upon the most commonly-made mistakes, whilst supplying students with correct answers and explanations on completion so that they could auto evaluate and see their score ranking with regard to their peers. For this reason, we recommended that students use this tool as often as they felt necessary, in order to limit errors and strengthen knowledge.

3.2 Results

The results presented below correspond only to the initial visualization test and the first obligatory standardization test. They were obtained subsequent to the implementation of the tools and tasks explained previously. Interpretation of results gained in the subject as a whole will be left for analysis at the end of the course, when the new resources have been applied to the totality of the course material.

In total, 96 students from groups G-311 and G-312 participated in the first visualization questionnaire, all of whom were in their first year of the degree in Electronics and Automatics at the University of Zaragoza. A global analysis of the results shows us that a little over half of the answers given were either incorrect or unanswered (see Fig. 1). Essentially, there was a slight majority of students who experienced some kind of difficulty in the execution of spatial reasoning exercises.

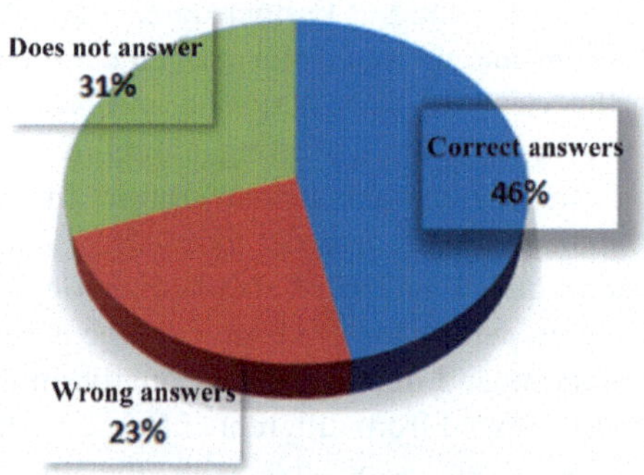

Fig. 1. Initial test, result questions 1–40.

This result could have been considered positive, had the participants involved not been selected from a skewed student demographic, i.e. from a range of different specialties. In fact, the participants belonged to an eminently scientific or technical background, having logically completed technical drawing courses in both the first and second years of their high school baccalaureate before commencing the degree in Electronics and Automatics. There would also have been further reinforcement for those who were repeating the course material.

On the other hand, if we focus on the percentage of correct answers given in single-answer questions, compared to those where 2 correct answers were possible from a selection of 4, we see a clear decrease in success rates in questions 31–40. These were the questions in which students were required to select the two correct options, i.e. AB, AC, AD, BC, BD and CD – six combinations of which only one was correct. The results could be due to a greater level of difficulty which student experience in spatially rotating and recognizing an object from various positions. Alternatively, it could also be that students ran out of time to complete the questions properly, especially given these appeared at the end of the test (see Fig. 2).

However, when we analyze results from the single-answer questions (1–30) separately from the global results of questions 1–40, we see that success rates rise notably.

Although the percentage of incorrect responses rose too, this was mainly due to questions which were left unanswered.

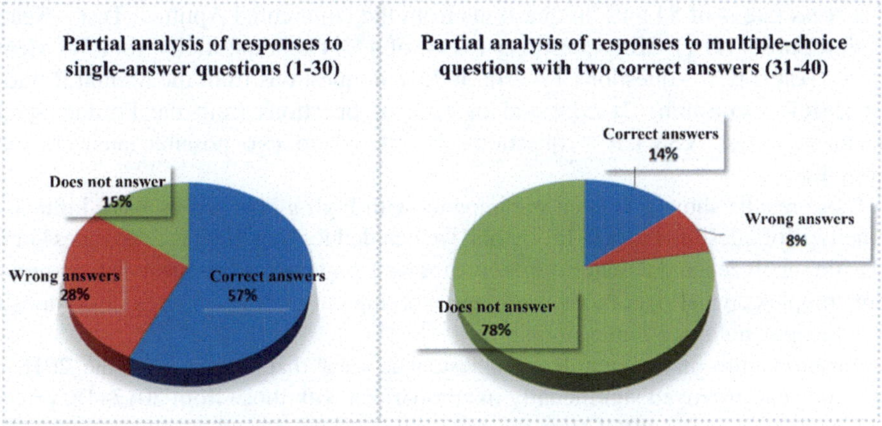

Fig. 2. Initial test, comparative analysis by answers.

In general, there was a diffusion of results, depending on the type of test used and whether the analysis of results was localized or global (see Fig. 3).

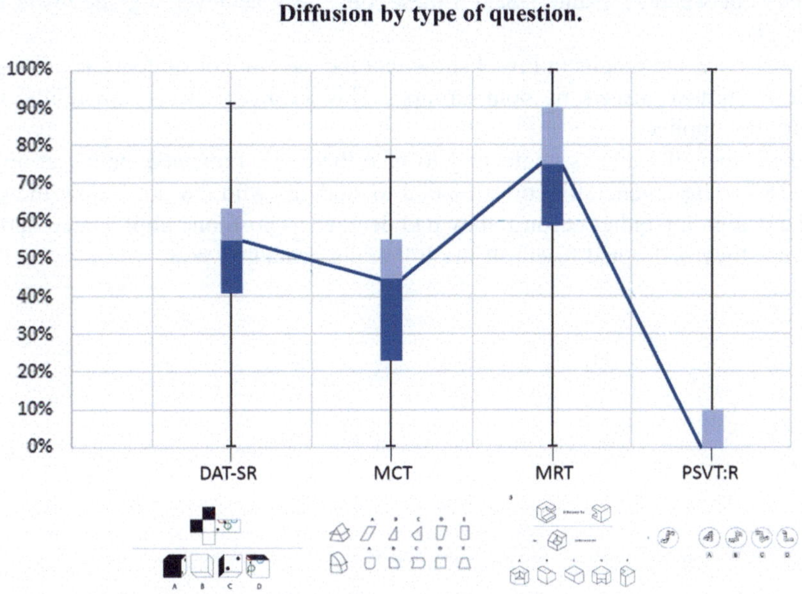

Fig. 3. Initial test, comparative analysis by question type.

This diffusion could be related to the wide-ranging differences of ability in terms of students' spatial perception (either natural or acquired), when compared with the different abilities necessary to complete each test successfully.

An analysis of the average number of correct answers revealed the following: that the success rate is of 54.54% in questions from the Differential Aptitude Test – Spatial Relation Subset (DAT-SR – questions 1–11); of 44.44% in questions from the Mental Cutting Test (MCT – questions 12–20); of 75% in questions from the Mental Rotation Test (MRT – questions 21–31); and of 10% of questions from the Purdue Spatial Visualization test (PSVT:R – questions 31–40, where two possible answers were required).

These results show that even participants with high global scores were influenced by the type of questions posed. From this, we can deduce that there is a weakness in the capacities analyzed and a major difficulty amongst students in those tests which require more complex spatial operations, such as revolving cut views of figures or composing 2D developments three-dimensionally.

Regarding the first standardization test, it is clear that results from the 2018–19 academic year improved significantly in comparison with those from 2017–18, prior to the introduction of the PIIDUZ in the tutor-supervised practical work.

The fail rate decreased in all of the groups by at least half. In courses prior to the implementation of the PIIDUZ, the rate of failure in the first standardization test was high. In fact, between 50% and 65% of students did not pass the test. From these percentages, only 10% of students achieved a fail + grade (between 4 and 4.9), which could be compensated with the other parts of the course. After the introduction of the PIIDUZ, with its interactive tools and tasks, student preparation improved, reducing fail rates to between 22% and 32%. Of these, only 18% achieved a grade lower than 4 (see Fig. 4).

Furthermore, the graphs in Fig. 4 show that the pass rate of students achieving over a grade 7 almost tripled in both groups. This clearly reflects favorably on the methodology applied.

The number of students neglecting to take the exam remained stable, at approximately 18%. These generally corresponded to students who did not follow the course from the beginning, either because they had decided to postpone until a later start-date, or because they had opted to enroll the following academic year.

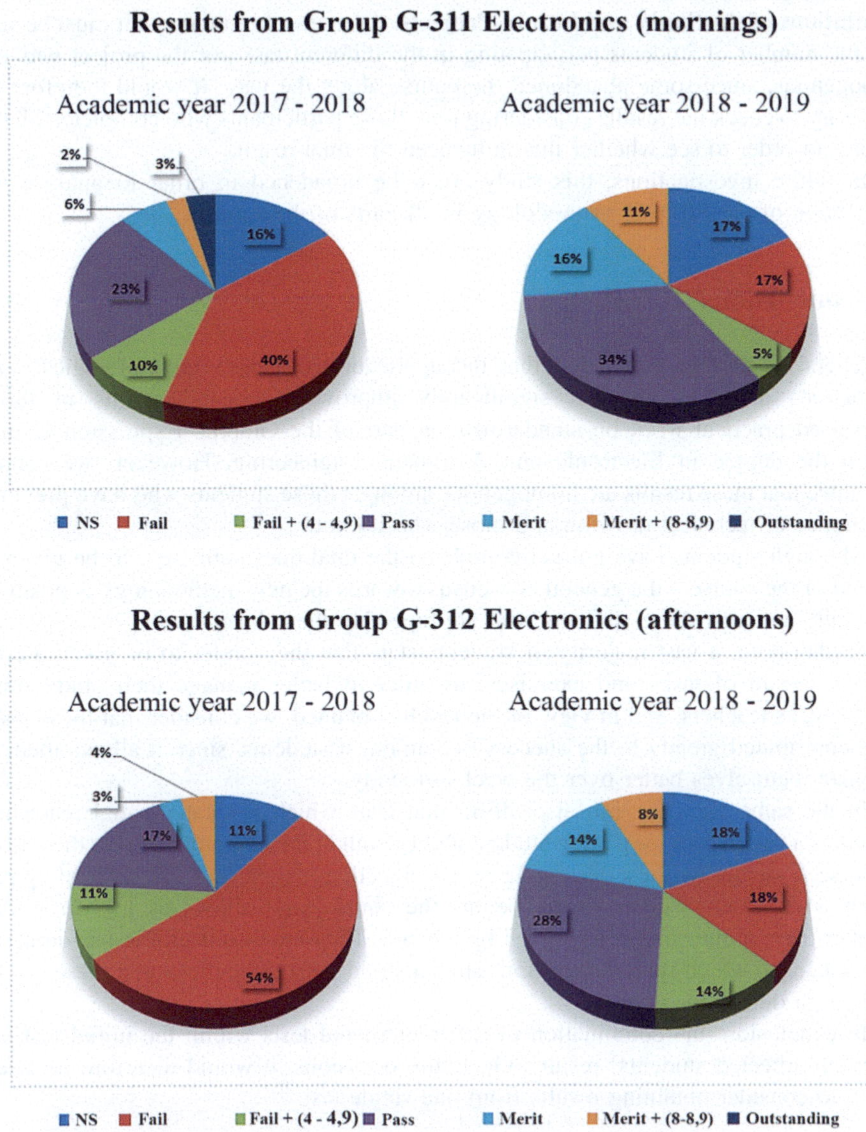

Fig. 4. Test 1 standardization, comparative analysis by results.

3.3 Discussion

The Influence of the Tasks and 3D-PDF on Academic Results. The implementation of weekly tasks in ascending order of difficulty clearly improves academic results amongst students, since it allows them to prepare gradually for the standardization part of the course.

Limitations of the Study and Future Projects. In terms of limitations, it must be said that the number of students participating in the different tasks of the project was not homogenous, since some abandoned the course along the way. It would therefore be necessary to check the results considering only those participants who completed all the phases, in order to see whether this influenced the final result.

In future investigations, this study could be broadened in order to analyze the application of the PIIDUZ methodology in all parts of the subject.

4 Conclusions

In general, it can be concluded that the application of active methodologies, auto evaluations and 3D-PDF has significantly improved students' results in tutor-supervised practical work on standardization part of the Graphic Expression Course within the degree in Electronic and Automatic Engineering. However, we cannot guarantee that these results are homogenous amongst those students who have previous experience of technical drawing and those who do not.

Although students have not yet completed the final questionnaire – to be given at the end of the course – the general consensus towards the new methodology is positive, especially amongst those students who are repeating the subject.

Furthermore, a vast majority of students state that they prefer to be given a continuous stream of tasks and exercises, in order to better manage their study time. Similarly, as teachers, and in view of the results obtained, we consider that these tasks have contributed greatly to the success rate amongst students, since it allows them to organize themselves better over the weeks of study.

In the same way, we can also affirm that tests which evaluate spatial reasoning served as a useful tool for gaging student ability both at the beginning and at the end of a course. Essentially, these tests give us an objective insight into the level of spatial vision with which a given group begins the course, and allows us to corroborate whether these abilities have improved by the end. They are also useful in providing us with data as to which spatial operations students find most difficult, so that these can be worked on during the course.

In conclusion, the combination of different spatial tests within the initial test has favorably affected students' results. On future occasions, it would therefore be interesting to consider obtaining results from one single test.

References

1. Gardner, H.E.: Intelligence Reframed: Multiple Intelligences for the 21st Century. Basic Books, New York (1999)
2. Arrieta, M.: La capacidad espacial en la educación matemática: estructura y medida. Educación matemática **18**(1), 99–132 (2006)
3. Sorby, S.A., Baartmans, B.J.: A course for the development of 3-D spatial visualization skills. Eng. Des. Graph. J. **60**(1), 13–20 (1996)

4. Strong, S., Smith, R.: Spatial visualization: fundamentals and trends in engineering graphics. J. Ind. Technol. **18**(1), 1–6 (2001)
5. Miller, C.L., Bertoline, G.R.: Spatial visualization research and theories: their importance in the development of an engineering and technical design graphics curriculum model. Eng. Des. Graph. J. **55**(3), 5–14 (1991)
6. Martín Gutiérrez, J.: Estudio y evaluación de contenidos didácticos en el desarrollo de las habilidades espaciales en el ámbito de la ingeniería. Tesis doctoral. Universidad politécnica de Valencia. España (2010)
7. Villa, A.: Desarrollo y evaluación de las habilidades espaciales de los estudiantes de ingeniería. Actividades y estrategias de resolución de tareas espaciales. Tesis doctoral. Univer-sidad politécnica de Cataluña. España (2016)
8. Melgosa, C.: Diseño y eficacia de un gestor web interactivo de aprendizaje en ingeniería gráfica para el desarrollo de la capacidad de visión espacial. Tesis doctoral. Universidad de Burgos. España (2012)
9. Saorín, J.L., Navarro, R., Martín, N., Contero, M.: Las habilidades espaciales y el programa de expresión gráfica en las carreras de ingeniería. In: ICECE 2005 (2005)
10. Gómez-Tone, H.C.: Impacto de la Enseñanza de la Geometría Descriptiva usando Archivos 3D-PDF como Entrenamiento de la Habilidad Espacial de Estudiantes de Ingeniería Civil en el Perú. Formación universitaria **12**(1), 73–82 (2019)
11. Manzanares, M.C.S., Dios, M.Á.Q., Sánchez, R.M., Llamazares, M.D.C.E., González, Á.A.: Cuestionarios de e-autoevaluación y e-feedback: una aplicación en Moodle. Eur. J. Health Res. **4**(3), 135–148 (2018)
12. Gumbau Albert, M., Garcia-Carceles, B., Marín, A., Pastor, J.M., Villagrasa, J.: Clickers y Exámenes Tipo Test, Herramientas Interactivas de Evaluación Contínua. In: In-Red 2016. II Congreso nacional de innovación educativa y docencia en red. Editorial Universitat Politècnica de València, July 2016

Experience in the Implementation of a Product Lifecycle Management System (PLM) for the Management of Projects in the Subject of Technical Office

Pedro Ubieto-Artur(✉), Cristina Royo-Sánchez, and César García-Hernández

Departamento de Ingeniería de Diseño y Fabricación, Universdad de Zaragoza, María de Luna, 3, 50018 Saragossa, Spain
pubieto@unizar.es

Abstract. This paper is the result of a teaching innovation project that aims to implement a commercial document management system for the administration of the technical drawings included in the projects that belong to the subject Technical Office of the Degree in Engineering in Industrial Design and Product Development.

The main objective of this work is to ensure that students learn to work with a commercial PLM system, and make them able to manage the graphic documentation of their projects, reduce the duration of the project and face them with the problems that occur when using different versions of project documents. It is also considered an objective of this work to bring students closer to the reality they will encounter when they leave the labour market.

Although the experience has been carried out in the subject of Technical Office and dealing with projects based on designs made with Inventor, its use is possible in other subjects of the area and even in subjects of other areas of knowledge. The software used is designed to manage any type of documentation that always accompanies a project.

Keywords: Product Lifecycle Management (PLM) · Oficina Técnica · Vault

1 Introducción

Following the guidelines of the European Higher Education Area (EHEA) for the development of a new educational context, education should focus on student learning [1, 2] and "more specifically on the learning outcomes expressed in terms of competence" [3].

Specifically, the subject of Technical Office of the Degree in Engineering in Industrial Design and Product Development of the University of Zaragoza, sets as one of its objectives that students acquire the basic and general competence (CG07): "Ability to use and master the techniques, skills, computer tools, information and communication technologies and design engineering tools necessary for the practice of it" [4] in order to achieve the goal that the student is able to perform the "planning and management of technical documentation: technical drawing" [4].

F. Cavas-Martínez et al. (Eds.): INGEGRAF 2019, LNME, pp. 386–394, 2020.
https://doi.org/10.1007/978-3-030-41200-5_42

One of the main characteristics of the current market is the globalization of competition and the shortening of the business cycle of a product, so in the mid-80s, in order to reduce time and cost in the development of a new product, a new strategy called product lifecycle management [5, 6] emerged. Product lifecycle management (PLM) is an information strategy that extends throughout the life of a product, from design processes until it ceases to be manufactured, recycled or eliminated [6, 7].

For the PLM to fulfill its objective, all product information must be managed efficiently and must always be available to all persons who need product information [8].

At present, to carry out the management of the product lifecycle, companies use specific computer-aided design software (CAD), computer-aided manufacturing (CAM), computer-aided engineering (CAE) and product data management (PDM) [6].

Autodesk Vault® is a commercial software which serves to centralize and manage all the documentation related to the design of a product, whether technical drawing generated with different Autodesk programs (AutoCad, Revit, AutoCad Civil 3D, etc.) or documents, Microsoft Office spreadsheets and presentations, pdf files, etc. [9]. In this way, all members of a project team or technical office can access and work together on this documentation [10].

In this paper, we will present the experience acquired in the training of students of the subject of Technical Office in the use of Autodesk Vault®, so as to make them able to use this software in the development of their projects, both in the management of technical drawings, generated with Autodesk Inventor, and in the management of the rest of the non-graphic documentation.

We begin by explaining the context in which the experience develops and the objectives set at the beginning. Next, we describe the implanted system, with the most important technical characteristics. Next, and as the most important part of the exercise, we develop the tasks performed in the practical classes and the competences worked in them.

Although the experience is carried out with a three-hour practical class, we have divided it into two levels of skills: basic and advanced. We should be able to divide this practice into two parts, since the basic skills can be explained in previous course subjects, where team projects are done with Autodesk Inventor®. We conclude by commenting on the conclusions we have reached, after the experience.

We hope that this paper can serve as a starting point for those who want to implement a PLM system in their subject, being all the team responsible for this work available to the teaching community, in order to share experiences and continue advancing in this discipline, considered crucial within the area of Graphic Expression in Engineering.

2 Context and Objectives

In the subject of Technical Office of the Degree in Engineering in Industrial Design and Product Development, two construction projects are carried out, following the Spanish standard "UNE 157001" [11]. These projects are developed in work groups of three students and take as a starting point the designs made in other subjects.

There are three roles for carrying out projects: writing of project documents, preparation of technical drawings with their associated information and project management. The students of each work group distribute the roles according to their criteria, being the only condition that nobody repeats the role in the two works to be performed.

Since course 2012–2013, a Technical Office Management system has been developed, within the Web 2.0 environment, which is the platform used by students to develop their projects. This system focuses on the non-graphic information of the project. The technical drawings are presented in PDF format, as a “final product”.

Generally, the student responsible for the role of “Technical drawing” develops them on his personal computer, which hinders teamwork and negatively influences project deadlines. Generally, the rest of the members of the work group do not have access to this information until the document containing the technical drawing has not been completed.

To solve this problem, the work groups establish temporary solutions to access this information: the technical drawings are sent by email, shared folders are used in Drive, etc.

But these solutions are nothing more than another source of problems: version errors, no change control, lost files and, more importantly, these solutions are far from what students’ experience will be, once they enter the market labor.

2.1 Objectives

Although the main objective of the experience is to ensure that students learn to work with a commercial PLM system, what is really intended is that students:

- Be able to manage the graphic documentation of your projects, using a commercial PLM system
- Be able to reduce project times, thanks to the access of all the members of the work group to the system information.
- Reduce version problems with the necessary change controls, improving the quality of their projects.
- Understand the importance of PLM systems in the development of engineering projects.
- Get closer to the reality they will find in the labour market.

3 Materials

Commercial Autodesk Vault® software has been used to carry out this experience. As the main client, Autodesk Inventor® has been used, a 3D modeling program that allows students to design their product and generate project technical drawings. This program is also used in other subjects of the same degree, such as Computer Aided Design I and Graphic Expression II, among others.

One of the main advantages of using this software is that, thanks to Autodesk’s policy for university centers, students have free licenses without any limitations. In addition, this software is widely used in technical offices around the world.

The first step to work with Vault would be to install the server version of the program on a computer. Students' computers must have the Vault client version installed, in addition to Inventor. It should be clarified that, although the Vault client version is used here as an add-in installed in the main Inventor program, it can also be used as a stand-alone application. In this case, it is about the Vault application itself on which the management of all product information can be performed: changes of state, versions, revisions, change orders, etc.

3.1 Server Configuration

Following the system requirements recommended by Autodesk [12], Microsoft Windows Server® 2016 and Microsoft SQL® Server 2017 were installed. This software is available for free, for teaching purposes, if the University is part of the Azure Dev Tools program for Teaching (formerly Microsoft Imagine) [13].

The computer used has an Intel Core i3 at 3.70 GHz and 8 Mbyte of RAM memory. It does not meet the requirement recommended by Autodesk regarding the processor (it should be at least one i5), but no performance problems have been detected.

Once the operating system and database manager are configured, Autodesk Vault Server 2019 is installed. The process is quick (no more than 15 min).

If it was not previously done, it is necessary to configure the license server, which is installed with Vault. Installing Vault Client on client computers requires activation through the license server.

To complete the installation of the Server, it is necessary to configure the system firewall, to open the necessary ports [14]. It should be kept in mind that Vault uses Microsoft Internet Information Server (IIS) for remote access that, by default, uses port 80. If the computer is used as a web server, it is advisable to configure IIS on another port, the 8080 for example, before installing Vault. Otherwise, it is necessary to reinstall it after changing the port in IIS.

3.2 Configuration of Client Computers

Client computers must meet the requirements recommended by Autodesk for Inventor [15] as they are more demanding than those recommended for Vault Client. The client computers will be the students' computers and those of the computer classrooms in which the practical classes will be taught.

It is advisable to install Inventor first, since when installing Vault Client the Vault Add-in for Inventor is added [16], which allows access to the warehouse without leaving it.

4 Method

In order to carry out the experience, three phases have been established. First, it is necessary to configure the server in order to work on the skills with the students. Two levels of competencies are defined below. The basic level allows the student to interact with Vault, from the point of view of file management. The second level relates Vault's functionalities to the concepts of project management and management, typical of the Technical Office subject.

4.1 Practice Preparation

As we want to work on students' skills, it is necessary to configure the server in advance. You have to create vaults and users, just as the work teams are organized. In the specific case of the subject of Technical Office, they were organized into 19 groups of three students each. The necessary tasks are broken down and detailed in Table 1.

Table 1. Server configuración.

Description	Remarks
Create vaults	• Create a varehouse for each work group • The vault name matches that of the work group • Approximate duration: 15 min for 20 vaults
Create users	• One user must be registered for each student • To facilitate the work, users are created with the Personal Identification Number, as an identifier, plus their ID as a password
Assign stores to users	• Each user will be assigned the warehouse of their work group • The teaching team has access to all stores
Assign permissions to users	• Each user is assigned the maximum level of permissions according to the role they play in each job • Approximate duration of management of 60 users: 90 min

4.2 Basic Skills

Working the basic skills described in Table 2, the student will be able to interact with Vault, both from the Vault Client and from Inventor, from the point of view of file management [17, 18].

So that each student can perform the proposed practical exercises, it has a compressed file that contains three sets of files (one for each member of the work group) composed of a couple of documents in Microsoft Word format, as well as a simple assembly, of about nine pieces, in Inventor format. Each set of files has a different name (ending in 1, 2 or 3) to avoid problems with Vault. Although duplicate file names are allowed in a vault, Autodesk recommends using unique file names.

With these files, each student will create their own folder in which they will solve the proposed problems.

Table 2. Basic skills

Description	Skills
Introduction to Vault	• Understand what Vault is • See its main features • Log in to Vault with the client program • Know the Vault client program interface
Document management	• Understand the relationship between the folder structure of the Vault project explorer and the folders on the local computer • Learn how to create folders in a vault • Add/get files to/from a vault • Understand the concept of applying check-out and check-in to the archives of a vault
Vault plug-in for Inventor	• Learn about the capabilities of the Vault add-in added to Inventor • Understand the scope of inventor projects of the Vault type to define a coherent project for all clients • Learn to transfer an existing project to Vault • Understand how Vault Browser works • Check the need to open or insert parts with the Vault plug-in for Inventor
File management	• Understand the importance of Vault's ability to maintain the relationship between files in an assembly • Learn to move, copy and rename files, as well as copy designs • Know the concept of labels and their relationship with the product life cycle

The exercise is planned for an hour and a half, approximately.

The content of this practice can be applied to any subject in which students who are organized in work groups must develop a multi-piece assembly with Inventor. In the degree of Industrial Design Engineering and Product Development, it could be useful in Graphic Expression II, Computer Aided Design I and Design Workshop IV and VI, among others.

This practice can be developed with the Basic version of Vault, although it is difficult to find the educational version.

4.3 Advanced Skills

When working on advanced skills, the student will learn and relate the operation of Vault with the management and project management processes [18] which, in the case of the Technical Office subject, are defined by the UNE ISO 21500 standard [19]. The proposed problems will be solved on the set of files provided in the previous practice, although this time the team will only work on one of the assemblies (Table 3).

Table 3. Advanced competencies.

Description	Remarks
Categories and rules	• Understand the concept of categories in Vault • Know the categories defined by default and their relationship with files and articles • Understand the concept of rules and their relationship with categories Be able to create custom categories and rules • Understand the relationships between categories, product life cycles, file reviews and associated properties • Be able to create custom categories and rules
Product life cycle State	• Understand the concept of life cycle associated with Vault files and articles • Know the meaning and properties associated with each state • Learn to change the states of articles and files • Know how to assign life cycles to categories
Versions and revisions	• Understand the difference between version and revision • Learn to change the revision number of articles and files • Know how to assign the revision number format to the categories
Articles	• Understand the concept of Article in Vault, as well as its associated properties • Learn to create articles from scratch or from a file • Know how to use the article master
Material's list	• Understand how to manage material lists in Vault • Understand the relationship between article properties and Inventor iProperties • Understand the correspondence of the position number and number (part number) properties with ISO 7573: 2008 • Learn to add non-CAD items to the bill of materials • Know how to export the bill of materials
Exchange orders	• Understand the concept of change order used in Vault • Relate the change orders with the results of the control processes defined in the UNE ISO 21500 standard • Learn to create change orders and relate them to the affected objects • Know the life cycle defined for change orders • Learn to change the status of a change order

The practice is planned for an hour and a half, approximately.

5 Conclusions

With this experience, we will introduce a very specialized commercial program to develop the practices of the subject of Technical Office. Students will be able to work as a team with a PLM system to develop their projects, optimizing the final result, avoiding version problems and reducing project deadlines. It is a good way to bring them closer to the professional environment, where it is increasingly necessary to use project documentation management systems.

Although the experience has been developed in the subject of Technical Office, with projects based on designs made with Inventor, it is possible to apply it to other subjects of the area and, even, in subjects of other areas of knowledge. Vault software is designed to manage any type of documentation that is always associated with a project, even if it loses the functionalities of tracking file relationships.

Acknowledgements. This paper is the result of the teaching innovation project PIIDUZ_18_312 with the title "Implementation of a commercial document management system for the management of technical drawings in the projects in the subject Technical Office". This project has been recognized by the University of Zaragoza within the 2018/2019 teaching innovation call.

References

1. de Miguel, M.: Modalidades de enseñanza centradas en el desarrollo de competencias. Orientaciones para promover el cambio metodológico en el espacio europeo de educación superior (2005). https://www2.ulpgc.es/hege/almacen/download/42/42376/modalidades_ensenanza_competencias_mario_miguel2_documento.pdf. Accessed 21 marzo 2019
2. Rodríguez, R.M.: Mejora continua de la práctica docente universitaria: una experiencia desde el proceso de convergencia del Espacio Europeo de Educación Superior (2007). https://dialnet.unirioja.es/servlet/articulo?codigo=2783317. Accessed 21 marzo 2019
3. Cano, M.E.: La evaluación por competencias en la educación superior (2008). https://www.ugr.es/~recfpro/rev123COL1.pdf. Accessed 21 marzo 2019
4. Universidad de Zaragoza: Oferta de estudios oficiales universitarios (2019). https://sia.unizar.es/documentos/doa/guiadocente/2018/25821_es.pdf. Accessed 29 marzo 2019
5. Wikipedia: Product lifecycle, 6 marzo 2019. https://en.wikipedia.org/wiki/Product_lifecycle. Accessed 12 abril 2019
6. Martínez Lebrón, L.: La gestión del ciclo de vida del producto: PLM, 15 febrero 2013. https://www.eoi.es/blogs/mtelcon/2013/02/15/la-gestion-del-ciclo-de-vida-del-producto-plm/. Accessed 10 abril 2019
7. Rouse, M.: Definition. Gestión del ciclo de vida del producto (PLM), diciembre 2018. https://searchdatacenter.techtarget.com/es/definicion/Copy-of-product-lifecycle-management-PLM. Accessed 10 abril 2019
8. Pérez González, J.: Diseño e implantación en una herramienta PLM de un prototipo aeronáutico, Sevilla (2016)
9. Autodesk: Vault (2019). https://www.autodesk.com/products/vault/overview. Accessed 2 abril 2019
10. Autodesk: Productos Vault, 6 agosto 2018a. https://knowledge.autodesk.com/es/support/vault-products/learn-explore/caas/CloudHelp/cloudhelp/2019/ESP/Vault-About/files/GUID-87D9CA09-9881-4506-9465-0677392BCD7E-htm.html. Accessed 2 abril 2019
11. AENOR: UNE 157001:2014 Criterios generales para la elaboración formal de los documentos que constituyen un proyecto técnico. AENOR, Madrid (2014)
12. Autodesk: Productos Vault, 2 mayo 2018b. https://knowledge.autodesk.com/es/support/vault-products/learn-explore/caas/sfdcarticles/sfdcarticles/ESP/System-requirements-for-Autodesk-Vault-2019-products.html. Accessed 2 abril 2019
13. Microsoft: Microsoft Azure (2019). https://azureforeducation.microsoft.com/devtools. Accessed 12 abril 2019

14. Autodesk: Vault Products, 17 abril 2018. https://knowledge.autodesk.com/support/vault-products/troubleshooting/caas/CloudHelp/cloudhelp/2019/ENU/Vault-Install/files/GUID-D77453BD-68A1-465F-8979-05F21E5A382B-htm.html. Accessed 2 abril 2019
15. Autodesk: Productos Inventor, 22 marzo 2018a. https://knowledge.autodesk.com/es/support/inventor-products/learn-explore/caas/sfdcarticles/sfdcarticles/ESP/System-requirements-for-Autodesk-Inventor-2019.html. Accessed 2 abril 2019
16. Autodesk: Productos Inventor, 18 abril 2018b. https://knowledge.autodesk.com/es/support/inventor-products/learn-explore/caas/CloudHelp/cloudhelp/2018/ESP/Inventor-Help/files/GUID-AEBB84F1-CFF9-4606-9F0E-FC6EBD2A364F-htm.html. Accessed 2 abril 2019
17. Center for Technical Knowledge Ascent: Autodesk® Vault Basic 2019 Essentials, 1st ed. Ascent, Center for Technical Knowledge, Charlottesville (2018)
18. Center for Technical Knowledge Ascent: Autodesk® Vault Professional 2019 Data Management for Autodesk® Inventor® Users, 1st ed. Ascent Center for technical Knowledge, Charlottesville (2018)
19. AENOR: UNE-ISO 21500:2013 Directrices para la dirección y gestión de proyectos. AENOR, Madrid (2013)

From Y to Z Generation, an Engineering Students' Spatial Skill Analysis

Vicente A. López Chao and Jorge Martín-Gutiérrez(✉)

Dpto. Técnicas y Proyectos en Ingeniería y Arquitectura,
Universidad de La Laguna, 38202 San Critóbal de La Laguna,
Tenerife, Spain
{vlopezch, jmargu}@ull. edu. es

Abstract. At the beginning of the 21st century, in Spanish engineering schools, research was carried out on the levels of spatial ability in engineering students and how they could improve these levels with different training. The results of those investigations revealed low levels in spatial ability of the students who had access to engineering degrees. In addition, they confirmed that specific training could improve these levels. A decade later, how is the level of spatial ability of students who have access to engineering careers? In this decade, society, and specifically youth (belonging to the Z-generation) has acquired habits that would be worth analyzing and that can influence their levels of spatial ability (mass use of mobile devices, virtual reality video games, use of Apps…). This research compares the level of spatial skill of first year engineering students in 2018–19 to those from 2008–09 at Universidad de La Laguna (ULL).

Keywords: Engineering · Engineering drawing · Spatial ability

1 Introduction

Spatial ability has been studied over the years at an international level, while in Spain, it was not until 2004 that a first approach was produced by Carrión, Serrano, Díaz, Tomas and Sentana [1] and by DEHAES group [2] since 2007, followed by other university researchers from the engineering area. However, during the last few years there has been a research drop, despite the existence of training interventions of these skills in university students.

The importance of science, technology, engineering, arts and maths (STEAM) and its relation with spatial ability has increased the interest towards looking forward new research evidences and results. Moreover, its application in primary and secondary education remains essential for its influence on the cognitive development of children. Simultaneously, technological advances and social changes of the last two decades have motivated new behaviours in the youth, which has been called as Z generation, referring to people born in a purely digital era (since the end of the 90s).

F. Cavas-Martínez et al. (Eds.): INGEGRAF 2019, LNME, pp. 395–402, 2020.
https://doi.org/10.1007/978-3-030-41200-5_43

2 State of the Art

2.1 Spatial Ability

The field of spatial ability (also known as spatial skill) has acquired great prominence in academic engineering environments. So we can find interventions that improve the spatial ability using: 3D modelling tools [3], 3D tangible models [4], or virtual reality [5], augmented reality [6] and videogames based on orienteering races [7]. In addition, longitudinal research indicates that young people have low levels of spatial ability is not an isolated happening [8], but it has continuity over time.

In recent years, several studies have focused on primary and secondary education, seeking for relationships between learning strategies, tools or technologies and the acquisition of cognitive skills. Thus, some studies have compared teaching methodologies based on the application of game-based learning [9], the use of mobile applications [10] or the introduction of geographic information systems [11].

Furthermore, gender predisposition towards developing spatial ability has proven to be vital fields of research, since students' beliefs can influence their choice to perform activities of spatial ability [12], although there are results in primary education (k-12) that there are no significant differences in this ability by gender [13].

The importance of STEAM subjects has led to mathematics research in primary education due to its relationship with spatial abilities [14]. This multidisciplinary nature has led to studies on its relation to variables such as creativity [15], problem solving competence [16] and digital fluency [17].

The importance of this component of intelligence is not limited to areas of engineering. Technological advances have focused interest on its importance for various trainings such as in the field of medicine for surgical operations [18, 19] or to manage a laparoscopic camera [20]. These studies indicate that there is a direct relationship with spatial abilities.

2.2 Engineering Students in 2008 vs 2018

In a decade university students have changed from the Y-generation (or millennials) to what we now call Z-generation (or postmillennials). In both temporary spaces, 2008 and 2018, these are young people who are accustomed to the use of digital technologies. While millennials were born in the late 80s and early 90s, they have grown up in the early days of the digital world. They arrived at the university classrooms as consumers of videogame technologies, use of communication tools such as chat and video chat, and Internet browsing.

A decade later, in 2018/19 the Z-generation access to the university, students who have been using digital technologies since they were born, which has generated a great dependence on them. Its most characteristic feature is immersion in the digital world. That is why they are attributed an attraction to visual information and ease in the management of multitasking [21]. This innate characteristic is linked to other parameters such as a greater potential for creativity, suggesting its relationship with the daily use of technologies [22].

Being a digital native is the main attribution of the Z-generation, understanding that they have great competences on technologies. However, there are studies that show that it could be an unreal conception, since despite their continuous exposure to digital tools, they have not developed digital competences [21]. This situation is directly related to the lack of digital training of teachers [23] as well as its application in the teaching of graphic engineering [24]. Likewise, the experience in ICT does not necessarily implies an improvement of capacities whether the person does not have a prior knowledge base, as in the case of reading and writing [25].

Social networks are their main means of communication, which implies the creation of online communities, where they evaluate and comment on: products they buy, videos they consume, etc., generating a way of life framed by the need for feedback. However, the critical capacity that they can have when receiving feedback is unknown [26], being a world of ephemeral consumption that does not last in their memory.

3 Methodology

This research follows a descriptive longitudinal design with the objective of comparing the spatial abilities of the engineering degree students of the University of La Laguna (ULL) of the academic courses 2008–09 and 2018–19, by measuring the components of spatial visualization and mental rotation, according to the classification of Carroll [27] both are the components of spatial ability. The tests used to measure these variables are DAT-5SR and MRT, validated and widely accepted in the spatial ability literature. DAT: Differential Aptitude Test; MRT: Mental Rotation Test).

The data collection of DAT in 2008 was performed with the test version with scale 0–60 points, and it has been transformed to scale 0–50 of the new version of DAT-5SR test, in order to be able to establish comparisons with the scale conducted in the year 2018. The MRT test version has not changed in time, scaled from 0 to 40 points.

The analysis of results begins with the description of the mean values. Subsequently, Mann-Whitney test was applied to determine whether there are significant differences between both results regarding the academic courses.

In addition, a comparative descriptive analysis of the *University Admission Tests* (UAT) (known as PAU, recently EBAU) will be carried out for the subject of technical drawing, with the purpose of triangulating information with the results of the student's abilities on both dates. In this temporary space, these tests have been related to three legislative stages: the LOGSE (PAU 2008-2009), LOE (PAU 2010-206) and LOMCE (EBAU 2017-2018). This technical drawing test is optional and implies that the student had prepared the subject a few months before their access to the university, so it is the reference framework to know the level of students.

The sample consists of a total of 493 students in the first year of university engineering degrees from the University of La Laguna, distributed in two different academic years 2008–2009 (n = 337) and 2018–2019 (n = 156).

The student sample of 2008–09 came from different engineering degrees: agricultural engineering (n = 44), quantity surveyor (n = 64), mechanical engineering (n = 59), electronic engineering (n = 78), chemical engineering (n = 19) and civil engineering (n = 73). The student sample of 2018-19 came from nautical engineering (n = 79), agricultural engineering (n = 36) and computer science (n = 40).

4 Results

4.1 MRT and DAT Longitudinal Analysis

The mean value of MRT test in 2008 was 18.53 while in 2018 it is 18.63. Likewise, the mean value of DAT test in 2008 was 29.05 and in 2018 the result was 27.03.

Subsequent, Kolmogorov-Smirnov test was performed to identify the distribution of the sample regarding DAT and MRT tests in both academic courses. The results presented in Table 1 show only a normal distribution in the DAT test of 2018, so a nonparametric test will be applied to find out the presence of significant differences between groups.

Table 1. Kolmogorov-Smirnov test.

Test	Academic year	Statistic	P-value
MRT	2008–09	.066	.001
	2018–19	.099	.001
DAT	2008–09	.056	.013
	2019–19	.050	.200

Finally, Mann-Whitney test was applied on both MRT and DAT tests to verify the existence of significant differences with respect to the academic year category. The results show significant differences for DAT, but not for MRT (Table 2).

Table 2. Mann-Whitney test.

	MRT	DAT
Mann-Whitney U	26149.5	22921.0
Wilcoxon W	38395.5	35167.0
Z	−.093	−2.289
P-value	**.926**	**.022**

4.2 Longitudinal UAT Results in Technical Drawing at ULL

Frequency results of students who have taken the UAT at ULL from 2008 to 2018 are shown in Fig. 1. Data follow a positive trend, from 2.939 in 2008 to 4.231 in 2018. The growth presents similarities with the number of students presented in the scientific-technical area, from 1.621 enrolled in 2008 to 2.338 in 2018.

The frequency analysis of students who have submitted to the technical drawing test can be categorized into three stages, PAU (2008–2009), PAU-14 (2010–2016) and

EBAU (2017–2018). In the first and second, the mean value for students presented is 402 and 306 respectively, which follow a negative trend. In the EBAU, the number increases to an average of 568.5 students.

The test result means for the different stages are 6.26, 6.13 and 6.21 correspondingly. While the percentage of students that have passed Technical Drawing test is 60.97, 69.49 and 74.79, following the same order.

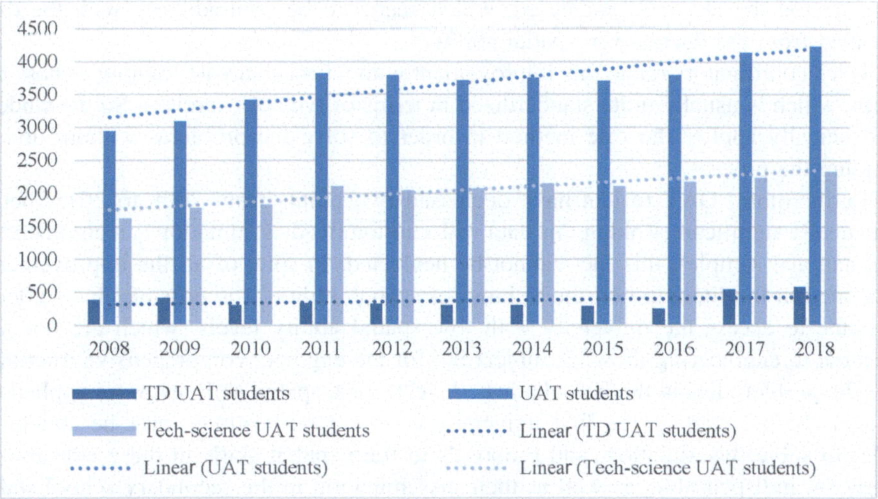

Fig. 1. Comparative bar graph of technical drawing UAT students, total UAT students, and tech-science UAT students

5 Discussion

Spatial ability is a factor increasingly assumed as necessary in different areas of knowledge and is related to skills and competencies such as creativity, problem solving and visual thinking [15, 16].

Many authors have revealed benefits of using ICT to develop spatial ability. This evidences caused the expectation of an increase of these capacities with the arrival of Z-generation to the university. However, this study provides results of a decrease in spatial visualization, while the capacity for mental rotation remains. These results support other studies that indicate a reduction of other competences in this generation of students [25], in addition to the lack of digital competence that is assumed for this population [21].

Technologies have diminished people's efforts on spatial visualization processes. Every time it is less necessary to generate a 3D vision from orthogonal views since it is increasingly common to visualize directly in three-dimensional environments. However, these students regularly use the smartphone and play videogames, tools that work under three-dimensional rotation patterns. This may explain why mental rotation has remained unlike the spatial visualization component.

In addition, in recent years, the perception of the university teaching staff on the student level of technical drawing agreed to follow a negative trend (according to discussion forums at conferences and congresses). This situation is supported by the analysis performed on the UAT, in which student attendance to technical drawing reduces over the years, despite having increased the amount of student candidates. However, in the last two years, both the population and the percentage of students who have passed the test have increased, which seems to be contradictory with the data obtained from the decrease in spatial ability.

UAT coordinators relate this improvement with a baccalaureate training to pass the exam, which is usually quite standardized in terms of type of exercises. So the student mechanically applies the rote method in order to solve the problems without understanding the process.

Furthermore, DAT results have decreased two points from 2008 to 2018, being statistically significant. Taking this data with caution (its dissemination is recommended to a national sample), this fact cannot be neglected. In spite of all the improvements generated in the literature on the training of spatial ability, it is possible that students continue to access the university with low spatial ability levels, which are not just essential to engineering drawing subject but for the engineer comprehensive training.

The problem lies in the baccalaureate levels, since spatial skill test were applied on students before starting the first university course. Clarifications must be sought in order to solve this situation, and proposals to train spatial skills in these educational levels are indispensable, as well as their dissemination in the secondary school staff.

One of the limitations of the study is the lack of information on ICT methods in the teaching of technical drawing. Likewise, it would be interesting for future studies to relate this information to the use and type of videogames.

6 Conclusions

XXI century world requires students with mastery of different ICT-related competences, with spatial ability being an increasingly interdisciplinary factor. This research tries to find out whether the technological development and its supposed application in secondary education, as well as the fact of having been born in a digital world, means an improvement of the spatial ability in the Z-generation with respect to a previous generation.

Despite the limitations of this study, it is possible to conclude that there is a decrease in spatial visualization in first-year engineering students of the university of La Laguna, while mental rotation capacity remains at similar levels.

Results prove a decrease in levels of spatial ability, which are essential in the training not only for engineers but also for many specialists that have digitalized their procedures in professional practice. Bearing in mind that this is not an isolated case in the Z-generation, it is essential to investigate the reasons that lead to these results and promote methods for the development of spatial ability in secondary education, as well as its dissemination.

References

1. Pérez Carrión, M.T., Serrano, M., Díaz, M., Tomás, R., Sentana, I.: El desarrollo de la percepción espacial de los alumnos de estudios técnicos universitarios. In: Actas del XIV Congreso Internacional de Ingeniería Gráfica. Santander (2002)
2. DEHAES Group: http://dehaes.webs.ull.es. Accessed 31 Mar 2018
3. Torner, J., Alpiste, F., Brigos, M.: Spatial ability in computer-aided design courses. Comput.-Aided Des. Appl. **12**(1), 36–44 (2015). http://doi.org/c32t
4. Huang, T.C., Lin, C.Y.: From 3D modeling to 3D printing: development of a differentiated spatial ability teaching model. Telematics Inform. **34**(2), 604–613 (2017). http://doi.org/gc5vnh
5. Molina-Carmona, R., Pertegal-Felices, M., Jimeno-Morenilla, A., Mora-Mora, H.: Virtual reality learning activities for multimedia students to enhance spatial ability. Sustainability **10** (4), 1074 (2018). http://doi.org/gdsd5v
6. Martín-Gutiérrez, J., Saorín, J.L., Contero, M., Alcañiz, M., Pérez-López, D.C., Ortega, M.: Design and validation of an augmented book for spatial abilities development in engineering students. Comput. Graph. **34**(1), 77–91 (2010). http://doi.org/b4nm3g
7. Roca González, C., Martín Gutiérrez, J., García Domínguez, M., Carrodeguas, M., del Carmen, M.: Virtual technologies to develop visual-spatial ability in engineering students. Eurasia J. Math. Sci. Technol. Educ. **13**(2), 441–468 (2017). http://doi.org/f9vttq
8. Veurink, N.L., Sorby, S.A.: Longitudinal study of the impact of requiring training for students with initially weak spatial skills. Eur. J. Eng. Educ. **44**(1–2), 153–163 (2019). http://doi.org/c32v
9. Newman, S.D., Hansen, M.T., Gutierrez, A.: An fMRI study of the impact of block building and board games on spatial ability. Front. Psychol. **7**, 1278 (2016). http://doi.org/f8zqkq
10. Wang, J.Y., Wu, H.K., Hsu, Y.S.: Using mobile applications for learning: effects of simulation design, visual-motor integration, and spatial ability on high school students' conceptual understanding. Comput. Hum. Behav. **66**, 103–113 (2017). http://doi.org/f9fkwk
11. Jadallah, M., Hund, A.M., Thayn, J., Studebaker, J.G., Roman, Z.J., Kirby, E.: Integrating geospatial technologies in fifth-grade curriculum: impact on spatial ability and map-analysis skills. J. Geogr. **116**(4), 139–151 (2017). http://doi.org/c32x
12. Vander Heyden, K.M., van Atteveldt, N.M., Huizinga, M., Jolles, J.: Implicit and explicit gender beliefs in spatial ability: stronger stereotyping in boys than girls. Front. Psychol. **7**, 1114 (2016). http://doi.org/c32z
13. Julià, C., Antolì, J.Ò.: Enhancing spatial ability and mechanical reasoning through a STEM course. Int. J. Technol. Des. Educ. **28**(4), 957–983 (2018). https://doi.org/10.1007/s10798-017-9428-x
14. Gilligan, K.A., Flouri, E., Farran, E.K.: The contribution of spatial ability to mathematics achievement in middle childhood. J. Exp. Child Psychol. **163**, 107–125 (2017). http://doi.org/c322
15. Cho, J.Y.: An investigation of design studio performance in relation to creativity, spatial ability, and visual cognitive style. Think. Skills Creat. **23**, 67–78 (2017). http://doi.org/f9wbtm
16. Buckley, J., Seery, N., Canty, D.: Investigating the use of spatial reasoning strategies in geometric problem solving. Int. J. Technol. Des. Educ. **29**(2), 341–362 (2019)
17. Tchoubar, T., Sexton, T.R., Scarlatos, L.L.: Role of digital fluency and spatial ability in student experience of online learning environments. In: Arai, K., Kapoor, S., Bhatia R. (eds.) Science and Information Conference, vol. 2, pp. 251–264. Springer, Cham (2019)

18. Schlickum, M., Hedman, L., Felländer-Tsai, L.: Visual-spatial ability is more important than motivation for novices in surgical simulator training: a preliminary study. Int. J. Med. Educ. **7**, 56–61 (2016). http://doi.org/f3t238
19. Henn, P., Gallagher, A.G., Nugent, E., Seymour, N.E., Haluck, R.S., Hseino, H., Traynor, O., Neary, P.C.: Visual spatial ability for surgical trainees: implications for learning endoscopic, laparoscopic surgery and other image-guided procedures. Surg. Endosc. **32**(8), 3634–3639 (2018). http://doi.org/gdxzdv
20. Roch, P.J., Rangnick, H.M., Brzoska, J.A., Benner, L., Kowalewski, K.F., Müller, P.C., Kenngott, H.G., Müller-Stich, B.P., Nickel, F.: Impact of visual–spatial ability on laparoscopic camera navigation training. Surg. Endosc. **32**(3), 1174–1183 (2018). http://doi.org/gc2n43
21. Pérez Escoda, A., Castro Zubizarreta, A., Fandos Igado, M.: La competencia digital de la Generación Z: claves para su introducción curricular en la Educación Primaria. Comunicar **49**, 71–80 (2016). http://doi.org/cxn8
22. Martini, A., Sotille, S.S., de Quadros Martins, A.R.: Process of creativity and innovation: a comparison between baby boomer generation and z generation. Braz. J. Educ. Technol. Soc. **10**(2), 141–152 (2017)
23. Fernández-Cruz, F.J., Fernández-Díaz, M.: Los docentes de la Generación Z y sus competencias digitales. Comunicar **46**, 97–15 (2016). http://doi.org/c323
24. Lopez-Pena, V., López-Chao, V.A., Lopez-Chao, A.M.: Analysis of teaching methods in graphic design in the galician university system (GUS) in Spain. Anthropol. **25**(3), 214–219 (2016). http://doi.org/c324
25. Amiama Espaillat, C., Mayor Ruiz, C.: Lectura digital en la competencia lectora: La influencia en la Generación Z de la República Dominicana. Comunicar **52**, 105–114 (2017). http://doi.org/c325
26. Dolot, A.: The characteristics of Generation Z. e-mentor **74**(2), 44–50 (2018). https://www.ceeol.com/search/article-detail?id=737237
27. Carroll, J.B.: Human Cognitive Abilities. A Survey of Factor-Analytic Studies. Cambridge University Press, New York (1993)

Fostering Outcomes; INGENIA Subjects and Digital Prototyping Laboratories

M. Luisa Mtz. Muneta(✉), Gregorio Romero Rey, Juan de Dios Sanz Bobi, and Antonio Carretero Díaz

Grupo de Ingeniería Gráfica, Departamento de Ingeniería Mecánica, ETSII, Universidad Politécnica de Madrid, C/José Gutiérrez Abascal, 2, 28006 Madrid, Spain
luisa.mtzmuneta@upm.es

Abstract. Since the adaptation of the different traditional engineering programs to the degree and master degrees according with the Bologna frame, the number of subjects with CDIO methodology has been increasing. This methodology aims that the student, from the beginning, conceive, design, build and implement improvements and operate the project. In the ETSII, and within the Master of Industrial Engineering, all students (362 students in the course 2018–2019) organized in teams must attend and pass a subject called INGENIA in which they must develop a product/project/service that must comply with a series of constraints. These subjects INGENIA are designed following the CDIO philosophy and has as its fundamental objective the development of outcomes according to the ABET accreditation model, and has been hindered by the limitation in the construction of final prototypes since there were not sufficient means available for its manufacturing.

Since the end of 2018, a digital manufacturing laboratory has been enabled at ETSII, offering students the opportunity to manufacture and build their prototypes with basic materials, but with functionalities similar to those of the designed artifact/project. Although the commissioning of this laboratory is recent, it has already allowed students to develop and validate projects and improve redesign according to user requirements.

Keywords: Outcomes · Skills · CDIO · ABET · Fablab · Digital prototyping

1 Introduction

The engineering study programs after their adaptation to the European Higher Education Area are based on the development of specific and transversal competences. This requires the adoption of new teaching strategies that allow students to develop knowledge and skills simultaneously. Within all of them, project-based learning stands out. This methodology aims to motivate students to participate and encourage their own learning process. Learning is done around realistic engineering problems where students apply the theoretical knowledge learned. The effectiveness of these PBL activities is contrasted in many studies, whether in basic education, higher education and university and within different fields, as engineering, medicine, law, informatics, etc. The

F. Cavas-Martínez et al. (Eds.): INGEGRAF 2019, LNME, pp. 403–410, 2020.
https://doi.org/10.1007/978-3-030-41200-5_44

University of Aalborg (Denmark, 1974) and the University of Maastricht (Netherlands 1976) are a point of reference in this type of teaching [1].

Not only are these type of activities suitable for the development of technical competences, generic o transversal competencies are necessary to successfully develop these learnings [2, 3]. The combination of teamwork and PBL promotes creativity according to the work of Zhuo and Kolmos [4] as well as the autonomous learning. The CDIO™ initiative [5] is surely the most ambitious approach in the development of PBL experiences. Student learning is established within a context of conceiving-designing-implementing and operating problems and designs of the real world [6]. Any university can join this initiative where the objective is describe with the phrase "Graduating engineers should be able to conceive-design-implement-operate complex value-added engineering systems in a modern team-based environment" [7] and that it can be used as a framework for curricular development and its results.

2 INGENIA Subjects

The degree of Industrial Technologies and the Master of Industrial Engineering of the Escuela Técnica Superior de Ingenieros Industriales are accredited by ABET [8] which is a certified organization that accredits college and university programs in applied and natural science, computing, engineering and engineering technology. The study programs must meet 8 criteria, among which the criterion 3 "Outcomes" and criterion 4 "Continuous Improvement" stand out. Therefore, a working framework has been defined where the students develop the 7 ABET competences where, in a planned way, results are measured, evaluated and improvements proposed [9]. The competences to be implemented into the programs are:

1. an ability to identify, formulate, and solve complex engineering problems by applying principles of engineering, science, and mathematics
2. an ability to apply engineering design to produce solutions that meet specified needs with consideration of public health, safety, and welfare, as well as global, cultural, social, environmental, and economic factors
3. an ability to communicate effectively with a range of audiences
4. an ability to recognize ethical and professional responsibilities in engineering situations and make informed judgments, which must consider the impact of engineering solutions in global, economic, environmental, and societal contexts
5. an ability to function effectively on a team whose members together provide leadership, create a collaborative and inclusive environment, establish goals, plan tasks, and meet objectives
6. an ability to develop and conduct appropriate experimentation, analyze and interpret data, and use engineering judgment to draw conclusions
7. an ability to acquire and apply new knowledge as needed, using appropriate learning strategies.

According to [9] one of the strategies followed is to design teaching strategies that allow the measurement of these competences. To this end, the Escuela Técnica Superior de Ingenieros Industriales (ETSII) of the Universidad Politécnica of Madrid

(UPM) implements a group of compulsory subjects in the first year of Master's Degree program at ETSII - UPM (a two-year program with 120 ECTS after a four-year Grade in Industrial Technologies with 240 ECTS) generically known as INGENIA, which is part of this CDIOTM initiative [10].

These subjects meet the CDIO™ requirements but may have different orientations and projects, so some of them are directed to electrical subjects, others to mechanical engineering or others to the development of business models. These subjects consist of 12 ECTS distributed in 2 semesters and organized in 120 face-to-face hours and between 180 and 240 h of personal work of the student in teams. The subjects occupy 4–5 h weekly of the teacher's timetable.

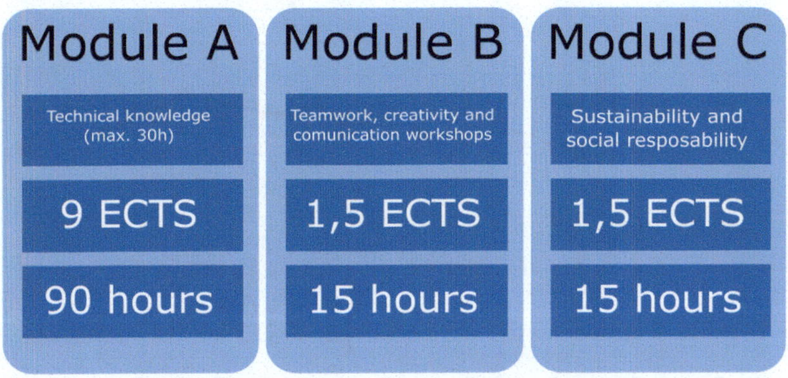

Fig. 1. Structure of contents of INGENIA subjects.

The contents of the INGENIA subjects (Fig. 1) are structured as follows:

- Module A (9 ECTS): a maximum of 30 h are established for the introduction of new theoretical content if necessary; the rest of the time the teachers will supervise the students' progress in their project.
- Module B (1.5 ECTS): it is dedicated to workshops on the skills of teamwork, creativity and communication necessary for the development of the project and its communication.
- Module C (1.5 ECTS): they are dedicated to contents of sustainability, ethics and social responsibility that they must implement in their project.

These subjects are taught annually to more than 300 students (in the 2018–2019 academic year they were 362) since the 2014–2015 academic year and represent a challenge for our college. Each of the INGENIA subjects have at least a team of 5 teachers responsible for monitoring the students and has a global cost of 10,000 euros/year that are used in the consumables necessary to build prototypes or final designs. The necessary machinery and tools are in charge of the teaching teams responsible of the different INGENIA subjects.

3 Satisfaction and Assessment of Outcomes by Students in the Subjects INGENIA

During the 2014–2015 academic year, 109 students and in the 2015–2016 academic year, 218 students from the first year of the Master's Degree in Industrial Engineering completed a series of surveys where their satisfaction with these INGENIA subjects was collected, obtaining the following results (see Fig. 2):

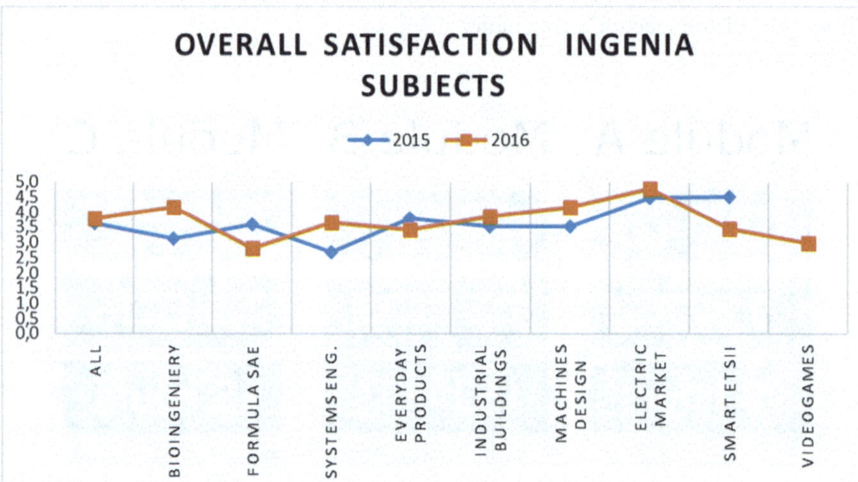

Fig. 2. Satisfaction of the students about INGENIA subjects

The global average in the 2014–2015 academic year was 3.62. In the 2015–2016 academic year it reached 3.82 out of 5 points.

Overall, the following average results are obtained for the four outcomes (Table 1):

Table 1. Student self-perception for teamwork, communication, creativity and sustainability outcomes.

	Teamwork		Comunicación		Creativity		Sustainability	
Year	2015	2016	2015	2016	2015	2016	2015	2016
All INGENIA subjects	3,5	3,64	3,32	3,45	3,33	3,52	3,33	3,48

On the other hand, the students were asked to evaluate their communication, creativity, sustainability and teamwork skills at the end of the assignments obtained (Fig. 3):

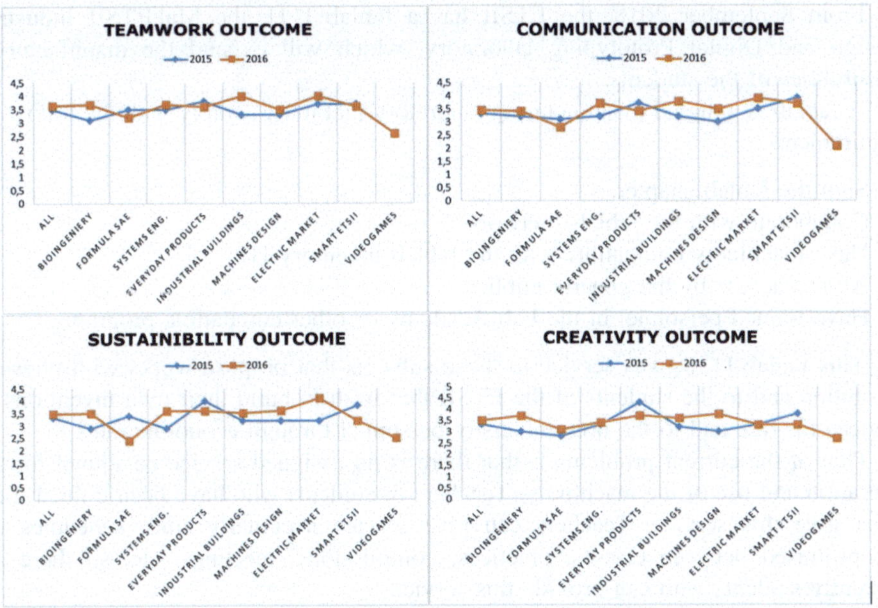

Fig. 3. Self-perception of students in the outcomes of teamwork, communication, sustainability and creativity

Although globally there is a slight improvement, both in the satisfaction and in the self-perception of the outcomes, in the year 2015–2016 three INGENIA subjects obtained worse results in all of them. The comments and suggestions of the students mostly referred to problems in planning and especially about the difficulty of implementing their ideas in physical prototypes. This problem also appeared in the following academic courses, both by means and materials have been one of the most frequent student complaints in all of them.

4 Industrial Design and Digital Prototypes Laboratory MakETSII

In order to make it easier for students to build prototypes in the 2018–2019 academic year, a laboratory has been established where students could implement their physical prototypes based on low-cost materials.

This laboratory has started from an initiative of the Graphic Engineering Group of ETSII that has provided the essential machinery for its operation and has been supported by the ETSII management.

From September 2018, the ETSII has a fablab [11], the MakETSII Industrial Design and Digital Prototyping laboratory, which will expand the manufacturing possibilities of the students.

A fablab is a digital manufacturing laboratory that must comply with the following requirements:

- Sign the fablab chapter.
- Collaborate with the fablabs network.
- Have machinery compatible with the fablab inventory [12].
- Allows access by the general public.
- Have trained personnel in the Fab Academy or other compatible programs.

This fablab [13] gives service to those subjects that propose works with physical resolution and to the students of the ETSII that wish to build their own inventions or devices for free and to the non-university personnel through economic rates.

One of the current problems is that there is no assigned service personnel for the operation and use of the machinery. Teachers or students who have been trained for it must give this service. Teachers can give service irregularly since schedules are compromised between classes, practices, commissions, meetings, etc. so there are currently students who can provide this service.

For this, a program of fablab monitors students has been designed. These students must learn the operation and use of the machines with their different technologies, as well as the use of other less sophisticated tools and the corresponding software; they will have to attend to the complete process of design, construction and testing of devices, in order to help and collaborate with inexperienced students later.

With this laboratory, we want in a short time:

- Promote the MAKER and DIY culture (Do It Yourself)
- Solve the problems derived from the construction of prototypes of the INGENIA subjects.
- Allow other subjects to build physical prototypes in their activities.
- Develop contents that allow online learning to the student before working with the FABLAB machinery
- Train a group of students who act as monitors for other inexperienced students
- Develop good practices in the laboratory with the different technologies of the machinery of the lab.
- Encourage entrepreneurship.

In order to verify its usefulness in the elaboration of the projects, a group of 46 students has been proposed the design of a 3D puzzle in the first semester of the 2017–2018 academic year. The elaboration, whose final result is a physical prototype, made with low-cost materials or a virtual prototype (Fig. 4) and their satisfaction with the work has been surveyed (Table 2).

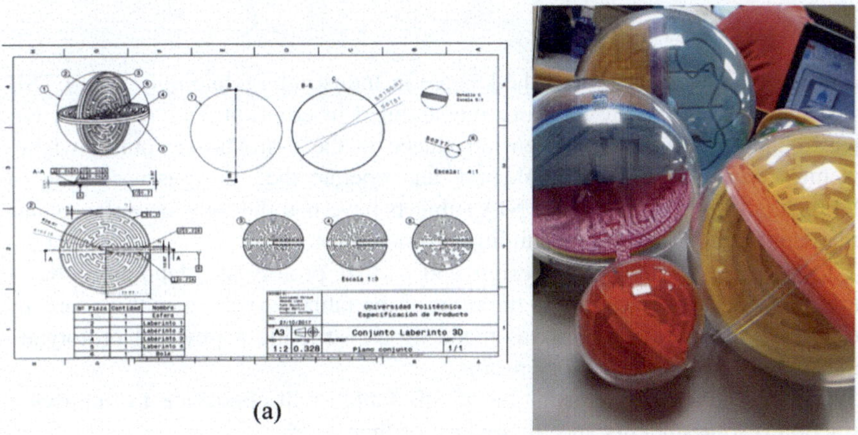

(a) (b)

Fig. 4. Results of the project (a) Virtual prototype and drawing (b) Physical prototype

Table 2. Descriptive data of the experience.

Type of work	Count	Mean	Standard Dev.	Var. Coef.	Min.	Max.
Low cost prototype	28	4,07143	0,766356	18,8228%	1,0	5,0
Virtual prototype	18	2,88889	1,18266	40,9383%	1,0	5,0
Total	46	3,6087	1,1051	30,6234%	1,0	5,0

An ANOVA analysis has been carried out (Table 3), obtaining that the ratio-F, which in this case is equal to 17,009, is the quotient between the estimated between-groups and the estimate within-groups. Since the P-value of the F-test is less than 0.05, there is a statistically significant difference between the mean of Satisfaction between one level of Type of work and another, with a level of 5% of significance.

Table 3. ANOVA for satisfaction by Type of work

Source	Sum square	Df	Mean square	F	P value
Between groups	15,3216	1	15,3216	**17,01**	**0,0002**
Intra groups	39,6349	44	0,900794		
Total (Corr.)	54,9565	45			

In other words, the elaboration of physical prototypes versus virtual prototypes increases project satisfaction for students; also improving the compression of other concepts such as tolerances and functional dimension.

Given the above data, it is expected that the construction of physical prototypes will improve student satisfaction within the INGENIA subjects; getting more quality jobs and simultaneously developing the competences of these subjects.

5 Conclusions

Since the ABET accreditation of the Degree in Industrial Technologies, the ETSII has worked hard to incorporate the outcomes defined in criterion 3 of ABET. A type of subjects called INGENIA has been introduced in the first Master course, where students must conceive, design, implement and operate their designs according to the CDIO methodology. Not all INGENIA subjects have had the same acceptance, and the main reason is the difficulty of building physical prototypes.

To this goal, an industrial design and virtual prototyping laboratory has been created, which works according to the rules of the fablabs to alleviate this problem. An analysis of student satisfaction was made by constructing a physical prototype as a result, obtaining very satisfactory results.

We hope that the implementation of this fablab will contribute to improving the satisfaction of our students and the quality of their work.

Acknowledgments. This work has received funding from the call for Educational Innovation Aids and the Improvement of the Quality of Education 2018–19 with reference IE1819.0509 and the support of the ETSII of the UPM.

References

1. Kjersdam, F., Enemark, S.: The Alborg Experiment: Project Innovation in University Education. Aalborg University Press, Denmark (1994)
2. Lee, R.M.K.W., Kwan, C.Y.: The use of problem-based learning in medical education. J. Med. Educ. **1**(2), 149–157 (1997)
3. Lofthouse, R., Cole, S., Thomas, U.: Creativity and enquiry in action: a case study of cross-curricular approaches in teacher education. Tean J. **2**(1). https://bit.ly/2ty5hYo. Accessed 6 Feb 2020
4. Zhou, C., Kolmos, A.: Interplay between individual creativity and group creativity in problem and project-based learning (PBL) environment in engineering education. Int. J. Eng. Educ. **29**(4), 866–878 (2013)
5. http://www.cdio.org
6. Crawley, E.F., Malmqvist, J., Östlund, S., Brodeur, D.R.: Rethinking Engineering Education: The CDIO Approach, pp. 1–286. Springer, Heidelberg (2007)
7. Crawley, E.: The CDIO syllabus a statement of goals for undergraduate engineering education (2001). http://www.cdio.org/files/CDIO_Syllabus_Report.pdf
8. https://www.abet.org/
9. Hernández Bayo, A., Ortiz Marcos, I., Carretero Diaz, A., del mar de la Fuente Garcia-Soto, M., Lumbreras Martin, J., Martinez Muneta, M.L., Riveira Rico, V., Rodriguez Hernandez, M.: Integral framework to drive engineering education beyond technical skills. Int. J. Eng. Educ. (IJEE) **30**(6), 1697–1706 (2014). ISSN 0949-149X
10. Tanarro, E.C., et al.: Teaching engineering design to a multidisciplinary audience at master's level: benefits and challenges of the CDIO approach. In: 11th International CDIO Conference, Chengdu, China (2015)
11. https://www.fablabs.io
12. https://docs.google.com/spreadsheets/d/1U-jcBWOJEjBT5A0N84IUubtcHKMEMtndQPLCkZCkVsU/pub?single=true&gid=0&output=html
13. http://www.fablab.etsii.upm.es

Introducing Parametric CAD in a First Year Course in Engineering Degree: A Case Study

Alba Roda-Sales(✉), Verónica Gracia-Ibáñez, Maria-Jesús Agost, Miquel Gómez-Fabra Gómez, and Mariana Núñez-García

Department of Mechanical Engineering and Construction, Universitat Jaume I, Av. de Vicent Sos Baynat s/n, 12071 Castelló de la Plana, Castelló, Spain
rodaa@uji.es

Abstract. Teaching in technical drawing is evolving towards new paradigms based on 3D modeling. This tendency is reflected in the evolution of the first course subject of technical drawing, common to different bachelor's degrees (Mechanical, Industrial Technologies, Electrical and Chemical) in Engineering at Universitat Jaume I. Nevertheless, although 3D modelling has been gaining importance, its introduction to students has always been preceded by lessons on projection systems. In this course, as a step further in a general methodological change, it was proposed to introduce students to the different projection systems and the achievement of engineering drawings from 3D models, improving their spatial vision and shape recognition. Aiming for a more intuitive introduction to 3D modeling, the use of a specific commercial parametric CAD software has been implemented. Moreover, its use will provide students a basis to work on contents of subsequent subjects (finite element analysis, computer-aided manufacturing, etc.). Beyond the above advantages, the effect of this implementation has been assessed on academic performance through scores comparisons. Furthermore, both students and teachers answered specific questionnaires, obtaining that students prefer the new software implemented in several aspects assessed. It was also obtained that both students and teachers perceive the methodological change as an improvement regarding the past course, what is coherent with the score increase reported this present course regarding the past one.

Keywords: 3D modelling · Parametric CAD · Learning experience · Engineering drawing

1 Introduction

Teaching in technical drawing has been challenging during the last decades since CAD technologies are constantly evolving. Become knowledgeable of 3D modeling can be considered as a requirement for future engineers properly prepared [1]. Training and teaching CAD students mean being in continuous evolution [2] and according to Rubio García et al. [3] both teaching methodologies and software used in CAD teaching in Spain are varied. However it appears to be a certain consensus of the 3D modeling introduction in Engineering learning, especially in some degrees as Mechanical

F. Cavas-Martínez et al. (Eds.): INGEGRAF 2019, LNME, pp. 411–419, 2020.
https://doi.org/10.1007/978-3-030-41200-5_45

Engineering [2] where its knowledge would be deeper promoted in higher courses [4]. At Universitat Jaume I, in the first course subject of technical drawing (common to Mechanical Engineering, Industrial Technologies Engineering, Electrical Engineering and Chemical Engineering bachelor's degrees) CAD 3D modeling has been progressively introduced throughout last years.

Nevertheless, the introduction of students to CAD 3D modeling in this subject has always been preceded by lessons on projection systems, what implied a particularly tough task for those students that lacked basic concepts on technical drawing or had more difficulties in spatial vision. Graphic communication seems to go through visualization, sketching, modeling and preparation of technical documentation [5], but for those with limited spatial visualization abilities can be a hindrance to reach the modeling part. After a cumbersome process of shape recognition from projected views by students, only those capable to interpret properly the object were able to apply 3D modelling module of AutoCAD in order to get the 3D model. Therefore, in this course, as a step further in the methodological change, it was proposed to switch the order of contents of the subject. The main idea was to introduce students to 3D modeling firstly, in order to improve their spatial vision and shape recognition. Then, once students have acquired a basis on 3D models and modelling techniques, they are introduced to projection systems and the achievement of engineering drawings from 3D models. Method is focused on students through their own practice in 3D modeling introduced by teachers, in opposition to the teacher-centered model, as it should be both from the European Higher Education and Research Area (Bologna Declaration) [6] as from a constructivism paradigm on education. Spatial reasoning should be considered a main competence for engineers [6, 7], but given the difficulties students have been revealing in previous years a change in methodology is proposed [8].

Not only was the methodological order decided to be changed, but also the software. AutoCAD is one of the most powerful tools to create engineering drawings, but its 3D modeling module may not be as intuitive as a parametric software one's. Therefore, in order to ease the students' introduction to CAD 3D modeling, this academic course was also implemented the use of the parametric CAD software Solid Works. This software, apart from being one of the most used in engineering 3D modeling, it is used in subsequent subjects of Industrial Technologies Engineering and Mechanical Engineering bachelor's degrees (to perform finite element analysis, computer-aided manufacturing, vibration analyses, etc.) [2, 4]. Furthermore, the use of a parametric CAD software provides them the view of its capabilities of design reusability [9].

The main purpose of the methodological change is to improve students' comprehension and perception of the subject, as well as their academic results. Therefore, the aim of this work is to assess the impact of the methodological change proposed by applying action-research [10] through three different indicators: students and teachers' perception measured with questionnaires, and the comparison of academic results regarding the past course.

2 Materials and Methods

The study comprised two phases, the online questionnaire and the academic results comparison. In the first phase, a total of 102 students out of the 295 enrolled students answered a specific questionnaire and 8 teachers volunteered to complete an-other questionnaire. In the second one, scores from specific parts of exams (theoretical and practical parts) were compared, taking theoretical parts of 242 and 214 exams (from past course and from the present one, respectively) and 206 and 192 practical parts.

2.1 Students' Questionnaire

Students were asked to complete an online anonymous questionnaire, in order to assess their perception towards the subject. This questionnaire was divided into four main areas: general aspects, experience during SolidWorks use, experience during AutoCAD use (if they used it), experience during the use of other CAD 3D software (if they used it) and perception of the change in the order of contents (if they were repeating course).

In the general part they were asked about their gender, the degree in which were enrolled, their previous knowledge on technical drawing and if they were repeating course or not. In the part where CAD software were assessed, students rated in a Likert 1 to 5 scale the aspects presented in Table 1. Regarding the perception of the change in the order of contents of the subject, they were asked if they perceived that this change has improved their spatial vision and eased their comprehension of the subject.

Table 1. CAD software aspects assessed by students.

Aspect	Description
Qs1	Intuitiveness of the software
Qs2	Contribution of the software to the improvement of their spatial vision
Qs3	Contribution of the software to the identification of principal views (front, top and right) and axonometric projections
Qs4	Feasibility to perform changes to the 3D model
Qs5	Future applicability of the 3D software knowledge in their respective careers

After collecting all the questionnaire data, descriptive analyses were performed and Kolmogorov-Smirnov and Wilcoxon tests applied to the scores obtained in the software assessment, in order to determine if differences were significant (Significant level = 0.05).

2.2 Teachers' Questionnaire

Teachers of the subject were also asked to complete a questionnaire, in order to assess their perception regarding the teaching experience and perceptions during the classes. The main areas of the questionnaire were general aspects, perception regarding the experience of the students while using SolidWorks and AutoCAD (teachers were asked depending on the academic course/courses they taught the subject), perception regarding the difficulty experienced by the students in several aspects (only for those teachers that taught the theoretical part of the subject) and perception regarding the impact of the change in the order of contents of the subject.

In the general part they were asked whether they taught the subject only the past course, only the present course or both courses. In the part where perception of teachers regarding students' experience while using CAD software was assessed, teachers rated in a Likert 1 to 5 scale the aspects presented in Table 2. In the assessment of difficulty observed in students in different theoretical parts, teachers rated in a Likert 1 to 5 scale the aspects presented in Table 3 for past and present course. Regarding the perception of the change in the order of contents of the subject, they were asked if they perceived that this change has improved students' spatial vision and eased their comprehension of the subject.

Table 2. CAD software aspects assessed by teachers.

Aspect	Description
Qt1	Intuitiveness of the software
Qt2	Contribution of the software to the improvement of students' spatial vision
Qt3	Contribution of the software to the students' identification of principal views (front, top and right) and axonometric projections

Table 3. Comprehension aspects assessed by teachers.

Aspect	Description
Qt4	Perceived level of difficulty experienced by students while obtaining principal views from a model
Qt5	Perceived level of difficulty experienced by students while obtaining a 3D model from its principal views

After collecting all the questionnaire data, descriptive analyses were performed to the scores obtained in the software assessment to compare results to students' perceptions.

2.3 Academic Results Analyses

In order to compare academic results regarding the past course, scores obtained in similar questions from exams regarding both past and present course were collected and compared. The types of questions compared and aim pursued are described in Table 4.

Kolmogorov-Smirnov tests showed if scores could be considered as normally distributed. Mann-Whitney U tests were applied in order to determine if significant differences in score distributions between courses could be identified. Besides, analysis of variance was applied to detect possible significant differences in means, when normally distributed scores could be considered (Significant level = 0.05).

Table 4. Theoretical and practical parts of exams compared.

Aspect	Description	Aim
Theoretical Question (TQ)	Obtaining front and top views from an axonometric representation of a part	Comparing students' spatial vision prior and post methodological change
Practical Exercise (PE)	Modeling a part from an axonometric drawing	Comparing students' modeling ability prior and post software change

3 Results and Discussion

3.1 Students' Questionnaire

From 295 enrolled students in the subject, 102 completed the online questionnaire (73.5% male and 23.5% female). Participants were enrolled in Mechanical Engineering (31.4%), Chemical Engineering (26.5%), Industrial Technologies Engineering (22.5%) and Electrical Engineering (19.6%).

Regarding the previous knowledge, students answered if they had coursed any subject related with technical drawing in any period before enrolling their respective degrees. The answers obtained are those detailed in Table 5.

Table 5. Percent of students that coursed subjects related with technical drawing in each period.

Students (%)	Period
25.5	Never
9.8	During first years of high school
52.0	During last two years of high school
8.8	Several times during the high school period
1.0	During last two years of high school and Professional Training Courses
2.9	Other periods

From all the participants, 30.4% were repeating course. These students rated the benefit of the order change of the subject contents regarding the past course with a mean score of 3.81 out of 5, showing their high level of acceptance with the methodological change applied.

As for the software usage, 37.3% answered that had used AutoCAD, and 13.7% said that had used other software (Autodesk Inventor, LibreCad, FreeCad, Cinema4D, Paint 3D, AutoSketch 3D, Blender, SketchUp, Autodesk 3ds Max). From the part where different aspects of software were assessed, the 95% interval of confidence of mean value (Likert scale) in scores given are those presented in Fig. 1.

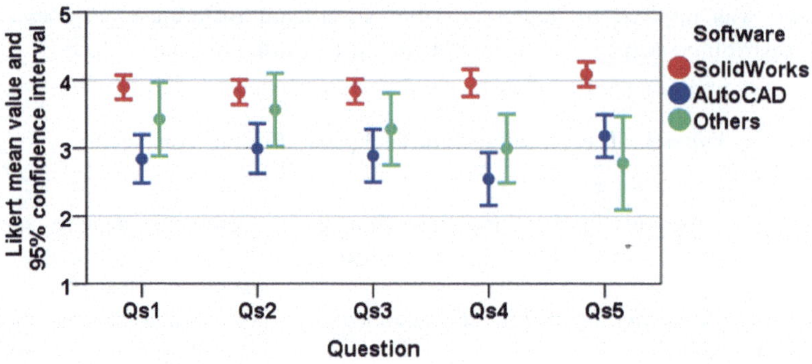

Fig. 1. 95% interval of confidence of mean value (Likert scale) in scores given by students to each aspect (Qs1: Intuitiveness, Qs2: Spatial vision, Qs3: Views, Qs4: Changes, Qs5: Applicability). First column for SolidWorks, second for AutoCAD and third for other software.

It can be seen that the mean rates obtained for all aspects assessed are higher for SolidWorks. Furthermore, scores obtained for the other software were also higher than AutoCAD ones in almost all the aspects.

Kolmogorov-Smirnov confirmed that answers were not normally distributed, thus Wilconxon test to repeated measures were applied to scores obtained for the students repeating course between SolidWorks and AutoCAD scores obtaining significant differences ($p < 0.05$) in all the aspects assessed. The same procedure was applied to scores given by all the students that have used SolidWorks and AutoCAD (students repeating course and non-repeating but having used it in other circumstances) and also significant differences ($p < 0.05$) were obtained in all the aspects assessed. Nevertheless, when applying the same analysis to scores given by those students that have used SolidWorks and another software, significant differences were only found for aspects assessed in Qs3 (identification of principal views), Qs4 (changes to the 3D model) and Qs5 (applicability). Nevertheless, these results can be attributable to the simplicity and the lack of technical aspects that the other software used by students present. For this reason, in scores from Qs1 (intuitiveness) and Qs2 (spatial vision) no significant differences were reported, but they were the worst rated in Qs5 (applicability).

3.2 Teachers' Questionnaire

All the teachers of the subject completed the online questionnaire. Four of them taught theoretical and practical parts of the subject in both academic courses. The rest of the teachers only taught practical part of the subject (one of them during past course, another during both courses, and two of them during present course).

Regarding the perception of the benefit of changing the order of the subject contents, they all gave a rate of 4 out of 5. This result is in accordance with students' perception (3.81 out of 5).

As for the teachers' perception of students' experience while using SolidWorks and AutoCAD, the 95% interval of confidence of mean value (Likert scale) in scores given are those presented in Fig. 2.

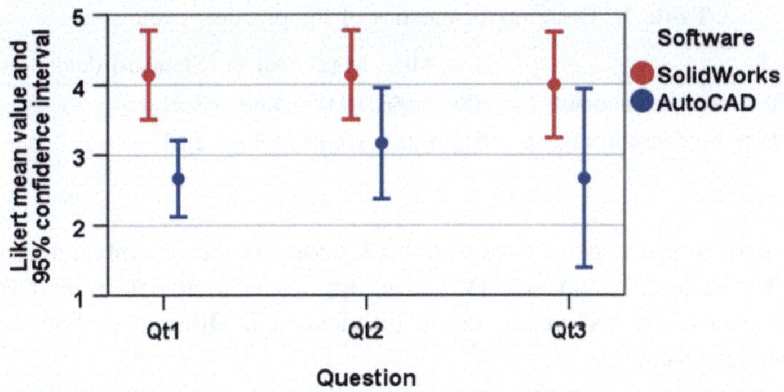

Fig. 2. 95% interval of confidence of mean value (Likert scale) in scores given by teachers to each aspect (Qt1: Intuitiveness, Qt2: Spatial vision, Qt3: Views). First column for SolidWorks and second for AutoCAD.

Questions asked to teachers Qt1 to Qt3 are in fact the same to those asked to students (Qs1 to Qs3). It can be observed that teachers' perceptions regarding the goodness of the software change are in accordance with the students' ones.

As for the rates obtained when assessing the difficulty observed in obtaining principal views from a model (question Qt4), the mean score was 2.5 (SD $\pm$ 0.577) in both courses. Nevertheless, in the difficulty when visualizing the object from principal views (question Qt5), the mean values for the past course and current one were 3.75 (SD $\pm$ 0.5) and 3.25 (SD $\pm$ 0.957), respectively.

3.3 Academic Results Analyses

Regarding the theoretical question (TQ), scores from 242 exams from the last course and 214 from the present one were compared. When appropriate, scores from both courses were transformed in order to make the maximum score uniform (1 point). General descriptive statistics are shown in Table 6.

Table 6. Descriptive statistics of the theoretical questions.

	N	Min	Max	Mean	Standard deviation
TQ score (last course)	242	0.00	1.00	0.64	0.32
TQ score (present course)	214	0.00	1.00	0.75	0.31

Scores could not be considered as normally distributed. Significant differences in score distributions were found between both courses ($p < 0.05$).

Two hundred and six practical exercises (PE) of last course and 192 of the present one were also compiled, and scores were compared. Scores went from 0 to 10 points. Table 7 shows descriptive statistics.

Table 7. Descriptive statistics of the practical exercises.

	N	Min	Max	Mean	Standard deviation
PE score (last course)	206	0.00	10.00	4.68	3.21
PE score (present course)	192	0.00	10.00	4.92	2.61

Kolmogorov-Smirnov tests showed that PE scores could be considered as normally distributed. Consequently, an ANOVA was applied, which determined that no significant differences between means could be identified. Differences between distributions neither were found.

Results obtained show an improvement regarding spatial visual abilities according to the theoretical questions. It is noticeable that teachers' answer to Qt4 did not show optimism in that improvement, although they perceived a spatial visual abilities improvement in their answers to Qt5. No differences have been seen in modelling abilities due to the software change, even though both teachers and students perceptions from questionnaires showed acceptance to the software change.

4 Conclusions

The experiment carried out has shown a clear acceptance on the methodological change implemented both from teachers' and students' perceptions, as well as from the academic results. Even though academic results did not show differences regarding the software change, the perception towards the change was positive, both from students and teachers. Results encourage teachers to keep on with these methodological changes, in the hope of helping students acquire improved spatial vision competences.

Acknowledgments. The research work reported here was made possible by the Innovative Education Project titled "Aprofundiment en conceptes de modelatge 3D paramètric" given by USE - Universitat Jaume I.

References

1. Ye, X.: Today's students, tomorrow's engineers: an industrial perspective on CAD education. Comput. Aided Des. **36**, 1451–1460 (2004)
2. Field, D.A.: Education and training for CAD in the auto industry. Comput. Aided Des. **36**, 1431–1437 (2004)
3. Rubio García, R.: Present state of CAD teaching in Spanish Universities. Comput. Educ. **44**, 201–215 (2005)
4. Faath, A.: Interdisciplinary and consistent use of a 3D CAD model for CAx education in engineering studies. In: Volume 5: Education and Globalization. ASME 2016 International Mechanical Engineering Congress and Globalization. ASME (2016)
5. Jeli, Z.: Usage of 3D computer modelling in learning engineering graphics. In: Virtual Learning. InTech (2016)
6. Contero, M.: Learning support tools for developing spatial abilities in engineering design. Int. J. Eng. Educ. **22**, 470–477 (2006)

7. Contero, M.: Improving visualization skills in engineering education. IEEE Comput. Graph. Appl. **25**, 24–31 (2005)
8. Gracia-Ibáñez, V.: Methodologies to improve academic results even when increasing the level of demand: an experience in cad teaching. In: International Congress on Education, Innovation and Learning Technologies Proceedings (ICEILT), Barcelona (2014)
9. Camba, J.D.: Parametric CAD modeling: an analysis of strategies for design reusability. Comput. Aided Des. **74**, 18–31 (2016)
10. Gracia-Ibáñez, V.: Applying action research in CAD teaching to improve the learning experience and academic level. Int. J. Educ. Technol. High. Educ. **13**, 9 (2016)

Augmented Reality and Mobile Devices as Tools to Enhance Spatial Vision in Graphic Representations

T. Polhmann, D. Parras-Burgos, F. Cavas-Martínez, F. J. F. Cañavate, J. Nieto, and D. G. Fernández-Pacheco(✉)

Technical University of Cartagena, C/Doctor Fleming s/n, 30202 Cartagena, Murcia, Spain
daniel.garcia@upct.es

Abstract. The application of new technologies in the teaching field provides a series of advantages in teaching-learning processes. This paper presents a tool based on augmented reality that enhances and develops the capacity of spatial vision in engineering students, and more specifically in subjects of the Graphic Expression area. To do this, the student is provided with a series of QR codes that accompany the drawings of parts or mechanical assemblies usually used in these subjects. The tool, called ARTID (Augmented Reality for Teaching, Innovation and Design), consists of a free mobile application that can be used on any mobile device with the Android operating system. In this way, students can scan the QR code that accompanies each drawing and obtain a virtual 3D model of the represented object. The operation of the application has been verified by several tests on different devices, obtaining satisfactory results in all cases and demonstrating its ease of use and great utility.

Keywords: Augmented Reality · Spatial vision · Teaching-learning · 3D model · Teaching

1 Introduction

Augmented reality is considered as a technology that combines three-dimensional real and virtual elements in real-time. Although this type of technology is being applied in different areas of knowledge, advertising is which has most boosted its development [1]. The use of augmented reality applications has been increased, above all, by the boom of mobile devices and Internet access, since information related to positioning can be shown to the user thanks to GPS location [2].

Nowadays, augmented reality has a high demand in many areas of society, including teaching, where it begins to be applied in different educational levels and subjects [3–7]. The visualization of 3D objects in space is one of the main challenges faced by engineering students during their academic training. The quick evolution of technology jointly with the growing popularity and easy access to smart mobile devices make possible the development of new opportunities that can serve to improve the teaching quality [8, 9]. In this sense, the use of this type of technology inside and

F. Cavas-Martínez et al. (Eds.): INGEGRAF 2019, LNME, pp. 420–427, 2020.
https://doi.org/10.1007/978-3-030-41200-5_46

outside the classroom can increase the understanding and motivation of students in technical drawing subjects taught in Industrial Engineering degrees [10].

This communication describes the creation of a mobile application to visualize, using augmented reality, 3D virtual models of figures used in graphic expression subjects.

2 Objectives

The main objectives for the creation and development of the ARTID (Augmented Reality for Teaching, Innovation and Design) augmented reality tool for its application in Graphic Expression subjects are, among others:

- Provide the teacher with a support tool for teaching the subject, inside and outside the classroom.
- Transform classic classes into more interactive and dynamic activities that encourage student motivation.
- Help students to develop their spatial vision and understand the exercises done in class.
- Create an open access tool available to all students interested in the subject.

3 Methodology

With the aim of obtaining an easy-to-use and open access tool, different software was used to obtain the ARTID application. Its development process can be divided into three main phases: (i) creation of QR stamps to scan and reproduce the pieces, (ii) 3D modeling of the pieces, and (iii) joining of stamps and pieces within Unity software.

3.1 First Phase: Stamp Creation

The stamp is defined as the image scanned through the mobile device that identifies the piece to be projected in augmented reality. It is important to consider a series of characteristics before its creation:

- The stamp must be unique. This is important to prevent a delay when showing the 3D model or another object appearing in its place. To this end, it is necessary to choose an abstract, but well defined, image. In the case of ARTID application, QR images were chosen to ensure that there are not two equal stamps.
- The stamp must have an optimal size for the projection of the model. On one hand, if the stamp has a small size it will be projected faster, but some details may be missing. On the other hand, if it is too large, the piece may not be generated or may take too long to appear. The optimum size defined for this tool is an A4 format size.

The user can resort to web pages and applications for the creation of the QR code. For this tool, a free and easy to use application obtained from the Microsoft store and named *Simple QRCode Creator* was used.

3.2 Second Phase: 3D Modeling of the Pieces

In a second stage, the 3D modeling of the pieces is performed with the aid of a design and 3D modeling program, as for instance Solidworks. Since Unity software requires 3D pieces to be in FBX format for importation, the following steps are carried out: (i) exportation of the 3D models obtained in Solidworks to STL format, (ii) importation of the STL file into Blender software to add colours or textures, and (iii) exportation in Blender to FBX format. If models are directly drawn in Blender, the first conversion is not be necessary. Figure 1 shows some modeled pieces to be used as 3D models in the ARTID tool.

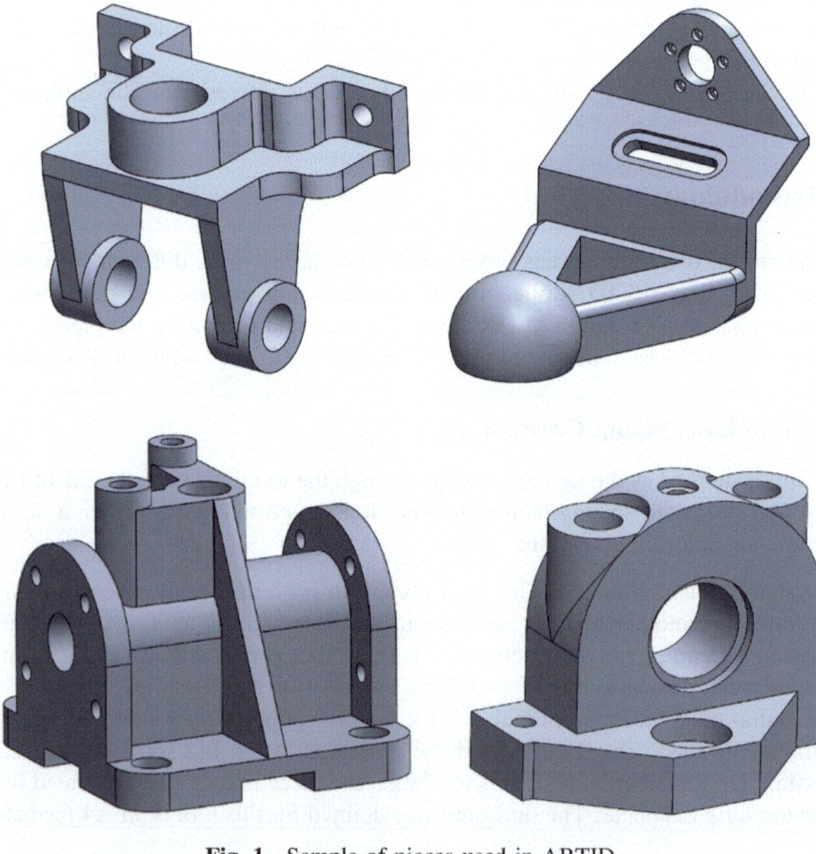

Fig. 1. Sample of pieces used in ARTID.

3.3 Third Phase: Creation of the Augmented Reality Tool

Vuforia is an extension of the Unity program that allows working with augmented reality. Both Vuforia and Unity are licensed software that can be used for educational and non-profit purposes, and whose authorization was requested for the realization of this work. Vuforia is divided into two work steps:

1. Vuforia website (https://developer.vuforia.com), which is available for developers. In this page, the stamps created in the first phase must be uploaded, generating a file that will need to be added to Unity.
2. The second part of Vuforia is located inside the *GameObjects* tab in Unity. The file generated in Vuforia must be executed while Unity running and the chosen stamps must be added. Vuforia is accessed using the *GameObjects* tab, then *ARCamera* has to be selected. In *ARCamera*, the stamps will be firstly added (ImageTarget), and then the 3D models (Fig. 2).

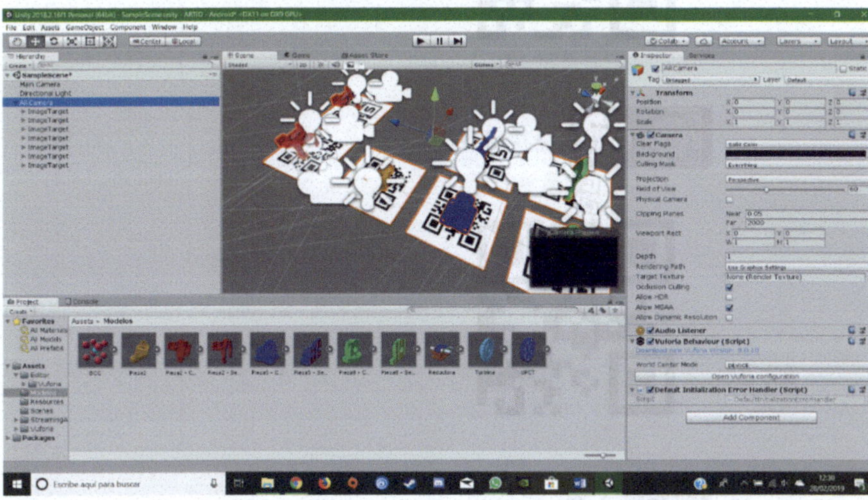

Fig. 2. Unity screenshot.

In order to generate the application a series of procedures within Unity are necessary to be carried out. Firstly, we must determine the platform on which the tool is going to be developed. For the present work, the Android platform was chosen. Once the platform has been selected, the settings for the mobile device are modified by clicking on the *Player settings* button, activating *Vuforia Augment Reality* and indicating the name of the application. Once all these steps have been done, the *Build* button is activated to generate the APK file of the final application that will be installed on the mobile device.

4 Results

In order to verify the suitability of the created application, several tests were performed using different mobile devices with Android Operating System (v4.2 or higher) and a database made up of twelve drawings with QR codes (Fig. 3).

Fig. 3. Some drawings with QR codes used in the test.

The QR codes can be scanned from any angle and position, showing the 3D model that corresponds to the selected piece/drawing (Fig. 4).

Fig. 4. Visualization of model 2 using the ARTID application.

The application is also endowed with the possibility of showing the sectioned model. For this purpose, an additional card with the letter S must be placed at the bottom-right position of the QR code (Figs. 5 and 6).

Fig. 5. Positioning the additional card for the visualization of sectioned models.

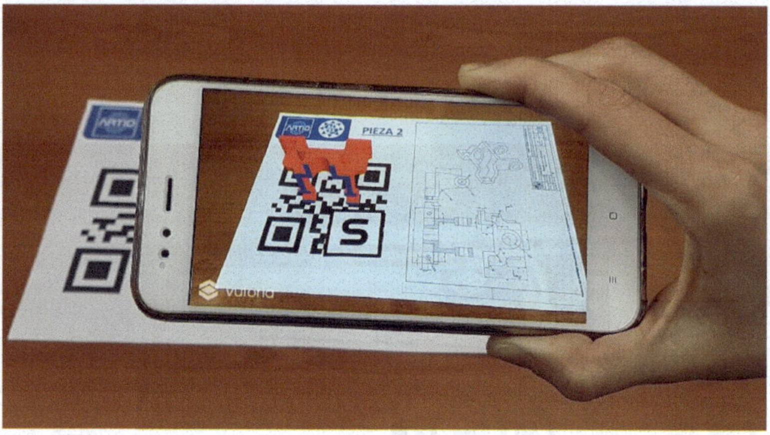

Fig. 6. Visualization of sectioned model 2 in the ARTID application.

To encourage the use of augmented reality in this type of subjects, a website (www.artid.es) has been developed. This site aims to provide a training tool for students that need support for the visualization of pieces in graphic expression subjects. The objective is to supply a model and figure library from which students can download the desired models, as well as the last version of the software application (Fig. 7).

Fig. 7. Screenshots from ARTID website.

5 Conclusions

This communication describes the process of creating a tool based on augmented reality and called ARTID. The main objective of this new tool is to obtain an open source application that allows to visualize, through augmented reality, different types of pieces and sets usually used in technical drawing subjects. The obtained results showed the proper operation of the application in the majority of the mobile devices used, as well as a very positive evaluation for the use of this type of technology as support for teaching.

References

1. Azuma, R.T.: A survey of augmented reality. Presence: Teleoperators Virtual Environ. **6**, 355–385 (1997)
2. Uribe, F.C.: Realidad aumentada aplicada a la enseñanza de la geometría descriptiva. AUS (Valdivia) **18**, 18–22 (2015)
3. Arora, R., Kazi, R.H., Grossman, T., Fitzmaurice, G., Singh, K.: Symbiosissketch: combining 2D & 3D sketching for designing detailed 3D objects in situ. In: Proceedings of the 2018 CHI Conference on Human Factors in Computing Systems, Montreal QC, Canada, pp. 1–15. ACM, Montreal QC, Canada (2018)
4. Atrio Cerezo, S., Guardado Moreno, E.: La realidad aumentada y su presencia en un modelo docente tecnológico para la didáctica de la química en bachillerato. Revista Educación y tecnología **1**, 9–38 (2012)
5. Chang, H.Y., Hsu, Y.S., Wu, H.K., Tsai, C.C.: Students' development of socio-scientific reasoning in a mobile augmented reality learning environment. Int. J. Sci. Educ. **40**(12), 1410–1431 (2018)
6. Ibáñez, M.B., Delgado-Kloos, C.: Augmented reality for stem learning: a systematic review. Comput. Educ. **123**, 109–123 (2018)
7. Suh, A., Prophet, J.: The state of immersive technology research: a literature analysis. Comput. Hum. Behav. **86**, 77–90 (2018)
8. Ferrer Torregrosa, J., Jiménez Rodríguez, M.Á., Barcia González, J.M., Torralba Estellés, J.: La realidad aumentada en la docencia de ciencias de la salud. Nuevos caminos para la comprensión. II Congreso Internacional de Innovación Educativa, Cartagena 2014, Cartagena, pp. 1–10 (2014)
9. Lineros, M.L., Jiménez, F.J.S., Carballo, F.M., Tatay, C.V.: Realidad aumentada como herramienta docente en la asignatura de construcción en las enseñanzas de ingeniería. XXIV Congreso de Innovación Educativa en las Enseñanzas Técnicas, Cádiz 2016, Cádiz, pp. 1–10 (2016)
10. Juan, C., YuLin, W., Tjondronegoro Dian, W., Wei, S.: Construction of interactive teaching system for course of mechanical drawing based on mobile augmented reality technology. Int. J. Emerg. Technol. Learn. **13**, 126–139 (2018)

Comparison and Analysis of Graphic Engineering in the Degree in Engineering in Industrial Technologies

Elidia Beatriz Blázquez-Parra(✉), Isidro Ladrón de Guevara-López, and Manuel Damián Marín-Granados

Departamento de Expresión Gráfica, Diseño y Proyectos Escuela de Ingenierías Industriales, C/Doctor Ortiz Ramos, s/n, 29010 Málaga, Spain
ebeatriz@uma.es

Abstract. The degree is the first level of university studies. It includes the basic teachings along with other subjects aimed at training the future engineer. This article describes a study carried out on the inclusion of the area of Graphic Expression in Engineering in the Degree in Engineering in Industrial Technologies at national level in both public and private universities. As a result, we can observe a difference in the involvement of this area in public universities with respect to private universities. Thus the Polytechnic University of Valencia offers this subject to a greater extent whilst private universities—such as those of Navarra and Deusto—only offer 6 credits. Finally, a graphical comparison is presented regarding the participation and status of this subject in the various universities. At the local level the relationship with different subjects of the curriculum of the University of Malaga is also analyzed, through its teaching plan, taking into account the course schedules and timing. The results reveal how subjects within the area of Graphic Expression in Engineering are most strongly linked to the study plans of Electricity, Structures and Materials and, finally, Mechanics and Machines.

Keywords: Graphic Engineering · Industrial Technologies · University

1 Introduction

In 1998, a declaration urging the development of a European Higher Education Area (EHEA) was signed at the Sorbonne in Paris, France, by the Ministers of Education of France, Germany, Italy and the United Kingdom. The La Sorbonne Declaration represented the first step towards a long-term change in higher education in Europe [1]. In 1999, the ministerial conference was held in Bologna (Italy), giving rise to the Bologna Declaration. This declaration, which was signed by 30 European states (not only the European Union), is the one that laid the foundations for the construction of the aforementioned EHEA, having six objectives, [2], including: a system of easily comparable degrees, two Higher education cycles (graduate cycle and postgraduate cycle), an ECTS credit system, promotion of student and teacher mobility, European

F. Cavas-Martínez et al. (Eds.): INGEGRAF 2019, LNME, pp. 428–437, 2020.
https://doi.org/10.1007/978-3-030-41200-5_47

cooperation for quality assurance, and the promotion of the European dimension of higher education.

The principle of university autonomy is included in the Spanish Constitution. This principle plays a fundamental role in allowing the educational institution to be both formative and critical of society. Autonomy includes academic freedom, and the management and administration of its own resources. The Spanish law states these powers in the Organic Law 6/2001, of December 21, of Universities (LOU) [3].

1.1 The Degree of Engineering in Industrial Technologies

The Degree in Industrial Technologies Engineering (GITI) is offered in both public and private Spanish Universities, distributed across 28 centers throughout the national territory (see Fig. 1). From this figure it should be noted that these centers belong to 26 universities; since two of these, the Polytechnic University of Catalonia and the University of Cádiz, offer the GITI degree in two of their centers.

With respect to the representation of private universities, there are five that are distributed from the center of the peninsula to the north, located in the Community of Madrid (Antonio de Nebrija University, Universidad Pontificia de Comillas), Catalonia (University of Ramón Llull) and País Vasco (University of Deusto, University of Navarra (San Sebastián). In these communities they share this presence with public universities that teach the same degree. As shown in Fig. 1, there is no center that offers this degree in La Rioja, Extremadura, Castilla-La Mancha, the Balearic Islands, or the Canary Islands. However, with the exception of the islands, the other autonomous communities are located near to or have good communication with centers that offer this degree.

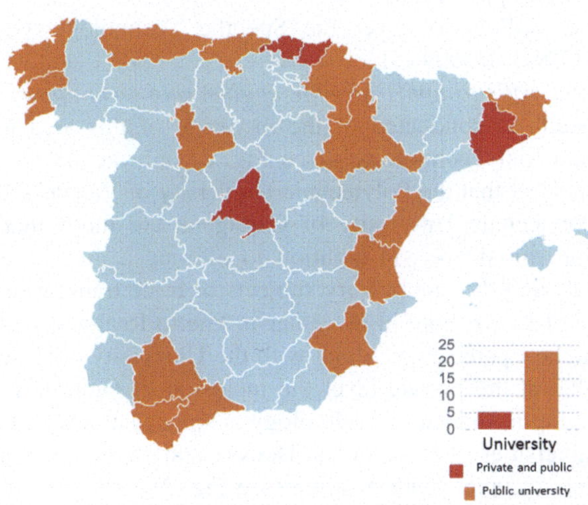

Fig. 1. Distribution of the GITI degree according to type of university.

The academic courses for the GITI degree offered by public universities in 2017, published on the page of the Ministry of Science, Research, and Universities constituted a total of 3,050 places [4]. In the case of UNED, since it employs a distance-based modality of learning, the number of places is not limited. It is well known that, although its scope of action covers the entire national territory, the headquarters are located in Madrid. The Autonomous Communities that offered the highest number of academic courses for the 2017–2018 academic year are Catalonia, Madrid, and Andalusia, which account for 19% of the courses offered in the national territory, followed by the Valencian Community and Aragón, with the lowest number of courses being offered by Asturias, Murcia, Basque Country, and Castilla & León.

2 The Area of Graphic Expression in Engineering and GITI

The main objective of the teaching activities in this degree is to focus on the acquisition of a set of skills in the subject programs, allowing the students to assimilate new knowledge, terminologies, and skills related to the disciplines under study. The subjects and their teaching methodologies must include both a scientific/technical component and encourage development of a way of thinking or acting when dealing with the practical problems that an engineer might expect to encounter. The area of Graphic Expression in Engineering is primarily included in the teaching and basic training involved in university qualifications at degree level. In the case of GITI, the role of this subject is to allow the student to acquire the skills of Basic Engineering Training from the Basic Training module described in the Ministerial Order CIN/351/2009 [5]:

"Capacity for spatial vision and knowledge of graphic representation techniques, both by traditional methods of metric geometry and descriptive geometry, as well as by computer-aided design applications."

However, these skills also cover the Specific Technologies described by the Ministerial Order CIN/351/2009 [5]:

"Knowledge and skills to apply graphic engineering techniques."

If a detailed study is conducted of the inclusion of this area in the Degree of Engineering in Industrial Technologies (see Fig. 2) only at the level of the simple taught degrees, it is clear that the Polytechnic University of Valencia, the University of Cádiz, and the Polytechnic University of Cartagena are those that have a greater presence in offering this degree, in addition to offering a greater range of optional courses. When analyzing the compulsory subjects or basic training of the degree, it is still the University of Cádiz, both in its center in Puerto Real and in Algeciras, which accounts for the highest percentage, along with the University of Oviedo. It should be noted, however, that at the private level the inclusion of Graphic Expression in the Degree of Engineering in Industrial Technology is somewhat low, accounting for only 6 credits at the Universities of Navarra and Deusto, and 15 credits at the University of Ramón Llull, which occupies tenth position (see Fig. 2).

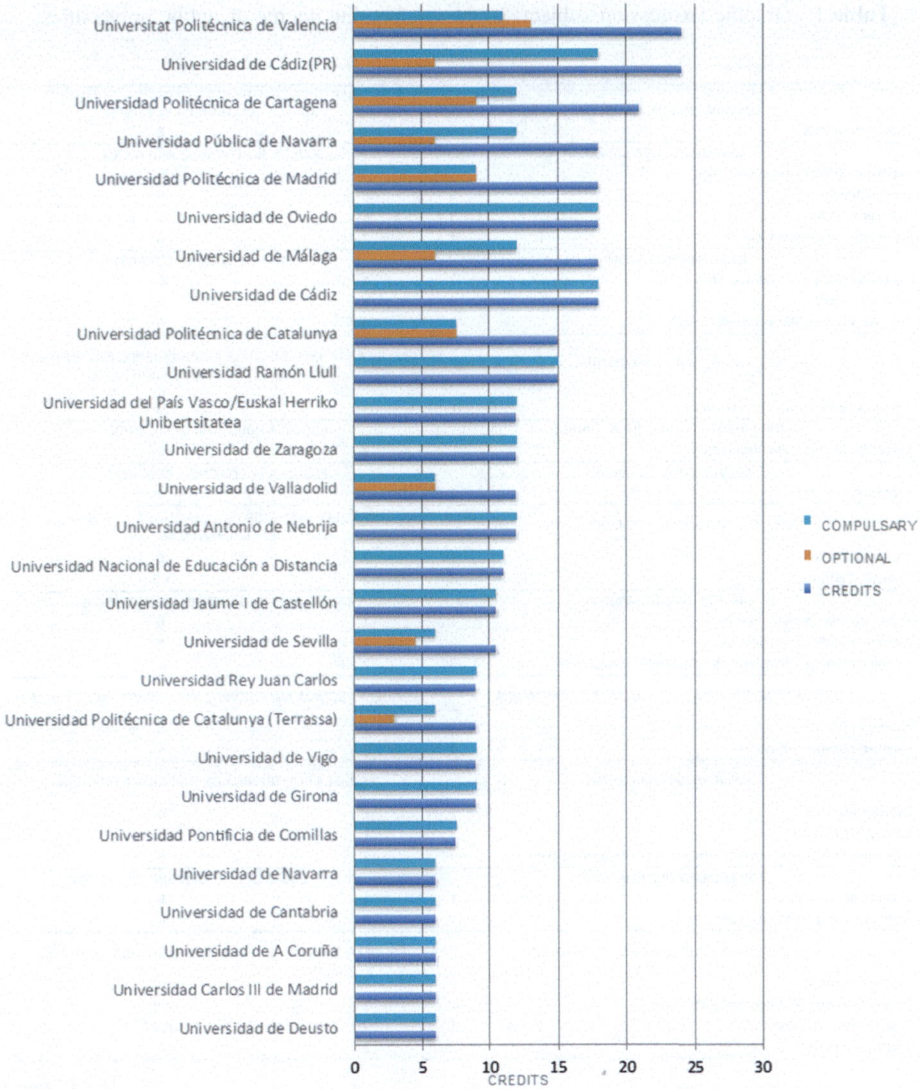

Fig. 2. Distribution of credits according to compulsory and optional subjects across the universities studied.

Table 1. Graphic Expression subjects in the engineering degree of public universities.

Subject	Year	Semester	Credits
UNIVERSIDAD DE A CORUÑA			ESCUELA POLITÉCNICA SUPERIOR
Expresión Gráfica	1	1	6
UNIVERSIDAD DE CÁDIZ			ESCUELA SUPERIOR DE INGENIERÍA
Expresión Gráfica y Diseño Asistido	1	1	6
Dibujo Industrial	3	5	6
Ingeniería Gráfica	3	6	6
Topografía (multidisciplinar)	4	7	6
UNIVERSIDAD DE CÁDIZ			ESCUELA POLITÉCNICA SUPERIOR
Expresión Gráfica y Diseño Asistido	1	1	6
Dibujo Industrial	3	5	6
Ingeniería Gráfica (itinerario Mecánica)	3	6	6
UNIVERSIDAD DE CANTABRIA			ESCUELA TÉCNICA SUPERIOR DE INGENIEROS INDUSTRIALES Y DE TELECOMUNICACIÓN
Técnicas de Representación Gráfica	1	1	6
UNIVERSIDAD CARLOS III DE MADRID			ESCUELA POLITÉCNICA SUPERIOR
Expresión Gráfica en la Ingeniería	1	2	6
UNIVERSIDAD DE GIRONA			ESCUELA POLITÉCNICA SUPERIOR
Expresión Gráfica	1		9
UNIVERSIDAD JAUME I DE CASTELLÓN			ESCUELA SUPERIOR DE TECNOLOGÍA Y CIENCIAS EXPERIMENTALES
Expresión Gráfica	1	2	6
Dibujo Industrial	3	2	4,5
UNIVERSIDAD DE MÁLAGA			ESCUELA DE INGENIERÍAS INDUSTRIALES
Expresión Gráfica en la Ingeniería	1	1	6
Ingeniería Gráfica y Topografía	2	1	6
Ingeniería Gráfica y Sistemas de Información Geográfica	3	2	6
UNIVERSIDAD NACIONAL DE EDUCACIÓN A DISTANCIA			ESCUELA TÉCNICA SUPERIOR DE INGENIEROS INDUSTRIALES
Expresión Gráfica	1	1	6
Ingeniería Gráfica Industrial	2	2	5
UNIVERSIDAD DE OVIEDO			ESCUELA POLITÉCNICA DE INGENIERÍA DE GIJÓN
Expresión Gráfica	1	2	6
Expresión Gráfica II	3	1	6
Ingeniería Gráfica y Sistemas de Información Geográfica (OP)	3	2	6
UNIVERSIDAD DEL PAIS VASCO			ESCUELA DE INGENIERÍA DE BILBAO
Gráficos de Ingeniería	1	1	6
Ampliación de gráficos de ingeniería	1	2	6
UNIVERSIDAD POLITÉCNICA DE CARTAGENA			ESCUELA TÉCNICA SUPERIOR DE INGENIERÍA INDUSTRIAL
Expresión Gráfica	1	1	6
Diseño Asistido por Ordenador (OP)	4	2	4,5
Fundamentos de Geomática	4	2	4,5
Diseño Industrial	2	2	6

(Continued)

Table 1. *(Continued)*

Subject	Year	Semester	Credits
UNIVERSIDAD POLITÉCNICA DE CATALUNYA	ESCUELA SUPERIOR DE INGENIERÍAS INDUSTRIAL, AEROESPACIAL Y AUDIOVISUAL DE TERRASSA		
Expresión Gráfica I	1	1	6
Expresión Gráfica II	1	2	3
UNIVERSIDAD POLITÉCNICA DE CATALUNYA	ESCUELA TÉCNICA SUPERIOR DE INGENIERÍA INDUSTRIAL DE BARCELONA		
Expresión Gráfica	1	2	7,5
Diseño Asistido por Ordenador (OP)	2	2	3
Simulaciones gráficas 3D (OP)			4,5
UNIVERSIDAD POLITÉCNICA DE MADRID	ESCUELA TÉCNICA SUPERIOR DE INGENIEROS INDUSTRIALES		
Dibujo Industrial I	1	1	6
Dibujo Industrial II	1	2	6
Diseño Asistido por ordenador (OP)	4	1	3
Sistemas de información Geográfica	4	2	3
Sistemas de representación en construcción			
UNIVERSIDAD PÚBLICA DE NAVARRA	ESCUELA TÉCNICA SUPERIOR DE INGENIEROS INDUSTRIALES Y TELECOMUNICACIÓN		
Expresión Gráfica	1	1	6
Dibujo Industrial	1	2	6
Diseño industrial (OP)	3	2	6
UNIVERSITAT POLITÉCNICA DE VALENCIA	ESCUELA TÉCNICA SUPERIOR DE INGENIEROS INDUSTRIALES		
Expresión Gráfica	1	2	6
Ingeniería Gráfica	4	1	4,5
CAD básico en ingeniería de la construcción (OP)	4	2	4,5
CAD para el cálculo de estructuras industriales (OP)	4	2	4,5
Product Design (OP)	4	1	4,5
UNIVERSIDAD DE SEVILLA	ESCUELA TÉCNICA SUPERIOR DE INGENIERÍA		
Expresión Gráfica	1		6
Diseño Asistido por Ordenador (OP)	4		4.5
UNIVERSIDAD REY JUAN CARLOS	ESCUELA SUPERIOR DE CIENCIAS EXPERIMENTALES Y TECNOLOGÍA. CAMPUS DE MÓSTOLES		
Expresión Gráfica	1	1,2	9
UNIVERSIDAD DE VALLADOLID	ESCUELA DE INGENIERÍAS INDUSTRIALES		
Expresión Gráfica en la Ingeniería	1	1	6
Dibujo Asistido por Ordenador	3	2	6
UNIVERSIDAD DE VIGO	ESCUELA DE INGENIERÍA INDUSTRIAL		
Expresión Gráfica	1	1	9
UNIVERSIDAD DE ZARAGOZA	ESCUELA DE INGENIERÍA Y ARQUITECTURA		
Expresión Gráfica y Diseño Asistido	1	1,2	6
Procesos de fabricación y dibujo industrial	2	2	6

Tables 1 (continued) and 2 display the subjects taught in the different centers of the Spanish territory, indicating the course, the semester, and the credits assigned.

Although the names of the subjects in Tables 1 and 2 are the same in some centers, it should be noted that following a detailed study of the different contents of the syllabus programs, it appears that in those centers in which the credit load of the subjects is inferior, the greatest content corresponds to industrial drawing, with descriptive geometry taking second place.

Table 2. Subjects of the area of private universities

Subject	Year	Semester	Credits
UNIVERSIDAD ANTONIO DE NEBRIJA			ESCUELA POLITÉCNICA SUPERIOR
Expresión Gráfica I	1	1	6
Expresión Gráfica II	1	2	6
UNIVERSIDAD DE DEUSTO			FACULTAD DE INGENIERÍA
Expresión Gráfica en la Ingeniería	1	1	6
UNIVERSIDAD DE NAVARRA			ESCUELA SUPERIOR DE INGENIEROS
Expresión Gráfica	2	1	6
UNIVERSIDAD PONTIFICIA COMILLAS			ESCUELA TÉCNICA SUPERIOR DE INGENIERÍA (ICAI)
Expresión Gráfica	1	2	7,5
UNIVERSIDAD RAMÓN LLULL			IQS SCHOOL OF ENGINEERING/ESCUELA TÉCNICA SUPERIOR IQS
Expresión Gráfica	1	1,2	9
Dibujo Industrial	2	2	6

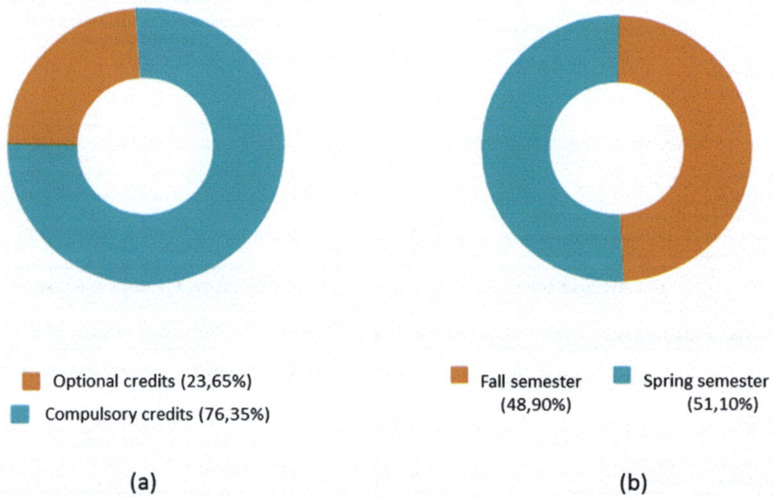

Fig. 3. (a) Proportion of optional and compulsory credits in all of the universities. (b) Temporal distribution of the teaching load of the subject for the GITI degree.

Figure 3(a) shows a summary graph that takes into account all universities, the distribution of optional and compulsory credits, with the fourth year offering optional subjects linked to graphic engineering. Figure 3(b) shows the temporal distribution of the subjects, with the teaching being equally balanced between the two semesters (50%) at a global level.

Finally, the subject area has the greatest weight in the first academic year, accounting for 56% of the subjects taught. This fact is in line with the comments made at the beginning of this section regarding the basic skills that are acquired in the first year.

3 Status of the Subject Area in the GITI of the University of Málaga

The area of Graphic Expression in Engineering is taught at the School of Industrial Engineering across 10 ordinary degrees and 4 master's degrees [6]. Most of the teaching in the area is taught in the Degree in Industrial Design and Product Development, with a total of 10 subjects, of which 3 are optional and the others forming part of the basic and compulsory training. The lowest presence at degree level is found in the Andalusian Tech qualifications (Degrees of Energy Engineering and Electronic Engineering, Robotics, and Mechatronics). The teaching curriculum that leads to the Degree in Engineering in Industrial Technologies by the Universities of Malaga incorporates the distribution of ECTS credits that is presented in Table 3 and has eight study plans [7], among which students have to choose one. Each plan consists of various modules, with specific skills (18 ECTS credits) and general course skills (24 ECTS credits). Depending on the chosen plan, the student will be required to take the basic, compulsory, and elective courses common to all of the plans, along with those corresponding to each unique plan.

Table 3. Distribution of the curriculum in terms of ECTS according to type of subject.

Subject Type	Credits
Basic training (BT)	60
Compulsory (C)	96
Optional (OP)	72
Final research project (FRP)	12
Total	240

In this degree, the subjects related to Graphic Engineering are shown in Table 4. As can be seen, there are 3 subjects of 6 credits each, of which 12 credits are basic and compulsory, and 6 optional (Graphic Engineering and Systems or Geographic Information).

If a detailed study is conducted of the subject guides for the GITI degree, paying special attention to their programs, it can be concluded that the subjects of this plan that have a link with the subject of Graphic Expression (GE) and/or Graphic Engineering and Topography (GET) are those indicated in Fig. 4, which shows the semester in which it is taught, with even numbers corresponding to the first semester and odd numbers to the second semester. For example, 5 s corresponds to the first semester of the third year. A green background shows the subjects corresponding to the different plans taught at the School of Industrial Engineering of Malaga.

Table 4. Subjects of the area of graphic expression in engineering.

Graduates of engineering in industrial technology				
51243	Graphic Expression in Engineering	BT	1	6
51248	Graphic Engineering & Topography	C	3	6
51271	Graphic engineering & Geographic Information Systems	OP		6

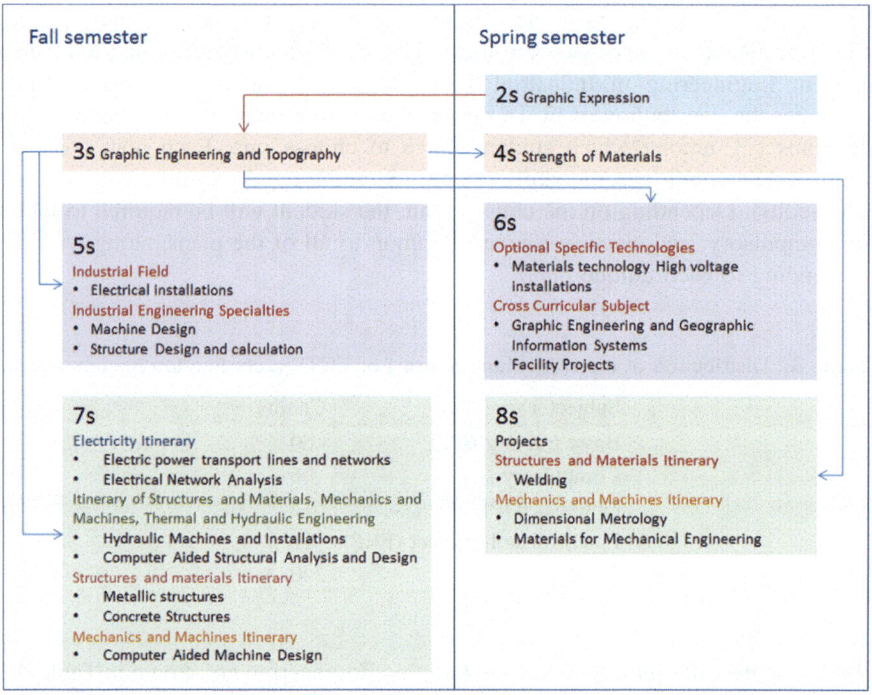

Fig. 4. List of Graphic Expression subjects and Graphic Engineering and Surveying along with others of the GITI curriculum in EIT.

The GE and GET subjects have contents that consist of basic training. After their completion, students acquire skills that will shape them as a future engineer, and this type of course content will be included during all semesters in certain study plans. However, it should be noted that the inclusion of these subjects is most strongly linked to courses in Electricity, Structures and Materials, and Mechanics and Machines, although such content will also be fundamental for projects in which the use of drawings and maps is inherent in the subject.

4 Conclusions

The Degree in Engineering in Industrial Technologies can be studied in most of the autonomous communities of Spain in various public and private universities. However, the proportion of academic courses offered in private universities is lower, because we must take into account the fact that the total number of private universities in Spain is 32, of which only 5 offer this degree, the latter being geographically located in the central, northern, and eastern regions of the country. On the other hand, there are a total of 51 public universities, with approximately 50% of these universities offering the Degree in Industrial Technology Engineering.

With respect to the inclusion of Graphic Expression in the Degree in Engineering in Industrial Technologies, this varies between a weight of 6 credits in universities such as the University of Navarra and 24 in the Polytechnic University of Valencia. By conducting an analysis of this distribution, it can be concluded that the average for all universities, taking into account all the subjects, is 11.04 credits. If this is compared with the average number of course credits offered by private universities, the resulting average is 9.3 credits, whilst in public universities the average is 11.38 credits. Therefore this area of study has a stronger presence in the degrees offered by public universities. As a future line of research it would be interesting to conduct a similar comparison of the presence of other areas of study within the degree whilst also comparing the subject guides for this course.

Acknowledgment. This work is partially supported by the University of Malaga, I Plan Integral de Docencia.

References

1. Allègre, C., Blackstone, T., Berlinger, L., Ruettgers, J.: Declaración de la Sorbona. Paris (1998)
2. Joint Declaration of the European Ministres of Education: Bologna Declaration. The Europan Higher Education Area (1999). http://www.ehea.info/cid100210/ministerial-conference-bologna-1999.html. Accessed 13 Feb 2019
3. Ley Orgánica 6/2001. (2001, diciembre 24). Ley Orgánica 6/2001, de 21 de diciembre, de Universidades. Boletín Oficial del Estado. https://www.boe.es/boe/dias/2001/12/24/. Accessed 15 Mar 2019
4. Ministerio de Innovación, Ciencia y Universidades. http://www.ciencia.gob.es/. Accessed 10 Mar 2019
5. Orden Ministerial CIN/351/2009. https://www.boe.es/diario_boe/txt.php?id=BOE-A-2009-2893. Accessed 03 Mar 2019
6. Departamento de Expresión Gráfica, Diseño y Proyectos. http://www.edp.uma.es/. Accessed 30 Mar 2019
7. Plan de Estudios Tecnologías Industriales. https://www.uma.es/grado-en-ingenieria-en-tecnologias-industriales/info/8101/plan-de-estudios-tecnologias-industriales/. Accessed 30 Mar 2019

Validation of the Use of Concept Maps as a Tool for the Teaching and Learning of Industrial Drawing

Edurne Iriondo-Plaza[1(✉)], Fernando Veiga-Suárez[2,3], Urko Eslava-Adot[2], and Alain Gil-Del-Val[2,3]

[1] Department of Mechanical Engineering, University of the Basque Country, Paseo Rafael Moreno 3, 48013 Bilbao, Spain
edurne.iriondo@ehu.eus

[2] Department of Engineering, Public University of Navarre, Los Pinos Building, 31006 Pamplona, Spain

[3] Industry and Transport Department – Advanced Manufacturing Tecnalia, Mikeletegi Pasealekua 1-3, 20009 Donostia-San Sebastián, Spain

Abstract. After some years of practice, common misunderstandings and lack of conceptual organization/structure have found from the learning results of Industrial Drawing students. This article presents an experimental work developed in order to measure the learning process through the concept map analysis. The main aspects to investigate are map complexity, the influence of mapping individual or collectively and the perception of the students on the usefulness of the activity. The complexity level of the maps is set extracting qualitative data and evaluating it through a rubric. The data studies the performance of individual or group working dynamics. In order to evaluate the perception of the students about the mapping activity, a survey presents the results obtained. The present work can conclude that: the students are able to identify and organize main and secondary concepts but their ability to represent a complex system is weak. Individually drawn maps show slightly higher average rate than the ones done in groups and parallel to this, the survey shows a higher usefulness appreciation of the tool from the students that have been mapping individually. The purpose with highest rate is the one related to Information Management and Collective Work. Finally, the use of concept maps improved overall from 0,8 to 1 point the tests scores, compared to the students that did not use this tool.

Keywords: Concept maps · Conceptual knowledge assessment · Industrial drawing · Teaching & learning

1 Introduction

Concept maps originated as a tool to achieve meaningful learning in students, as support for the organization and representation of knowledge (ideas and concepts) of the students. This tool helps and facilitates the transmission of information between the professor and the student. Concept maps are graphic tools to organize and represent knowledge. They include concepts, generally enclosed in circles or squared, and a

F. Cavas-Martínez et al. (Eds.): INGEGRAF 2019, LNME, pp. 438–445, 2020.
https://doi.org/10.1007/978-3-030-41200-5_48

connecting arrow with its word or sentence, called link. The link specifies the relationship between the two concepts.

Conceptual maps are the visual representation of the relationships between the hierarchy of mental concepts and the connections between the concepts in the mind of each person. When we speak and/or read, this hierarchy becomes linear. For someone to achieve meaningful learning with the information collected in a linear way (through reading or orally), a hierarchical structure must be given to the information in order to add and adapt it to the previously internalized and learnt data. Conceptual maps work as "translators" to move to linear mode and vice versa, information categorized in a hierarchical way (Fig. 1).

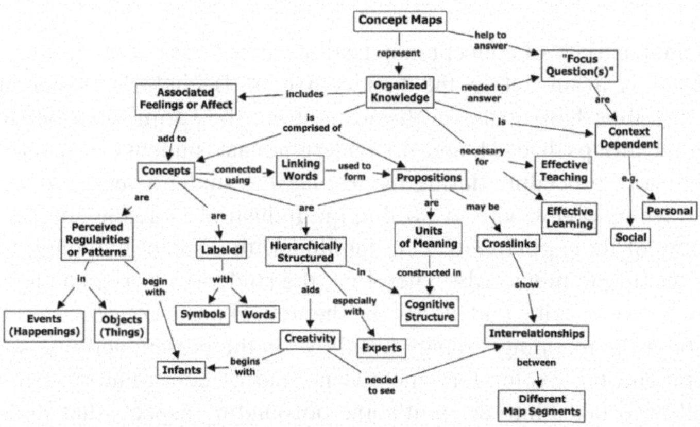

Fig. 1. Elements and links of the set up of a concept map [1].

Humans think through concepts and over the time, the meaning of the concepts are changing depending on the events experienced and the new information acquired. Over time, more and more examples and cases are associated with each concept and we will create new ways of relating them. The representations of the relationships between the concepts of the concept map are based on a "from the general to the specific" model. In each map, the most general or inclusive concepts are on the top and below, more specific concepts or examples. Another feature of the conceptual map is its conceptual hierarchy, when two or more concepts given under a broader and compatible concept, and finally, "cross connections" connect different hierarchical branches with each other. Concept maps help professors to identify, understand and organize the concepts intended to teach. They help to specify the relationships (propositions) necessary for the understanding of concepts and ideas. As a basic principle in education, in this theoretical field the students must begin to take into account what they know previously and to consider that knowledge is a basic and essential first step. Concept maps help students to understand previously internalized knowledge and to relate it to new ones to acquire.

The main objective of this study is the validation of the use of concept maps as a tool to improve the teaching and learning of Industrial Drawing. Under the main target, this work sought to research:

- The relationship between the complexity representation of the students and the individual or collective mapping.
- The perception of the students about the usefulness of the concept mapping as a tool to achieve the learning results and skills of the subject.
- The effect of the use of concept maps on the scores of the students.

2 Research Design

The experimental activity of concept map testing carried out in two groups of Industrial Drawing, which is a subject in the first course of Bachelor's degree in Industrial Engineering at Public University of Navarre, during the spring semester in 2019. The first and second groups had 41 and 45 undergraduate students, respectively. Before explaining the work procedure during the testing, the students recruited with a general invitation to develop a new activity, within the Industrial Drawing subject framework. However, a few of the participants were familiar with concept mapping, but most had not used this technique previously. Therefore, the students received a class on concept mapping in order to describe and clarify the theoretical basic concepts and the mapping structures. About the mapping strategy to follow on the present activity, some research studies recommend the exploratory method instead of confirmatory method [2]. The procedure followed takes into account some outstanding aspects that students have to manage inside and outside the classroom: complexity of the working system and individual/team work. Regarding to the complexity analysis from the concept maps, the qualitative date is studied. Focusing on measurements, first group worked individually, and the students selected the method and generated a concept map per chapter (Removable and fixed joints, Geometric tolerances and Movement transmission). However, the second group followed a different method. In this case, they developed a concept map in groups of five students taking into account the entire whole subject. The third aspect to evaluate, through a survey, is the perception of the students about the concept mapping and its contribution to the achievement of the skills related to the Industrial Drawing subject. The last aspect to analyze is the level of the conceptual knowledge of the students though the test score.

The assessment tool consists of four stages of the learning process: the learning, the knowledge, the motivation and the effect of the use of concept maps on the scores of the tests carried out. Beginning with the first stage, the learning assessment requires assessing analytically the concept maps using a rubric [3]. Table 1 shows these evaluable criteria. The assessment tool consists of seven concepts to evaluate.

Table 1. Rubric of concept map evaluation.

Evaluable criteria	Score	Leverage
1. Leading concept	From 3 to 0	X 3
2. Subordinate concepts	From 3 to 0	X 10
3. Propositions Is the relationship of meaning between two concepts indicated by the joining line and by their link word(s)? Is a valid relationship?	From 3 to 0	X 10
4. Presence of dynamic propositions They involve: (1) movement, (2) action, (3) change of state, or (4) dependency relationships	From 3 to 0	X 10
5. Cross connections and creativity Does the map show significant connections between the different segments of the conceptual hierarchy? Is a valid relationship?	Number of cross connections and creative connections	X10
6. Graphic examples that support the explained concepts	Number of graphic examples	X 10
7. Hierarchy Does the map present a hierarchical structure? Is each of the subordinate concepts more specific and less general than the main?	Number of valid hierarchical levels	X5

The main function of concept maps is to explore the knowledge and understanding level of the students in a specific topic [4]. This work analyzes the results from an individual and collective mapping perspective and its influence on the global mapping score. The surveys evaluate the student motivation by analyzing the general proficiencies and the learning outcomes. The student assessment has a main question: "What overall assessment does the methodology use? However, four purposes divide the survey to be easier for the students. Table 2 illustrates the parts of this assessment tool. The students have five levels of interest (1. Strongly disagree, 2. In disagreement, 3. Indecisive, 4. In agreement and 5. Strongly agree).

Table 2. Assessment survey of concept map methodology.

Purposes
1: Student spotlight, evaluation of teaching methods and improvements of teaching/learning processes
2: Acquisition of general skills of industrial drawing
3: Industrial drawing learning results
4: Information management and collective work

The first purpose evaluates the teaching methodology, learning process, student spotlight and improvements. Therefore, this point includes four questions (e.g., Does the methodology that include the performance of conceptual maps encourage the participation of students in the classroom?). To assess the acquisition of general skills, the survey proposes these themes from the guide of Industrial drawing. Moreover, the third purpose focuses on learning results published in the same teaching guide of the subject. Finally, the survey studies how the students have recollected and managed the information. The survey ends by asking to the students the main question related to the overall assessment the concept map methodology. As described before, this question summarizes the aim of this survey and, consequently, these results of this issue will be so much conclusive.

The scores of three tests done by students that used concept maps will be compared to the ones that did not used them as a teaching and learning tool. Test 1 is a theoretical exam and its main target is to measure the conceptual learning of the students. Test 2 is a practical activity where students represent graphically all the elements of an assembly, as well as their tolerances and adjustments. Test 3 is a practical teamwork.

3 Results

The results obtained on concept map metrics evaluate by means of the rubric. The surveys were the tool used to measure the perception of the students about the concept map tool and the scores of the tests of the students are the indicator to measure the effect of the use of the maps.

3.1 Concept Map Metrics

One evaluator, the professor of the subject, scored each map. The maps do not grade for course credit. With the amount of map metrics analyzed in this study, it was important to use the explicit assessment guideline, the rubric on Table 1 to ensure consistent scoring.

Table 3. Results of concept map metrics evaluated through the rubric: Mean Value and Standard Deviation (SD).

	Group 1- completed individually (37 samples)		Group 2- completed in groups (31 samples)	
	Mean value	SD	Mean value	SD
1. Leading concepts [/9 points]	8,48	1,74	9,00	0,00
2. Subordinate concepts [/30 points]	25,78	6,12	17,78	6,67
3. Propositions [/30 points]	9,69	7,12	8,89	7,82
4. Dynamic propositions [/30 points]	6,09	7,04	0,00	0,00
5. Cross connections & creativity [X10 per element]	7,50	9,09	0,00	0,00
6. Graphic examples [X10 per element]	5,86	4,41	12,22	3,63
7. Hierarchy [X5 per element]	7,03	4,43	4,44	3,00
Total	70,44	23,75	63,17	4,92

Table 3 shown the results on concept map metrics evaluated through the rubric and distributed in the different sections. The rubric evaluates seven criteria and the results split in two groups: Group 1 (students that mapped individually) and in Group 2 developed the maps collectively.

Individuality/collectivity results on map complexity: Group 2 presents a slightly lower value of conceptual map complexity; it seems that carrying out the maps collectively does not stimulate the origination of more complex maps.

Individuality/collectivity results on map configuration: The maps obtained individually show a great dispersion in their results, while when they done them in a group the results are more homogeneous and uniform.

Leading and subordinate concepts: The main and subordinate concepts appear in the optimal way; practically all the students who make the conceptual maps are able to identify the concepts of the subject and its hierarchy.

Complex propositions and connections: The students show difficulties to make complex propositions and connections. Most of them are beginners in concept mapping and this can be a limitation because it requires some training and experience to acquire skills to map complex systems. In addition, it is necessary to take into account that the time they have to make the maps is limited and they do not have any additional support material to the instructions of the professor of the subject.

Graphic examples: Due to the nature of the subject, some students include graphic examples that support the acquisition of concepts of graphic representation. The following Fig. 2 shows a sample.

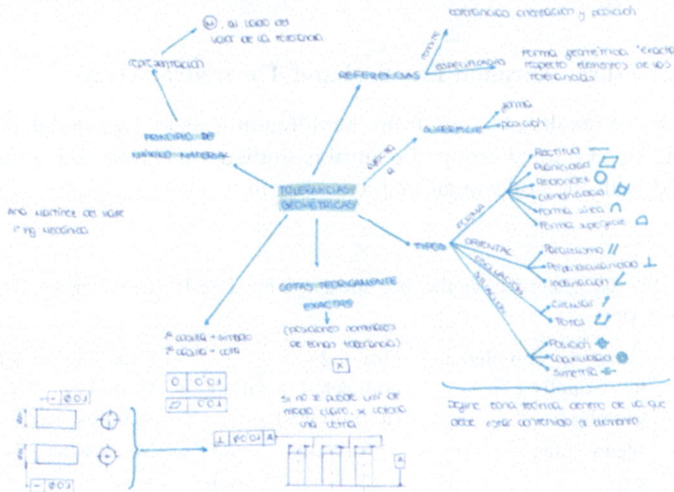

Fig. 2. Example of a concept map showing the Dimensional Tolerances subject matter.

3.2 Students Perception Though the Surveys

The results obtained from the survey conducted in the classroom, show the degree of student satisfaction with the concept mapping activity.

Table 4. Results of the survey according to previously defined purposes.

	Group 1 - completed individually (37 samples)		Group 2 completed in groups (31 samples)	
	Mean value	SD	Mean value	SD
Purpose 1	3,68	0,84	3,49	0,89
Purpose 2	3,72	0,75	3,43	0,79
Purpose 3	3,78	0,80	3,35	0,85
Purpose 4	3,92	0,79	3,62	0,81
Total	3,78	0,80	3,47	0,84

The results indicated in Table 4 show that the concept mapping activity is satisfactory to the students. Group 1 considers it more positive than Group 2, with a slightly lower satisfaction value (3,47 from 5, versus 3,78), but it can be considered positive for both.

It is verified that the concept maps are perceived most useful for purpose 4 (Information Management and Collective Work), being this appreciation same for both groups.

3.3 Students Scores Through Practical and Theoretical Tests

Table 5 shows, the results obtained from the different tests of knowledge conducted in the classroom. An additional group of control is studied; this group did not perform any activity related with the conceptual map experiment.

Table 5. Scores obtained on the previously defined tests (considering 10 as the maximum score).

	Group 1 - completed individually (37 samples)		Group 2 – completed in groups (31 samples)		Group 3 - no activity carried out (79 samples)	
	Mean value	SD	Mean value	SD	Mean value	SD
Test 1	9,83	0,51	7,80	0,94	5,93	1,19
Test 2	6,94	1,36	7,61	1,89	5,98	1,04
Test 3	6,53	1,04	6,30	1,63	6,59	1,84
Total	7,19	0,88	6,98	1,51	6,15	1,01

The results, Table 5 show that the concept mapping activity leads to better results in all of the performed tests. Being especially significant, the improvement obtained in the results of the Test 1 considered as a test of theoretical concepts. Group 1 leads this score with a high value (9,83 from 10, versus 7,8 and finally the group that did not perform any activity related with this work with 5,93). Finally, it is possible to confirm that the concept maps are useful to improve the teaching and learning of new concepts. The overall valuation of the knowledge improves from the previously obtained 6,15 to both group involved in this experimental work (Group 2–6,98 and Group 1–7,19).

4 Concluding Remarks

Based on the results presented, we identify several implications of the application of concept maps as a teaching and learning improvement tool. The main research questions outlined for this study has an answer: the use of mapping as a complementary activity is convenient to identify the understanding level of the subjects. Individual maps obtain better overall results: the scoring of the maps through the rubric is higher, as well as the perception of the students captured by means of the survey. The reason why this tool is more convenient for individual work/assessment is due to its capacity to represent the mind hierarchy of each human being on a specific topic. This is reinforced with the results obtained from the 3 tests carried out to students that used concept maps and others that did not. The students that completed their maps individually obtained the highest scores.

Concerning to the results obtained from the survey, the students consider that the most valued purpose for concept mapping is the Information Management and Collective Work.

The results presented show the lack of the skill on managing complex conceptual relationship of the students. This outcome is a very useful information for the professors in order to redefine some activities and to facilitate the acquirement of this skill.

Finally, the overall valuation of the knowledge after using concept maps improves overall from 0,8 to 1 point on the tests scores, compared to the students that did not use this tool.

References

1. Cañas, A.J., Carff, R., Hill, G., Carvalho, M., Arguedas, M., Eskridge, T.C., Lott, J., Carvajal, R.: Concept maps: integrating Knowledge and Information Visualization. In: Tergan, S.O., Keller, T. (eds.) Knowledge and Information Visualization. Lecture Notes in Computer Science, vol. 3426, pp. 205–219. Springer, Heidelberg (2005)
2. Walker, J.M.T., King, P.H.: Concept mapping as a form of student assessment and instruction in the domain of bioengineering. J. Eng. Educ. **92**(2), 163–179 (2003)
3. Lucas-Molina, B., Pérez-Albéniz, A., Fonseca-Pedrero, E., Ortuño-Sierra, J., Urraca, M.L., Santarén-Rosell, M.: Fiabilidad y evidencias de un instrumento para la evaluación de la calidad de los mapas conceptuales. Contextos Educativos, Extraordinario **2**, 119–130 (2017)
4. Watson, M.K., Pelkey, J., Noyes, C.R., Rodger, M.O.: Assessing conceptual knowledge using concept map scoring. J. Eng. Educ. **105**(1), 118–146 (2016)

Interactive and Lightweight Step-by-Step-Based Environment for Self-learning in Basic Graphic Expression Skills

Francisco L. Sáez-Gutiérrez(✉), Laura S. C. Albaladejo, J. S. Velázquez-Blázquez, Dolores Parras-Burgos, Francisco J. F. Cañavate, Daniel G. Fernández-Pacheco, and Francisco Cavas-Martinez

Graphic Expression Department, Technical University of Cartagena, C/Doctor Fleming, s/n, 30202 Cartagena, Murcia, Spain
francisco.saez@upct.es

Abstract. A key factor in Graphic Expression teaching is the previous knowledge acquired on high-school studies. The Spanish education system has created a high degree of uncertainty due to the legislative reforms that have established different educational pathways for university access. Those successive reforms have affected the previous knowledge of the Engineering students. In relation to Graphic Expression, a lack of the fundamental geometrical knowledge applies most specifically to those students who have not studied technical drawing in their previous educational stage. For this reason, the implementation of an interactive and self-learning tool has a great interest to solve this problem. The aim of this work was to design and implement a learning environment based on modern techniques (gamification, flipped classroom, etc.) and computer-aided design and digital publishing tools.

Keywords: Teaching innovation · Self-learning · CAD

1 Introduction

In the last 30 years, the changes experienced by the Spanish education system are unprecedented. During these years, there were eight educational reforms. This had generated different pre-university educational pathways [1]. In these pathways, the technical drawing subject is considered an optional subject. This means that some students have not received any technical drawing lessons before the university [2].

In order to solve this issue, in the Department of Graphic Expression at the Technical University of Cartagena (UPCT), we are carrying out a teaching innovation work to design an interactive environment for self-learning in basic Graphic Expression skills. This paper attempts to document this work. This paper is organized as follows: the first section gives an overview of the study framework. The second section examines the methodology. The next chapter analyses the results. Our conclusions are drawn in the final section.

F. Cavas-Martínez et al. (Eds.): INGEGRAF 2019, LNME, pp. 446–453, 2020.
https://doi.org/10.1007/978-3-030-41200-5_49

2 Study Frameworks: Legal Review and Teaching Innovation Methods

In the last 30 years, successive Spanish governments have made eight educational reforms. The LOE law [3], in 2006, and the LOCE law [4], in 2002, had given the authority to high-school centers to fix the range of optional subjects. The LOMCE law [5], in 2013, fixed the range for all high-school centers. In Sciences and Technologies knowledge branch, the student can choose two subjects between the follows one: Biology-Geology, Physics-Chemistry, and Technical Drawing II. As we can see, Technical Drawing subject is always an optional subject. For this reason, it is possible that a student starts his Engineering academic grade without receive technical drawing lessons.

The use of Information Technologies (IT) has changed our information and communication modes. These changes are also translated into our teaching systems [6]. The teaching system has evolved toward a two-way communication model and away from the traditional one-way communication model. Actually, the teacher's role is to guide the student in his self-learning, monitoring and fostering the available information and the information sources. A study on IT technologies by Coscolla and Agustó [7] concludes that 90% of teachers considered that IT application improves learning and academic performance and 60% of teachers considered that the IT technologies increase collaborative work and responsibility.

There are plenty of teaching innovation methodologies: In the Flipped Classroom, students have access to the teaching material outside the classroom. In classroom sessions are where the exercises are solved [8]. Collaborative learning allows students to reach objectives in groups. This methodology reinforces teamwork but hinds the individual evaluation [9]. In Project-based learning and Service Learning, students solve a project into a final product. These methodologies are perfect for final degree projects and at the beginning of the engineering grade [10]. Gamification involves the use of video game mechanics into the teaching system. This motives learning and exercise solving [11]. The Self-learning method is a technique that many people have been practised with success throughout history [12]. This methodology together with a virtual environment, also called e-learning, is our ideal choice for the following reasons [13]:

- Allows to access to the teaching material in the needed time (just-in-time learning)
- Provides an interactive and flexible teaching
- It is possible to access anywhere by IT technologies
- It could be merged with others teaching innovation methodologies as Gamification and Flipped Classroom.

3 Objectives and Methodology

Our main goal is to design and implement a self-learning teaching material in basic Graphic Expression skills. To meet this objective, we structure the work in the following stages those are based on Instructional Systems Design theory (ISD) [14]:

1. Teaching material analysis and selection
2. Implementation with Computer-Aided Design (CAD)
3. Digital export of the learning material
4. Digital publishing on an e-learning platform

3.1 Teaching Material

The Graphic Expression subject is a core subject in most of the UPCT Engineering academic grades. The UPCT study plans, those are approved by the national certification agency (ANECA) [15], sets some linked competences for the subject. On the other hand, the guide for the university admission test [16] sets the subject requirements knowledge. With these two documents, we prepare the basic Graphic Expression teaching material. Focused on the previous aspects, the developed material is shown in Fig. 1.

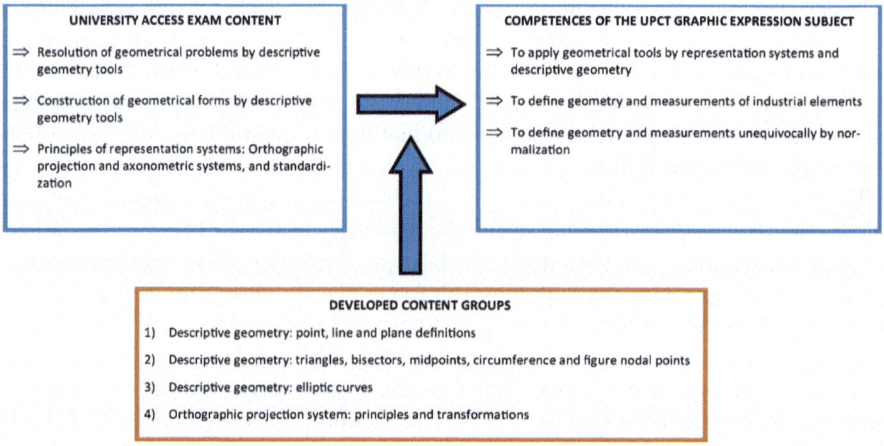

Fig. 1. Mainstreaming between the university and pre university Graphic Expression subject content.

3.2 E-Learning Platform: UPCT Virtual Classroom

Chosen methodology requires an IT online platform. In 2000, Technical University of Cartagena (UPCT) introduced, for the first time, a virtual learning environment. This platform was based on web course tools technology (Web CT). This initial platform includes the first functions as web directory and webmail. In 2007, the platform was migrated into a MOODLE system. This migration improved system capacities and interoperability. Nowadays, the university uses a new version, called UPCT Virtual Classroom (Fig. 2), with advanced features as online tutoring and digital publishing

[17]. UPCT Virtual Classroom is the proper e-learning platform to our work with the following advantages:

- The platform is owned by the University. For this reason, it is not needed a third-parties cloud storage service
- Students have instant access with their course registration. It is not needed any registration on a third-party e-platform or webpage
- The teacher has total control over the access and the teaching material
- The platform is operated and supported by the university. For this reason, it is not needed any third-party support service

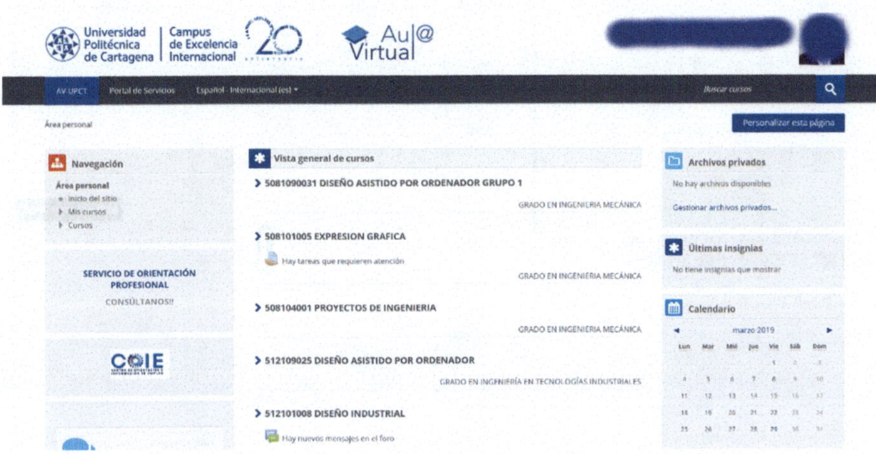

Fig. 2. UPCT Virtual Classroom e-learning platform.

3.3 Step-by-Step-Based Teaching Drawing by Computer Aided Design

Actually, there are many innovative formats in teaching material: e-notes and e-books, non-interactive tutorials, video tutorials and interactive tutorials. E-notes and e-books are the conversions to IT technologies of the traditional teaching material [18]. This format depends on the readability and comprehension of the student. Non-interactive step-based tutorials can reach a very large extension [19]. For example, if one exercise divided into 15 resolution steps and it is represented in one step by page, it could reach 15 pages of extension. Another format is the video tutorial [20]. In this case, the student can see the step-based resolution in a single document. However, in this case, the student-document interaction is limited to pause/resume the visualization. The interactive tutorial format is a similar format to the non-interactive tutorial, but each resolution step has a certain level of interaction with the student. In our case, each step is overlaid in the same space representation (Fig. 3). The student can activate each step for convenience. We have named "layers" to the resolution steps. In this format, the interactive features are: turn on/off layers, zoom in/out, move, turn and navigate.

We have used the computer-aided design (CAD) to design the teaching material. This tool allows drawing different layers in a single representation space. We have used AutoCAD. To draw the layers, we have set the following guidelines:

- Each layer should represent only one resolution step, unless a resolution step can be represented by a previous exercise or example
- Each layer is drawn in one color in order to be able to stand out from the rest
- The number of layers should be the necessary minimum for the proper learning of the exercise or example

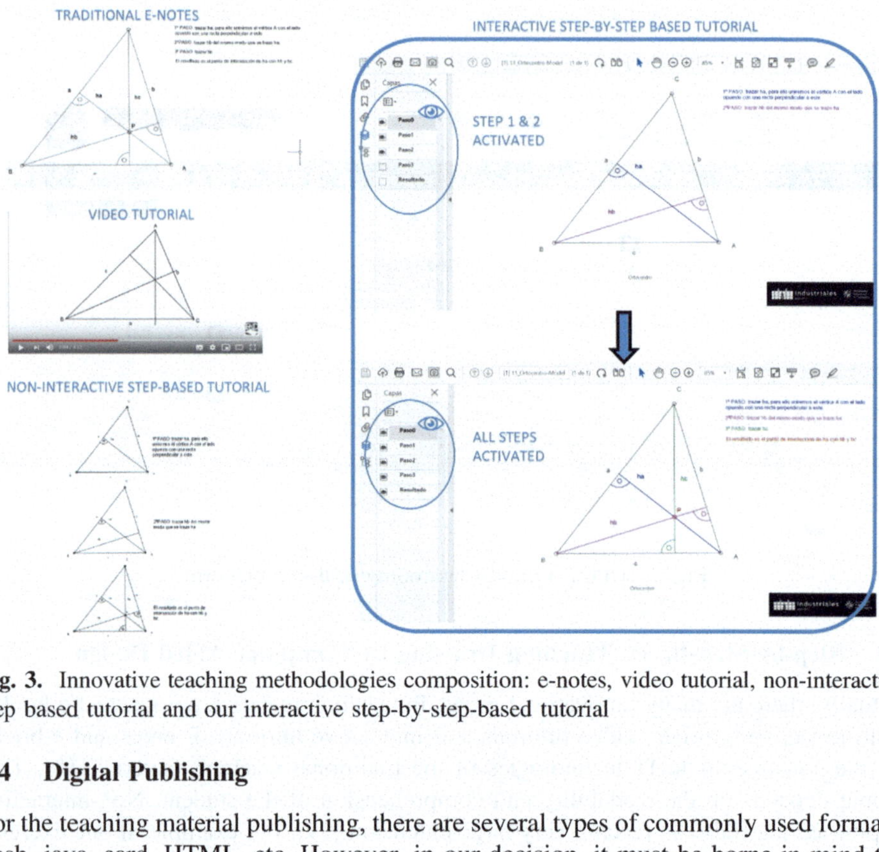

Fig. 3. Innovative teaching methodologies composition: e-notes, video tutorial, non-interactive step based tutorial and our interactive step-by-step-based tutorial

3.4 Digital Publishing

For the teaching material publishing, there are several types of commonly used formats: flash, java, card, HTML, etc. However, in our decision, it must be borne in mind the following requirements:

- It should be a standard format and compatible with different operating systems: windows, MacOS, Linux and Android
- It should allow AutoCAD layers exporting
- It should have a freeware reader

Considering these requirements, our proper decision is the portable document file (PDF) format. Since the 7.0 version, the Adobe Acrobat software allows the CAD layer exporting and reading (Fig. 4).

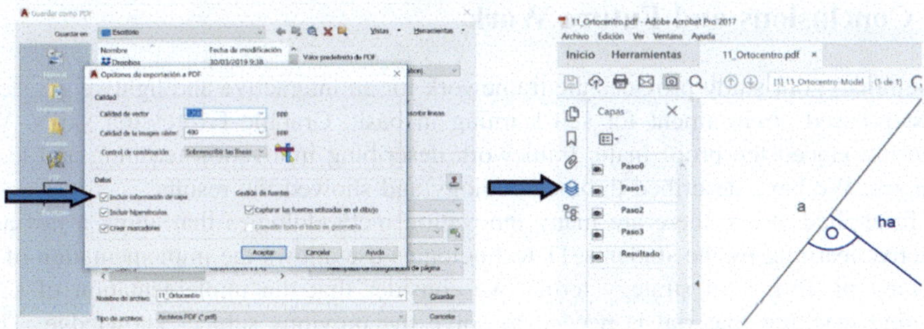

Fig. 4. AutoCAD export to PDF dialog box and Adobe Acrobat layer options menu.

4 Results

As a result of this work, we obtain different exercises and example from the collection described in Sect. 3.1 those are published on an e-learning platform in a lightweight and interactive format (Fig. 5). The teaching material has the following features:

- It is published on an IT e-learning platform. Available in anywhere in anytime
- It comprises the basic Graphic Expression skills that are needed to start engineering university studies
- It is lightweight, and it has a proper interaction level with features as turn on/off layers, zoom in/out, move, turn and navigate
- It allows students the self-learning in basic Graphic Expression skills

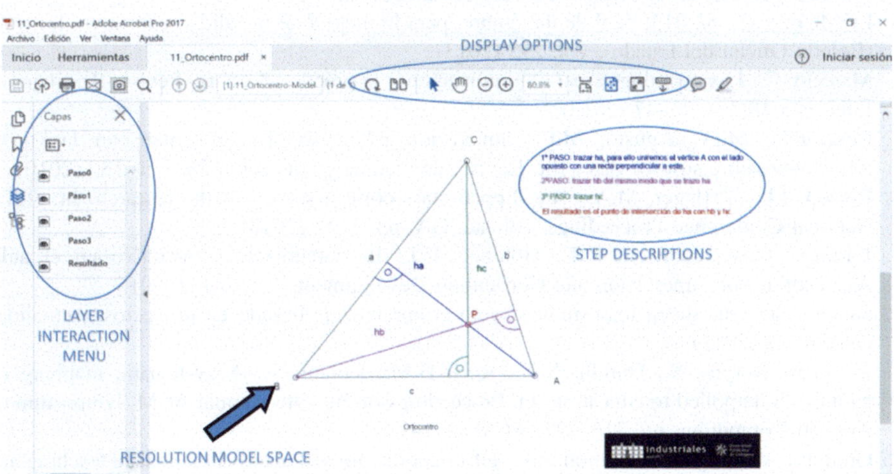

Fig. 5. Step-by-step based interactive format.

5 Conclusions and Future Work

In summary, our study provides the framework for an interactive and lightweight step-by-step-based environment for self-learning in basic Graphic Expression skills. We have introduced the problematic framework describing innovation teaching and legal changes. We have described the methodology and showed the results.

In applied terms, there are many innovation methodologies that allow a fast and effective learning by the students. IT technologies use allows the implementation of e-learning platforms. In strategy terms, we consider that the implementation of self-learning teaching material is needed to solve the previous subject knowledge. This implementation should be evaluated by quality systems, for example, surveys. In addition, we consider that this environment should be tested in a pilot experience.

References

1. Llorent-Bedmar, V., Palma, V.C.D.: Spanish educational legislation reforms during the current democratic period: a critical perspective. Education Policy Analysis Archives, 26 (2018)
2. Guirao Sánchez, A.: Análisis comparativo de los programas oficiales de dibujo técnico en la enseñanza media y su implicación en las tecnologías de la información y de la comunicación (tic) como recurso metodológico. Non-published Thesis, Universitat Politècnica de València (UPV) (2010)
3. Loe, L.O.: 2/2006, de 3 de mayo, de educación. In: Boletín Oficial del Estado, vol. 106, pp. 17158–17207 (2006)
4. Boletín Oficial del Estado, B.: Ley orgánica 10/2002, de 23 de diciembre, de calidad de la educación (loce). In: Boletín Oficial del Estado (2002)
5. LOMCE, L.O.: 8/2013, de 9 de diciembre, para la mejora de la calidad educativa. In: BOE (Boletín Oficial del Estado), vol. 295 (2013)
6. Marcelo, C.: Las tecnologías para la innovación y la práctica docente. Revista Brasileira de Educação **18**(52), 25–47 (2013)
7. Coscollola, M.D., Agustó, M.F.: Innovación educativa: Experimentar con las tic y reflexionar sobre su uso. Píxel-Bit. Revista de medios y educación **36**, 171–180 (2010)
8. Bishop, J.L., Verleger, M.A.: The flipped classroom: a survey of the research. In: ASEE National Conference Proceedings, Atlanta, GA, pp. 1–18 (2013)
9. Johnson, D.W., Johnson, R.T., Holubec, E.J.: El aprendizaje cooperativo en el aula. Association For Supervision and Curriculum Development, Virginia (1999)
10. Sánchez, J.: Qué dicen los estudios sobre el aprendizaje basado en proyectos. Actualidad pedagógica (2013)
11. de Sousa Borges, S., Durelli, V.H., Reis, H.M., Isotani, S.: A systematic mapping on gamification applied to education. In: Proceedings of the 29th Annual ACM Symposium on Applied Computing, pp. 216–222. ACM (2014)
12. Gremmo, M.J., Riley, P.: Autonomy, self-direction and self access in language teaching and learning: the history of an idea. System **23**, 151–164 (1995)
13. Cabero-Almenara, J.: Bases pedagógicas del e-learning. RUSC. Univ. knowl. Soc. J. **3**, 1 (2006)

14. Huang, C.H., Chiu, C.F., Chin, S.L., Hsin, L.H., Yu, Y.P.: A sports e-learning platform: teaching and learning by using multimedia contents. In: 2010 3rd IEEE International Conference on Ubi-Media Computing, U-Media 2010, pp. 222–226 (2010)
15. UPCT: Memoria de titulos aneca. http://www.upct.es/estudios/grado/5081/documentos/5081_memoria_01-06-2016.pdf. Accessed 30 Mar 2019
16. UPCT-UMU: Evaluación de bachillerato para el acceso a la universidad. https://www.boe.es/boe/dias/2018/01/26/pdfs/BOE-A-2018-984.pdf. Accessed 30 Mar 2019
17. Silva Pérez, M.M., Miguel Hernández, B., León Albert, G., Sánchez, M., Ignacio, J., Fernández López, J.A.: El aula virtual de la universidad politécnica de cartagena. Repsitorio digital UPCT (2010)
18. Mitsuhara, H., Shishibori, M.: Hybrid textbook: fusion of digital textbooks and traditional learning tools. Int. J. Intell. Syst. Technol. Appl. **14**, 237–255 (2015)
19. Chi, M., Jordan, P., Vanlehn, K.: When is tutorial dialogue more effective than step-based tutoring? In: International Conference on Intelligent Tutoring Systems. Lecture Notes in Computer Science (including subseries Lecture Notes in Artificial Intelligence and Lecture Notes in Bioinformatics). LNCS, vol. 8474, pp. 210–219 (2014)
20. Paliokas, I.: Reinforcing metacognition in cad education using video tutorials. Comput.-Aided Des. Appl. **6**, 613–623 (2009)

TDT-L0 a Test-Based Method for Assessing Students' Prior Knowledge in Engineering Graphic Courses

Barbara Motyl[1(✉)], Gabriele Baronio[2], Domenico Speranza[3], and Stefano Filippi[1]

[1] University of Udine, Via delle Scienze 206, 33100 Udine, Italy
barbara.motyl@uniud.it
[2] University of Brescia, Via Branze 38, 25123 Brescia, Italy
[3] University of Cassino and Southern Lazio, via G. Di Biasio, 43, 03043 Cassino, Italy

Abstract. Since students attending basic engineering courses are numerous and show different background regarding high school, experience, skills and attitudes, it has become important for educators to assess the level of their prior knowledge. In Italy, in order to access university engineering courses (BS degree level), students usually must take an entrance test; it assesses their knowledge about basic subjects such as mathematics, physics and chemistry, as well as their logical and linguistic skills. However, this test does not refer to technical subjects, e.g., the technical drawing. The two university research units of Udine and Brescia collaborate in developing the Technical Drawing Test-Level 0 (TDT-L0), a test-based method to assess students' prior knowledge about basic engineering graphics topics like axonometric and orthographic projections, sectional views, basic dimensioning standards, etc. This paper describes the definition of the test contents - the selection of the questions to ask to the students - as well as the way to analyze the collected information and exploit the results to improve the learning and teaching experience of both students and educators. An adoption of the TDT-L0 at the University of Udine is reported as first validation.

Keywords: Engineering education · Technical drawing · Prior knowledge assessment · Test

1 Introduction

Since our information society makes massive use of images, diagrams, and animations to spread messages, especially when it comes to providing technical information quickly and immediately, visual and graphics information represents an effective form of communication. In this context, the study and research on "graphicacy" represent an interesting and still rather unexplored topic. As defined in [1], graphicacy represents the basic skill of communicating through visual images; together with literacy, articulation,

The original version of this chapter was revised: The surname of the author has been corrected. The correction to this chapter is available at https://doi.org/10.1007/978-3-030-41200-5_67

F. Cavas-Martínez et al. (Eds.): INGEGRAF 2019, LNME, pp. 454–463, 2020.
https://doi.org/10.1007/978-3-030-41200-5_50

and numeracy, it can be considered as the fourth basic skill of any school [1, 2]. On the other hand, engineers and technical professionals have always used and preferred the graphics language to convey technical information and represent design ideas. Moreover, especially in this moment, the engineering education must be carefully prepared in order to allow the integration between hard and soft skills as well as to better face the challenges of the new industrial revolution [3]. In this context, as engineers use graphical methods for communication and analysis, it is important that engineering students possess spatial visualization skills, graphicacy, and Technical Drawing - TD competences next to the use of specific tools like CAD systems [4]. This way, from the authors' point of view, these competencies must develop and improve during higher studies [4]. This paper describes the definition of the Technical Drawing Test - Level 0 (TDT-L0), a test-based method to assess students' prior knowledge about basic engineering graphics topics like axonometric and ortho-graphic projections, sectional views, basic dimensioning standards, etc. The paper is organized as follows. Section 2 illustrates the background and the authors' previous research work. Section 3 introduces the TDT-L0 design and development starting from the motivations and describing its structure highlighting the desired assessment logic afterwards. Section 4 describes an early adoption of the TDT-L0 at the University of Udine and proposes a discussion about the results, all of this as a first validation and tuning of the method. Finally, Sect. 5 proposes the conclusions and some future developments.

2 Background and Our Previous Research

As reported in [5], since 2014 the Design and Methods Research Groups of the University of Udine and Brescia have been working together to develop a specific framework for the innovation and the improvement of TD teaching and learning [6–8].

This framework is based on the Technical Drawing Evaluation Grid - TDEG [9], the reference grid used by the authors to classify and divide TD topics into eight levels of knowledge, creating in this way a correspondence between specific topics and students' knowledge, skills, and competences (Fig. 1).

	Knowledge (STANDARDS)	Skills (to ...)	Competence (to be able to ...)	
Level 1	General principles of representation (UNI EN ISO 3098-0/2, UNI EN ISO 128-20/24, UNI EN ISO 5455, UNI EN ISO 5457, UNI 8187, UNI 938); Projection methods, representations and orthographic views (UNI EN ISO 5456-2, ISO 128-30/34); Cuts and sections (ISO 128-40/44/50, UNI 3972)	Carry out a "Wizard" (following indications provided by thirds parties) representation in views and sections of a part by following these basic rules: - Lines of text and characters, dimensional scales, sizes and folding of sheets, specification box; - Orthographic views and representations in sections and cuts in industrial and mechanical design.	Interpret the morphology of a part through its representation in views and sections	1-A
		Know by heart how to select appropriate views and sections needed to represent a part, autonomously applying relating standards	Independently carry out the representation in views and sections of a part	1-B

Fig. 1. An excerpt of the TDEG considering Level 1 specifications

In addition, a Level 0, as explained in [6] must be considered to specify and evaluate the prior knowledge that students entering the basic level classes of TD must possess in order to adequately address the study of this subject. In fact, since students attending these basic engineering classes are numerous and present different background regarding high school, experience, skills, and attitudes (as previously observed also in [10]), our research group is developing the specific TDT-L0 test to assess students' prior knowledge about basic engineering graphics topics.

3 TDT-L0 Design and Development

Starting from the consideration that large TD classes involve students with different educational backgrounds and different skills, the authors, based on their teaching experience, became aware of the need to analyze students' prior knowledge on TD basic topics to provide a better quality and more focused teaching, considering also the reduced number of hours (usually from 50 h to 60 h) reserved for frontal lessons and in classroom exercises. This analysis, performed on data collected using tests, must start during the first lesson, using some of the two hours available. In addition, the test will be used to verify both the knowledge of the TD key concepts and the abilities in hand drawing and sketching on paper. A literature search about the use of specific TD tests allowed discovering some interesting examples concerning the evaluation of graphicacy in elementary or middle/high schools [1, 2], the use of tests such as the Purdue Spatial Visualization Test: Rotations (PSVT: R, Mental Rotation Test (MRT), mental Cutting Test (MCT) and the differential aptitude test: space relation test used for the evaluation of spatial abilities of students at the university level [2] or other descriptive geometry based test [11]. Finally, some interesting references were found regarding the development of a TD inventory concept by [12], of a nationally normed Engineering graphics concept and skills test by [13, 14] and of an e-assessment test for engineering drawing in [15]. Unfortunately, these tests do not match the authors' intentions. In fact, our need is to get, the level of students' prior knowledge on some specifics and basic TD concepts at the beginning of the courses, as quickly and precisely as possible; unfortunately, all the cited references do not consider these aspects.

3.1 TDT-L0 Contents and Purpose for an Assessment Procedure

Starting from the analysis of the TDEG grid [9] and discussing with other TD instructors, the basic concepts to assess have been defined: the knowledge and the use of axonometric and orthographic projections, the creation of sectional views, the knowledge of dimensioning principles and the knowledge of specific terminology related to the drawing of mechanical parts such as screw, nut, or bolts. Table 1 summarizes the six questions used (column two) and the kind of knowledge/skills to verify (column three). These questions come from refinement both in the language and in the range of the proposed exercises; they are the result of three years of experimentation at the University of Udine, from 2014 to 2017.

The "Additional considerations" column of Table 1 reports further information about question assessment and answer evaluation. In fact, in the "Requirements"

Table 1. TDT-L0 questions about knowledge and skills.

#	Questions	Skills to verify	Requirements for correctness	Additional considerations
Q1	Draw freehand a three-dimensional cube	Spatial visualization skills Knowledge of axonometric or perspective projections Knowledge of descriptive geometry basics	Drawing the correct number of visible sides of a three-dimensional cube using axonometric or perspective projections	Use of an axonometric projection? Of what kind (Isometric, Cavalier, …? Use of a perspective projection? Drawing of only one visible side of the cube?
Q2	Draw freehand a parallelepiped using three views: top, front and side	Spatial visualization skills Knowledge of multiview drawing (rules and standards) using orthographic projections	Drawing of the three requested views placed according to orthographic projection standards (European projection method)	Use of non-correctly placed orthogonal views (for example lined up horizontally)? Use of other projection methods (axonometric prj.)? Drawing of less than three views?
Q3	Draw the missing orthogonal view 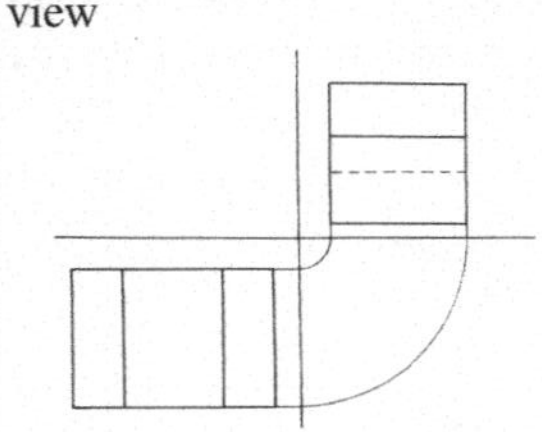	Spatial visualization skills Knowledge of multiview drawing (rules and standards) using orthographic projections.	Drawing the right shape (different possible solutions) and placing of the view using the European projection method	If he/she has drawn the correct view (shape & placing), what was the most used view shape? If he/she hasn't drawn the correct view, did he/she draw the incorrect shape? did he/she use the in-correct placing? If he/she hasn't drawn the correct view, did he/she read and recognized the different types of lines?

(*continued*)

Table 1. *(continued)*

#	Questions	Skills to verify	Requirements for correctness	Additional considerations
Q4	Draw the indicated sectional view 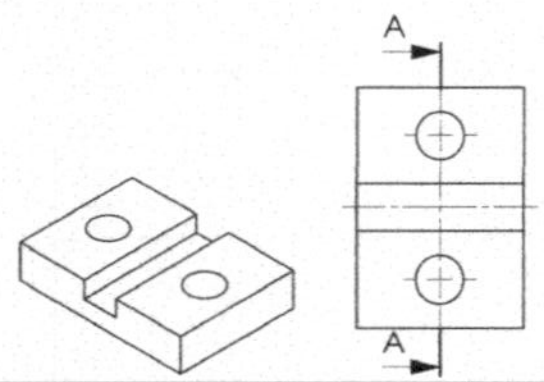	Spatial visualization skills Knowledge of sectional views standards and rules (lines, symbols, hatching, and positioning)	Drawing of the right shape, hatching, and placing of the sectional view according to rules and standards	Did he/she not understand where to "cut" the solid? Did he/she not place correctly the view according to orthographic projection standards? Did he/she not use the cross-hatching lines?
Q5	Dimension the perforated plate in figure 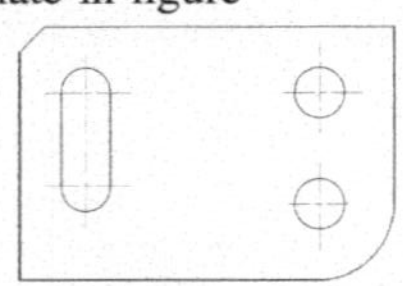	Knowledge of the dimensioning rules and standards (both writing in the appropriate way the dimensions and defining and measuring size and position information). Use of conventional dimensioning of mechanical features (hole, fillets, …)	Placing the right number of dimensions to geometrically define the shape of the object Use of mechanical conventions for dimensions of holes, fillets, and chamfers Dimensioning in mm	Has he/she used the correct unit of measurement? Has he/she used the appropriated placing of dimensions? Did he/she use an alternative dimensioning method?
Q6	Draw freehand a screw and a nut	Knowledge of specific language and terminology of TD and mechanical parts Conventional representation of mechanical parts	Drawing of a "screw" and "nut" using mechanical conventional representation	Did he/she use a pictorial representation? Did he/she have prior knowledge of mechanical conventions? Did he/she do a partial or incomplete answer?

column, the authors describe the requirements that answers must possess to be considered as processed correctly, while in the "Additional considerations" column, they propose possible different aspects that can be taken into consideration during answer correction and evaluation to obtain further information on the student knowledge and preparation level. The next section describes the practical application of TDT-L0 at the University of Udine.

4 Application of TDT-L0

During the 2017–18 academic year, the TDT-L0 was administered to 241 students (198 males and 43 females), enrolled in the TD course of the degree in Mechanical and Industrial Engineering at the University of Udine. For what concerns the female students in each of the two courses who took part in the test, it corresponds to the 15% in the case of Mechanical Engineering and about 20% in the case of Industrial Engineering respectively. The TDT-L0 test is proposed to the students as printed on a single A3 sheet, leaving the space for answering. The students are asked for inserting their personal information, the high school of origin and the address of their course of study, in order to use this information in the subsequent results evaluation phase (see Table 2).

Table 2. Characteristics of the students' high school provenance.

High School (HS) or Technical Institute (TI) of provenance	Mech. Eng. [%]	Ind. Eng. [%]
Scientific HS	34.5	56.7
Classical HS	0	3.7
Mechanical TI	41.1	9.0
Aeronautical TI	9.3	3.0
Electrotechnical/Electronic TI	5.6	7.5
Telecommunications/Information TI	0.9	5.2
Commercial TI	2.8	9.7
Land Surveyors TI	0.9	2.2
Professional HS	2.8	0
Other	0	0.7
Abroad HS	1.9	0

For the execution of the test, the first hour of the first lesson was used. After a brief introduction on the test purposes and the explanation on how to answer, the A3 sheets were distributed. Students were asked to answer using a pencil and a rubber. It was also possible to use rulers and squares if the students thought it as necessary. To answer the questions a time of 5–10 min is left, managed by the instructor. The instructor also projects each question on the screen in the meantime. At the end of the time, the sheets were collected and evaluated collectively by the members of the Udine research group. The main observations and results relating to the assessment of the answers of the six questions are reported in the next paragraph.

4.1 Data Analysis and Discussion of Results

To get a first validation of the proposed TDT-L0 test, the answers were evaluated following the trace proposed in the "Requirements" column of Table 1. This way, the results obtained for the six questions are reposted in Table 3. The values are in percentages and indicate the number of correct, incorrect and not given answers.

Table 3. Questions and results.

#	Mech. Eng. students			Industrial Eng. students			Total students		
	Correct [%]	Incorrect [%]	Not answer [%]	Correct [%]	Incorrect [%]	Not answer [%]	Correct [%]	Incorrect [%]	Not answer [%]
Q1	94.4	5.6	0	78.4	21.6	0	85.5	14.5	0
Q2	72	27.1	0.9	63.4	34.3	2.3	67.2	31.1	1.7
Q3	66.4	29	4.6	41	55.3	3.7	52.3	43.6	4.1
Q4	48.6	38.3	13.1	10.5	64.9	24.6	27.4	53.1	19.5
Q5	64.5	14	21.5	44	28.4	27.6	53.1	22	24.9
Q6	45	42	13	8.2	83.6	8.2	24.5	65.1	10.4

Only the answers that met exactly the correctness requirements indicated in Table 1 were considered as correct. All the other possibilities and "nuances" of response, from completely wrong to partially given or incomplete, were included in the incorrect answers. The analysis of the given answers confirmed, in the first instance, the hypothesis, which concerns the influence of the high school of provenance on the level of TD knowledge owned by the students [9], in fact, also, in this case, it was noted a different behavior in the two courses: Mechanical and Industrial. Going now to detail the analysis o considering the single questions, while there is a similar trend (considering the percentage of questions correctly answered) for Q1 and Q2 (related to the basic concepts of orthogonal projections), all the other questions indicate different trends. Such tendencies are probably due to the massive presence of students coming from technical institutes enrolled in Mechanical Engineering and from Scientific high schools in Industrial Engineering. Particularly significant are the data that can be obtained by analyzing the answers to question Q3 in reference to the assessment of spatial visualization skills. Instead, the analysis of the trend for Q4 (use of sectional views) leads to the conclusion that this topic will have to be carefully revised and deepened in the course. In fact, more than half of the students (53.1%) made mistakes in the execution of the section and almost the 20% did not respond. Other significant information concerns question Q5, which again highlight a lack of knowledge of more technical subjects such as dimensioning, with 47% of total students who provided incorrect answers or did not reply (25%) to the question. The same tendency lasts also for question Q6, in which only 25% of the total students seem to possess notions related to the use of a specific technical language and the knowledge of the conventional representation of mechanical pieces. About this question, however, it should be highlighted that the answers containing an "illustrative" representation of the screw and nut, which correspond to the 57.3% out of the 65.1% indicated for the total of students, are also included among the incorrect answers. In order to obtain a meaningful, visual representation of the level of a priori knowledge possessed by the students at the time of entrance to the TD course, the graphic representation proposed in Fig. 2 is used. In particular, with the use of this radar diagram instructors can visualize what is the trend of the correct answers obtained for the six questions of the TDT-L0 test, visualizing the coverage areas of the knowledge given for the two different engineering courses (Mechanical and Industrial) or the general one. This information allows instructors to get an immediate picture of which topics are the most known/unknown.

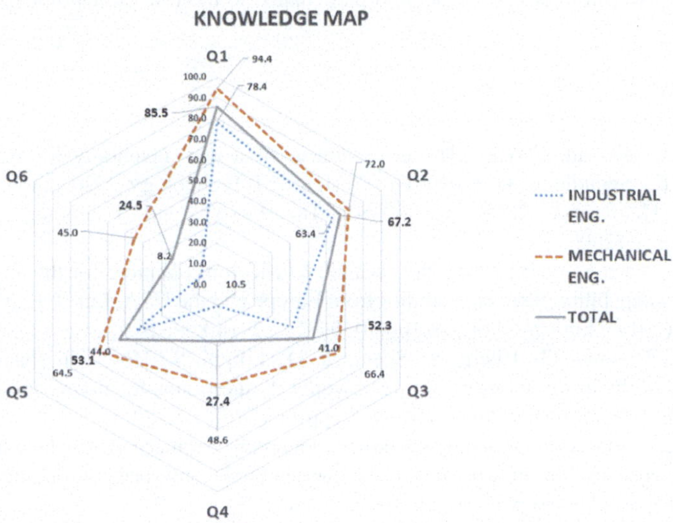

Fig. 2. Knowledge map related to TDT-L0 test at the University of Udine.

5 Conclusions and Future Developments

The first validation of the TDT-L0 test provided encouraging results regarding the reliability of the answers obtained to verify the preliminary knowledge on TD of engineering students. Furthermore, the feasibility of the test was also verified with very large classes of students. Regarding the evaluation of the results and the rapidity in the correction of the answers, it was observed that the definition of the evaluation objectives - through the drafting of the requirements for the correctness of the answers, and an efficient teamwork of the people involved in the correction, are fundamental. For example, the corrections and the analysis of the results related to the test performed in Udine took a week. This allowed catching an almost immediate picture of the students' knowledge and deciding which topics would have required to be deepened during the lessons. These results also enabled the instructors to plan the exercises and topics to be consolidated and/or strengthened during the course. It should be also emphasized that the test was deliberately proposed in paper format rather than online, in order to evaluate also the students' sketching abilities. It is not planned to repeat the test at the end of the lessons; however, the authors are reasoning about the opportunity to carry out a comparison of the test results with those obtained by the students as final exams. In the future, the authors will consolidate the evaluation requirements proposed here, also using the experience from the adoption of the TDT-L0 by other universities. In fact, it is planned to administer the test, in its current version, also at the Universities of Brescia and Cassino and Southern Lazio, as well as to carry out an evaluation of the prior knowledge on TD of students of different Universities.

Acknowledgments. The authors want to acknowledge Dr. Daniela Barattin for her help and all the students at the University of Udine who participated to the test validation over the years.

References

1. Danos, X., Norman, E.W.L.: The development of a new taxonomy for graphicacy. In: Norman, E., Spendlove, D. (eds.) The Design and Technology Association International Research Conference 2009, pp. 69–84. The Design and Technology Association, Wellesbourne (2009)
2. Danos, X., Barr, R., Górska, R., Norman, E.: Curriculum planning for the development of graphicacy capability: three case studies from Europe and the USA. Eur. J. Eng. Educ. **39**(6), 666–684 (2014). https://doi.org/10.1080/03043797.2014.899324
3. Motyl, B., Baronio, G., Uberti, S., Speranza, D., Filippi, S.: How will change the future engineers' skills in the Industry 4.0 framework? A questionnaire survey. Proc. Manuf. **11**, 1501–1509 (2017). https://doi.org/10.1016/j.promfg.2017.07.282
4. Metraglia, R., Baronio, G., Villa, V.: Issues in learning engineering graphics fundamentals: shall we blame cad? In: Proceedings of the International Conference on Engineering Design, ICED 2015 (DS 80-10), vol. 10, pp. 31–40 (2015)
5. Speranza, D., Baronio, G., Motyl, B., Filippi, S., Villa, V.: Best practices in teaching technical drawing: experiences of collaboration in three Italian Universities. In: Eynard, B., Nigrelli, V., Oliveri, S.M., Peris-Fajarnes, G., Rizzuti, S. (eds.) Advances on Mechanics, Design Engineering and Manufacturing. LNCS, pp. 903–913. Springer, Cham (2017). https://doi.org/10.1007/978-3-319-45781-9_90
6. Villa, V., Motyl, B., Paderno, D., Baronio, G.: TDEG based framework and tools for innovation in teaching technical drawing: the example of LaMoo project. Comput. Appl. Eng. Educ. **26**(5), 1293–1305 (2018). https://doi.org/10.1002/cae.22022
7. Baronio, G., Motyl, B., Paderno, D.: Technical drawing learning tool-level 2: an interactive self-learning tool for teaching manufacturing dimensioning. Comput. Appl. Eng. Educ. **24** (4), 519–528 (2016). https://doi.org/10.1002/cae.21728
8. Baronio, G., Bodini, I., Copeta, A., Dassa, L., Grassi, B., Metraglia, R., Motyl, B., Paderno, D., Uberti, S., Villa, V.: Integrated approach to the innovation of technical drawing teaching methods. In: Cavas-Martínez, F., Eynard, B., Fernández, C.F., Fernández-Pacheco, D., Morer, P., Nigrelli, V. (eds.) Advances on Mechanics, Design Engineering and Manufacturing II. LNCS, pp. 705–713. Springer, Heidelberg (2019). https://doi.org/10.1007/978-3-030-12346-8_68

9. Metraglia, R., Baronio, G., Villa, V.: Learning levels in technical drawing education: proposal for an assessment grid based on the European Qualifications Framework (EQF). In: Proceedings of ICED 11 - 18th International Conference on Engineering Design - Impacting Society Through Engineering Design, vol. 8, pp. 161–172 (2011). SCOPUS code: 2-s2.0-84858825431. ISBN: 978-190467028-5
10. Metraglia, R., Villa, V., Baronio, G., Adamini, R.: High school graphics experience influencing the self-efficacy of first-year engineering students in an introductory engineering graphics course. Eng. Des. Graph. J. **79**(3), 16–30 (2016)
11. Cobos-Moyano, A., Martín-Blas, T., Oñate-Gómez, C.: Evaluating background and prior knowledge: a case study on engineering graphics learning. Comput. Educ. **53**(3), 695–700 (2009). https://doi.org/10.1016/j.compedu.2009.04.010
12. Nozaki, S.Y., Sorby, S.A., Study, N.E., Steinhauer, H.M., Sadowski, M., Miller, R.: The development of a concept inventory for engineering graphics. In: Proceedings of 2016 IEEE Frontiers in Education Conference (FIE), pp. 1–4 (2016). https://doi.org/10.1109/fie.2016.7757430
13. Croft, F.M., Demel, J.T., Meyers, F.D.: A proposal for a nationally normed engineering graphics concepts and skills test. In: Proceedings of the 2001 American Society for Engineering Education Annual Conference & Exposition, Paper # 1124 Session # 1138 (2001)
14. Demel, J.T., Croft, F.M., Meyers, F.D.: Development of nationally normed engineering graphics concepts and skills tests. In: Proceedings of the 2002 American Society for Engineering Education Annual Conference & Exposition, Session 1338 (2002)
15. Valentine, R.: An e-assessment for engineering drawing. In: Proceedings of the International Conference on Engineering and Product Design Education, E&PDE 2018, pp. 358–362 (2018)

Design of Emergency Vehicle Lighting Integrated in Fire Fighting Trucks

María Alonso-García[1(✉)], Aída Solís-Mellado[2], Cristina Gómez-Vázquez[2], Manuel Fernández-Rubio[3], Óscar D. de-Cózar-Macías[4], and Elidia Beatriz Blázquez-Parra[4]

[1] Dpto. Ingeniería Mecánica y Diseño Industrial, Escuela Superior de Ingenierías, Universidad de Cádiz, Av. Universidad de Cádiz, 10, 11519 Puerto Real, Cádiz, Spain
maria.alonso@uca.es
[2] Universidad de Cádiz, Av. Universidad de Cádiz, 10, 11519 Puerto Real, Cádiz, Spain
[3] Departamento I+D+I, Feniks Cleaning & Safety SL, Seville, Spain
[4] Dpto. de Expresión Gráfica, Diseño y Proyectos, Escuela de Ingenierías Industriales, Universidad de Málaga, c/Doctor Ortiz Ramos s/n, 29071 Málaga, Spain

Abstract. This research focuses on the design and development of emergency vehicle lighting integrated in fire fighting trucks. The Project uses the Design Thinking methodology, and has been developed at University of Cádiz, in collaboration with Feniks Cleaning and Safeting, S.L. It has been part of a teaching innovation project developed during the first semester of the academic year 2018–2019, being the responsible, two students from "Industrial Design Engineering and Product Development Degree" (IDEPDD), who have had the opportunity to work on a real product design, under the University-Business co-tutorization. The students have combined practical ECTS of the subject "Creation of New Products" with two internships offered by the Company, which is dedicated to special vehicles customization. This company looks for a versatile product adaptable to the range of vehicles they work with, in order to differentiate itself from the competition. A challenge is not only for the formal and aesthetic characteristics, but also for the need to guarantee the viability and manufacturability of the product, with an estimated production of less than 200 vehicles per year.

Keywords: Industrial design · University-Business · Dual training · Product design

1 Introduction and Background

According to the report prepared in 2015 by the Catalonian Association of Public Universities (ACUP), the high level of youth unemployment registered in Europe since 2006 has made educational policies directed to the labor insertion of young people [1]. Through the European Higher Education Area (EHEA), curricula have been modified in order to achieve European convergence to the acquisition of adequate professional

F. Cavas-Martínez et al. (Eds.): INGEGRAF 2019, LNME, pp. 464–471, 2020.
https://doi.org/10.1007/978-3-030-41200-5_51

skills and facilitate the student transition to the labor market, through what is regulated by the Ministry of Education, Culture and Sport (2003), in RD 1393/2007, modified by RD 861/2010, which establishes the guidelines for the implementation of the current Degrees, by demanding a greater proximity between University, Society and Enterprise. In the specific case of Engineering, many researches require the change of curricula, focusing it on "employability", "entrepreneurship" and "innovation" [2]. On the other hand, according to the Accenture report [3] almost 50% of graduates from university degrees are dissatisfied with their job. Personal and professional skills are the key to finding a job, as established by the companies surveyed in the investigation. The report highlights motivation and enthusiasm skills.

In the same way, the Company recognizes not being able to cover certain jobs because it does not find qualified workers. As previous studies set [4], professionals in industry are flexible, adaptable, of lateral thinking and creative, but normally, they lack essential professional skills or do not find the connection between what they have learned and what they need to work. More and more universities are locating in these degrees different levels and strategies of collaboration with companies for academic purposes, which suppose a bridge to professional training and to the insertion of students in the current labor market [5, 6].

University of Cádiz and Feniks Cleaning and Safeting, S.L. (from now on "FENIKS" or "the Company"), have worked together on the R+D Project "Design of emergency vehicle lighting integrated in fire fighting trucks", being part of the Teaching Innovation Project "Development of academic Works co-tutored between the University of Cadiz and companies under Dual training components" (sol-201800112282). This Teaching Innovation Project consists in the development of real projects in the classroom, under the co-supervision and monitoring of both, the professor and the Company. This Project, which has had the financial support of the University of Cádiz, and two student internships offered by the Company, has been carried out within the fourth grade subject "Creation of New Products" of IDEPDD during a full semester.

The Project has counted also with the collaboration of different companies and professionals for the product manufacturability analysis and the prototype development.

FENIKS is a company dedicated to the bodywork of urban cleaning vehicles and firefighting trucks, and cleaning and maintenance of solar thermal power plants. This company proposes to the University of Cádiz the design of emergency vehicle lighting composed by an external case (tulip) and a quick fixing system (platform) which should be perfectly integrated into the vehicles chassis. The lighting must have a standardized design enabling the use of the product in any brand and any single or double truck cabin. The project objectives are, in addition to offering this differentiating value to the brand, minimizing manufacturing and assembly costs, and ensuring easy maintenance and repair, taking into account the limited number of units and the wide variety of models (vehicles) where the product could be installed.

The Company has the supervening need to adapt the lighting to the new requirements of the Vehicles General Regulation (CEPE/ONU number 65). This regulation is modified through the PCI/810/2018 order by the Annex XI, regarding the signal V-1 of the "First Priority vehicle". The order requires Spain to adhere to the existing regulations in most of the Union Europe States. This modification establishes that the emergency service lighting devices of first priority vehicles must be blue, while the yellow is reserved to the vehicles that constitute an obstacle in the road.

On the other hand, First Priority vehicles and Heavy Goods vehicles regulations and administrative processing, set a maximum height of 4 m. At this length, it is necessary to add 0.20 m in order to absorb vertical movements, obtaining a minimum height of 4.20 m.

Fire fighting vehicles are built according to their own service needs, as the EN 1846 regulation [7] is specified. The regulation sets also the requirements for the homologation of vehicles, defining, classifying, categorizing and codifying the different types of fighting trucks and auxiliary services in a designation common system. In addition, EN 1846 identifies the main characteristics of any motor vehicle from an alphanumeric code, serving as reference for making supply requests and identifying the main characteristics of any vehicle.

2 Method

The methodology used to design the product has been based on the one created by Tim Brown: "Design Thinking", focused on the user experience. This methodology is constituted by 6 phases (see Fig. 1). Through these, the design team must understand the problem in all its facets (1), in order to empathize (2) with users through an observation process. Later, all the information is organized and synthesized to reframe and define (3) the specific problem, and to generate as many ideas as possible (4) that will be previously selected by professors, and afterwards, by the company.

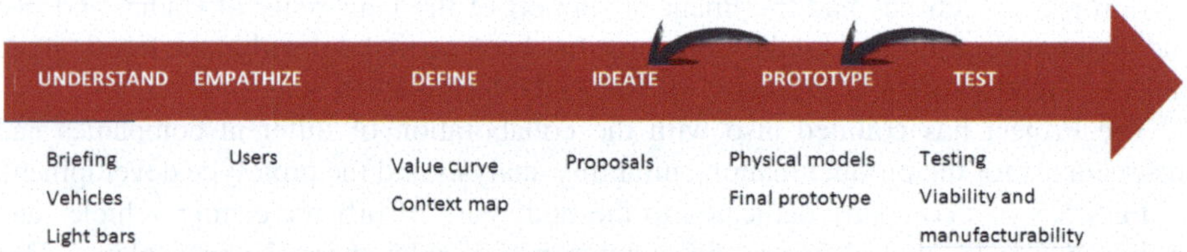

Fig. 1. Methodology employed in research.

Finally, the prototype phase is carried out (5) in order to present a selection of product (6), which will be accompanying a manufacturing budget.

During the semester, four meetings are considered to discuss the briefing of the Project, the analysis of solutions, the optimization of the selected proposal and the viability and manufacturing study respectively.

3 Results

To reach the final solution, and working under this methodology, it has been necessary to study both the dimensional product context (relating the surface of the vehicle and the lighting) and the formal context. After the selection of the proposal, it has been developed also a study of materials, and a viability and fabricability analysis. In this sense, to guarantee its correct production, all the appropriate modifications are made to the concept.

3.1 Dimensional and Formal Study

The product, to be designed, as mentioned before, must be positioned and adapted quickly to any truck cabin. To achieve this, the binomial lighting-vehicle's surface has been analyzed, studying the different models of bodywork that can be qualified as "firefighting truck", and the light bars most used by Feniks. In order to guarantee the correct installation of the tulip on any kind of cabin truck, the dimensions of the main brands are established, taking into account the different types of existing truck cabins, and especially their maximum and minimum dimensions. It is concluded that the length of the analyzed truck cabins is between 900 and 1600 mm and the width between 1200 and 2500 (see Fig. 2).

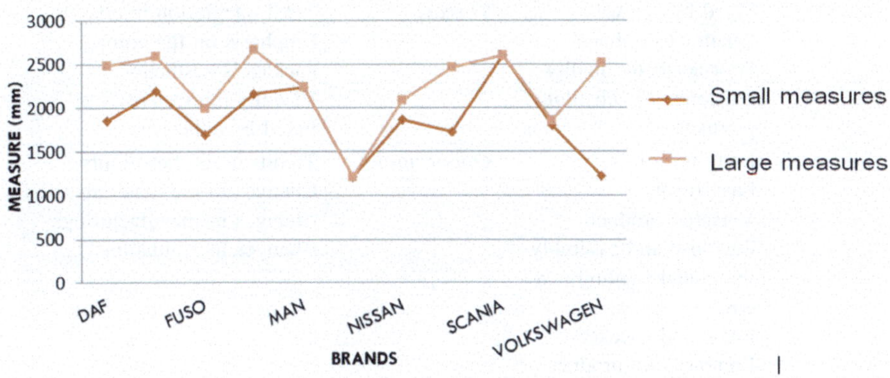

Fig. 2. Comparative widths of possible fighting vehicle cabins

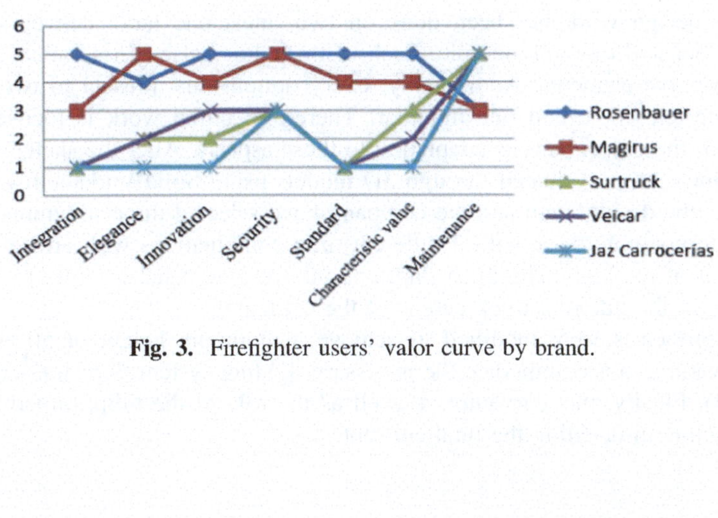

Fig. 3. Firefighter users' valor curve by brand.

The elements most valued by the potential client must be understood. These are: integration (fusion of design and functionality), elegance (simplicity and design beauty), and innovation (safety, standardization, brand value and easy to maintain and to repair any degradation). In order to evaluate the application grade the competition has, these parameters have been analyzed, each one, in a different value curve, placing each competitor according to the grade that corresponds to them (see Fig. 3).

The study of the different light bars used by Feniks, and those most demanded by users, has been developed through a SWOT analysis (Table 1). In a comparative mode, taking into account the previously established design requirements, the weaknesses have been obtained, as well as the threats, strengths and opportunities presented by the models used by the Company, against those competing products. This supposes the starting point for the formal product design.

Table 1. SWOT analysis "Lighting bridge".

Internal origin (brands used by Feniks)		*External origin (other brands)*	
Weaknesses	Fixed by screws Unattractive shape Poor aesthetic quality Existence of substitute products	*Threats*	Total integration in chassis Emphasis on the compact package's aesthetic Visually innovative Fixed by adhesive
Strengths	Long useful life Easy repair Versatile product Easy use and assembly Integrated lighting low cost Integrated speakers Emphasis on product technology	*Opportunities*	Trends in car headlights Changes in the materials used Changes in manufacturing Changes in regulations

Formal design work has been done on two lines; one more conventional, and another riskier and more ergonomic. Each of these lines also has a variety including others decorative elements. Additionally, every option must present to two different models: one telescopic and one modular. Therefore, initial work is focused on the semantics of the object, and its adaptation to these aspects. After the sketching phase, proposals have been evaluated through 3D models using SolidWorks software. From 3D models, the design team and the company have selected three structures from the same line (ergonomic one), with 3 different tulip combinations with greater or lesser involvement of the logo. The three final options were modeled in order to check the proportions in the different truck cabins of the company.

These proposals were modified to achieve a minimum height of 50 mm in the tulip's curvature, to accommodate the necessary lighting system to be housed inside it (see Fig. 4). Finally, this curvature, as well as the role of the tulip, turned out to be decisive factors in deciding the final concept.

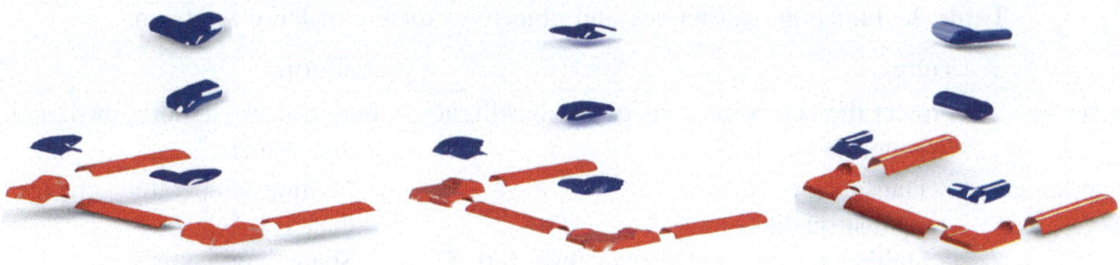

Fig. 4. Explosion of selected concepts, with 50 mm minimum arch.

3.2 Materials and Product Manufacturability

The final result is a product adapted, ergonomic, identifiable and easy to assemble. Its telescopic structure permits to fit each piece in the previous one, reducing its length and allowing its quick adaption to any vehicle in the market. It presents an ergonomic design, providing continuity with the truck cabin thanks to the use of curved and soft lines. It provides identity to the brand through subtle details of its own logo. The tulip-platform union by screws and adhesive fixation achieves an easy assembly (see Fig. 5).

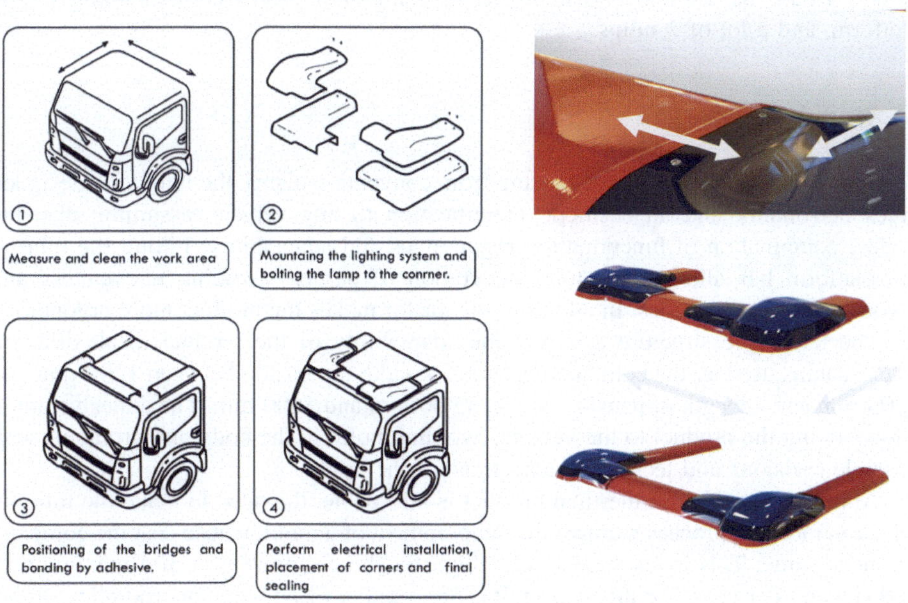

Fig. 5. Assembly process (left), anchor detail (upper right) and telescopic action (lower right).

For the material and manufacturing process selection, it was necessary to analyze the functions and requirements that each component of the product must have, in addition to the objectives that are to be achieved (Table 2).

Table 2. Function, restrictions and objectives for the material selection.

	Tulip	Platform
Function	Protect the component inside the headlight (wiring, lamp, etc.)	Integrate the lighting designed to the vehicle
Restriction	– Transparent – Optical qualities – Melting temperature greater than 100 °C – Shape: curve surface – Number of pieces per lot: 320 – Specific dimensions	– Melting temperature greater than 100 °C – Shape: curve surface – Number of pieces per lot: 160 – Specific dimensions
Objectives	Maximize tenacity, elastic modulus, resistance to fatigue and traction, minimize the price, weight and environmental footprint, and ensure easy maintenance	

Once completed, it is concluded that the materials suitable for the design are polycarbonate for the tulip and the PRFV for the platform. This is because polycarbonate is a material that meets all requirements and it is one of the most used in lighting. PRFV is already used by Feniks, so it is the easiest to work with; it is an economic material and complies with the requirements. The estimation per lighting system would be 1.487,6 euros, including a lot of 5 pieces corresponding to the platform, and a lot of 2 tulips.

4 Conclusions

Searching for innovation and distancing from conventionalisms, the luminaire designed gives personality and an aesthetic identification to any vehicle, assuming also the perfect combination of functionality, ergonomics and adaptability. Fitting the tulip in the platform has allowed the total integration of parties, avoiding unevenness and favoring continuity shapes. In addition, this factor makes the product more ergonomic.

The telescopic structure achieves the adaptability of the product to all different trucks cabins used by the company, giving it a width dimension between 1200 mm and 2200 mm and a length dimension between 900 mm and 1600 mm. The adhesive union allows fixing the product to the vehicle, avoiding holes in the body and problems with possible revisions and technical inspections of the truck.

Although the price of the final product is affordable, the need to make the tulip by injection molding process requires the development of a mold, whose cost the company should assume. This investment is very high for the customization of a maximum of 160 vehicles per year. For this reason, it is proposed to FENIX the incorporation of this luminaire to other types of vehicles: cleaning and forestry (Fig. 6), as well as the continuity on the research for the manufacture of the tulip by other cheaper manufacturing techniques, as it could be additive manufacturing techniques.

Fig. 6. Cleaning, forest and fire fighting vehicles.

On the other hand, the collaboration has been highly satisfactory for both parties, entailing for the students an opportunity to have a first contact with the labor market, as well as the development of a first real project, getting involved in issues not only of design, but also in contact with suppliers, the communication with different departments of the company and several visits to the factory, which helps their training as engineers. The company confesses to having managed to carry out a project, which, after being approached on several occasions, had never gone ahead due to lack of time. "Having this program gives us the opportunity to count on the help of young people with an external point of view, with the ability to innovate and contribute with new ideas that we would not have thought of".

References

1. ACUP: Promoción y desarrollo de la formación dual en el sistema universitario catalán. http://www.acup.cat/sites/default/files/promocion-y-desarrollo-de-la-formacion-dual-en-el-sistema-universitario-catalan.pdf. Accessed 10 May 2019
2. Kolmos, A., Hadgraft, R.G., Holgaard, J.E.: Response strategies for curriculum change in engineering. Int. J. Technol. Des. Educ. **26**(3), 391–411 (2016)
3. Accenture Universia: Las competencias profesionales de los titulados. Contraste y diálogo Universidad-Empresa. Centro de Alto Rendimiento de Accenture y Universia. http://www.unizar.es/ice/images/stories/calidad/EstudioCompleto.pdf. Accessed 11 May 2019
4. Ball, L.: Preparing graduates in art and design to meet the challenges of working in the creative industries: a new model for work. Art Des. Commun. High. Educ. **1**(1), 10–24 (2002)
5. Evyapan, N., Korkut, F., Hasdogan, G.: Implications of collaboration with industry for educational strategies in industrial design: the graduation project course, pp. 137–159. METU, Faculty of Architecture Press, Ankara (2005)
6. Yang, M.Y., You, M., Chen, F.C.: Competencies and qualifications for industrial design jobs. Implications for design practice, education, and student career guidance. Des. Stud. **26**(2), 155–189 (2005)
7. European Union: European Standard EN 1846 Firefighting and rescue service vehicles (2012)

Engineering and Construction

3D Semantic-Rich Modelling of Underground Utility Networks

Adrian Preciados Royano[1], José Andrés Díaz Severiano[1], Valentín Gómez-Jáuregui[2], Cristina Manchado del Val[2], Julen López Iglesias[1], Olmo Fernández García[1], and César Otero González[2(✉)]

[1] School of Civil Engineering, Universidad de Cantabria, Avda. Los Castros s/n, 39005 Santander, Spain
[2] EgiCAD Research Group, School of Civil Engineering, Universidad de Cantabria, Avda. Los Castros s/n, 39005 Santander, Spain
cesar.otero@unican.es

Abstract. In this paper, we seek to document the state of art of the 3D semantic-rich modelling of underground utilities. To address this challenge, we develop a review of most common platforms used for 3D data modelling, mostly based on GIS or XML. We plan to detail what each platform is capable of: the information they can provide and their potential of displaying 3D models and information associated in portable devices for maintenance or any other field of work. Next, we carry out a case study where both capabilities and limitations of these kind of systems are described. In order to overcome limitations, we remark the relevance of automation and programming techniques as an auxiliary resource.

Keywords: 3D model · Data model · Utility infrastructure · Utility network · Sematic-rich model · BIM · GIS

1 Introduction

Nowadays society has total dependence on utility networks; at the same time, we ask for more efficient distribution networks in term of energy consumption, carbon balancing and water saving, among others. Because of this, the stress on utility networks is becoming higher and higher, so a reliable and well-maintained network is very necessary.

However, the availability of highly detailed and accurate geo-referenced map is usually very limited. In the last decades, most urban places have suffered a rapid growth and density of utilities has gone up without a detailed planning; as a result, networks are not completely catalogued. When detailed maps are available, they are mostly 2D models and usually incomplete or vague [1]. They don´t provide information other than very basic data.

These situations often result in poor preservation, waste of resources (i.e. wasting freshwater) and delays or/and cost-overruns in any kind of construction performed in the public space, when finding unexpected networks. Thus, the knowledge of precise

F. Cavas-Martínez et al. (Eds.): INGEGRAF 2019, LNME, pp. 475–481, 2020.
https://doi.org/10.1007/978-3-030-41200-5_52

location of underground services is vital to properly maintaining the existing networks or to plan new parts of any network [2], which must be designed depending on existing components. This includes the Z coordinate, since it is very important to know the depth of the network for digging ditches or similar duties.

In Sect. 2, methods for generating these semantic-rich 3D models will be analyzed. In Sect. 3, a case of study using Autodesk Infraworks in a real city will be detailed.

2 Methods for 3D Semantic-Rich Modelling of Utility Networks

Methods for modelling utility networks can be classified in two kinds: those based on standards and those based on specific software for modelling.

2.1 Frameworks

CityGML – Utility Network ADE. "CityGML (Geography Markup Language) is an open data model and XML-based format for the representation, storage and exchange of virtual 3D city models" [3]. These models contain 3D objects with their geometry, topology, semantics and appearance. However, these standardized models lack a rich information model for underground utilities.

The Utility Network ADE is and application domain extension for CityGML developed in 2012 in the Technical University of Berlin that "*provides the required concepts and classes for the integration of multi-utility networks into the 3D CityGML models*" [4], such as network connectivity and other functional aspects. As well, it can represent the topography of networks in the model.

A very useful feature is the integration with the urban 3D model itself. This way, it is possible to have not only a highly detailed and semantic-rich 3D model of utility networks, but also a part of the complete urban 3D model. This makes it possible the analysis of the different networks together with the rest of the city: energy consumption, carbon balancing, disaster management, etc.

Utility Network ADE provides classes and concepts to multiple utility networks regarding most common features of these networks. These features consist on a set of parameters, such as type of network, transported fluid, type of element, etc. All kinds of components in the network might be represented: distribution elements (pipe, channels, etc.), storage elements (water tanks, batteries, etc.) terminal or protective elements, among others. In addition, hierarchy and connectivity might be easily established, as well as the topology of the networks.

The 3D visualization of the models depends upon specific software admitting CityGML models. Commercial software supporting CityGML might be found in its Wiki. Most software for 3D representation allows web-visualization for the end user, not depending on any software, and some of them (i.e. CityGRID) are also suitable for portable devices (iOS or Android). However, this framework is appropriate for management more than for real time enquiring for maintenance or similar (Fig. 1).

Fig. 1. CityGML 3D visualization by using SIMKAS-3D. Source: Semantic 3D modeling of multi-utility networks in cities for analysis and 3D visualization [2]

IFC. IFC (Industry Foundation Classes) is an ISO standard for modelling buildings among their utilities developed by buildingSMART, and it is one of the mostly used semantic models in buildings [5]. Models are based on the relationship and connectivity between network objects. It might be used for public networks, but it is more useful for building management (e.g. BIM). However, integration between IFC and CityGML is currently being researched, in cases such as Stockholm.

IMKL. IMKL is a Dutch standard for 3D models [6] which focuses on the georeferencing of the network components and its topology; it was meant to avoid damaging utility networks when working in the public space. Although it was not conceived to be semantic-rich, it is compatible with CityGML Utility Network ADE.

2.2 Software

Here below methods based on commercial software will be shown. There are numerous utility network modelling software, but mostly 2D [5] (such as Bentley OpenUtilities), so only software with 3D capability will be mentioned.

ArcGIS Utility Network Management. ArcGIS Utility Network Management is an extension for ArcGIS Enterprise that enhances the capability of network management in GIS providing specific tools of representation and analysis of utility networks [7]. It also allows the generation of 3D models, despite of GIS being traditionally a 2D system.

This way, it is possible to have a 3D model that contains traditional information in the attached database but also provides the possibility of being represented regarding its properties.

In addition, it is possible to analyze the network and simulate different cases that might happen, such as a pipe being connected to a valve, a source of water being cut, etc. Models can also include elements other than distribution ones, such as transformers, buildings, etc. 3D visualization is possible either via web services or on portable devices.

Autodesk Infraworks 360. Autodesk Infraworks 360 is a software for designing, modelling and analyzing most of civil infrastructures. Utility networks might be modeled within Infraworks, by using the tools and 3D models provided (for light poles, manholes, pipes, etc.), or might be loaded from a range of different files, e.g. a GIS or Autodesk Civil3D file. It is able to read information attached to those input files, such as attributes from GIS, so Infraworks models can be semantic-rich.

Infraworks 360 is based on collaborative work: models might be stored in the network, so an entire team is able to work simultaneously on the same model. Additionally, end-user will be able to always access an up to date model.

Same way as in ArcGIS Network Model, components of the networks can be visualized according to their physical characteristics (i.e. a pipe may be represented according to its material and diameter, which is vital for visual understanding of the network by introducing depth perception [8]). This performance, combined with digital terrain models, orthoimages, 3D buildings or other similar urban elements, produces a model with high fidelity to the reality, which is very useful for getting oriented in the model itself.

3 Case of Study

The authors of this communication have been working on a 3D semantic-rich model of the wastewater network in the city of Leon, north of Spain. The aim of the model is to improve and update the current 2D representation by generating a 3D model including both GIS and BIM capabilities. It will be extremely useful for tasks as checking network structure and errors, interferences analysis, integration with other network components as sewage treatment plants or real-time monitoring.

3.1 Input Data

The input of the model consists on GIS data regarding the entire network of a portion of the city. This includes several shapefiles regarding both punctual and linear components of the network (mainly manholes and pipes) (Fig. 2).

Raw data had to get through a process to be imported to Infraworks. Coordinates system was transformed from ED50 datum into ETRS89. GIS file had to be analysed to find anomalies, and Z coordinates of pipes were extracted from the punctual components of the network, which had height information stored as an attribute.

Fig. 2. 2D GIS model of Leon wastewater network

After processing brute data, it was imported into Autodesk Infraworks. The generated 3D model is focused on the wastewater network, but it includes precise data of the terrain, orthoimages and 3D models of buildings with highly detailed geometry. This provides a straightway process to get easily orientated for any end-user that might not be familiar with the network model but be familiar with the environment (i.e. maintenance staff).

The result is a semantic-rich 3D model generated from GIS files. This way, pipes are represented taking into consideration their size; in addition, textures are applied to the pipes according to the material they are built of. Figure 3 shows the 3D model completely built. However, this software lacks of more in-depth analysing tools and integration with the rest of the model.

Fig. 3. Result of the 3D model in Infraworks 360

This lack of analysis capacity in Infraworks was solved by means of a preprocessing of the data in the Civil 3D programming environment. Civil 3D elements are imported directly into Infraworks. Figure 4 shows the data model of this application, which allows to show that the elements of the analyzed network (alignments and pipe elevations) are accessible numerically via software.

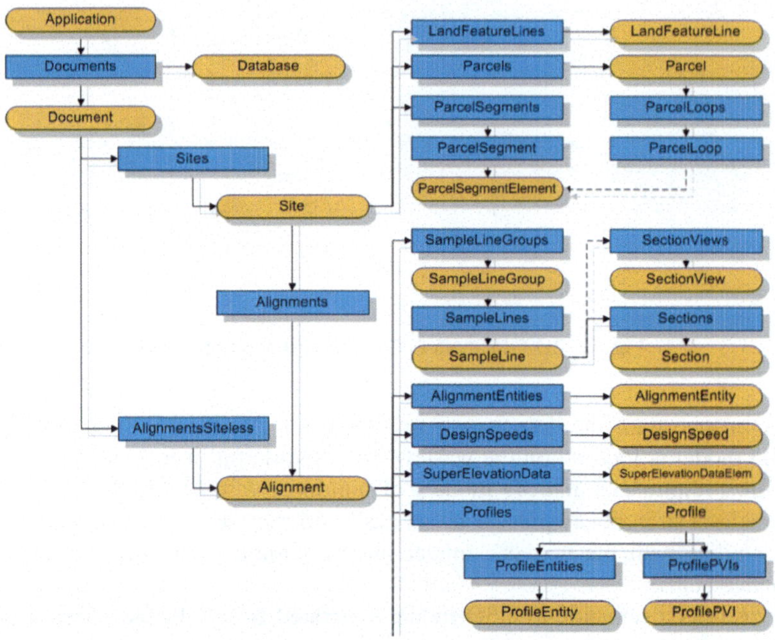

Fig. 4. Civil 3D data model

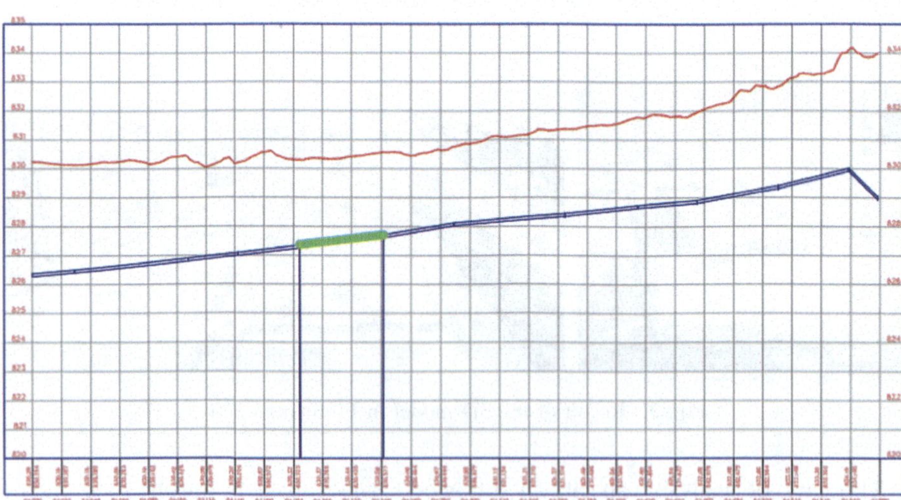

Fig. 5. Solution to a portion of the network containing faulty Z data

Through this pre-process it has been possible to verify the quality of the data collection of the network conductions and to interpolate those sections in which the existence of defective portions was revealed. Figure 5 illustrates a typical case in which a portion of the network contained faulty Z data (which was 0) and how easily can be solved when looking at the cross section of that portion of the network.

4 Conclusions

3D semantic rich models are very useful for utility management. Interest in this topic is growing up as it appears new standards (Utility Network ADE for CityGML) or software (extensions for ArcGIS, Bentley).

Having a geo-referenced model is very valuable for any construction job whereas a semantic-rich model is truly useful for management, by using analysis methods provided, or for maintenance labours, using portable devices for field work. It would also be possible to combine accurate geo-reference of the model with GPS positioning of the device itself, so these maps would be not only much more complete than traditional ones, but also easier to use.

In addition, obtaining a 3D semantic-rich model of networks is obviously easier, as many features have already been defined. Therefore, enhancing BIM models through mixed reality, where the user can interact and visualize the virtual world in a more realistic way can improve communication and problem solving between stakeholders.

References

1. Yan, J., Jaw, S.W., Son, R.V., Soon, K.H., Schrotter, G.: Three-dimensional data modelling for underground utility network mapping. Int. Arch. Photogram. Remote Sens. Spat. Inf. Sci. **42-4**, 711–745 (2018)
2. Cazzaniga, N.E., Carrion, D., Migliaccio, F., Barzaghi, R.: A share database of underground utility lines for 3D mapping and GIS applications. Int. Arch. Photogramm. Remote Sens. Spat. Inf. Sci. **40-4/W1**, 105–108 (2013)
3. Open Geospatial CityGML Frontpage. https://www.opengeospatial.org/standards/citygml. Accessed 29 Mar 2019
4. Becker, T., Nagel, C., Kolbe, T.H.: Semantic 3D modeling of multi-utility networks in cities for analysis and 3D visualization. In: Progress and New Trends in 3D Geoinformation Sciences, pp. 41–62. Springer, Berlin (2013)
5. El-Mekawy, M., Östman, A., Hijazi, I.: A unified building model for 3D urban GIS. ISPRS Int. J. Geo-Inf. **1**, 120–145 (2012). https://doi.org/10.3390/ijgi1020120
6. Den Duijin, X., Agugiaro, G., Zlatanova, S.: Modelling below- and above-ground utility network features with the CityGML utility network ADE: experiences from Rotterdam. ISPRS Ann. Photogram. Remote Sens. Spat. Inf. Sci. **4-4/W7**, 43–50 (2018)
7. ESRI Utility Network Management Frontpage. https://www.esri.es/producto/utility-network. Accessed 30 Mar 2019
8. Guerrero, J., Zlatanova, S., Meijers, M.: 3D visualization of underground pipelines: best strategy for 3D scene creation. ISPRS Ann. Photogram. Remote Sens. Spat. Inf. Sci. **2-2/W1**, 139–145 (2013). https://doi.org/10.5194/isprsannals-ii-2-w1-139-2013

Revision of Automation Methods for Scan to BIM

Julen López Iglesias[1], Jose Andrés Díaz Severiano[1], Piedad Eliana Lizcano Amorocho[1], Cristina Manchado del Val[2(✉)], Valentín Gómez-Jáuregui[2], Olmo Fernández García[1], Adrian Preciados Royano[1], and César Otero González[2]

[1] School of Civil Engineering, Universidad de Cantabria, Avda. Los Castros s/n, 39005 Santander, Spain
[2] EgiCAD Research Group, School of Civil Engineering, Universidad de Cantabria, Avda. Los Castros s/n, 39005 Santander, Spain
cristina.manchado@unican.es

Abstract. Nowadays in the BIM (Building Information Modelling) world there are several functions that are not fully developed. It is well known that, during the last years, BIM methodology, is being implemented more and more around the world and we are witnessing the benefits that can provide. Despite having plenty of benefits for engineering, there is a problem that it has not been solved yet: the correct extraction and/or recognition of the physical space represented by a point cloud, to create an accurate BIM model, which is better known as Scan to BIM. Scan to BIM, is a handwork methodology where the user tries to convert a point cloud, which can be obtained using different reality capture devices, into a 3D model by means of a BIM software. The problem is that the process is almost entirely manual, so taking into consideration the significant amount of time spent to represent complicated geometries, Scan to BIM can become a very hard work and time-consuming work. We will show in this paper some methods that are currently available to automate the process of Scan to BIM, which allows to partially reduce the manual work of the person responsible of this task. Additionally, this will make it possible to obtain a much more complete product than if it had been made without any automation, as well as to reduce consumed time to make the model.

Keywords: BIM · 3D model · Point cloud · Automation · Scan to BIM

1 Introduction

BIM (Building Information Modelling) methodology is growing in AEC (Architecture, Construction and Engineering) community. To perform a BIM 3D model for an existing project there is a very important step such as Scan to BIM.

This article summarizes some existing procedures to automate Scan to BIM process. Basically, there are two approaches, on the one hand, indoor modelling, and on the other hand, outdoor modelling. About indoor modelling, algorithms have been developed to segment the point clouds in planes to identify the different elements

F. Cavas-Martínez et al. (Eds.): INGEGRAF 2019, LNME, pp. 482–490, 2020.
https://doi.org/10.1007/978-3-030-41200-5_53

(walls, floors and ceilings). Once the plans are correctly detected, the boundaries of the elements are drawn to complete indoor modelling. In order to represent windows and doors, some algorithms have been developed to detect gaps in the segmented planes and assimilate them to doors and windows. Another approach allows to model the secondary elements of indoor modelling, creating regions of potential interest and comparing them with elements of a database (Fig. 1).

Although for outdoor modelling, previous algorithms can also being applied, a parametric procedure is presented to model a building façade following some basic architectural rules.

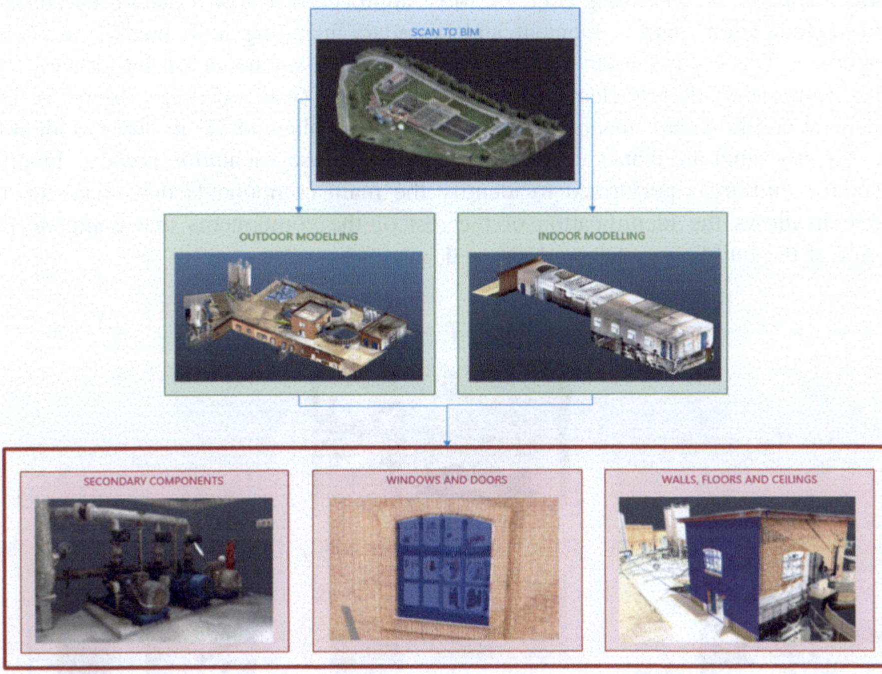

Fig. 1. Approaches for generation of BIM elements depending on their indoor/outdoor properties

2 Modelling

The outdoor modelling of a building is an important concept for improving the BIM 3D model, but much research has been done on the indoor modelling because it is the most difficult area to be represented, mainly due to occlusions. However, there is a semi-automatic approach [1] to model the exterior façade of a building from architectural rules based on the classical façades of buildings.

The methodology begins by developing a library of parametric objects that make up the different parts of a building's façade. These objects are a solid block (known as TD), a solid block with a hole inside (TW), a window (W) and finally a horizontal solid block (BL). The user can also manually modify certain parameters such as the number of floors, position of the doors, etc.

In order to be able to modify the façade in a more efficient way, it is also possible to interact graphically with the façade for moving its elements. The customized parameters are divided into three groups: the first group consists of the basic parameters of the façade (number of floors, wall thickness, etc.); the second group includes the parameters that allow to modify the classic architectural proportions of the building; and finally, the third group is used to modify any specific object of the façade in a graphic way.

About indoor modelling research, there are different methodologies to model the interior elements of a building (Fig. 2). Most approaches follow a pattern based on a point segmentation, data refinement and boundary mapping and, finally, an object recognition. The segmentation process tries to make a segmentation into planes that will constitute the different elements within the building (walls, ceilings, floors…). The refinement and boundary mapping aims to project the plans in 2D in order to identify and correctly label the planes extracted in the previous segmentation process. Finally, object recognition is performed to identify the main components that make up the model; it allows the identification of the rest of the components that complete the interior of the buildings, such as doors and windows.

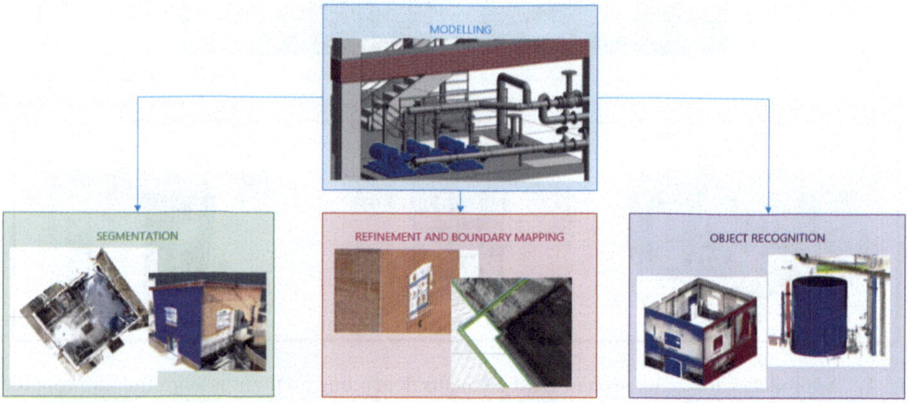

Fig. 2. Steps for modelling automatization

2.1 Walls, Floors and Ceilings

Segmentation. To perform the correct segmentation of the point cloud, numerous algorithms have been developed. However, one of them, RANSAC [2], is the one most commonly used today to identify the planes that make up the structural elements (Fig. 3). This algorithm tries to find parameters that fits the point data and filter the outlier ones.

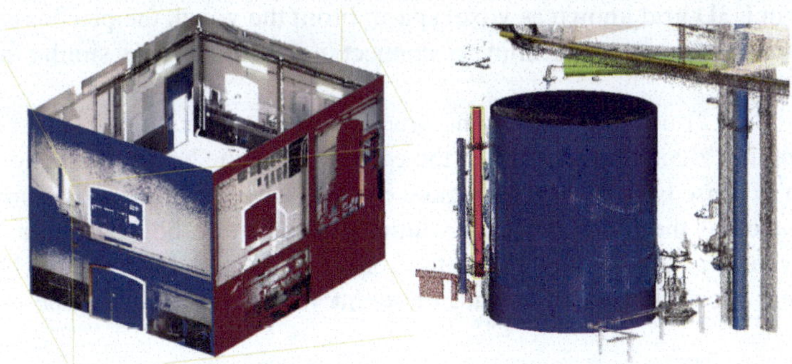

Fig. 3. Segmentation and object recognition for detection of walls and elements using RANSAC algorithm

Other approach [3] performs the segmentation of the point cloud by defining the level of abstraction desired by the user as the sole parameter. It is a region growth algorithm, which re-evaluates local characteristics during segmentation to describe the roughness and curvature of the surface of each point making a surface estimation based on the PCA (Principal Component Analysis).

The process continues searching for a seed point that will grow progressively. Once the process is finished, another seed point is necessary to start with another segment. The level of abstraction increases until all possible radius is exhausted; then, the threshold of roughness increases until the moment that the roughness equals the value of the level of abstraction, which would conclude the process. The approach presents a challenge because as the threshold of roughness increases, low density point clouds may not be detected correctly, so this approach is valid for high density clouds.

In order to specify the segmentation, several authors have added procedures that make it easier to determine the components. Huber et al. [4], Khoshelham and Díaz-Villariño [5] and Oeseau et al. [6], determine floors and ceilings using a histogram of the distribution of points along the z-axis. Ochmann et al. methodology [7], is based on the probability that a point belongs to a room in order to segment it.

All these procedures refer to the complete segmentation of the point cloud; however, there are other techniques where several segmentations are performed sequentially such as the one by Thomson and Boehm [8] and Valero et al. [9], which perform a horizontal segmentation for the ceiling and floor and another vertical segmentation for the wall planes.

Jung et al. [10] performs the segmentation from a hypothesis where a room is a segment surrounded by vertical walls and linked to other rooms by open doors. Segmentation is performed using a circular-shaped detection window that checks in 2D whether the cells of a binary map are occupied or not. In the case of occupied cells, the algorithm incorporates them to the segment. The approach is very limited because sloped walls are not correctly detected and does not consider elements such as stairs.

There is a more automatic algorithm [11] for segmentation of the point cloud, automatically detecting the extracted planes and classifying them as floors, ceilings or walls. The algorithm considers the rest as noise and excludes it. To do that, the data is discretized in a 3D grid known as voxel space. From the voxel, the patches are detected using a regional growth algorithm to connect points that have similar normal flat surfaces.

Hong et al. [12] use a pseudo-point structure for segmentation. The pseudo-points have the implicit z-coordinates, so that the estimation of the height of the floors and the elimination of noise is carried out in a more efficient way. Each pseudo-point has stored the coordinates of the points that fall within a defined threshold. The maximum and minimum values of z would represent the ceiling and the floor respectively. If points with different z-coordinates are found, the point would automatically be classified as noise.

Refinement and Boundary Mapping. After the segmentation process, there may be points that are not classified correctly and may be misleading. In order to avoid it, a refining and boundary mapping process is carried out to eliminate the noisy points for the correct labelling of the extracted segments.

Most methods are based on a binary image for specifying whether the cells are occupied or not. For labelling, the process checks how many cells are interconnected to determine the set of them that make up the segment. The refining method depends on the level of precision that the user wants, bearing in mind that a greater accuracy in the refining takes an increase in calculation time. Conversely, lower accuracy may lead to errors as it may include noisy points in the segment. The segment limit is carried out by starting with a point on a corner previously labelled as occupied in the binary image and connect adjacent occupied cells to form the boundary of the element.

Noisy points are processed by Huber et al. [4], Previtali et al. [13], and Xiong et al. [14] using ray tracing labelling. The algorithm tries to create a voxel space for each scanning station. In order to know the state of the voxel (occupied, occluded or empty) it is necessary to analyse the position of the points in relation to the scanner and the surface. Finally, they apply an interior painting algorithm to fill in the occluded regions.

Refining in segmentation can also be done using Douglas-Peucker's algorithm [15], which begins with segmentation of the extracted boundary between the wall and the floor into separate lines from the wall surface. The lines are represented by Hesse's line model and the algorithm automatically extracts the corners of the wall assuming the intersection of two lines at a right angle.

2.1.1 Object Recognition

Most automatic modelling methods include this stage in the segmentation process, but there are methods that are not able to do it after the segmented and refined point cloud to establish the relationship between the extracted geometry and the object to be modelled. This is the reason why there are different processes that try to assign an element (wall, floor or ceiling) to each extracted geometry.

The most used method establishes the relationship between the plane and the object by its topology. Thus, Nguyen et al. [16] presented a methodology that was able to automatically deduce the existing topological relationships of the elements (walls, ceilings and floors). The relationships include separation, intersection, adjacency, containment and connectivity. In this way, a 3D computer-aided design system (3D

CAD) or a solid modeler is used to detect the topological information of the elements and by means of algorithms and a prototype system with the relationship between them is generated.

2.2 Windows and Doors

The modelling of windows and doors is a less developed process than that of the main components (doors, windows and ceilings). The problem is that when taking data with the scanners, the transparent surfaces of the window generates noisy points taken from the outside. As a result, the segmentation of points for the windows modelling can become erroneous. Therefore, the modelling of windows and doors is considered a separate operation.

For the modelling of windows and doors, the binary image obtained in the segmentation of the point cloud can be used. Thus, the interior rectangular areas in the walls that are considered empty are the areas where there may be a door or window. To do this, there are different options: Douglas-Peucker's algorithm [15] detects the limit of the area and generates a segmentation in order to extract the plane of the window. Other softwares such as Libsvm [17] automatically detect openings such as doors and windows through a supervised learning process.

Pu and Vosselman's approach [18] is based on a triangulation method that detects the boundaries of scattered regions on walls and adjusts them to rectangles to form windows or doors. This approach is more appropriate for point clouds with similar density in all areas. About window modelling, Böhm et al. [19], use edge detection based on the density of points on the façade to look for horizontal and vertical lines that form the window.

About door modelling, Faber and Fischer [20] use an algorithm to adjust parametric models of door point clouds from a previously segmented model. It is a method that requires a lot of point density and is valid for rectangular and arched doors.

All the procedures described in this section can also be used in outdoor modelling to detect windows and doors on the façade.

2.3 Secondary Components

The approaches presented so far, have focused on the modelling of main components but inside buildings there are secondary components (such as switches, extinguishers, etc.) that can distort the segmentation of the main elements.

In order to detect them, there is an algorithm [21] that is based on the colour and depth of the objects. It starts by converting each previously segmented element into a 4D orthoimage (J_{CD}) in which each pixel has colour and depth information. This orthoimage is divided into colour image (J_C) and depth image (J_D) in order to work independently.

On the one hand, the detection of the secondary components using the depth filter is done by means of strategies of recognition of the corresponding potential regions of interest, which are calculated using the Canny algorithm [22] that can detect objects with geometric discontinuities. The detected objects are compared with the models available in the database and in case of finding a similarity the process would have finished successfully.

On the other hand, using the colour filter, colour discontinuities are detected in the elements to define the regions of potential interest. To do this, the previously calculated colour image is transformed into a binary image in order to find compact sets of pixels that can be regions of potential interest. To finalize the process, the potential regions of interest are compared with the available database using global descriptors.

However, despite witnessing what the method does, we must also highlight its limitations, for example, for inclined walls or indoors with many objects the method is not so accurate.

3 Conclusions and Future Work

This paper presents different approaches for reconstruct building elements from a point cloud but they have some shortcomings that still need to be improved. Figure 4 shows a summary of them.

Authors / Automation methods	Dore C. et al	Schnabel R. et al	Dimitrov A. et al	Huber D. et al	Oesau S. et al	Khoshelham K. et	Ochmann S. et al	Thomson C. et al	Valero E. et al,	Junj J. et al.	Hong S. et al	Previtali M. et al	Xiong X. et al.	Douglas D.H. et al	Nguyen T.H. et al	Chan C.C. et al.	Pu S. et al	Böhm J. et al.	Faber P. et al.	Adan A. et al.	Canny J.
Walls detection	X	X					X				X	X									
Floors and ceilings		X		X	X	X	X	X	X		X			X	X						
Windows, doors		X	X											X		X	X	X	X	X	X
Refinement (noise)				X								X	X								

Fig. 4. Summary of different approaches presented

About indoor modelling, all the approaches exposed start from a simple building where the rooms to be modelled are simple rectangular areas surrounded by vertical walls. But buildings usually have more complex shapes such as pillars in the middle of the rooms or protruding pillars in the walls.

There is an analogous problem related to modelling windows and doors: all studied approaches are based on modelling indoor movable panels with simple rectangular shapes, which can be far from reality as there is a lot of variety of windows and doors in buildings. In the future, these procedures will need to be improved to obtain a more accurate 3D BIM model. The algorithm for modelling secondary objects shown in Sect. 2.3 can be of great help in solving the problems discussed above. If the database of objects to be recognized by this procedure increases, it will be possible to detect elements such as pillars, beams, etc. and will be able to develop a more complete interior modelling.

In terms of outdoor modelling, the current approach is very limited, since it is established upon basic architectural rules; moreover, the façades of the buildings are nowadays complex elements. As a result, using the proposed approach will make it very difficult to represent the façades correctly.

In next steps, it is essential to implement the different algorithms previously reviewed on a BIM software in order to compare them and evaluate the detection methods in terms of efficiency and precision among others. Also to investigate about a better segmentation method of the point clouds, maybe artificial intelligence or new genetic algorithms could be developed to perform the Scan to BIM more accurately and reducing the consumed time.

References

1. Dore, C., Murphy, M.: Semi-automatic generation of as-built BIM façade geometry from laser and image data. Electron. J. Inf. Technol. Constr. **19**, 20–46 (2014)
2. Schnabel, R., Wahl, R., Klein, R.: Efficient RANSCAC for point-cloud shape detection. In: Computer Graphics Forum, pp. 214–226. Blackwell Publishing Ltd., Oxford (2007). https://doi.org/10.1111/j.1467-8659.2007.01016.x
3. Dimitrov, A., Golparvar-Fard, M.: Segmentation of building point cloud models including detailed architectural/structural features and MEP systems. Autom. Constr. **51**, 32–45 (2015). https://doi.org/10.1016/j.autcon.2014.12.015
4. Huber, D., Akinci, B., Stambler, A., Xiong, X., Anil, E., Adan, A.: Methods for automatically modelling and representing as-built building information models. In: Proceedings of the 2011 NSF Engineering Research and Innovation Conference, Atlanta, Georgia (2011)
5. Oesau, S., Lafarge, F., Alliez, P.: Indoor scene reconstruction using feature sensitive primitive extractions and graph-cut. ISPRS J. Photogramm. Remote. Sens. **90**, 68–82 (2014). https://doi.org/10.1016/j.isprsjprs.2014.02.004
6. Khoshelham, K., Díaz-Vilariño, L.: 3D modelling of interior spaces: learning the language of indoor architecture. In: Proceedings of the ISPRS Technical Commission V Symposium, Riva del Garda, Italy, vol. XL-5, pp. 321–326 (2014). https://doi.org/10.5194/isprsarchives-XL-5-321-2014
7. Ochmann, S., Vock, R., Wessel, R., Tamke, M., Klein, R.: Automatic generation of structural building descriptions from 3D point cloud scans. In: Proceedings of the 9th International Conference on Computer Graphics Theory and Applications (GRAPP), Lisbon, Portugal (2014)
8. Thomson, C., Boehm, J.: Automatic geometry generation from point clouds for BIM. Remote Sens. **7**(9), 11753–11775 (2015). https://doi.org/10.3390/rs70911753
9. Valero, E., Adán, A., Cerrada, C.: Automatic method for building indoor boundary models from dense point clouds collected by laser scanners. Sensors **12**(12), 16099–16115 (2012). https://doi.org/10.3390/s121216099
10. Jung, J., Stachniss, C., Ju, S., Heo, J.: Automated 3D volumetric reconstruction of multiple-room building interiors for as-built BIM. Adv. Eng. Inform. **38**, 811–825 (2018). https://doi.org/10.1016/j.aei.2018.10.007
11. Ochmann, S., Vock, R., Wessel, R., Klein, R.: Automatic reconstruction of parametric building models from indoor point clouds. Comput. Graph. **54**, 94–103 (2016). https://doi.org/10.1016/j.cag.2015.07.008
12. Hong, S., Jung, J., Kim, S., Cho, H., Lee, J., Heo, J.: Semi-automated approach to indoor mapping for 3D as-built building information modelling. Comput. Environ. Urban Syst. **51**, 34–46 (2015). https://doi.org/10.1016/j.compenvurbsys.2015.01.005

13. Previtali, M., Barazzetti, L., Brumana, R., Scaioni, M.: Towards automatic indoor reconstruction of cluttered building rooms from point clouds. In: Proceedings of the ISPRS Technical Commission V Symposium, Riva del Garda, Italy, vol. II-5, pp. 281–288 (2014). https://doi.org/10.5194/isprsannals-II-5-281-2014
14. Xiong, X., Adan, A., Akinci, B., Huber, D.: Automatic creation of semantically rich 3D building models from laser scanner data. Autom. Constr. **31**, 325–337 (2013). https://doi.org/10.1016/j.autcon.2012.10.006
15. Douglas, D.H., Peucker, T.K.: Algorithms for the reduction of the number of points required to represent a digitized line or its caricature. Int. J. Geogr. Inf. Geovis. **10**(2), 112–122 (1973). https://doi.org/10.3138/FM57-6770-U75U-7727
16. Nguyen, T.H., Oloufa, A.A., Nassar, K.: Algorithms for automated deduction of topological information. Autom. Constr. **14**(1), 59–70 (2005). https://doi.org/10.1016/j.autcon.2004.07.015
17. Chang, C.C., Lin, C.J.: LIBSVM: a library for support vector machines. ACM Trans. Intell. Syst. Technol. (TIST) **2**(3), 1–27 (2011). https://doi.acm.org/10.1145/1961189.1961199
18. Pu, S., Vosselman, G.: Extracting windows from terrestrial laser scanning. In: ISPRS Workshop on Laser Scanning, pp. 320–325 (2007)
19. Böhm, J., Becker, S., Haala, N.: Model refinement by integrated processing of laser scanning and photogrammetry. In: Proceedings of 3D Virtual Reconstruction and Visualization of Complex Architectures, Zurich, Switzerland (2007)
20. Faber, P., Fischer, B.: How can we exploit typical architectural structures to improve model recovery? In: 3D Data Processing Visualization and Transmission, pp. 824–833 (2002)
21. Adán, A., Quintana, B., Prieto, S.A., Bosché, F.: Scan-to-BIM for ‘secondary’ building components. Adv. Eng. Inform. **37**, 119–138 (2018). https://doi.org/10.1016/j.aei.2018.05.001
22. Canny, J.: A computational approach to edge detection. IEEE Trans. Pattern Anal. Mach. Intell. **8**(6), 679–698 (1986). https://doi.org/10.1109/TPAMI.1986.4767851

Modelling As-Built MEP Facilities in a BIM Environment

Olmo Fernández García[1], Jose Andrés Diaz-Severiano[1], Valentín Gómez-Jáuregui[2(✉)], Cristina Manchado del Val[2], Julen López Iglesias[1], Adrian Preciados Royano[1], and César Otero González[2]

[1] School of Civil Engineering, Universidad de Cantabria, Avda. Los Castros s/n, 39005 Santander, Spain
[2] EgiCAD Research Group, School of Civil Engineering, Universidad de Cantabria, Avda. Los Castros s/n, 39005 Santander, Spain
valen.gomez.jauregui@unican.es

Abstract. BIM (Building Information Modeling) is a working methodology that centralizes all the information of a project in a single digital environment. The data are shared between all the agents of the project and contains not only the physical characteristics of the facility, but also the functional ones.

BIM is also being pushed by European governments; it is already a requirement for public projects in several countries, such as the United Kingdom or Denmark among others. In Spain, BIM is required for architectural projects since 2018 and it will be compulsory for all infrastructure projects in 2019. When applying the BIM methodology to MEP (Mechanical Electrical and Plumbing) projects, it makes maintenance easier, as it is a digital model of the installation that considers the operation period. Other benefits include clash detection and easier connections. Due to all these factors, there is a demand for making BIM models for facility management in already built installations, either to comply with legal requirements or to cheapen the operational costs.

Modelling the piping or finding and identifying the equipment of these installations can be a tedious process; therefore, this study aims to present a summary of the state of the art and a case study in modelling as-built MEP facilities in a BIM environment.

Keywords: BIM · MEP · Point cloud · As-built modeling

1 Introduction

Mechanical, Electrical and Plumbing (MEP) systems are vital for the normal functioning of buildings and facilities. They also require maintenance and imply operation costs during all their lifetime. As the operation phase is the longest in a facility's lifetime, those costs are, according to previous research [1], up to 60% of the total project cost.

One of the advantages of the BIM methodology is making the operation and maintenance of a facility easier, as the MEP information is already in a digital

F. Cavas-Martínez et al. (Eds.): INGEGRAF 2019, LNME, pp. 491–498, 2020.
https://doi.org/10.1007/978-3-030-41200-5_54

document. In the event of an incidence, the BIM model is much easier to manage than the individual documents where the information is traditionally stored, saving time searching among them. In the event of a leak in a pipe, visualizing a 3D model is a faster and more intuitive way to localize it, helping to close the upstream valve as soon as possible. In the event of modifications or expansions of the facility, having the as-built information minimizes the risk of unexpected problems.

BIM models of new public facilities are, in fact, compulsory in all public projects in the UK, the Netherlands, Denmark, Norway, Sweden and Finland. There are also plans to reach that level of exigence in Spain, France, Germany, Russia and Australia. In other countries, such as the USA, Canada, Brazil and China, the usage of BIM is widespread in most projects [2]. However, the BIM models are not always as accurate as they should; some studies have advocated for requiring an actual as-built model, when the construction phase ends, that contains all the changes and modifications that have been made to the initial planning [3].

However, many already existing facilities were not built using BIM methodology, so there is not a digital model that can be used for maintenance coordination. This paper studies the different methodologies currently available for creating an as-built BIM model for already existing facilities. An added benefit of creating a digital model of the MEP system is finding any discrepancies between the plans and the actual ubications of the elements that could have been modified/overlooked during the building phase or changed in later maintenance operations (Fig. 1).

Fig. 1. As-built BIM model with the overlaying point cloud used to build it.

2 Geometrical Data Management

2.1 Obtaining Geometrical Data

In the construction phase of projects, there is already a need of accurate tracking of the as-built structure, to find any differences between the already built elements and the layouts of the original project. The same methodology used for that purpose can be used to obtain a digital model of an executed facility. The technologies used include photogrammetry and three-dimensional terrestrial laser scanning (TLS) [4].

From all those different options currently available, the one widely considered as the most suitable for modelling as-built MEP facilities is the TLS, as it offers high precision and speed. This method outputs a point cloud that must be processed to obtain an as-built model that can be used in a BIM environment.

However, TLS has limitations; some areas are inaccessible for the equipment and this fact produces gaps in the point cloud, requiring some other method to be defined, such as handheld scanners.

2.2 Processing Geometrical Data

The process of building manually all the BIM objects of a facility, from the point cloud data, is a tedious and time-consuming process; however, there are some techniques that can automate the process to a certain degree [5, 6].

Scan-to-BIM is an automated process that can create a BIM geometrical model from the TLS scanning. Some software companies have started to include semi-automatic modelling capabilities in their Scan-to-BIM software, but the results that are obtained from them still need a lot of manual supervision and corrections [7].

If the original construction plans of the facility are available and there have not been major deviations from them after the construction or in subsequent modifications, Scan-vs-BIM methodologies can be applied to the project. While Scan-to-BIM works by processing the point cloud to find shapes and building upon them, Scan-vs-BIM compare the shapes found in the point cloud with a BIM representation of the facility that is known or suspected to be inexact. This extra step allows Scan-vs-BIM to be more precise (Fig. 2).

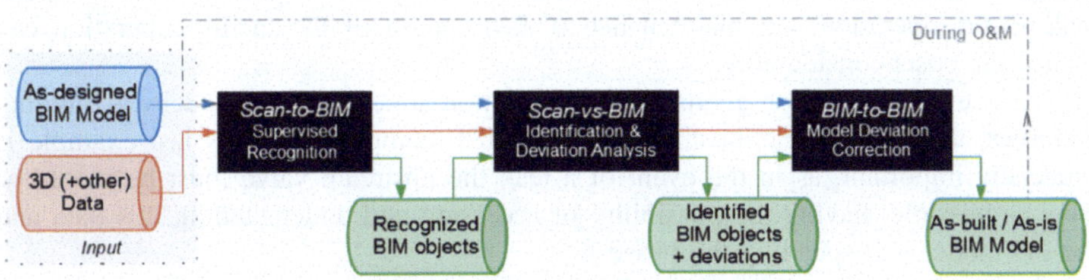

Fig. 2. Scan-vs-BIM process [8].

The original blueprints are used as a base for a BIM model of the facility. This model is then used as a base for a comparison between the planned layout of the facility and the actual position of all its elements, given by the laser scan.

When there is no reliable data about the facility, for example due to outdated or unreliable construction plans, the scan-vs-BIM step cannot be applied.

The Hough transform [9] is a technique used for detecting arbitrary shapes in greater images or pictures. With some modifications, it can be utilized for finding parametric features in raw point clouds, automatically detecting the pipes. The methodology assumes that the pipes are parallel to the main axes of the facility. Then, a series of parallel planes are created, and the program tries to find circles of the expected diameter (the diameter of the pipes) [10].

A curvature-based shape descriptor can be used as an alternative to the Hough transform to find pipe-shaped objects in the point cloud. It works by estimating the curvature of the pipe and then applying clustering and bag-of-features matching to locate shapes similar to a given example [11].

Other option is using the geometric primitives of the point cloud. With the information from the construction plans it is possible to obtain the radius and the axis position. Then, the program searches for geometric primitives that correspond with those values in the point cloud [12].

In any case, automated scan-to-BIM and scan-vs-BIM are still limited in their scope. Some frequent problems that will require manual refinement are:

- Occlusion of pipes. When part of a pipe is not visible in the point cloud, automated detection usually recognizes the previous and next segments of the pipe as different ones.
- Elbows. Again, detection of elbows as a part of one pipe is problematic, tending to consider it as two different ones.
- T-connections. Pipes are recognized as different separate segments instead of as a connected network.

2.3 Non-geometrical Data

Storing geometrical data from the facility is, however, far from being the only thing expected from a BIM model. One of the biggest strengths of the BIM methodology is the ability to have a detailed 3D map of the facility with precise and easily available non-geometrical information. In the case of MEP systems, information that helps making the operation and maintenance (O&M) phase of the facility operation can reduce the costs considerably.

Some examples of non-geometrical information are equipment capacities, building materials, and logical relationships between MEP components. This last example is especially important, as in the event of a leak the upstream valve must be closed as soon as possible; having the possibility of accessing and understanding this data and the 3D model will speed up the process.

This input data must be entered manually, usually by assigning codes to each element and connecting them to a database with all the pertinent information.

3 A Case Study of a WWTP

3.1 Introduction

The wastewater treatment plant (WWTP) object of this study has a total equivalent population of 300.000. It does pretreatment, primary treatment and secondary treatment with an average flow of approximately 4500 m^3/h.

The main objective of the project is to create a BIM model of the plant that can be used for Facility Management. The current management uses a CMMS (computerized maintenance management system) that will be linked from the BIM model to make the change as seamless as possible for the maintenance personnel.

Geometrical data of this facility was obtained by a combination of aerial photogrammetry and TLS capture. Related to TLS, the scanner was located at key points to avoid blind spots as much as possible. The points that could not be reached by the scanner were photographed as a way to obtain additional data.

3.2 Building the Model

The BIM software used in this project is Revit, with Autodesk Recap as a supplementary for processing the point cloud, checking pipe diameters and having spherical real view images of the facility as references. As the project is a relatively small one, most of the BIM model creation will be done manually, but automatic modelling systems will be considered for locations with long, straight pipes such as corridors.

There are construction plans of the facility available, but some complications during the building phase and a couple of modifications at a later stage meant that they are not reliable enough to be used as a base for the BIM model. Nevertheless, these construction plans are useful for identifying the different equipment and assigning them their corresponding codes and properties in the final model.

The current working methodology is as follows:

- First, auxiliary planes are located parallel to each building's walls, as those directions are the most commonly followed by the piping. This will be done for every building in the plant.
- Those planes are either parallel or perpendicular to the pipes, something that makes much easier placing a pipe object following the point cloud data. Locating a plane perpendicular to a pipe also facilitates calculating the pipe diameter, although the diameters were cross-checked with the pipe snap tool of Autodesk Recap (Fig. 3). This tool is intended for measuring distances between the surface of a pipe and another point, but its automatic recognition of pipes and their diameters is useful for this project. At this point, it is already possible to model every pipe in the facility in its place.

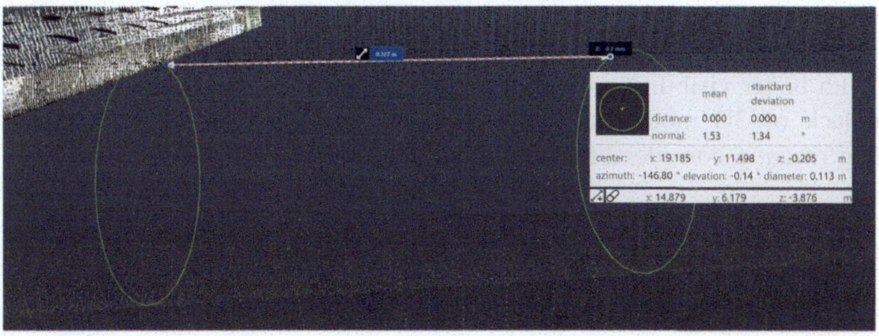

Fig. 3. Using the Pipe Snap Recap tool to obtain the diameter of a pipe

- For the pumps, pressurizers, boilers and other specialized equipment it is necessary to build a custom 3D model for each one of them, as the manufacturers did not have models available at the time of the project. An extensive web search was not enough to find models for WWTP-specialized equipment, either. These equipment models are made in situ, using the point cloud data as a reference.
- Finally, each element of the model is given its corresponding code to make them more easily identifiable, and they are linked to the CMMS database with all the pertinent information to make operation and maintenance easier.

3.3 Problems and Difficulties

There are several processes that proved to have some special complications in their execution. Modelling the elements that were occluded from the TLS with not completely reliable project plans and some photography, as well as some leftovers of the laser scanning is possible, but much slower due to the need of checking the photographs several times to make sure that the position of the missing parts was right (Fig. 4).

Fig. 4. A shadow produced by a staircase. The modelling was performed assisted by photographs.

In some cases, especially when horizontal pipes are very close to both the wall and the ceiling, calculating their diameters is more complicated. Comparing several different planes, checking the construction plans and searching for a point up or downstream with better visibility proved to be enough to solve this problem (Fig. 5).

The lack of 3D models of the equipment and the subsequent need of in-site modelling meant a considerable time investment.

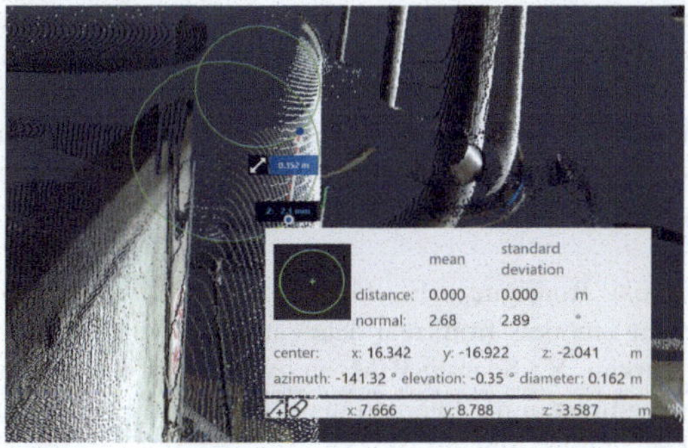

Fig. 5. A pipe difficult to measure that required cross-checking the diameter.

4 Conclusions

While automated systems for pipe detection are being developed, they still do not work reliably outside some very favorable cases, requiring straight line pipes parallel to the walls of the facility and high-quality renders without occlusions. Even if those requisites are met, human supervision of the process is still needed.

There is a promising future for these methods, however; especially for the scan-vs-BIM methodology, which gives the program more data to work with and can be much more precise, once the problems with occlusions and intersections are solved.

For smaller scope projects that cannot afford the required computational power or for those cases where there is not a reliable initial BIM model, however, manual modelling of the facility will still be needed. Even in bigger projects, some especially conflictive zones, due to inaccessibility or complicated geometry, may require manual intervention.

The problem that signified the biggest time loss, however, was the need of building individual models for each of the equipment of the WWTP; although some manufacturers are already providing 3D models of their machines, this is not yet a widespread practice. It would benefit both the as-built industry and the new built facilities, as they are or will be required to provide a BIM model in more countries each year.

References

1. Hu, Z.-Z., Tian, P.-L., Li, S.-W., Zhang, J.-P.: BIM-based integrated delivery technologies for intelligent MEP management in the operation and maintenance phase. Adv. Eng. Softw. **115**, 1–16 (2018)
2. Building Smart Spain. https://www.buildingsmart.es/bim/qu%C3%A9-es/. Accessed 01 Apr 2019
3. Lin, Y.-C., Lin, C.-P., Hu, H.-T., Su, Y.C.: Developing final as-built BIM model management system for owners during project closeout: a case study. Adv. Eng. Inform. **36**, 178–193 (2018)

4. Bosché, F., Ahmed, M., Turkan, Y., Haas, C.T., Haas, R.: The value of integrating Scan-to-BIM and Scan-vs-BIM techniques for construction monitoring using laser scanning and BIM: the case of cylindrical MEP components. Autom. Constr. **49**, 201–213 (2015)
5. Manchado, C., Varela, E., Gómez-Jáuregui, V., Otero, C.: A BIM model from reality capture data. In: Proceedings of the International Joint Conference on Mechanics, Design Engineering & Advanced Manufacturing, JCM 2018, 20–22, Cartagena, Spain, pp. 60–61, June 2018
6. Manchado, C., Varela Rodriguez, E., Gómez-Jáuregui, V., Otero, C.: BIM (building information modelling) models from real data capture. Dyna **94**(2), 137–139 (2019)
7. Son, H., Kim, C., Turkan, Y.: Scan-to-BIM–an overview of the current state of the art and a look ahead. In: Proceedings in the International Symposium on Automation and Robotics in Construction, IAARC, Oulu, Finland (2015)
8. Ahmed, M., Haas, C.T., Haas, R.: Automatic detection of cylindrical objects in built facilities. ASCE J. Comput. Civ. Eng. **28**, 04014009 (2014)
9. Rabbani, T., van den Heuvel, F.: Efficient Hough transform for automatic detection of cylinders in point clouds. In: Workshop "Laser scanning 2005", Enschede, The Netherlands, June 2005
10. Ahmed, M, Haas, C.T., Haas, R.: Autonomous modeling of pipes within point clouds. In: Proceedings of the 30th ISARC, Montréal, Canada, pp. 1093–1100 (2013)
11. Czerniawski, T., Nahangi, M., Haas, C., Walbridge, S.: Pipe spool recognition in cluttered point clouds using a curvature-based shape descriptor. Autom. Constr. **71**, 346–358 (2016)
12. Maalek, R., Lichti, D.D., Walker, R., Bhavnani, A., Ruwanpura, J.Y.: Extraction of pipes and flanges from point clouds for automated verification of pre-fabricated modules in oil and gas refinery projects. Autom. Constr. **103**, 150–167 (2019)

Geometric Analysis of the Windows of Cordoba's Minarets (Spain)Contribution Title

Rafael E. Hidalgo Fernández[1(✉)], Rafael Ortiz-Cordero[2], Jerónimo Sanz Cabrera[1], Pilar Carranza Cañadas[1], Enrique Burgos Ladrón De Guevara[1], and Paula Triviño Tarradas[1]

[1] Dpto. de Ingeniería Gráfica y Geomática, Universidad de Córdoba, Campus de Rabanales, 14071 Córdoba, Spain
iglhifer@uco.es

[2] Ingeniero Cabildo Catedral de Córdoba, C/Cardenal Herrero, 1, 14003 Córdoba, Spain

Abstract. The main objective of this work is developing step by step the geometric drawing of the windows of Cordoba's minarets (Spain). For this purpose, we have used basic geometric elements such as circumferences. Furthermore, the measures used were always in cubits, the type of units used during the Islamic period. This paper shows that there is a sketch design for minarets' windows using circles and cubits as some basic elements for the design. Thus, from circles, intersections and re-markable points we are able to carry out the design of a window which will have the same dimensions as those currently exiting in the minarets of the Church of Santiago, Córdoba and the Church of St. Juan de los Caballeros, Córdoba.

Keywords: Minarets · Geometry · Cubit · Relations · Al-Andalus · Windows

1 Introduction

After the fall of the Roman Empire the mathematical knowledge was reduced to its basics, Euclidean geometry and Pythagorean mathematics. In this era, few people could study geometry and mathematics [1]. From the IX to the XII centuries, the classical mathematical knowledge was spread along Al-Andalus. Islamic constructions were characterized by the use of integers and the incorporation of basic geometric forms on their designs and constructions. In fact, the base and height's measures of many minarets are integers, for instance, Hishām I minaret or Madinat al-Zahra' minaret, Cordoba; both of them have a base of ten cubits and a height of forty cubits, [2, 3 and 4].

Likewise, the minarets architecture is very simple; it has a square ground floor, with a circular or square inner disposition and a second sub-construction with the same design. The first study about geometric analysis and proportions at Cordoba's Mosque-Cathedral was written by Emilio Camps Cazorla [5]. His study followed an important premise, analysis and proportions must be simple. Only two sketches from the Islamic

F. Cavas-Martínez et al. (Eds.): INGEGRAF 2019, LNME, pp. 499–506, 2020.
https://doi.org/10.1007/978-3-030-41200-5_55

era are kept; the first was found by Ricardo Velázquez Bosco, architect of Cordoba's Mosque-Cathedral, in Madinat al-Zahra', which was found incomplete in a wall. The second sketch was found by Manuel Gómez Moreno during the restoration of the Holy Ark of Oviedo, Fig. 1. The Ark suffered the 1934 riots when a bomb destroyed part of the Church. This sketch shows a lobed arch with the compass layout in the wood. It is dated back in the XI century [5].

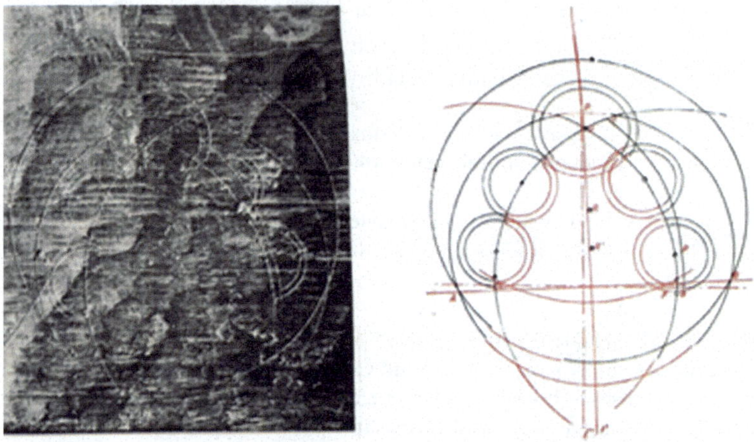

Fig. 1. Sketch of the Holy Ark of Oviedo [5].

During the constructions carried out in the Middle Ages diverse units of measurement were used such as the rod, the finger, the span, the toesa, and the cubit [6]. Arab writers wrote about the measures of the old mosque of Cordoba and the rest of constructions carried out during the Islamic occupation. The architect Félix Hernández studied all the measures of Cordoba's Mosque Cathedral and wrote a book about this study [7]. As might be expected, the work of Hernández is a great help to know the design and construction of this building and other built around the same era.

An important study of the proportions of the minarets built in Al-Andalus was conducted by Pavón Maldonado [8]. First, the author studied of proportions in the merlons found by him and Félix Hernández in Madinat al-Zahra'. Maldonado classifies merlons into two groups, big and small. Big merlons were used for the lower part of the minarets and small merlons were used for the top part of the minarets. In both cases merlons have a proportion of 1/2 between the base and the height. By contrast, the merlons located in the main zones of the mosques have a proportion of 0.88, a proportion very close to 7/8 or 0.875.

Pavón obtained in his study important conclusions about the proportions be-tween the base and the height of minarets in different zones and different construction eras. All of these relations and proportions studied before by different re-searchers in architecture, archaeology and art, enable us to study the potential existence of a model for the design of the minarets' windows.

At present the oldest window of the minarets is located in the minaret of Ibn Adabbas' Mosque, now Church of Salvador, Seville. This window was studied by Hernández [9]. The architect studied the different measures and relations between the elements that are part of the window. Furthermore, Hernández analyzed the windows of the minaret of the actual Church of San Juan de los Caballeros, Córdoba, its relation between sagitta and diameter is 7/8 (0,875), the same relation and dimensions of the windows of the Church of Santiago's minaret, Córdoba [10]. In his study [8] Pavón found the same relation in the merlons discovered in the Mosque of Madinat al-Zahra', Córdoba.

2 Methods

In this chapter we are dealing with the process of drawing step by step the windows of 7/8 typical of the Islamic architecture. To this end, Emilio Camps studied the geometry of different parts of Córdoba's Mosque-Cathedral, making use of geometric relations between circles, rectangles and lines.

The drawing has been carried out using an educational version of Autodesk Autocad 2016 software [11]. Furthermore, the montage of the images have been done with GIMP 2, a free software [12].

Firstly we begin drawing a circle with center A and a diameter of 4 mamuni cubits, (47.14 cm/cubit), 188.56 cm in total. It has been chosen this type of cubit because it is the most common cubit used along the Islamic constructions in Al-Andalus [7], Fig. 2, step 1.

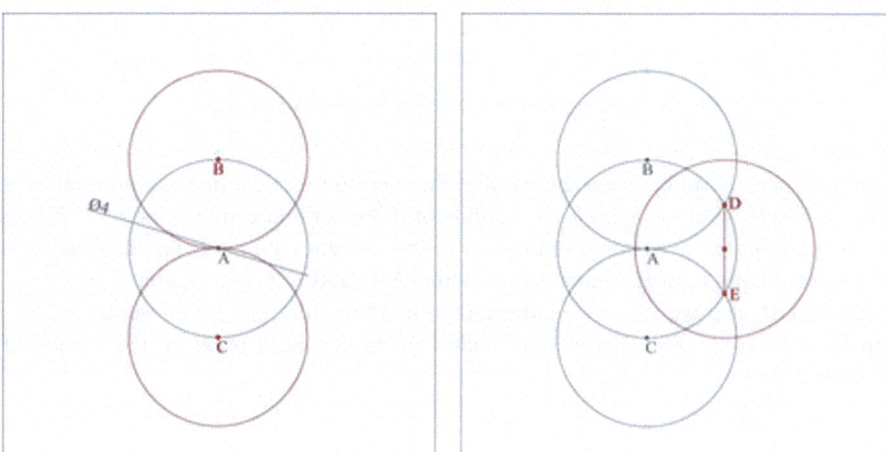

Fig. 2. Tracing an Islamic window. Step 1, step 2 and step 3.

Secondly we trace two circles with equal diameters to the first circle, in this case we use the points B and C for the center of the two new circumferences. These points are located above and below the intersection between the first circle and the vertical line from point A, Fig. 2, step 2.

Then we trace a line connecting points D and E, which are the intersection point of circumferences with centers B and C with first circle. We draw a circle of 4 cubits of diameter and center in midpoint of segment DE, Fig. 2, step 3.

In fourth step we draw a rectangle FGHI which upper and lower midpoints (Fo and Go) are the intersection points between the first circle and the circle number four. Furthermore, we trace a parallel line and tangent to point B, point of tangency of the first circumference. Figure 3, step 4.

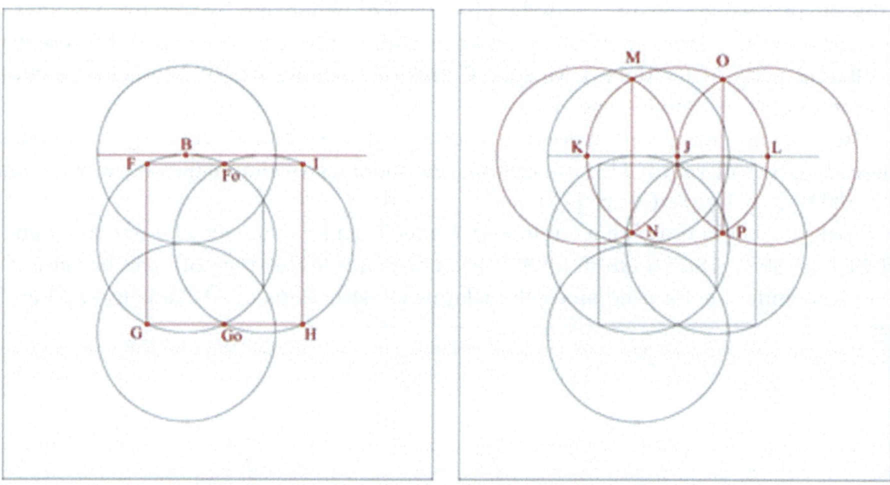

Fig. 3. Tracing an Islamic window. Step 4 and step 5.

In the intersection between the parallel line and the vertical line to midpoint of the rectangle FGHI we draw a circle of 4 cubits of diameter and center in point J. We also draw two circumferences equal to the first in the tangency points K and L. Finally, we plot two segments between intersection points MN and OP, Fig. 3, step 5.

In the sixth step we draw two circles, the first in the midpoint of the segment DE, in point Q and radius QG. The second circle had its center in point A and radius AH, Fig. 4, step 6.

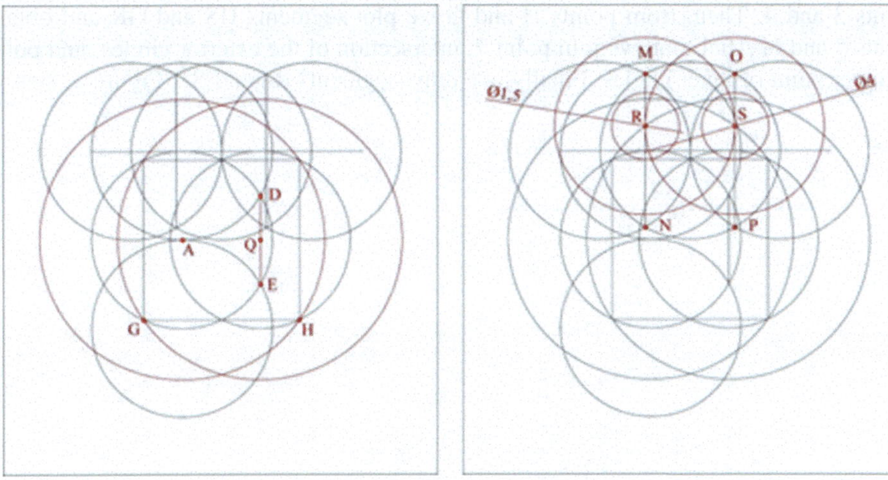

Fig. 4. Tracing an Islamic window. Step 6 and step 7

In the intersection between the circumferences and the segments MN and OP we draw two circles respectively, two of them of 1,5 cubits diameter (70,71 cm) and the other two circles of 4 cubits diameter with center R and S. Figure 4, step 7. After that, we plot a circle with center in point T, central point of the rec-tangle FGHI and radius TF. Figure 5, step 8.

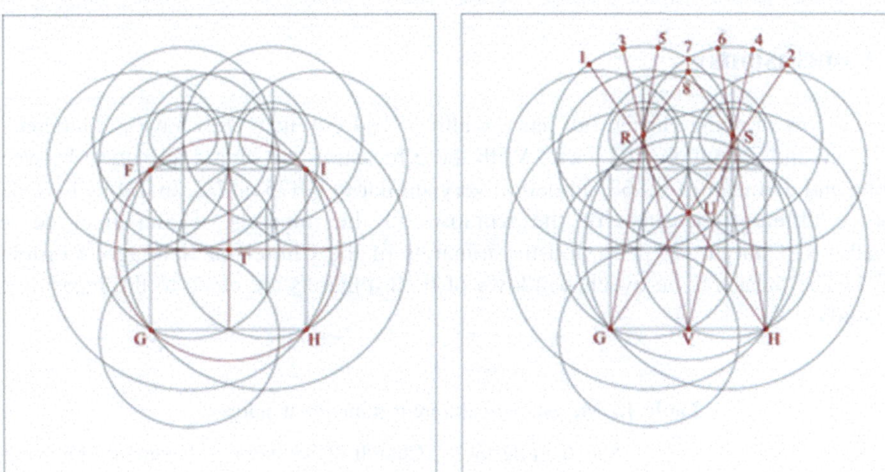

Fig. 5. Tracing an Islamic window. Step 8 and step 9

Finally, we trace two segments HU and GU and we continue them until their intersection with the upper circles. We obtain points 1 and 2. Plotting from point V, midpoint of segment GH, to point R and S, center of circles, and continue it we obtain points 3 and 4. Then, from points H and G we plot segments HS and GR and obtain points 5 and 6. After that, we join point 7, intersection of the exterior circles, and point 8, upper point of inner circles. Finally, we draw segments R8 and S8. Figure 5, step 9.

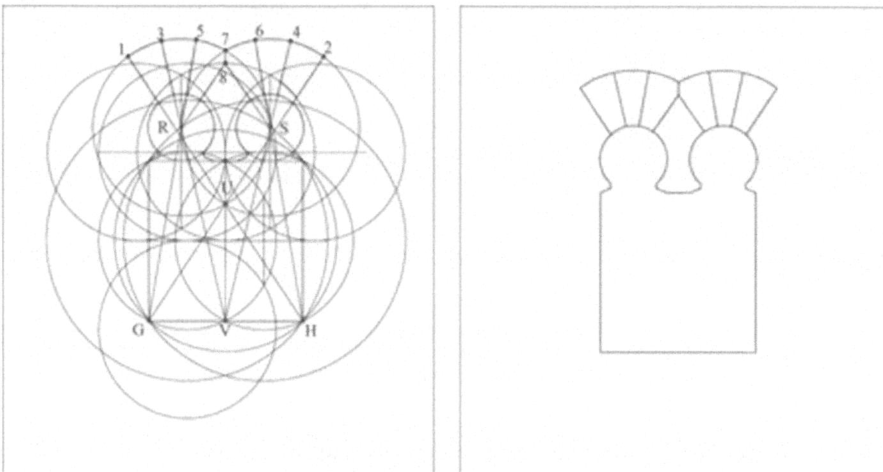

Fig. 6. Tracing an Islamic window. Step 10 and step 11.

In Fig. 6, step 11, we show the final layout of the windows. For achieving it we trace only the parts obtained before, Fig. 6, step 10.

3 Conclusions

The window drawing carried out has a width of 164 cm, its arches have a diameter of 70.71 cm and a sagitta of 61.4 cm. With these measures we obtain a relation between sagitta and diameter of 0,868, a relation very similar to 0,875 or 7/8. In Table 1 we can observe different measures of the windows of the minarets of San Juan de los Caballeros' Church, Córdoba and the minarets of the Church of Santiago, Córdoba, Fig. 7. The measures be-tween windows of both minarets are close to the measure of our study.

Table 1. Measures of different minarets windows.

	San Juan de los Caballeros		Church of Santiago	Geometric Drawing
Width	170	165	169	164
Diameter	75.2	71	75.86	70.71
Sagitta	65.8	60.8	66.5	61.4
Relation Sagitta/Diameter	0.875	0.87	0.877	0.868
Error deviation		2,94%	0,58%	3,53%

Fig. 7. Windows of Santiago and San Juan de los Caballeros, Córdoba

Table 1 shows different measures of the Windows of San Juan de los Caballeros, we can see the measure of the larger window and the smaller window. Santiago's Church minaret only has a window.

Likewise, it has been calculated the existing error deviation between windows, based on the larger window of San Juan de los Caballeros. In this case we can observe that the smaller window has a deviation of 2.94% compared to the larger window, while our geometric drawing has a deviation of 3.53%.

The deviation existing between our geometric drawing and the real examples is minimal, especially when it is checked that there is a similar deviation in other windows of the same construction, which demonstrate the existence of errors and deviations during the construction.

In this paper it has been demonstrated the existence of a geometric drawing for the design of the occidental Islamic windows with a relation between sagitta and diameter of 0.875. Architects designed the ground floor of their mosques but they also designed the decoration and the windows of the minarets too, making use of circles and notable points of it.

References

1. Ruiz de la Rosa, J.A.: Traza y Simetría de la Arquitectura: En la Antigüedad y Me-dievo. Servicio de Publicaciones de la Universidad de Sevilla, Sevilla (1987)
2. Idari, I.: Kitab al-Bayan al-Mugrib fi ajbar al andalus wa l-magrif, Historias de al-Andalus. Traducción de Francisco Fernández González, Granada (1860)
3. al-Jusaní, M.: Historia de los jueces de Córdoba. Kitáb al-qudá bi-Qurtuba. (970)
4. al-Makkarí, I.: The History of the Mohammedan Dynasties in Spain. Traducido al inglés por Pascual de Gayangos. Oriental Translation Fund, London (1840)
5. Camps Cazorla, E.: Módulo, proporciones y composición en la arquitectura califal cordobesa. CSIC Instituto Diego Velázquez, Madrid (1953)
6. Vallvé Bermejo, J.: Notas de metrología hispano-árabe. El codo en la España mu-sulmana. Al-Andalus **41**(2), 339–354 (1976)
7. Hernández Giménez, F.: El codo en la historiografía árabe de la mezquita mayor de Córdoba. Maestre, Madrid (1961)

8. Pavón Maldonado, B.: Alminares cordobeses. Boletín de la Asociación Española de Orientalistas **XII**, 181–210 (1976)
9. Hernández Giménez, F.: El alminar de 'Abd al-Rahman III en la mezquita mayor de Córdoba: génesis y repercusiones. Granada, Patronato de la Alhambra (1975)
10. Ortiz Cordero, R.: Reconstrucción del alminar de Hisham I de la Santa Iglesia Ca-tedral de Córdoba, Antigua Mezquita. Trabajo Fin de Máster. Universidad de Córdoba (2014)
11. Autodesk.com.: Free Student Software Downloads—Autodesk Education Community (2016). http://www.autodesk.com/education/free-software/all. Accessed 9 Apr 2016
12. GIMP: GIMP (2016). https://www.gimp.org/. Accessed 9 Apr 2016

The Analysis of Pathologies in the Church of San Ildefonso in Salorino (Cáceres) Using Geomatic Techniques

José Juan de Sanjosé Blasco(✉), Manuel Sánchez-Fernández, Alan D. J. Atkinson, and Angel M. Marra Recuero

INTERRA Research Institute, NEXUS Research Group, University of Extremadura, Avenida de La Universidad s/n, 10003 Cáceres, Spain
jjblasco@unex.es

Abstract. Buildings and other infrastructures suffer degradation over time and rehabilitation work to make improvements or for a change in use poses different challenges than those that are built anew. Appropriate measures must be taken to ensure that surrounding buildings and infrastructures are not altered. This paper studies the pathologies found in the parish church of San Ildefonso in Salorino (Cáceres) during remodeling work carried out in 2005 on the square surrounding it. The study uses geometric auscultation of the building using the comparison of geomatic techniques (terrestrial laser scanner, short-range scanner, high precision levelling and fissurometer) to assess the dynamic of the pathology. Geomatic measurements were first taken (November 2016 and April 2017) and later applied to the restoration work (July 2018). No significant movements were detected.

Keywords: Pathologies · Geomatic techniques · 3D modeling

1 Introduction

On the site where the town of Salorino (Cáceres) now stands (Fig. 1a and b) settlements are considered to have been established after the XIII th century during repopulation following the reconquest. At the end of the XV th century, the town is believed to have been a rest area frequented by travellers who stayed at its inns en route along the paths of the Order of Alcántara. The first inhabitants may possibly have arrived at the time of the repopulation by the Order of Alcántara.

The parish church of San Ildefonso is located in the town's main square. It is the largest building in the town and has the peculiarity of being free of any attached buildings of any kind. It is a voluminous barroque construction of the XVIII th century based on masonry with ashlar reinforcements on corners and façades. It has a single nave divided into four sections with a crossing and a rectangular apse. All of the crossings are covered by a barrel vault with lunettes except the crossing, which is covered by a dome over pendetives [1]. On the side of the epistle, by the apse, the small sacristy is located, dated by inscription in 1714 (Fig. 2a).

F. Cavas-Martínez et al. (Eds.): INGEGRAF 2019, LNME, pp. 507–517, 2020.
https://doi.org/10.1007/978-3-030-41200-5_56

(a) (b)

Fig. 1. Map of the location of Salorino and the site of the church. Source: https://commons.wikimedia.org/w/index.php?curid=16712100 y http://centrodedescargas.cnig.es

In 2005 the square was remodeled, mainly for the convenience of traffic circulation. A wall around the church of less than 1 m in height was removed along with the land infill between the walls (2 m estimated width). The structure removed formed a stabilizing force for the structural system of the church. Some years after the work on the square, visible cracks appeared together with collapses within the walls and vaults of the church. The remodeling work was considered to be a possible cause of these structural problems (Fig. 2b and c).

In 2016 the pathologies of the church were evident, and the research group NEXUS of the University of Extremadura in collaboration with the Diocesis of Coria-Cáceres (the owners) decided to examine its pathological state and whether this was evolving. For the case study it was decided that the full geometry of the site be determined using auscultation by terrestrial laser scanner and any structural movements of the building tracked using a level, fissurometer and short-range scanner.

The study of movements was made at three different times: November 2016, April 2017 and July 2018.

The following techniques were used for the study:

- High precision leveling to determine the settlement at one corner of the church.
- The assessment of the evolution of some cracks using a fissurometer. The cracks that appear in the vault of the central nave were measured from the over vault.
- The evolution of a crack over the main dome was analyzed using a hand scanner.
- The building was modeled in 3D (interior and exterior) using a terrestrial laser scanner.

These data were used to detect and quantify any changes that had taken place in the structure.

As a result of the state of deterioration of the church in December 2017, the Junta de Extremadura went ahead with the restoration and consolidation work, which began in January 2018 and ended in July of the same year. The last geometric auscultation survey was carried out in July 2018 when the work had been completed.

(a) (b) (c)

Fig. 2. Main façade of the church of San Ildefonso in Salorino and examples of its pathologies

2 Data Collection and Results

Leveling was used in the analysis of the settlement of some areas of the church, and a terrestrial laser scanner was used for the study of the collapses of the frontages. For the millimetric determination of cracks a hand-held scanning technique was used and supported by measurements using a crack monitor. The following is a description of the use of each of these techniques.

2.1 High Precision Leveling

The point chosen for geometric leveling and checking settlement was the corner of the church where the sacristy is located, since this is where the largest cracks are. Four points were measured on the eaves of the sacristy (Fig. 3).

The digital level used was a Leica NA 3000, which has a precision of a tenth of a millimeter on a barcode optical sight.

A reference point was chosen far from the area of influence of the church, and taking this point as a reference, high precision leveling was performed in which each point of the facade was measured. The results obtained in the three measurements are shown in Table 1. No significant movements were observed in the annual (2016–2017

Fig. 3. High precision leveling.

Table 1. Results of high precision leveling (values in meters).

	Measurement			Differences found		
	10/11/2016	05/04/2017	17/07/2018	2016–2017	2017–2018	2016–2018
Point 1	0,2500	0,2504	0,2503	−0,0004	0,0001	−0,0003
Point 2	0,2482	0,2479	0,2477	0,0003	0,0002	0,0005
Point 3	0,2660	0,2655	0,2660	0,0005	−0,0005	0,0000
Point 4	0,2859	0,2861	0,2863	−0,0002	−0,0002	−0,0004

and 2017–2018) and bi-annual differences (2016–2018). Any changes in the points leveled were of less than a millimeter.

2.2 Monitoring of Cracks and Fissures Using the Crack Monitor

The digital fissurometer used was the FIPSG150, which has a precision of 3 microns. This instrument must be calibrated using a pattern measurement, which is given a value of zero. For each crack measured, two reference signals are placed (one on each side of the crack) with a separation similar to that of the pattern (Fig. 4). The measurements obtained are the difference between the pattern and the real measurement, and so they have values close to a millimeter (Table 2).

Points 7 and 8 (Table 2) disappeared in the 2018 observation, but for the remaining signals the annual and bi-annual differences (2016–2018) are of less than a tenth of a millimeter. The cracks are interpreted as not having altered in this period.

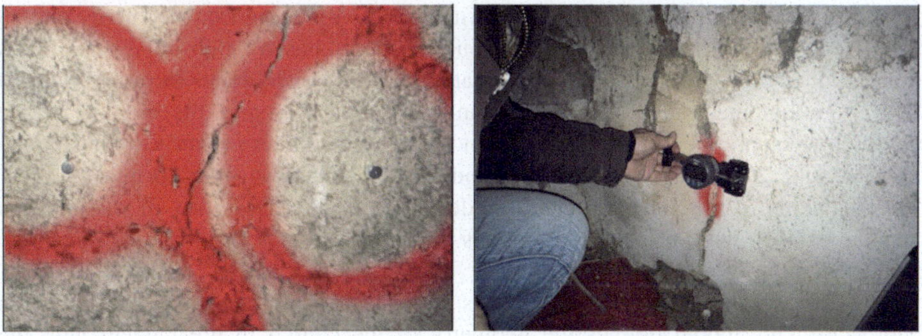

(a) (b)

Fig. 4. Placement of reference and measurement signals with the crack monitor.

Table 2. Results of measurement using the crack monitor (values in millimeters).

	Date of measurement			Differences		
POINTS	10/11/16	5/4/17	17/7/18	2016–2017	2017–2018	2016–2018
Point 1	−1.116	−1.163	−1.186	0.047	0.023	−1.233
Point 2	−0.190	−0.117	−0.170	−0.073	0.053	−0.097
Point 3	−0.120	−0.171	−0.105	0.051	−0.066	−0.156
Point 4	0.115	0.114	0.118	0.001	−0.004	0.117
Point 5	0.114	0.155	0.206	−0.041	−0.051	0.247
Point 6	−1.390	−1.357	−1.447	−0.033	0.090	−1.414
Point 7	−1.040	−1.084	–	0.044	–	–
Point 8	1.085	1.020	–	0.065	–	–
Point 9	−0.176	−0.186	−0.134	0.010	−0.052	−0.144

2.3 Terrestrial Laser Scanning

Terrestrial laser scanning is a technique that provides the highly detailed real geometry of an element or infrastructure. There are many examples of the use of terrestrial laser scanning for the documentation of heritage sites and geometric analysis as a tool for pathological purposes [2–4]. When an attempt is made to analyze the structural state of a building, the quality of its geometry is of fundamental importance to the precision of the calculation [5, 6].

In this case a geometric survey was performed using a terrestrial laser scanner (FARO Focus 3D X330, [7]). The equipment used has a range of 330 m and a data collection speed of up to 1 million points per second. For the entire survey of the church including exteriors, interiors and overvaults, 28 collections had to be made. These were obtained with a 6 mm resolution (determined by the mesh of points in a collection from a parament perpendicular to the scanner placed at 10 m). Given the density of the measurements and how they overlap one another, the resolution of the survey was ≤ 5 mm.

A closed polygon around the outer perimeter was drawn up for the survey and an open one to survey the interior and overvault. This was based on absolute coordinates in accordance with the system of coordinates ETRS89 UTM H29. In order to do so these were obtained from 4 control points distributed around the outside of the building, which are identifiable in the 3D model. These points were measured using the GNSS Leica 1200 equipment in static mode with readings of over 20 min duration. Georeferencing of the 3D point cloud model opens up the possibility of studying the position of this with respect to previous maps or any later studies that may be conducted.

A 3D model of 500 million points was obtained with a precision of union of less than 2 cm. This model defines the geometry of the church in great detail with the exception of the roof, which is difficult to access with this equipment. The geometry of the church was completed by an overhead drone flight.

The geometric study was carried out to examine the structural deficiencies that had been seen. In building structures, as in this case, geometry is highly relevant in terms of stability and structural safety. The geometric study centres on the analysis of:

- Walls - the study of collapses and thicknesses.
- Vaults - the study of deformation.
- Buttresses - location, thickness and height.

For the analysis of the collapses and thicknesses of the walls, sections horizontal and transversal to the 3D model were taken. The horizontal sections were taken every 2 meters from the ground upwards, and the transversal sections in the centers of the vaults and buttresses of the nave of the church (Fig. 5), where collapses of between 6 and 30 cm were observed (Table 3).

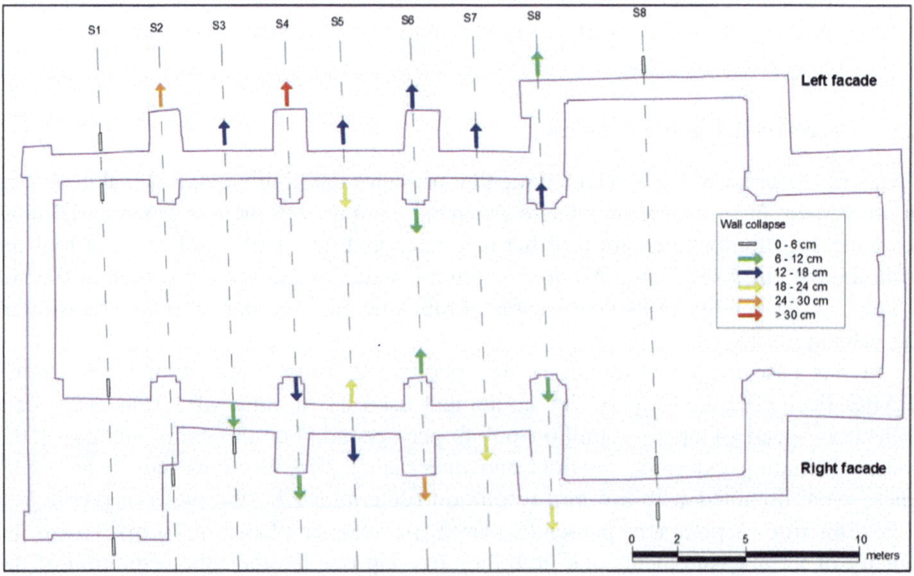

Fig. 5. Distribution of collapses in walls and buttresses, represented at a height of 8 m above the ground.

Table 3. Magnitude in centimeters of collapses in walls and buttresses. Data located and oriented in Fig. 5.

	Right		Left	
	Extrados	Intrados	Extrados	Intrados
Section 1 (S1)	0	0	0	0
Section 2 (S2)	0	–	–	23
Section 3 (S3)	0	6	–	15
Section 4 (S4)	8	13	–	30
Section 5 (S5)	14	18	20	11
Section 6 (S6)	27	18	20	11
Section 7 (S7)	19	–	–	11
Section 8 (S8)	6	13	12	6
Section 9 (S9)	0	–	–	0

Figure 6 shows the magnitude and direction of the collapses on the top of the walls and buttresses of the church. The largest collapses are in the buttresses. The movements estimated in the buttresses of Sects. 2 and 3 (S2 and S3) on the left façade correspond to cracks in nozzle of vault 5 on the same side.

The vaults were studied from horizontal sections of 15 cm equidistance (level curves). The work was conducted on the extrados and intrados of the vaults. Small deviations were found in the geometry of the latter (Fig. 6a and b).

The most relevant deformations appear in vaults 3 and 2.1. The determination of their cause is complex. They may be a result of movements in the foundations or walls, or of construction defects. From the survey of the intrados and extrados of the vaults of the nave, their thicknesses can be studied. Figure 6c shows a map of elevations representing the thicknesses of the vaults of the central nave. Vaults 4 and 5 are the least thick in the central part. The largest cracks can be seen in these vaults, and in addition they are located in the area in which the walls and buttresses show the greatest collapses.

2.4 Hand-Scanning

The short-range scanner is a tool that facilitates high resolution and precision 3D models. It was used in this case to monitor movements in structural cracks in elements of the church of Salorino. The equipment used was the GoScan 50 of CREAFORM®, with a maximum spatial resolution working capacity of 0.5 mm and precision of 0.2 mm. Movement is manual, so it is the user who, in real time, moves the scanner to determine the geometry of the object to be documented. The measuring range is 0.20 m to 0.60 m [8].

The scanner works from a calibrated system made up of a structured light projector and three photographic cameras, two for the acquisition of the geometric information and one to collect the texture. The 3D model is obtained from the reading of the light pattern by the cameras. The different positions of the equipment are referenced during scanning and can be carried out by different patterns: the geometry of the element,

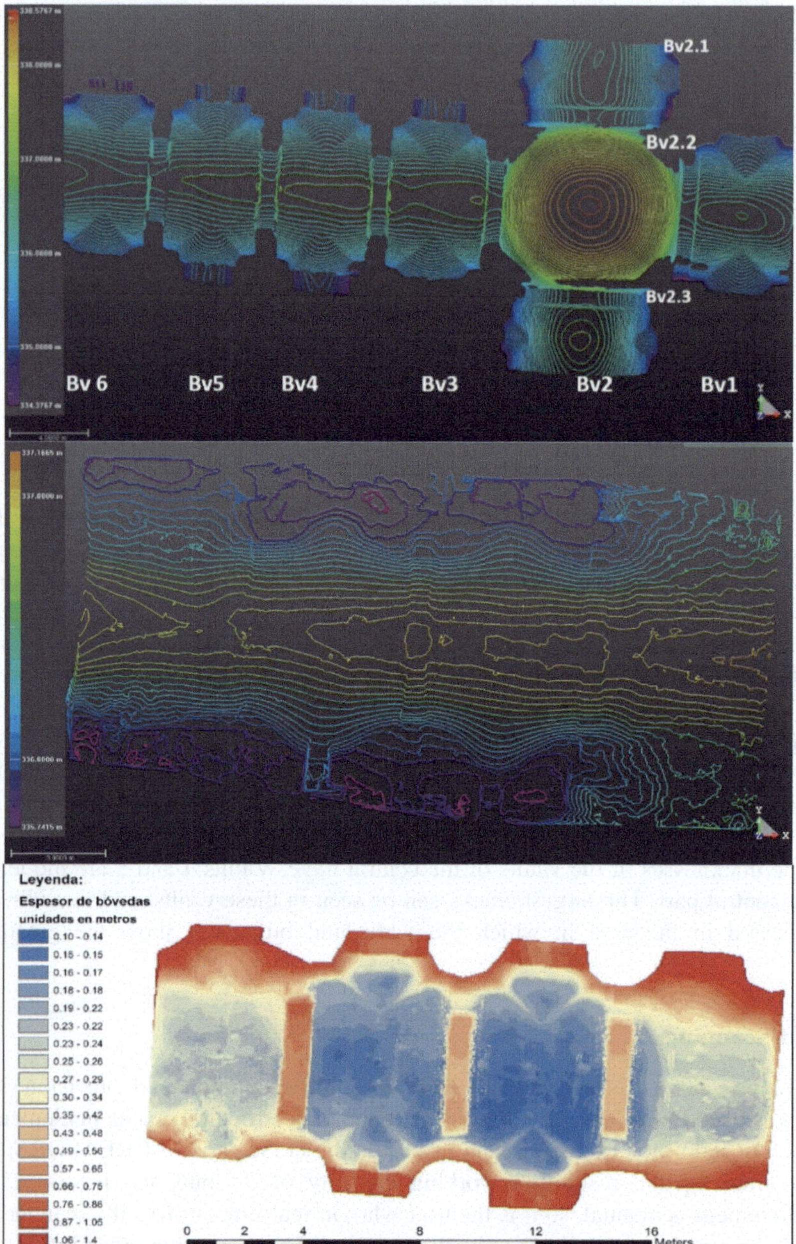

(a)

(b)

(c)

Fig. 6. Geometric study of the current state of the vaults. (a) Level intrados curves of the vaults; (b) Level extrados curves of vaults 3 to 6; (c) Map of thicknesses of vaults 3 to 6.

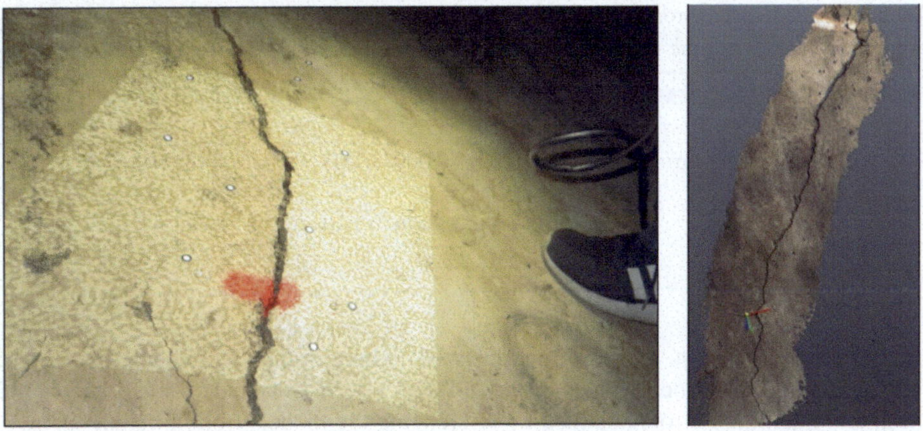

(a) (b)

Fig. 7. Data collection process using the short-range scanner for monitoring fissures and/or cracks in vaults. (a) Disposition of self-recognizable signal in a crack in the central vault of the apse; (b) model of the crack scanned.

difference in color or "target" self-recognizable signal set on the object. The signals used belong to the manufacturer and are specific to their range of short-range scanners. This method provides the most precise georeferencing.

Two cracks were analyzed, one in the apse vault, and the other located in the wall of the third vault of the nave.

As in the case of the other instruments, three data collection surveys were conducted in November 2016, April 2017 and July 2018. The study of cracks in the vault was performed by signals, since the surface could be cleaned and the markers stuck, ensuring their adherence and permanence over time. The signals were placed along the crack (Fig. 7). In this case the number and position of the signals in the fissure were sufficient for the study of its movement in multiple directions, unlike the difference of the fissurometer, for which we could only observe the movement of the crack in the direction in which the markers had been placed and with the spacing determined by instrument dimensions.

For the study and follow-up of the possible movements in the crack in the vault a total of 29 signals were used. The coordinate system of each of the models is relative and therefore different in each case. In order to study the geometric evolution of the cracks in the vault and wall, the system of coordinates used in the first survey (November 2016) was established as the system of reference. The remaining surveys were oriented according to this system of coordinates, thus ensuring that the data were comparable. Five points common to the reference model and the oriented one are selected. If the common points are "target" signals, the human factor is eliminated since identification and selection are automatic. When signals are not used, part of the error obtained is a result of human error in selecting the point. Tables 4 and 5 show the results of georeferencing and the movements observed in the cracks analyzed.

Table 4. Referencing errors (in mm) obtained in the study of the crack in the wall, errors expressed according to the coordinates and the 3D error (ε3D) as $\sqrt[2]{\Delta X^2 + \Delta Y^2 + \Delta Z^2}$

		April 2017			July 2018		
		X	Y	Z	X	Y	Z
Referencing error	Mean	0.001	0.000	0.003	0.001	0.000	0.003
		ε3D 1,938			ε3D 0,374		
	Maximum	1.470	1.970	1.080	0.360	0.320	0.070
		ε3D 2,800			ε3D 0,483		

For the analysis of the crack in the wall, where signals could not be placed, referencing and movement values of around 2 mm were obtained. Table 4 shows the mean and maximum errors obtained by referencing the models from different surveys. The average and maximum errors of the points used as the reference of the measurements were calculated. The average of the maximum deviations and errors in x, y, z and the 3D error are also shown. From the analysis made there is not considered to be any movement in the crack in the wall.

Table 5. Referencing and movement errors (in mm) obtained in the study of the crack in the vault. Errors expressed according to coordinates and the 3D error, such that: $\sqrt[2]{\Delta X^2 + \Delta Y^2 + \Delta Z^2}$

		November 2016			April 2017			July 2018		
		X	Y	Z	X	Y	Z	X	Y	Z
Referencing error	Mean	−0,002	−0,002	0,001	−0,002	−0,002	0,000	−0,012	0,795	0,646
		ε3D 0,003			ε3D 0,003			ε3D 1,025		
Movements in the vault	Mean	0,074	0,249	0,204	−0,118	0,072	0,451	−0,341	1,256	0,929
		ε3D 0,555			ε3D 0,877			ε3D 1,778		
	Maximum	0,398	6,224	3,152	0,289	6,655	3,261	0,426	8,970	2,253
		ε3D 7,108			ε3D 7,526			ε3D 9,386		
	Minimum	−1,362	−0,197	−0,239	−1,308	−0,555	−0,689	−7,141	0,003	−0,773

For the crack in the vault (Table 5) the analysis was performed in the same way as in the wall. Of the 29 signals scanned 5 were used for referencing and the remainder for studying movements in the vault. The referencing errors were smaller than the accuracy of the scanner (0.2 mm) in the data collections of 2017, and of 1 mm in the collection made in 2018. The movements in the vault are similar in the survey of 2017, with maximum displacements of 7 mm. In spite of having a greater referencing error, the movements detected in the crack in the vault in the survey of 2018 did not greatly differ from those found in the previous surveys, with a maximum of 9 mm and a mean of 1.7 mm.

3 Conclusions

By means of the geomatic methodologies applied, it can be concluded that during the period in which the church of Saint Ildefonso in Salorino was measured there were no movements within the building. Structural deformation was detected in different places, thus verifying the most representative pathologies of the building.

The techniques used were complementary to one another. In the case of the values obtained with the fissurometer and the short-range scanner a logical relationship was found, showing that both techniques can be used for the same purpose.

The consolidation works carried out in the first half of 2018 will prevent these pathologies from recurring.

Acknowledgements. This study was made possible thanks to financing granted by the European Fund for Regional Development of the European Union and the Junta de Extremadura to the NEXUS research group through the GR18053 support.

References

1. Salorino (2018). http://www.salorino.es/
2. Berenguer, F.: Aplicación de técnicas geodésicas y técnicas avanzadas de captura, escáner láser y termografía, en el estudio patológico de la estructura y la cimentación en la restauración integral de la Catedral de Santa María de la Asunción de Coria. In: Expresión Gráfica, Extremadura, p. 258 (2014). http://hdl.handle.net/10662/1622
3. Borrazás, P.M., Blancorotea, R., Rodríguez Paz, A.: La documentación geométrica de elementos patrimoniales con láser escáner terrestre. La experiencia del Lapa en Galicia. Cuadernos de Estudios Gallegos **56**(122), 33–65 (2009)
4. Serrano, F., Saumell, J., Berenguer, F.: Análisis de resultados métricos de una nube de puntos y una medición directa en el patrimonio edificado. El Santuario de La Montaña en Cáceres. Informes de la Construcción **66**(534), 016 (2014)
5. Acikgoz, S., Pelecanos, L., Giardina, G., Aitken, J., Soga, K.: Distributed sensing of a masonry vault during nearby piling. Struct. Control Health Monit. **24**(3), e1872 (2017)
6. Monserrat, O., Crosetto, M.: Deformation measurement using terrestrial laser scanning data and least squares 3D surface matching. ISPRS J. Photogramm. Remote Sens. **63**(1), 142–154 (2008)
7. Faro Focus 3D X330 (2017). http://www.faro.com/es-es/productos/topografia-3d/laser-scanner-faro-focus-3d/informacion-general
8. GoSCAN 50 (2018). https://www.creaform3d.com/es

HBIM. Parametric Families from Point Clouds in Heritage Elements

María José Marín Miranda(✉), Francisco Javier Chorro Domínguez, Manuel Sánchez-Fernández, and Juan Pedro Cortés-Pérez

INTERRA Research Institute, NEXUS Research Group, University of Extremadura, Avenida de La Universidad S/N, 10003 Cáceres, Spain
mjmarin@unex.es

Abstract. The conservation of heritage sites requires precise knowledge of the elements to be preserved. To this end, tools and methodologies must involve the integration of as much accessible and ordered information as possible while at the same time permitting the ongoing update of data.

Defensive elements have an important presence in the Iberian Peninsula and owing to both how widespread they are and the presence of recurring elements, the use of adaptative families on these characteristic elements offers an important step forward in their introduction into BIM methodology (Building Information Modelling).

BIM technology allows the generation of virtual models using parametric information of the physical and functional characteristics of a building while setting up a reliable database for decision taking during its lifecycle [1]. This technology was first used in construction projects over 15 years ago [2].

This communication presents a set of parametric families in defensive elements of the wall of Cáceres generated on the base of a geometric BIM study using point clouds obtained from a scanner laser survey of the site.

Keywords: HBIM · BIM families · Scanner laser

1 Introduction

Heritage is considered to be everything that has cultural, artistic, social, economic or historic value. All built heritage is under threat both from disasters and environmental effects, such as water, pollution and wind. To preserve this heritage, it must be maintained and protected to reduce such impacts over time.

The Charter of Venice [3] in its article 16 states that any heritage preservation works must be documented recording all phases of the process as well as the technical and formal elements identified during the course of the works. Later, the Charter of Krakow [4] updates the preservation principles in that it must be contemplated as a long-term strategy that reflects, in its restoration project, that all the information collected during the process be integrated.

One of the tasks required to perform these preservation works is the documentation of historic heritage including a geometric definition of the element to be preserved. Different techniques can be used to obtain the geometric information such as laser scan

F. Cavas-Martínez et al. (Eds.): INGEGRAF 2019, LNME, pp. 518–526, 2020.
https://doi.org/10.1007/978-3-030-41200-5_57

or photogrammetry, which have a high degree of automation and collect and process precise spatial data very quickly. From these three dimensional data the graphic two dimensional documentation of the survey is generated, including floors, elevations and sections, usually created in CAD files losing geometric data in the process of transformation to two dimensions.

In addition to the geometric data, there is a great deal of information that requires the preservation of a monument that cannot be made compatible with CAD files, and so in order to meet the requirements of the Charters of Venice and Krakow, other means are needed to include this information within the geometric model.

The application of BIM methodology to historical heritage is known as HBIM, the H referring to historic or heritage [5].

The platforms used in the generation of virtual models present limitations regarding their application to heritage. One of these is that there are no specific libraries of HBIM objects, sometimes with complex geometries associated with the geometric rules of the elements of the construction to be modelled [6].

For these reasons this research presents the development of parametric models of heritage elements from point clouds for the generation of HBIM families that can be introduced in complete models.

2 Materials and Methods

The parametrization of characteristic heritage elements facilitates the creation of a library of singular families for their incorporation into models, thus facilitating the modelling process.

In the methodology followed for modelling the families a selection of the most singular and characteristic elements is first made, defining the level of development for modelling the elements and proceeding to 3D data collection. A formal and constructive study of the elements is then made for their parametrization using this geometric information.

2.1 Selection of Elements

Military architecture is important in the Historic Patrimony of Spain as it has played an important part in the defense of the territory over the centuries, though its abandonment due to its loss of any function has in many cases led to deterioration or even destruction [7].

The strategic location of defensive elements substantially affects the perception of the landscape it is set in by adding considerable landscape value which, together with its historical value, makes its preservation necessary.

Defense heritage has been protected since the decree on protection of Spanish castles was published on the 22nd April 1949, in which all were protected regardless of their state of conservation and any interventions altering them were prohibited, as well as providing as detailed an inventory as possible in documentary and graphic form [8]. As the protection of this architecture was not considered enough, the Charter of *Baños de la Encina* for the Conservation of Defensive Architecture in Spain put forward a series of recommendations for the preservation of this heritage, one of which was the detailed knowledge of both the site and its surroundings for the purposes of any intervention [7].

The purpose of this paper is to make a formal constructive study of some characteristic elements of walled cities, particularly defensive elements from the times of the Almohads, which present characteristic features that clearly identify them [9]. In the Iberian Peninsula there are numerous Almohads city walls, among which the Alcazaba de Badajoz in Extremadura, those of Cáceres and of the Alcazaba de Reina stand out.

The towers and city walls of Cáceres between the Puerta de Arco del Cristo and the Torre Torremochada together with Torre del Horno were analyzed to define their identifying elements. This work focuses on the Torre del Horno as it contains a representative number of elements also present in the rest of the walled city, defining doors, windows, crenellations and loopholes.

2.2 Levels of Development LOD

Prior to the geometric data collection and modelling process the levels of development associated with the model must be defined and adapted to the result required, given that a very dense point cloud involves a great amount of information that may be surplus to the requirements of modelling a family for its introduction in BIM models.

The *American Institute of Architects* establishes five levels of development, LOD 100, LOD 200, LOD 300, LOD 350, LOD 400 and LOD 500. In its application to heritage using HBIM, based on publications of BIMForum, *Level of Development Specification* [10], LOD100 was established for those elements whose geometry can be defined on a basic level with no decorative nor constructive details. For LOD 200 a greater number of geometric details must be added, such as cornices, imposts or sunken ditches which, from level LOD 300, require the incorporation of information additional to the geometry, with materials also being defined. Level LOD 400 provides more precise knowledge, with detailed geometric information as well as auxiliary materials and systems that make up the family, such as closure systems in doors and windows or anchorage systems of elements. Finally, LOD 500 would represent the element just as it is built with all of its component parts and its precise geometry, including both constructive deformations and those due to deterioration. This latter case would be the only level not applicable to parametric families, as this personalization excludes it from the concept of the parametric family.

Regarding modelling heritage elements using BIM methodology, the developmental level established will depend on the needs of the project, with the establishment of different levels even within a single element. While a stretch of a wall can be suitably defined at LOD 100 due to its state of conservation and geometric integrity, another of the same canvas may require LOD 400 due either to its need for intervention or to its historical or artistic value. For this reason, it is difficult to establish a level of development applicable in generic terms to modelling families for their introduction in BIM.

For the case of this study LOD 200 was established with a geometric definition that allows its simplification to a level of visualization depending on the scale or information necessary and assigning information specific to certain elements to higher levels. A very high developmental level reduces the scope of application of the families created, whereas lower levels increase their chances of integration, as they are building blocks for their modification, thus permitting their adaptation to specific situations.

2.3 Data Collection

The parametrization of BIM objects requires precise geometry, which may be provided from existing maps or surveys. The cartographic source used depends on the typology of the element to model and the LOD that the work requires. In the case of heritage elements and even singular elements of them, they often need a degree of definition that makes an architectonic survey necessary.

The degree of definition and the precision of the survey to carry out depends on the level of development needed in the modelling. In this case, for surveying the families of singular elements located in defensive systems such as doors, windows, crenellations and loopholes, there are several techniques that provide suitable geometric documentation. Classical topography, using total station or laser distance meter, obtains the geometry of an element discretely, i.e. a simplified geometry of the element from the collection of individual points. Land photogrammetry and terrestrial laser scanner (TLS) are massive data collection tools, which acquire thousands or millions of points of the element studied in a short time period. The appearance of high resolution photograph scanners and digital photographic cameras, together with the development of photogrammetric software, make terrestrial photogrammetry a vanguard technique in obtaining precise 3D models of both small and large volumes. Classical topography, though it is a technique with which complex 3D models can be generated, is eclipsed by terrestrial photogrammetry and laser scanner [11].

Terrestrial scanner and photogrammetry techniques offer solutions of the same kind from different working processes. One of the significant differences between photogrammetry and TLS is the precision of the scale of the model. Photogrammetry requires topographical support to obtain metric models, and so the precision of these will depend on both the quality of the photogrammetric model and the topographic support. The TLS generates metric point clouds the precision of which depends on the technical specifications of the equipment and the methodology of data collection. This technique is used by numerous authors with the aim of carrying out geometric studies on historical buildings [12–14].

For the architectonic survey in question and its singular elements the TLS Faro Focus 3D X330 [15, 16] was used. This equipment has a measuring range of 330 m and a data collection speed of up to 1 million points per second. Data are collected with a resolution of 6 mm (this resolution is determined as the mesh of points obtained from a facing perpendicular to the scanner placed at 10 m). Given the density of the points collected and their overlap, the resolution of the survey was ≤ 5 mm.

For the survey of the Torre del Horno using the laser scanner an open polygon was used, the beginning of which was the patio of the upper floor, then downwards through the building including the two chambers and the patio of the first floor, the access stairway and finally the exterior. Given the dimensions and geometry of the entire survey 20 collections had to be made (Fig. 1). Owing to the 17.5 m height of the tower from top to bottom, the resolution of the wall and singular elements such as the outside of the windows, the top was not as well documented in terms of resolution as the lower chambers. This is due both to the height of the walls and the position of the scanner with respect to these. In this case, solutions are proposed such as increasing the number of collections by raising the scanner on tripods of greater height to improve the distance and relative position or collecting data from the roofs or balconies of nearby buildings.

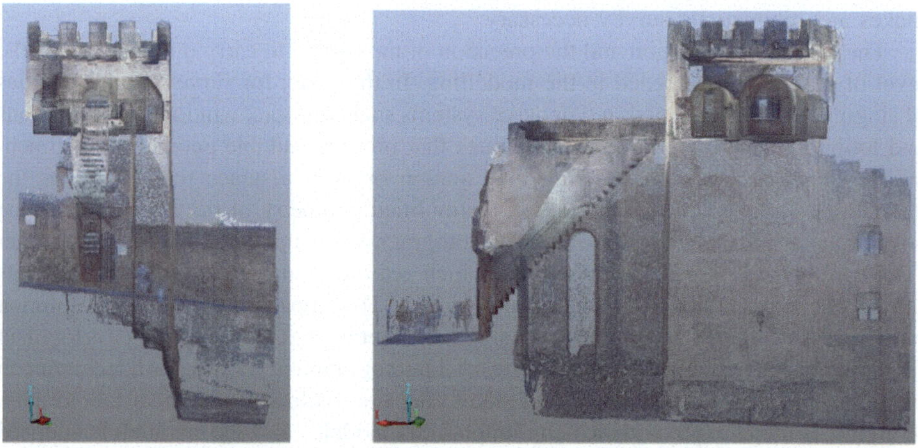

Fig. 1. Results of the survey of the Torre del Horno (Cáceres)

2.4 Modelling of Parametric Families from a Point Cloud

Parametrization of the families has been carried out using Revit 2019 by taking the point cloud as the basis of the laser survey. From this we define the geometry of the different elements to parametrize, which establishes the degree of development of the families in an LOD 200, providing a geometric idealized definition of the element by ignoring deformations due either to damage or the constructive process itself.

The point cloud is simplified using ReCap, and only the areas subject to the study are kept for the modelling of the families.

Taking the point cloud of the family to parametrize as the basis, the next step is the study of the geometry of the elements, and the floor layouts and elevations are defined before later modelling the geometry of the family (Fig. 2).

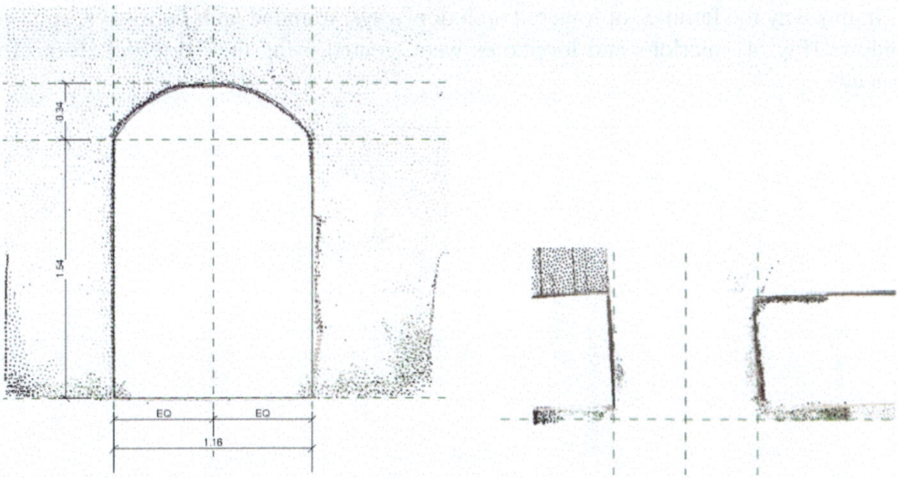

Fig. 2. Geometry of the family of the lowered arch from the point cloud.

From the study and formal modelling, the geometric rules that will define the parameters are taken, such that the family obtained might adapt to the greatest possible number of assumptions. In the case of the door of the lowered arch, only the total width and height of the door are defined, together with the start level of the arch, while the remaining parameters are calculated (Fig. 3), so it is possible to repeat the elements quickly and simplify field data collection at the same time as offering information to understand its geometry.

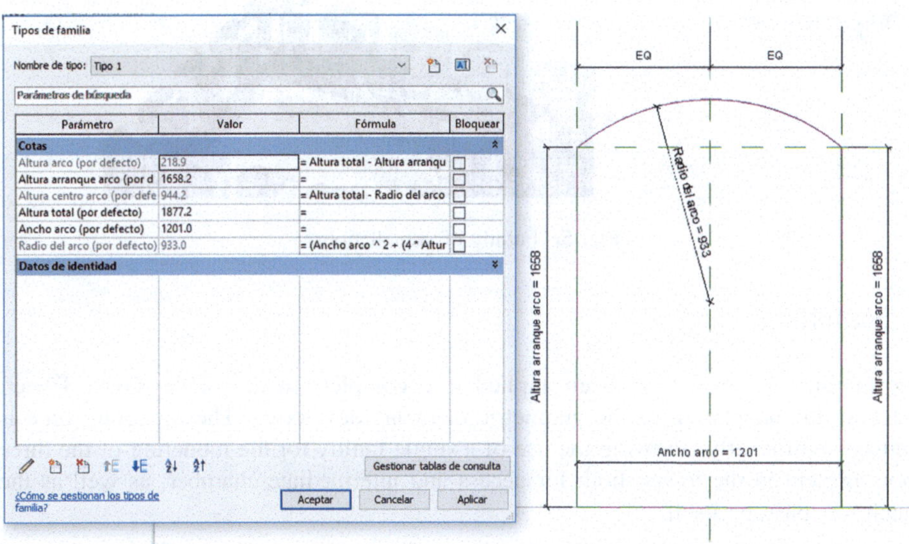

Fig. 3. Parameters of the family of the lowered arch.

In this way the families of lowered arch doorways, rounded arch doorways, splayed windows (Fig. 4), merlons and loopholes were created using the Torre del Horno for the study.

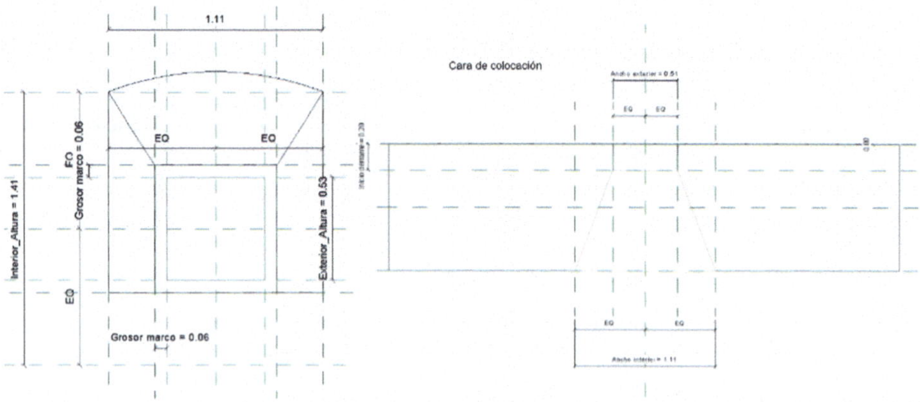

Fig. 4. Family of loophole window.

In the case of the crenellation (Fig. 5) we set out from families associated with matrices, such that direct modelling is possible from the whole crenellation instead of from element to element, thus defining the dimensions of the merlons.

Fig. 5. Family of crenellations.

3 Results

The families obtained have been applied in a complete model of the Oven Tower, checking the adaptation of the geometric elements developed. The modeling time is reduced significantly, allowing the use of a single family for the modeling of the three doors present in the tower, both for access and intermediate chamber, as well as the watchover passage itself.

Fig. 6. Complete LOD 200 modelling of the Torre del Horno.

In this way, all the modeling has been developed without the use of on-site family tools, which present limitations when are used in other models, also increasing the possibilities of integrating new parameters apart from geometric ones.

The result is a BIM model with an LOD 200 level of development (Fig. 6), which offers a geometric idealization of the tower and in all its components. This facilitates both its study and introduction of additional information, such as the comparison with real geometry offered by the point cloud of the scanner laser.

4 Conclusions

The existence of specific families for heritage elements represent a major advance in the introduction of these into BIM methodologies, which in the case of defensive elements, with large extensions and presence of recurring elements are particularly interesting, reducing runtimes and offering tools for geometric analysis with great value to establish relationships between the different elements, as well as the comprehension of the constructive processes used.

Use of scanner laser leads to high geometric precision and, above all, it facilitates the job since it encompasses the three dimensions from the survey itself, thus simplifying the job of defining the geometric parameters.

Acknowledgements. The present study has been made possible thanks to financing awarded by the European Regional Development Fund and the Junta de Extremadura to the research group NEXUS under grant GR18053 and to the project of investigation IB16115 I+D+i thanks to European Regional Development Fund and the Junta de Extremadura.

References

1. Kassem, M., Succar, B.: Macro BIM adoption: comparative market analysis. Autom. Constr. **81**, 286–299 (2017)
2. Kiviniemi, M., Sulankivi, K., Kähkönen, K., et al.: BIM-based safety management and communication for building construction. VTT Res. Notes (2597), 1–123 (2011)
3. de Venecia, C.: Carta Internacional sobre la conservación y la restauración de monumentos y sitios. In: II Congreso Internacional de Arquitectos y Técnicos de Monumentos Históricos (1964)
4. de Cracovia, C.: Principios para la conservación y restauración del patrimonio construido. In: Compend. leyes sobre la protección del Patrimonio Cultural en Guatemala, pp. 1–6. UNESCO (2000)
5. Murphy, M., McGovern, E., Pavia, S.: Historic building information modelling (HBIM). Struct. Surv. **27**(4), 311–327 (2009)
6. Murphy, M., McGovern, E., Pavia, S.: Historic building information modelling - adding intelligence to laser and image based surveys of european classical architecture. ISPRS J. Photogramm. Remote Sens. **76**, 89–102 (2013)
7. Instituto Andaluz de Patrimonio Histórico: Carta de Baños de la Encina actualiza la protección legal del patrimonio arquitectónico defensivo. In: PH Boletín, vol. 61, pp. 4–23 (2006)
8. Decreto de 22 de abril de 1949 sobre protección de los castillos españoles. BOE, 125, 5 de mayo de, pp. 2058–2059 (1949). http://www.boe.es
9. Bueno, S.M., Daza, P.G.: La muralla almohade de Cáceres: aspectos constructivos, formales y funcionales. AyTM **10**(1), 57–118 (2003)
10. Forum, B.: LOD Spec 2019 Part I, pp. 12–13, November 2019
11. Fernández, M.S., Cortés, J.P., Naranjo-Gómez, J.M.: Análisis geométrico para evaluación estructural. TLS vs Photogrametría terrestre: Aplicación a puente de fábrica ex–100 pk-8. In: Presented at the REHABEND 2016, Burgos, Spain (2016)
12. Ramos, L., Marchamalo, M., Rejas, J., Martínez, R.: Aplicación del Láser Escáner Terrestre (TLS) a la modelización de estructuras: precisión, exactitud y diseño de la adquisición de datos en casos reales. Informes de la Construcción **67**, 074 (2015)
13. Serrano, F., Saumell, J., Berenguer, F.: Análisis de resultados métricos de una nube de puntos y una medición directa en el patrimonio edificado El Santuario de La Montaña en Cáceres. Informes de la Construcción **66**(534), e016 (2014)
14. Mancera Taboada, J., Rodríguez Gonzálvez, P., González Aguilera, D., Muñoz-Nieto, Á., Gómez Lahoz, J., Herrero-Pascual, J., et al.: On the use of laser scanner and photogrammetry for the global digitization of the medieval walls of Avila. IAPRS **38**, 1–3 (2010)
15. SCENE, FARO, Ed., 5.4 ed. (2017)
16. F. D. X330, FARO, Ed., 1.4 ed. (2016). https://www.faro.com/es-es/

Data Management in "Scan to HBIM" Projects

Manuel Sánchez-Fernández(✉), Francisco Javier Chorro, María-José Marín, José Juan de Sanjosé-Blasco, and Luis Francisco Martínez

INTERRA Research Institute, NEXUS Research Group, University of Extremadura, Avenida de La Universidad S/N, 10003 Cáceres, Spain
msf@unex.es

Abstract. Heritage management and conservation is one of the challenges facing society today. The management and conservation model has evolved in recent years thanks to technological evolution. When carrying out a heritage conservation project, there are fundamental elements for a correct development, one is the geometry of the element and the other is the conceptualization of the project from the geometry. In the last decade, techniques have been developed that allow architectural surveys of buildings and infrastructures with a high level of detail. Among the techniques mentioned are terrestrial photogrammetry and terrestrial laser scanners, both with the ability to generate a high-precision, high-resolution 3D metric model. The 3D model obtained sows the geometric base for the development of a conservation or intervention project using the Building Information Modeling (BIM) methodology. By means of the execution of a project the management of the information of the data obtained with terrestrial laser scanner is studied for its later modeling in the software BIM Autodeskt Revit®, applied to the mode-side of two cloths of wall in the cities of Cáceres and Badajoz for the management and conservation of the patrimony.

Keywords: Laser scanning · Architectural survey · Conservation · Heritage · Heritage Building Information Modeling (HBIM)

1 Introduction

In today's society there is a need to rehabilitate, conserve and maintain the built historical heritage. The heritage built in Spain is distributed throughout the territory. The work of conservation and rehabilitation involves the study and analysis of the built element for the knowledge of the various that define it. The variables to be controlled are different depending on the type of intervention or conservation to be carried out or the purpose or use of the element being studied. When studying built heritage, it is interesting to know variables of the type of resistance of materials and structural elements, time of construction, construction material, geometry, historical phases of remodeling, etc. For the knowledge of these variables there are multiple techniques that have the capacity to offer values that characterize the patrimonial element. Within the possible study techniques, it is possible to discriminate between destructive and non-destructive techniques. The ideal in heritage intervention is to carry out analyses using non-destructive techniques, seeking the way to alter the heritage in the least possible

F. Cavas-Martínez et al. (Eds.): INGEGRAF 2019, LNME, pp. 527–534, 2020.
https://doi.org/10.1007/978-3-030-41200-5_58

degree. It is natural to find functional buildings dating back more than 10 centuries, this implies that the construction has gone through different phases in their state of construction. In these cases, it is essential to differentiate the constructive stages, as these will generally be treated differently.

The work presented is carried out in the Spanish region of Extremadura. The application of BIM methodologies associated with the conservation of built historical heritage (HBIM) is being studied. The research is based on the study of the number of heritage elements framed as defensive constructions and their current state for the development of a HBIM project aimed at conservation work. The BIM methodology is based on the ability to have information of different nature ordered in a single digital environment. Works [1–3] refer to methods used by other researchers for the development of HBIM projects of existing buildings or infrastructures. These works are based on the geometry of a heritage element obtained from a laser scanner or terrestrial photometry, from which a solid model is obtained by different paths [1–3]. In this way, it is intended to link the geometry of the infrastructure with pathological damage, sensor readings, classification according to phase and time of construction or improvement, etc. All this is based on the premise of studying the state of conservation through nondestructive techniques. Techniques such as terrestrial laser scanners, temperature and humidity sensors, thermal cameras, sclerometers, ultrasounds, fisurometers or hyperspectral cameras are used. Each of the values obtained characterized the infrastructure according to one parameter or another.

The object of study of the project is the application of the HBIM working method for the conservation and protection of heritage on defensive infrastructures in the region of Extremadura. The work process involves a historical-pathological study of existing infrastructures, which is a decision-making tool when it comes to focusing the project's action [4]. Figure 1 shows on cartography the walls dated in Extremadura (a) and the walls selected for study in the HBIMEX project (b). This is a research project being carried out by the NEXUS research group at the University of Extremadura in which the HBIM methodology is evaluated in the task of conservation and management of walls. The basic data of wall inventories have been extracted from the database [5], which has been verified and completed for an in-depth study of the situation of the infrastructures [4].

For the development of the project, attention has been focused on walls of Mozarabic origin. This type of wall contains geometric elements such as the dimensions of walls, doors, windows, arrow slits, etc. that keep geometric relations. The walls dating back to earlier times, such as the Roman walls, are mostly in ruins. The later, Christian walls are made up of masonry walls, which are less regular than the walls of mudwall than those given in Mozarabic walls and their singular elements such as doors or windows are not geometrically related, being completely different from each other. The conceptual basis of the BIM is the standardization of construction, build and model from preconceptualized elements, which is why Mozarabic walls have been selected or elements that meet geometric rules of construction.

The basis of the BIM is geometry, in heritage elements, mostly singular edifications, it is difficult to have a precision geometry of these infrastructures. In order to carry out the project, land laser scanning campaigns have been carried out on wall cloths and singular elements of Mozarabic walls in the region of Extremadura. The

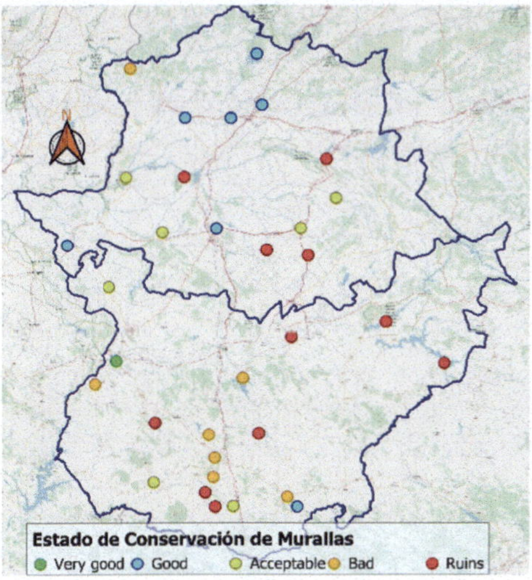

(a)

(b)

Fig. 1. (a) Distribution of walls dated in Extremadura classified according to state of conservation [4]; (b) Walls studied in the HBIMEX project.

purpose of the surveys carried out is the modeling of these elements in BIM through the Revit® software. Prior to modeling, the point cloud has been optimized so that point density and file size are appropriate for the modeling work being performed. From the point clouds we obtain the digital model of the terrain that is introduced in the software and the geometry of elements.

2 Materials and Methods

The terrestrial laser scanner is a technique that allows to obtain the real geometry, of an element or infrastructure, with high level of detail. There are many examples where terrestrial laser scanners are used for heritage documentation and geometric analysis from a pathological point of view [6–8].

In this case, a geometric survey has been carried out with a FARO Focus 3D X330, [9]. This equipment has a measuring capacity of 330 m range and data acquisition speed of up to 1 million points per second. A cloth from the wall of Badajoz has been lifted and access gates distributed in it, a cloth from the wall of Cáceres and isolated elements, the gate at a bend in the wall of Regina and the gate of the castle of Trujillo. Table 1 shows the number of shots taken for each documented element. The images are taken with a resolution of 6 mm (this resolution is determined as the mesh of points obtained in a shot on a wall perpendicular to the scanner located at 10 m). Given the density of the shots and their overlap, the resolution of the survey was ≤ 5 mm. In [1] point cloud acquisition and processing methods are evaluated for optimal modeling.

The different surveys are carried out from open polygonals and have been oriented in absolute coordinates, establishing the coordinate system ETRS89 UTM H29. For

Table 1. List of scanned elements, number of necessary shots and weight in GB of the files RawData

Elemento documentado	Número de tomas necesarias	Peso del RawData (GB)
Paño de muralla de Cáceres (250 m)	106	17,700
Aljibe del museo de Cáceres	5	0,875
Torre del Horno	18	2,960
Puertas de Muralla de Badajoz	19	3,440
Paño de Muralla de Badajoz (300 m)	90	17,000
Puerta en recodo de muralla de Regina	13	1,760
Puerta de acceso al castillo de Trujillo	6	0,920
Sumatorio	*257*	*44,655*

georeferencing, control points distributed in the external areas of the infrastructure and identifiable in the 3D model have been used. These points have been measured with Leica 1200 GNSS equipment in cinematic mode. The georeferencing of the 3D point cloud model allows the possibility of studying the position of the model with respect to previous cartographies or subsequent studies that may be carried out.

The architectural surveys carried out have been of singular elements in the walls of Extremadura, the object of the HBIMEX project. The aim of the work is the development of an HBIM methodology for the conservation of the built historical heritage. The different scanned elements present singularity in the data collection. Architectural lifting of doors or cisterns is a simple task as it involves the connection of no more than 4 to 6 shots executed according to an open polygonal. The most complex cases at the time of data collection and union of the same has been the architectural survey of cloths of walls of Cáceres and Badajoz, it is obvious that the more shots taken the more complex is a survey. In the case of Cáceres the wall is sheltered by constructions of intrados houses and extradós of wall. The existence of dwellings complicates the completeness of the documentation, as there are sections of walls or singular elements that belong to or are accessed through private dwellings. The documentation carried out was as complete as possible depending on the technique used, the terrestrial laser scanner. In the case of Badajoz, the wall is free and accessible in all its area, except for some sections of wall that are buried inside the wall. In this case the task of data collection is simpler since from a land point of view you manage to document the wall in its completeness.

The result of the architectural survey is a 3D point cloud model. This model can have a high point density, depending on the configuration of the data collection. When introducing the point cloud into Revit®, the optimum resolution/size ratio for the model was studied. Prior to modeling the point clouds have been processed in order to optimize them in RealWorks® software [10], Fig. 2 shows the workflow developed in this software. The point clouds have been classified according to the algorithms of the software for outdoor stays, in this case it allows to extract dreams, walls, streetlights, electrical lines and vegetation. With this step and a manual review, the point cloud is broken down into different clouds sorted by element type.

Fig. 2. Workflow performs RealWorks® software with point clouds.

From the resulting point clouds two main products are extracted, the architectural elements of the wall and the terrain. The point cloud of architectural elements is loaded into a project created with ReCap® software and this is directly integrated into Revit®. For the terrain model in RealWorks®, a mesh is created from the ground points and contour lines are generated, these contour lines are exported in CAD format and directly integrated into Revit®.

3 Results

From the work carried out, 7 point clouds were obtained, one from each of the elements studied, these clouds are georeferenced in the coordinate system ETRS89 UTM U29. Both the terrain and the geometry of the element are extracted from the clouds obtained. When it came to extracting the terrain model from the point cloud, certain difficulties were encountered for the two largest surveys. In the case of Badajoz, the point cloud of the terrain was not continuous, making it difficult to obtain the contour lines. This has been solved by using the point cloud surface of the IGN's LIDAR coverage, from the second rebound the terrain is extracted. In the case of Cáceres, given the buildings attached to the wall and the resolution of the IGN's LIDAR flight, it has been chosen to work with the surface obtained from the laser scanner survey.

The point clouds of the elements have been sampled according to the mesh spacing of resolution points 0.5–2.5–5–10–15–20 cm, Fig. 3 shows the volume of points for each of the resolutions in the different scanned elements. It is observed that from a resolution of 10 cm the difference in numbers of points in the different clouds is minimal.

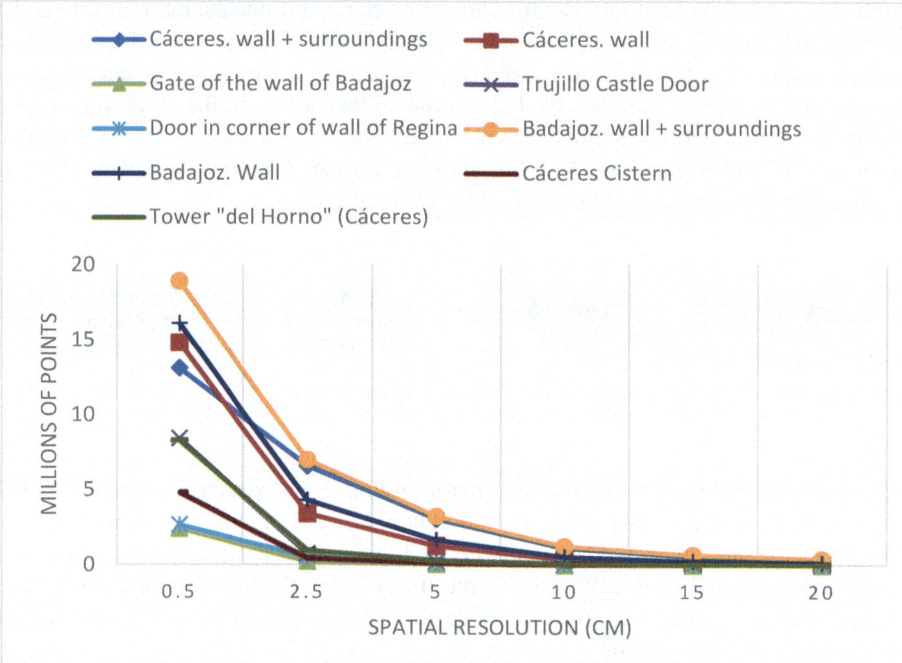

Fig. 3. Ratio of number of points and calculated spatial resolution

Figure 4 shows the graphical results of reducing the density of points for the lifting of one of the gates of the wall of Badajoz. It is noted that both the full resolution and the 0.5 cm resolution are detailed at a glance in such a way that you can model small objects such as your own factory parts. The resolutions of 2.5 and 5 cm offer a geometric definition of the coarsest element; the details of the model are lost. The optimal resolution level depends on the purpose of the modeling to be performed.

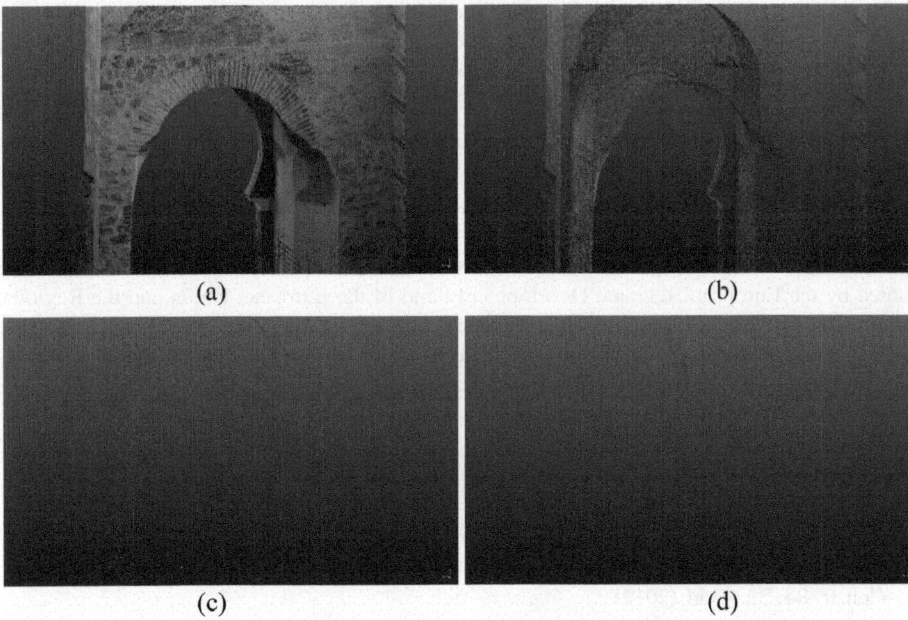

Fig. 4. Results of cloud of gate points in the wall of Badajoz. (a) maximum resolution taken; (b) resolution of 0.5 cm; (c) resolution of 2.5 cm; and (d) resolution of 5 cm.

These are levels of detail that vary depending on the objectives of the modeling. The American Institute of Architects [11] establishes the following: LOD 100, LOD 200, LOD 300, LOD 350, LOD 400 and LOD 500, the first level being equivalent to a representation of the object to be modelled, without the need for geometric precision and the LOD 500 a verification of the element that has a precise geometric definition in terms of shape, location, orientation and dimensions. In [12], the process of creating a parametric family of Almohade door façade is shown, in this case starting from a point cloud with a resolution of 2.5 cm that is sufficient for a LOD200 modeling, the work is developed in the Revit® software. In most cases, this resolution will be sufficient for LOD 200 and LOD 300 levels, while a LOD 350 level will be conditioned by the need for more precise geometric information, where the construction systems, materials and auxiliary elements have to be defined in greater detail, in this case precision levels of around 0.5 cm are recommended. From a LOD 400 level onwards, as much geometric information as possible will be required, as the resulting model is a faithful reproduction of the original, including both the material and constructive information of the element, as well as the deformations themselves, both associated with its construction process and with injuries.

4 Conclusions

For the SCAN to BIM work carried out the resolutions necessary to carry out a LOD200 and LOD300 modeling are far above what is necessary, i.e. more data is taken than is really necessary. The scan resolution could be reduced if the data collection and processing methodology allows it.

For modeling works that require a LOD350 or higher the necessary resolution is at least 0.5 cm, in these cases it would be convenient to study the relations nº of points/resolution for higher resolutions such as 0.1–0.2–0.3–0.4 cm.

The work done is sufficient to cover any level of geometric modeling in BIM. The knowledge and management of information facilitates the work of modeling at the computer level. Working with clouds of suitable resolution points does not reduce the performance of the PC.

Acknowledgements. The presentation of this work has been possible thanks to the financing granted by the European Regional Development Fund of the European Union and the Regional Government of Extremadura to the NEXUS research group through the GR18053 aid and to the IB16115 research project in public R&D&I centres thanks to the European Regional Development Funds and the Regional Government of Extremadura.

References

1. Rebolj, D., Pučko, Z., Babič, N.Č., Bizjak, M., Mongus, D.: Point cloud quality requirements for Scan-vs-BIM based automated construction progress monitoring. Autom. Constr. **84**, 323–334 (2017)
2. Baik, A.: From point cloud to jeddah heritage BIM nasif historical house–case study. Dig. Appl. Archaeol. Cult. Heritage **4**, 1–18 (2017)
3. Khodeir, L.M., Aly, D., Tarek, S.: Integrating HBIM (Heritage Building Information Modeling) tools in the application of sustainable retrofitting of heritage buildings in Egypt. Procedia Environ. Sci. **34**, 258–270 (2016)
4. Fernández, M.S., Blasco, J.J.S., Mogollón, F.-J.G., García, E.M.: Sistemas de información geográfica aplicados al inventario, catalogación y gestión de infraestructuras patrimoniales en Extremadura. In: IV Reunión del Grupo de Arte Contemporáneo del GEIIC: (Grupo Español del International Institute of Conservation), pp. 358–363 (2018)
5. www.amigosdeloscastillos.es (2017)
6. Berenguer Sempere, F.: Aplicación de técnicas geodésicas y técnicas avanzadas de captura, escáner láser y termografía, en el estudio patológico de la estructura y la cimentación en la restauración integral de la Catedral de Santa María de la Asunción de Coria. Doctoral Thesis, Expresión Gráfica, Extremadura (2014). http://hdl.handle.net/10662/1622
7. Borrazás, P.M., Blanco-Rotea, R., Rodríguez Paz, A.: La documentación geométrica de elementos patrimoniales con láser escáner terrestre. La experiencia del Lapa en Galicia. Cuadernos de Estudios Gallegos **56**(122), 33–65 (2009)
8. Serrano, F., Saumell, J., Berenguer, F.: XXX Análisis de resultados métricos de una nube de puntos y una medición directa en el patrimonio edificado. El Santuario de La Montaña en Cáceres. Informes de la Construcción 66 (2014)
9. Faro Focus 3D X330, ed. (2017). http://www.faro.com/es-es/productos/topografia-3d/laser-scanner-faro-focus-3d/informacion-general
10. RealWorks, TRIMBLE, Ed., 10.4 ed. (2017). http://www.trimble.com/3d-laser-scanning/realworks.aspx
11. BIM Forum: Forum: LOD Specification Volume Part I (2019)
12. Chorro-Domínguez, F.J., Marín-Miranda, M.J., Cortés, J.P.: Revit family from point cloud in heritage elements. the almohade gate. In: Presented at the APEGA 2019, Sevilla (2019)

Plates Pre-dimensioning for Injection Molds by Using Computational Design and Inclusion of the Parts in BIM Methodology

Daniel Diaz-Perete, Abelardo Torres-Alba, Jorge Manuel Mercado-Colmenero, and Cristina Martin-Doñate(✉)

Department of Engineering Graphics, Design and Projects, University of Jaén, Campus Las Lagunillas, S/N, 23071 Jaen, Spain
cdonate@ujaen.es

Abstract. This paper presents an algorithm for pre-dimensioning of plates of injection molds based on the measures of the part which will be manufactured. The pre-dimensioning plates are: the cavity plate, the core plate and the support plate. These plates will be pre-dimensioned according to the pressure loads generated in each case, derived from the specific requirements needed for the injection and manufacture of each plastic part. Additionally, the used equations will be introduced into computational design software linking each variable with his relative parameter in order to generalize the process of injection mold designing for simple plastic parts. After calculus process and the pre-dimensioning of the plates the final geometry will be introduced into BIM methodology using specifics BIM software in order to measure the quantities of material required, to collect dimensions and number of parts needed in an automatic way to get other interesting attributes like budget control. The implementation of BIM methodology in injection molds industry to project development gives engineering field many new possibilities of co-working between companies, making easier the communication of their design specialists optimizing the process of creation of new injection molds. This methodology will contribute to save resources improving the quality of the projects final results

Keywords: Injection molding · Computer aided manufacturing · CAD

1 Introduction

An injection mold is a high precision mechanical system designed to produce plastic parts [1]. Nowadays plastic injection molding is the most widespread manufacturing method in the industry, this process transforms the melt plastic flow into the geometry of the final part designed, located between the cavity and core plates [2, 3]. Though the majority of plastic parts are produced by injection molding method, injection molds design is still performed empirically, usually based on the experience of the mold designer [4]. The dimensioning tasks of the different mold components are highly dependent upon the knowledge of how the different geometrical and technological

F. Cavas-Martínez et al. (Eds.): INGEGRAF 2019, LNME, pp. 535–543, 2020.
https://doi.org/10.1007/978-3-030-41200-5_59

mold parameters interact with each other. Mold design requires qualified engineers from different disciplines and, in order to simulate the mold, each designer requires specifics software which the company must acquire [5, 6].

The globalization of the markets forces the producers to decrease the product development time, increasing at the same time the diversity of high quality products. The short-term economy of the company could depend on the award of the contract. This means that companies cannot afford to waste so much time to form new specialists in injection molding designing so it is a current trend to integrate knowledge and technology into automated design applications that engineers can use, reducing design and modelling time at same time that possible human errors during the design time. The applications developed so far required a lot of interaction on the part of the user and needed to the mold dimensioning values that have to be previously calculated outside the application using traditional parameters. These calculations require an advanced knowledge of rheological mold design, simulation, refrigeration and structural analysis [7].

This paper presents some algorithms to pre-dimensioning the cavity plate, the core plate and the support plates, introducing several geometrical parameters related with the number of cavities, the part dimensions and the injection pressure. The analytical equations proposed are able to calculate the minimum thickness of the mold plates and the rest of parameters. After the plates pre-dimensioning is introduced into the BIM methodology [8] the control of some interesting attributes linked with the mold design are obtained.

Geometric algorithms have been developed in the Matlab R2013 software [10] and the automated geometric modeling module has been performed Generative Components v8i. All of this from an MSI notebook with an Intel (R) Core (TM) i-77700HQ CPU @ 2.80 GHz

2 Dimensioning of the Main Plates of the Injection Mold

In this section we proceed to the dimensioning of the cavity plate, core plate and the support plate see Fig. 1. In a first phase, an analysis of the mechanical strength of the cavity and core plates is carried out in order to determine the separation H_x and H_y (in the coordinate axes X, Y) see Fig. 2 between each one of the mold cavities and between the mold cavities and the guide columns see (1), (2) [9, 11]. To do this, the maximum deflection allowed by the plates has been analyzed;

$$H_x = \sqrt[3]{\frac{12 \cdot P_{injection} \cdot l^4}{384 \cdot E_{steel} \cdot d_{max}}} \tag{1}$$

$$H_y = \sqrt[3]{\frac{12 \cdot P_{injection} \cdot w^4}{384 \cdot E_{steel} \cdot d_{max}}} \tag{2}$$

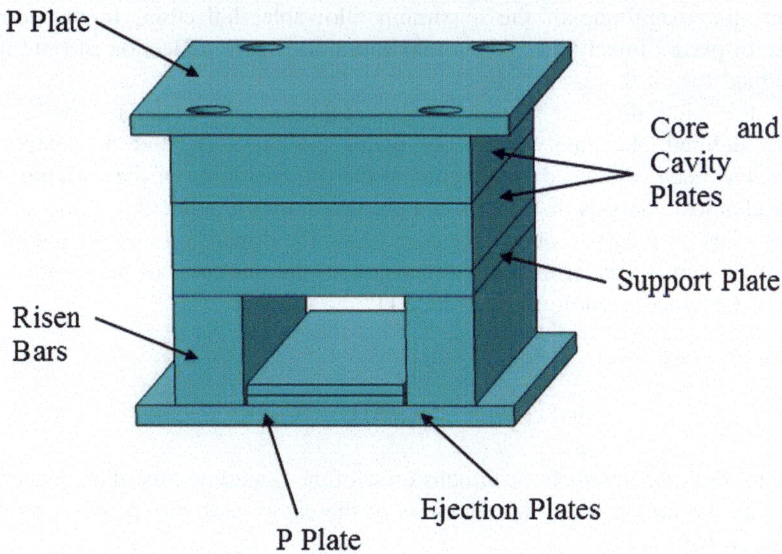

Fig. 1. Mold plates location scheme

Where $P_{injection}$ [MPa] represents the injection pressure, l [mm] represents the length of the bounding box of the plastic part on the X coordinate axis, *w* [mm] represents the length of the bounding box of the plastic part on the axis Y coordinate, E_{steel} [MPa] represents the elastic modulus of the plastic mold material and d_{max} [mm]

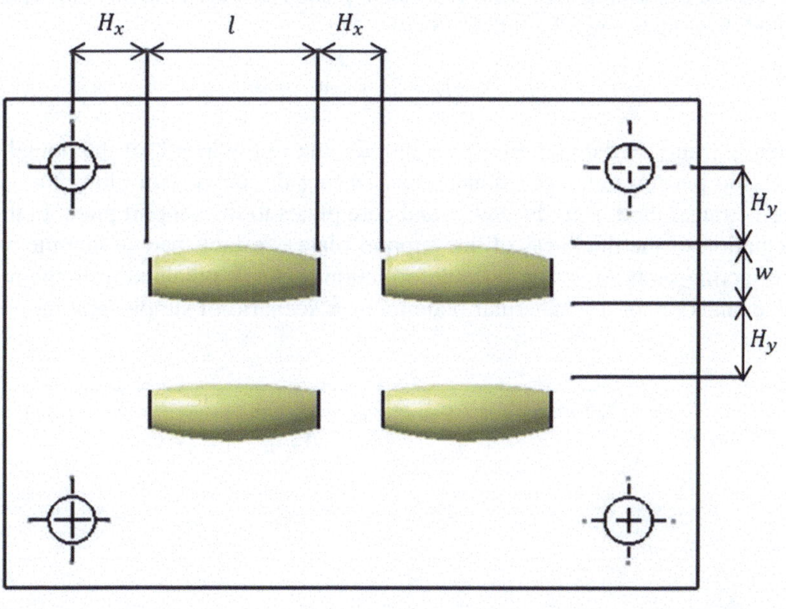

Fig. 2. Geometric parameters for dimensioning the main plates of the injection mold

represents the magnitude of the maximum allowable deflection. In the admissible tolerance in plastic injection molds, a maximum allowable deflection of 0.02 mm has been defined for all the case studies.

Once the safety distance between cavities and between cavities and guide columns has been defined, the cavity and core plates thickness H_z and the support plate thickness has been calculated. Analogous to the dimensioning of the variables H_x and H_y, a mechanical analysis of the resistance and maximum deflection of the cavity and core plates has been carried out. In the case where the dimension l of the bounding box of the plastic part is greater than the dimension w, the thickness of the cavity and core plates is established according to (3) [9, 11].

$$H_{z0} = \sqrt[3]{\frac{5 \cdot P_{injection} \cdot l^4 \cdot w}{32 \cdot E_{steel} \cdot d_{max} \cdot (H_x + n_{cavities} \cdot (l + H_x))}} \quad (3)$$

And for the case in which the dimension w of the bounding box of the plastic part is greater than the dimension l, the thickness of the cavity and core plates is established according to (4) [9, 11].

$$H_{z0} = \sqrt[3]{\frac{5 \cdot P_{injection} \cdot w^4 \cdot l}{32 \cdot E_{steel} \cdot d_{max} \cdot (H_y + n_{cavities} \cdot (w + H_y))}} \quad (4)$$

Where $n_{cavities}$ represents the number of injection mold cavities, defined by applying the economic criteria of mold theory [9]. It should be noted that in order to determine the final thickness of the cavity and core plates, the length or depth (depending on the demolding direction D_d) of the plastic part area that is demolded by the cavity and core plate must be added (5) [9, 11].

$$H_z = H_{z0} + v \quad (5)$$

Where v [mm], is the mentioned magnitude due to the length of the bounding box of the plastic part on the Z coordinate axis. During the process of filling the injection pressure is transmitted from the cavity and core plates to the support plate. In this way, the calculation of the thickness of this support plate is established so that the injection pressures acting on it do not produce deformations. This is obtained from the materials resistance analysis for a rectangular plate under a scenario of simple bending loads (6) [9, 11].

$$H_{support} = \sqrt{\frac{1,5 \cdot n_{cavities} \cdot A_{cavity} \cdot P_{injection} \cdot \left(H_y + \frac{n_{cavities}}{2} \cdot (w + H_y)\right)}{4 \cdot E_{steel} \cdot \left(H_x + \frac{n_{cavities}}{2} \cdot (l + H_x)\right)}} \quad (6)$$

Where A_{cavity} [mm^2] represents the area projected onto cavity and core plates, according to the parting direction D_d of the plastic part (Tables 1 and 2).

Table 1. Input variables for the dimensioning of the main plates of the injection mold

Input variables for the calculation and dimensioning of the mold plates		
Description	Input variables	Units
Bounding box	l, w, v	[mm]
Injection pressure	$P_{injection}$	[MPa]
Steel Young's modulus	E_{steel}	[MPa]
Minimum admissible tolerance	d_{max}	[mm]

Table 2. Output variables of the dimensioning of the main plates

Output variables for the calculation and dimensioning of the mold plates		
Description	Output variables	Units
Distance between cavities, X axis	H_x	[mm]
Separation distance between cavities, Y Axis	H_y	[MPa]
Minimum thickness of cavity and core plates	H_z	[MPa]
Minimum thickness of the support plate	$H_{support}$	[mm]

3 Implementation and Results

The case of study is based in the mold design for a bucket of vertical wind rotor see Fig. 3. Table 3 shows the input values for the main plates variables of the injection mold. Table 4 shows the output values for the main plates variables of the injection mold. Table 5 presents the results of the mold plates BIM Classification after Data-Set and Data-Group definition using Bentley AECOSim Software [12]. Figure 5 shows the configuration of the mold design based in obtained results. Figure 5 shows the mold design obtained by traditional procedure (designed manually by using catalogs) [12] (Fig. 4).

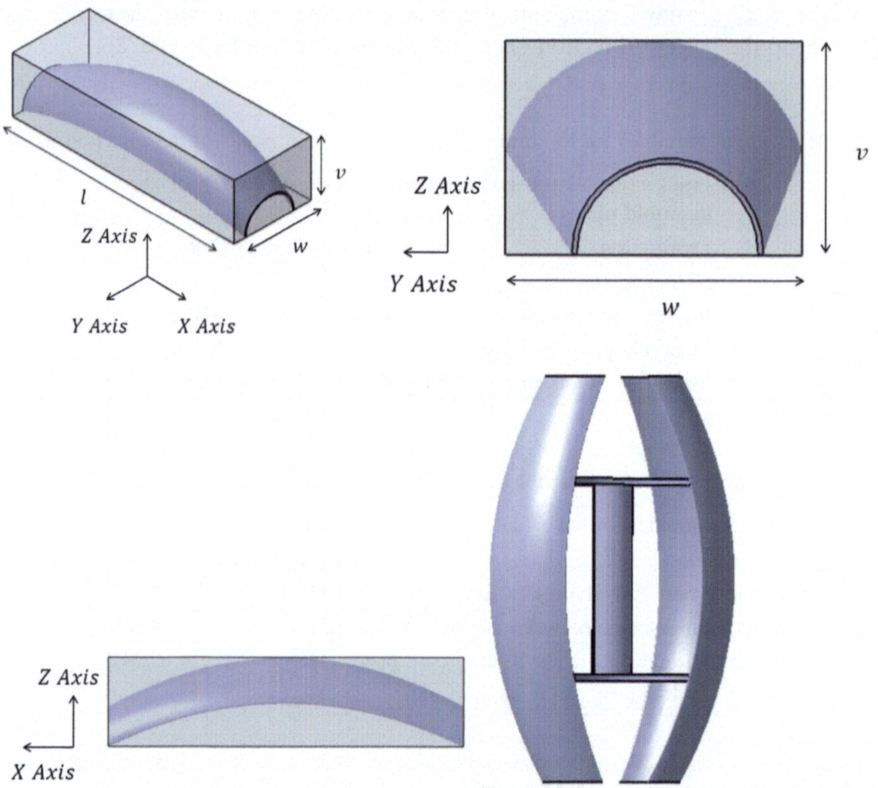

Fig. 3. Case of study

Table 3. Input values for the main plates variables for the injection mold

Input variables for the calculation and dimensioning of the mold plates			
Description	Input variables	Value	Units
Bounding box	l	320	[mm]
	w	115	
	v	57.5	
Injection pressure	$P_{injection}$	600	[MPa]
Steel Young's modulus	E_{steel}	2e5	[MPa]
Minimum admissible tolerance	d_{max}	0.02	[mm]

Table 4. Output values for the main plates variables of the injection mold

Output variables for the calculation and dimensioning of the mold plates			
Description	Output variables	Value	Units
Distance between cavities, X axis	H_x	366.31	[MPa]
Separation distance between cavities, Y Axis	H_y	93.59	[MPa]
Minimum thickness of cavity and core plates	H_z	449.32	[mm]
Minimum thickness of the support plate	$H_{support}$	6.97	[mm]

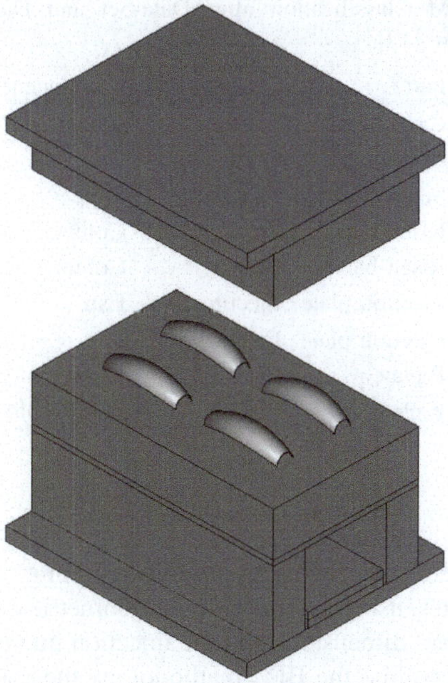

Fig. 4. Configuration of the mold design based in obtained results.

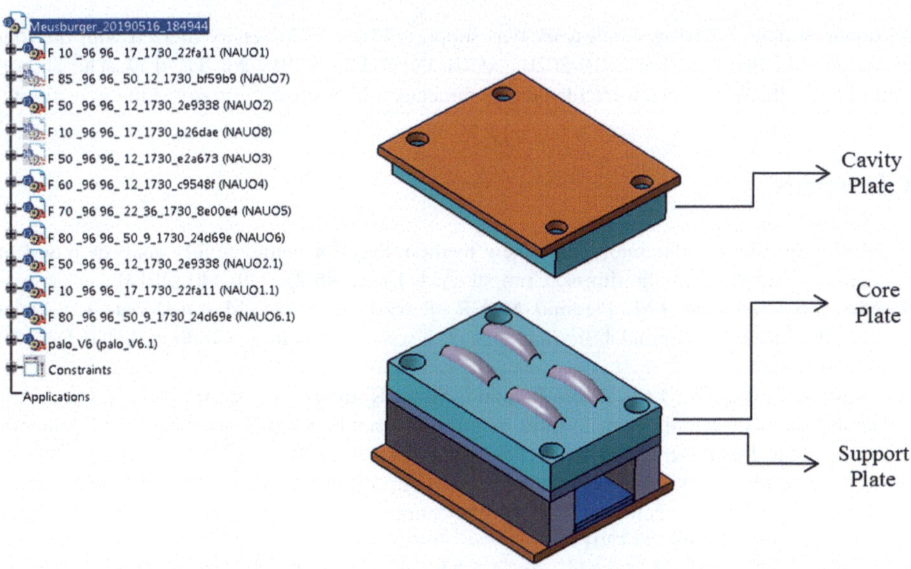

Fig. 5. Mold design manually using catalogs [13]

Table 5. Mold plates BIM Classification after Data-Set and Data-Group definition using Bentley AECOSim Software [12].

ID	Family	Part	Descrip. Part	Component	Quantity	Unit
198368	Mold1	Cavity plate	Cavity	Cavity	0.39909	m^3
198292	Mold1	Core plate	Core	Core	0.39909	m^3
198648	Mold1	Support plate	PlateSupp	Supp	0.06194	m^3
198512	Mold1	Risen bars	Bar	Colum	0.13896	m^3
198544	Mold1	Risen bars	Bar	Colum	0.13896	m^3
198448	Mold1	Ejection plate	Ejection	Exp	0.03264	m^3
198480	Mold1	Ejection plate	Ejection	Exp	0.03264	m^3
198260	Mold1	P plate	PlateP	PlatePmold	0.16301	m^3
198608	Mold1	P plate	PlateP	PlatePmold	0.16301	m^3

4 Conclusions

This paper presents several algorithms to pre-dimensioning the cavity plate, the core plate and the support plates, using for this purpose geometrical parameters related to the number of cavities, the part dimensions and the injection pressure. After the plates pre-dimensioning is introduced into the BIM methodology the control of some interesting attributes linked with the mold design are obtained. The parametric design of the mold parts shortens even more the duration of the design stage within the entire project that involves the creation of a complete injection mold.

Acknowledgments. The research work was supported by the University of Jaen with the Plan de Apoyo a la Investigación 2019-2020- ACCION1 PAIUJA2019-20: TIC-159. The authors would like to thank the reviewers for their comments, which greatly improved the exposition.

References

1. Martin Doñate, C., Paramio, M.R.: New methodology for demoldability analysis based on volume discretization algorithms. Comput.-Aided Des. **45**(2), 229–240 (2013)
2. Mercado-Colmenero, J.M., Paramio, M.A.R., Perez-Garcia, J.M., Martin-Doñate, C.: A new hybrid method for demoldability analysis of discrete geometries. Comput.-Aided Des. **80**, 43–60 (2016)
3. Mercado-Colmenero, J.M., Rubio-Paramio, M.A., Karlinger, P., Martin-Doñate, C.: A new procedure for calculating cycle time in injection molding based on plastic part geometry recognition. Int. J. Adv. Manuf. Technol. **98**, 441 (2018)
4. Mercado-Colmenero, J.M., Rubio-Paramio, M.A., Vizan-Idoipe, A., Martin-Doñate, C.: A new procedure for the automated design of ejection system in injection molds. Rob. Comput. Integr. Manuf. **46**, 68–85 (2017)
5. Mercado-Colmenero, J.M., Muriana, J.A.M., Paramio, M.A.R., Martín-Doñate, C.: An automated manufacturing analysis of plastic parts using faceted surfaces. In: Eynard, B., Nigrelli, V., Oliveri, S., Peris-Fajarnes, G., Rizzuti, S. (eds.) Advances on mechanics, design engineering and manufacturing. Lecture Notes in Mechanical Engineering. Springer, Cham (2017)

6. Mercado-Colmenero, J.M., Martin-Doñate, C., Rodriguez-Santiago, M., et al.: Int. J. Adv. Manuf. Technol. **102**, 1719 (2019). https://doi.org/10.1007/s00170-018-03235-3
7. Park, H.S., Dang, X.P., Nguyen, D.S., Kumar, S.: Design of advanced injection mold to increase cooling efficiency. Int. J. Precis. Eng. Manuf.-Green Technol., 1–10 (2019)
8. Eastman, C., Teicholz, P., Sacks, R., Liston, K.: BIM Handbook: A Guide to Building Information Modeling, for Owners, Managers, Designers, Engineers, and Contractors. Wiley, Hoboken (2008)
9. Kazmer, D.: Injection Mold Design Engineering, 1st edn. Hanser, Munich (2007)
10. Matlab, R2013a. https://mathworks.com
11. Menges, G., Mohren, P.: How to Make Injection Moulds, 3rd edn. Hanser, Munich (2001)
12. https://www.bentley.com/es/products/product-line/building-design-software/openbuildings-designer
13. https://www.meusburger.com/

Analysis of the Application of the Superblocks Model in the Context of the PEMUS in the City of Málaga

Francisco J. Ortiz-Zamora(✉), E. Beatriz Blázquez-Parra, Francisca Castillo-Rueda, Patricia Mora-Segado, and Francisco Benítez-Villaespesa

Universidad de Málaga, c/ Doctor Ortiz Ramos s/n, 29071 Málaga, Spain
fortiza@uma.es

Abstract. Episodes of atmospheric pollution are increasingly frequent in urban environments and metropolitan areas of large cities such as Madrid or Barcelona. By 2020, many cities will prohibit entry to the most polluting vehicles and will also impose specific actions on traffic. Based on the measures included in the Special Plan for Sustainable Urban Mobility (PEMUS) of the municipality of Málaga to meet the objectives set out by the EU in its 20-20-20 strategy, we will analyze the feasibility of the Superblocks proposal for Málaga. The methodology used for modeling is based on environmental capacity, which consists of removing space for private vehicles and assigning this to pedestrians and other means of transport, as proposed by the Urban Ecology Agency of Barcelona for the development of the concept of Superblocks, using the information published in the form of Open Data. With the modeling of the Superblocks proposal, some of the actions included in the PEMUS have been simulated and their impact has been assessed on the planning and management of the road infrastructure and transport networks operating in the city of Málaga.

Keywords: GIS · Superblocks · Mobility · Contamination · Open Data

1 Introduction

In the context of the preliminary draft of the Law on Climate Change and Energy Transition [1], the sixth edition of the Global Environment Outlook report (GEO-6) [2], recently published by the United Nations, confirms that the objectives of the European Union are not being met when it comes to matters of sustainability [3]. Some of the actions established as priorities affect the field of transport, regarding urban mobility as a key element in making cities more sustainable. In large and medium-sized populations, urban mobility is related to increases in energy consumption, the quality of the air we breathe, the diseases and deaths caused by pollution, noise, traffic congestion and lost hours, and road accidents, among others [4, 5].

Many cities have proposed strategies to achieve more sustainable development within the framework of the action plan proposed by the UN, known as Agenda 21 [6]. In the case of Málaga, the proposed measures have been established within the Special

F. Cavas-Martínez et al. (Eds.): INGEGRAF 2019, LNME, pp. 544–553, 2020.
https://doi.org/10.1007/978-3-030-41200-5_60

Plan for Sustainable Urban Mobility (PEMUS), which is a specific plan developed by the City's Urban Planning department (PGOU) [7–9]. In the PEMUS, a set of actions is established, the main objective of which is to limit the increasing environmental impacts of mobility linked to the use of private vehicles. The proposal of Superblocks is one of the most advanced measures to reduce the use of private vehicles [10, 11]. As shown in Fig. 1, the concept of the Superblocks consists of the reorganisation of surface mobility by structuring the road network into a network of basic roads. This network is made up of urban areas where private vehicle traffic is only allowed outside the area.

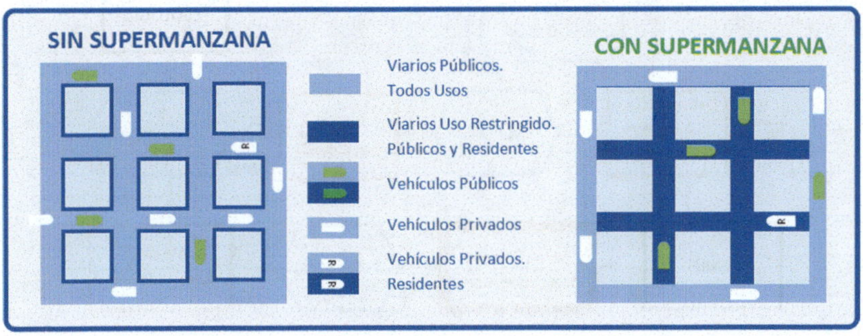

Fig. 1. Graphic analysis of Superblock model.

The current article has the following aims:

- To vectorize Superblock model which has been implemented by the Barcelona Ecology Agency for the city of Málaga [12] through Geographic Information Systems (GIS).
- To apply the concept of Superblock as a tool for organizing transport networks and mobility planning to give preference to alternative transportation modes over private vehicles [13–15].
- To evaluate the indicator *of the public roadway for car transit and public surface transportation (Vtm)* according to Eq. (1) using Superblocks proposal, along with its impact on planning. This indicator measures the percentage of road space for motorized transport over the road length [16–20].

$$Vtm = \frac{\sum Lvtm}{Ltv} \times 100 \tag{1}$$

Where Lvtm is the length of public road used for motorized transport and Ltv is the total length of the public road.

Evaluation parameter:

- Minimum value <40% of public road used for motorized transport.
- Desirable value <25% of public road used for motorized transport.

2 Methodology

The methodology used is based on environmental capacity, and consists of removing space from private vehicles and reallocating this space to pedestrians and other means of alternative transportation [9]. Figure 2 shows the set of operations carried out. To begin the construction of the model we will proceed to identify the necessary geographic elements.

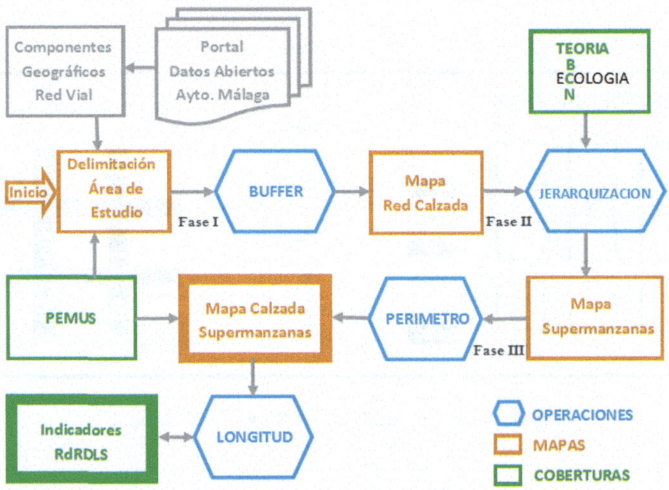

Fig. 2. Diagram of spatial operations developed in the GIS.

2.1 Data

The data used come from the various sources of Open Data that exist in the administration, primarily the open data portal of the City Council of Málaga [21] and the open data catalog of the Andalusian Regional Government [22]. The starting point is the georeferenced data and base cartography that contain the required isolated or discrete geographic components, such as lines and zones.

The files used have a Shapefile exchange format and represent the delimitation data of the roadways and sidewalks of urban roads, the building volumes located inside plots, the surfaces occupied by different water components, the margins of the roads, highways, and motorways, the delimitation of cadastral and non-cadastral parcels, the public land areas between the maritime-terrestrial zone and the urban maritime façade, the identification and geographical location of the stretches of road, and the public spaces of the Municipality of Málaga, all of which use the ETRS89 datum coordinate system.

2.2 Preparation of the Data

For the characterization of the geographic data, the Superblocks Zone 30, and Industrial Zone regions were initially constructed, as shown in Fig. 3, delimiting the study area to only Superblock region. Both Zone 30, which is composed of streets where there is a speed limit, and the Industrial Zone, which is the area corresponding to commercial and industrial estates, are areas where the use of the road is not restricted to any type of vehicle, and therefore the indicator used would not be applicable.

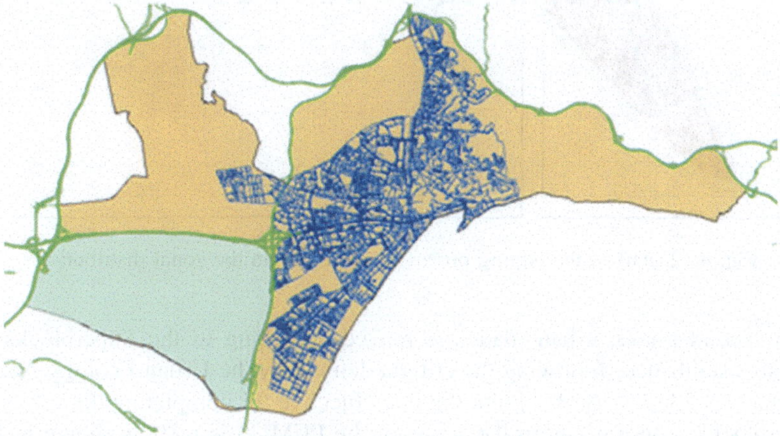

Fig. 3. Delimitation of the study area.

2.3 Spatial Data Management

To obtain the necessary operators of the motorized transport road indicator (Vtm), a process was followed for editing the spatial data.

In a first phase, the geographical components of the types of lines that make up the road are transformed into zone form. The process is conducted by editing all the open lines to close them and assign the polygon according to the example shown in Fig. 4.

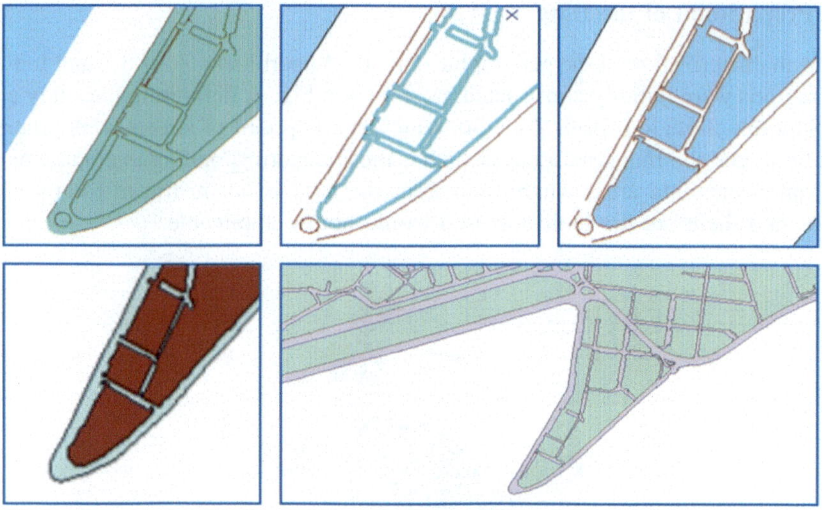

Fig. 4. Detail of the editing process used to obtain the zonal distribution.

In a second phase, urban roads are ranked according to the Superblocks model previously established, following the criteria defined by the Urban Ecology Agency of Barcelona [12, 23, 24]. At this point the first objective for completing the vectorization of Superblocks proposal within the scope of the PEMUS is met, as shown in Fig. 5.

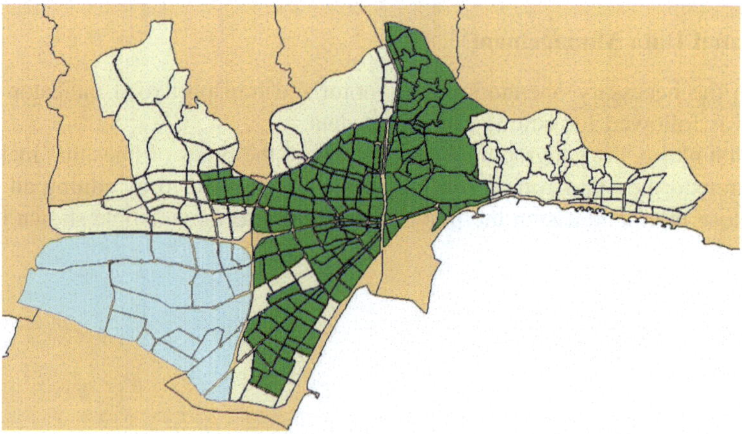

Fig. 5. Proposed vectorization of PEMUS Superblocks.

Finally, in a third phase, once the distribution into zonal units has been completed, the road network forming the roads surrounding the Superblocks can be obtained, as indicated in Fig. 6. By integrating and combining all the information obtained, it is possible to distinguish between the road network corresponding to all the urban roads and the road network of the perimeter roads of the Superblocks for the delimited study area.

Fig. 6. Detail of road network perimeter roads of the Superblocks.

3 Results

On the basis of the attributes that can be managed by the Geographic Information Systems, the geometry of the road network is calculated, both for all urban roads and for the perimeter roads of Superblocks within the delimited study area, as shown in Table 1.

Table 1. Attributes of road length.

Tabla

region supermanana calzada

Shape_Area	FID_da_car	RASANTE	CODFACTORE	longitud
56511568,1341	1386	1	3	13,908007
56511568,1341	1432	1	3	134,2845
56511568,1341	3430	1	15	238,228547
56511568,1341	3674	1	3	11,784401
56511568,1341	4224	1	8	22,163174
56511568,1341	4337	1	15	133,793671
56511568,1341	5088	1	15	278,345852
56511568,1341	5116	1	15	111,337143
56511568,1341	5538	1	7	114,909308
56511568,1341	5770	1	7	164,223404

1 (0 de 23808 Seleccionado)

region supermanana calzada

From these attributes we can obtain the data of the total length of public road and that of the Superblocks perimeter road, which provide the determinant operators of the Vtm indicator, as shown in Table 2.

Table 2. (a) Total length of public road and (b) Roadway perimeter Superblocks.

Campo: longitud_2

Estadísticas:

Conteo:	6750
Mínimo:	0,004849
Máximo:	661,413619
Suma:	281300,406014
Media:	41,674134
Desviación estándar:	44,820488
Valores nulos:	0

(a)

Campo: longitud_2

Estadísticas:

Conteo:	2757
Mínimo:	0,144894
Máximo:	623,335962
Suma:	92877,035715
Media:	33,687717
Desviación estándar:	45,314737
Valores nulos:	0

(b)

Table 2(a) calculates the total road length of the current basic road network integrated in the region of future Superblocks, while Table 2(b) calculates the length of the basic road network, which will be generated by the formation of Superblocks.

According to Eq. (2)

$$Vtm = \frac{92877\,\text{m}}{281300\,\text{m}} \times 100 = 33\% \tag{2}$$

As can be seen in Fig. 7, with the application of the Superblock model, the road space for motorized transport reaches 33%, remaining below the minimum value of the evaluation parameter, although there is still scope for improvement to reach the desired value of <25%.

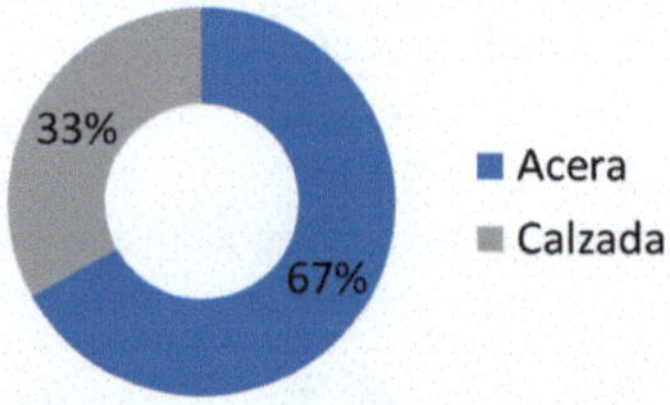

	Longitud (Km)	Incremento
Actual	281,3	-67%
Propuesta	92,9	

Fig. 7. Resulting proportion of sidewalk and roadway.

4 Conclusions

It is clear that spatial planning influences mobility patterns. A plan of the city using criteria for reducing the dependence on motorized vehicles would facilitate the transition towards more sustainable mobility. Several municipalities both in Spain and other countries have developed similar proposals during the drafting or revision of their respective Sustainable Urban Mobility Plans (PMUS), implementing the Superblocks scheme as a formula for territorial planning to help make cities more sustainable. Particularly noteworthy is the city of Vitoria-Gasteiz (winner of the European Green Capital Award 2012), where its application (recognized as Best Practice by Un-Habitat) is expected to reduce the public space allocated to private vehicles by up to 29%. A further example can be observed in the city of Barcelona, where pilot tests of application are being conducted in different districts such as Gràcia, with results showing a reduction of public space allocated to private vehicles of between 32.2% and 19.8%. The analysis of the proposed Superblocks model suggests the possibility of reducing the number of vehicles in circulation, without compromising the functionality of urban mobility in the city. In a certain sense, for the road hierarchy in the city of Malaga, with the creation of a basic network that delimits the large streets of the city and the traffic that circulates through them and Zones 30 for those places where a restriction of high volume traffic is necessary, its basic lines resembles and correspond to the concept of Superblocks.

References

1 Anteproyecto de Ley de Cambio Climático y Transición Energética (LCCTE). https://www.lamoncloa.gob.es/consejodeministros/Paginas/enlaces/220219-proyecto.aspx. Accessed 15 Mar 2019
2. UN Environment. Global Environment Outlook – GEO-6: Healthy Planet. Healthy People, Nairobi (2019). https://doi.org/10.1017/9781108627146
3. Normas sobre emisiones de CO2 para turismos y furgonetas: el Consejo confirma el acuerdo sobre unos límites más restrictivos. https://www.consilium.europa.eu/es/press/press-releases/2019/01/16/co2-emission-standards-for-cars-and-vans-council-confirms-agreement-on-stricter-limits/. Accessed 16 Apr 2019
4. Comisión Europea, Dirección General de Movilidad y Transportes, Dirección de Inversiones, Innovación y Sostenibilidad, Transporte, Unidad B4 - Transportes sostenibles e inteligentes; Movilidad urbana sostenible: Política europea, práctica y soluciones (2017). ISBN 978-92-79-66655-1
5. Ministerio para la Transición Ecológica; Evaluación de la calidad del aire en España 2017. https://www.miteco.gob.es/es/calidad-y-evaluacion-ambiental/temas/atmosfera-y-calidad-del-aire/informeevaluacioncalidadaireespana2017_tcm30-481655.pdf. Accessed 19 Jan 2019
6. Agenda 21. The United Nations Program of Action from Rio. https://web.archive.org/web/20090420073232/http://www.un.org/esa/sustdev/documents/agenda21/spanish/agenda21sptoc.html. Accessed 14 Mar 2017

7. IDEA: Instituto para la Diversificación y Ahorro de la Energía; PMUS: Guía práctica para la elaboración e implantación de planes de movilidad urbana sostenible (2006). ISBN 978-84-86850-98-2
8. Ayuntamiento de Málaga; Plan Especial de Movilidad Urbana Sostenible (PEMUS). https://movilidad.malaga.eu/opencms/export/sites/movilidad/.content/galerias/Documentos-del-site/PEMUS.pdf. Accessed 30 Jan 2019
9. Ayuntamiento de Málaga, Gerencia Municipal de Urbanismo; Plan General de Ordenación Urbana (PGOU). https://www.malaga.eu/recursos/urbanismo/pgou_ap2/PGOU2011AD1.html. Accessed 30 Jan 2019
10. Rueda, S.: Las Superblocks: reinventando el espacio público, reinventando la ciudad. Capítulo del libro de Usón, E.: "La Nueva Sensibilidad Ambiental. Arquitectura y Sostenibilidad en España. 2000–2005", Capsúnion Ediciones (2006)
11. Mehdipanah, R., Novoa, A.M., León-Gómez, B.B., López, M.J., Palència, L., Vasquez, H., Díez, È., Borrell, C., Pérez, K.: Effects of Superblocks on health and health inequities: a proposed evaluation framework. J. Epidemiol. Community Health **73**, 585–588 (2019)
12. Agencia de Ecología Urbana de Barcelona: Elaboración de una propuesta de Superblocks en el marco de la revisión del PEMUS de Málaga. https://www.omau-malaga.com/agendaurbana/subidas/archivos/arc_221.pdf. Accessed 05 Apr 2019
13. Rueda, S.: Superblocks for the design of new cities and renovation of existing ones: Barcelona's case. In: Nieuwenhuijsen, M., Khreis, H. (eds.) Integrating Human Health into Urban and Transport Planning, pp. 135–153. Springer, Cham (2019)
14. Delso, J., Martín, B., Ortega, E.: A new procedure using network analysis and kernel density estimations to evaluate the effect of urban configurations on pedestrian mobility. The case study of Vitoria Gasteiz. J. Transp. Geogr. **67**, 61–72 (2018)
15. Gyurkovich, M., Poklewski-Koziell, D., Duarte, C.M.: Supermanzana in practice: ability to create people friendly spaces upon the example of selected Barcelona-based projects. IOP Conf. Ser. Mater. Sci. Eng. **471**(9) (2019). Article Number 092010
16. Red de Redes de Desarrollo Local Sostenible; Sistema Municipal de Indicadores de Sostenibilidad. https://www.miteco.gob.es/es/calidad-y-evaluacion-ambiental/temas/medio-ambiente-urbano/SISTEMA_MUNICIPAL_INDICADORES_SOSTENIBILIDAD_tcm30-181856.pdf. Accessed 05 Apr 2019
17. Red de Redes de Desarrollo Local Sostenible; Sistema de indicadores y condicionantes para ciudades grandes y medianas. https://www.miteco.gob.es/es/calidad-y-evaluacion-ambiental/temas/medio-ambiente-urbano/INDICADORES_CIUDADES_GRANDES_Y_MEDIANAS_tcm30-181855.pdf. Accessed 05 Apr 2019
18. Agencia de Ecología Urbana de Barcelona: Plan Especial de Indicadores de Sostenibilidad Ambiental de la Actividad Urbanística de Sevilla. https://www.urbanismosevilla.org/areas/sostenibilidad-innovacion/Sostenibilidad/otros-documentos/borrador-del-plan-especial-de-indicadores-de-sostenibilidad-ambiental-de-la-actividad-urbanistica-de-sevilla/view. Accessed 05 Apr 2019
19. Agencia de Ecología Urbana de Barcelona: Plan de Indicadores de Sostenibilidad Urbana de Vitoria-Gasteiz. https://www.vitoria-gasteiz.org/docs/wb021/contenidosEstaticos/adjuntos/es/89/14/38914.pdf. Accessed 17 May 2019
20. Agencia de Ecología Urbana de Barcelona, Ajuntament de Barcelona: Dossier "Superblocks de Gràcia". https://www.bcnecologia.net/sites/default/files/prensa/dossier_Superblocks_gracia_es.pdf. Accessed 05 Apr 2019
21. Portal de Datos Abiertos del Ayuntamiento de Málaga. https://datosabiertos.malaga.eu/. Accessed 30 Jan 2019

22. Catálogo de Datos Abiertos de la Junta de Andalucía. https://www.juntadeandalucia.es/datosabiertos/portal.html. Accessed 30 Jan 2019
23. Agencia de Ecología Urbana de Barcelona; Modelo conceptual, Nuestras ideas: Supemanzana. https://www.bcnecologia.net/es/modelo-conceptual/supermanzana. Accessed 14 Mar 2017
24. Rueda, S., De Cáceres, R., Cuchí, A., Brau, Ll.: El urbanismo ecológico: Su aplicación en el diseño de un ecobarrio en Figueres. Agencia de Ecología Urbana de Barcelona, Barcelona (2012). ISBN 978-84-615-6947-2

An Accurate Structural Reconstruction in 3D of the Hypostyle Hall at the Ancient Egyptian Temple Complex of Karnak

I. A. Liarte Fernando, M. C. Juárez Castelló(✉), and F. J. Marrodán Esparza

Departamento de Ingeniería Mecánica, Universidad de La Rioja, C/ San José de Calasanz 31, 26004 Logroño, La Rioja, Spain
manuel.juarez@unirioja.es

Abstract. This paper presents an accurate structural reconstruction in 3D of the Hypostyle Hall at the ancient Egyptian temple complex of Karnak. Numerous reconstructions have already been produced, but they are approximations and seek to show what the Hall might have looked like. Here we seek to obtain a dimensionally accurate reproduction that can provide structural details of the elements that make up the Hall. The data obtained here will subsequently enable a full structural analysis of the Hall to be conducted using suitable calculation methods, as they provide the dimensions of its component elements. That analysis will be based on the dimensions given and the characteristics of the materials.

Keywords: Karnak · Hypostyle Hall · High fidelity graphics · Virtual archaeology · Ancient Egypt · 3D reconstruction

1 Introduction

There is currently great interest in virtual archaeological reconstructions. As indicated below, several 3D reconstructions of the Hypostyle Hall at Karnak have been produced, but they are merely descriptive in nature, with approximate data. None of them is dimensionally reliable.

The temples of Karnak in the city of Luxor, dedicated to the god Amun, are one of the greatest achievements of the civilization of ancient Egypt in terms of temple architecture. They are still the biggest complex of religious buildings ever constructed. Within that complex stands the Great Hypostyle Hall. In the time of the pharaohs the city was called Thebes, and it was the capital of ancient Egypt. Construction of the Hall was begun under pharaoh Seti I (1294–1279 BC) and completed under his son Rameses II (1279–1213 BC). It is one of the most colossal known works of antiquity [1].

The structure of the Hall is made up mainly of columns raised on foundation plinths, with architraves and slab ceilings. It is delimited by Pylons II and III and by the north and south walls, which were already standing when the ancient Egyptians built the Hall itself [1].

F. Cavas-Martínez et al. (Eds.): INGEGRAF 2019, LNME, pp. 554–567, 2020.
https://doi.org/10.1007/978-3-030-41200-5_61

Modern industrial workshop buildings are made mostly of reinforced and/or prestressed prefabricated concrete, with their structures simply resting on top of load-bearing elements so that the stress of the whole assembly can be transferred to the foundations in the ground. The construction of the Great Hypostyle Hall at Karnak resembles this modern prefabricated structure technique to a truly striking extent.

Unfortunately, the Hall is currently in a highly deteriorated state: many parts of it have collapsed, probably due to earthquakes.

This paper seeks to contribute to the task of maintaining the archaeological heritage of ancient Egypt by providing data that can help maintain, reconstruct and conserve the site.

2 Purpose

This publication seeks solely to obtain an accurate 3D structural model of the Great Hypostyle Hall at the Karnak temple complex. That entails obtaining the dimensions and structural details of the component elements of the Hall and any other information that may prove useful in furthering our understanding of its dimensions, of what it once looked like and of what it looks like now. These data can be found in the doctoral thesis by Liarte [2] which provides a complete bibliography and gives the dimensions of several structural elements that are no longer standing, enabling the full dimensions of the Great Hypostyle Hall to be worked out.

Once the full dimensions of the Hall are worked out, a 3D model can be created that should prove highly useful in any potential reconstruction or restoration work and for conducting structural studies.

3 Review of Previous 3D Reconstructions from the Relevant Literature

As mentioned above, several 3D reconstructions of the Great Hypostyle Hall at Karnak have been produced. Chief among them are those drawn up by Marc Mateos [3] the Digital Karnak Project at UCLA [4].

Both these projects have produced 3D reconstructions of the whole Karnak Temple Complex, but they are concerned with what Karnak might have "looked like" in the time of the ancient Egyptians and do not go into details, for instance as regards the way in which each structural element of the Hypostyle Hall was worked. Here we seek to go beyond appearance and clearly define each structural component of the Hall.

Other scale models produced by researchers to show what Karnak might have looked like at the peak of its splendour include the models of Gérard Homann [5] and the scale model on display at the entrance to the complex itself.

4 Software Used

The 3D models of the Hall presented here were made with 2 programs: AutoCAD by Autodesk [6], a specific program for designing 2D and 3D digital plans, and 3ds Max Design, also by Autodesk [7], which was used to give texture to the 3D model and for the final rendering process.

The method consisted of drawing up plan views and cross-sections with AutoCAD based on dimensions and drawings already published, and then using the resulting 2D models as the basis for drawing up 3D models. Once the 3D models were completed, they were exported from AutoCAD to 3ds Max Design, where textures were applied, and they were finally rendered. The model was oriented to match the actual orientation of the Hypostyle Hall, whose walls face north and south, with Pylon II facing west and Pylon III facing east. This enables the play of light in the Hall to be shown accurately, with the sun simulated to recreate any hour of the day and season of the year.

5 The Structural Elements of the Hall

This section sets out to describe the structural elements that make up the Hypostyle Hall at Karnak as a basis for the production of the 3D model. The construction materials used in the Hall and their mechanical properties also need to be identified for the model proposed here to be used for drawing up a structural analysis.

Figure 1 shows the structural parts of the Hall, based on the conclusions of the authors of the reports used. As can be seen in the figures, the Hall is made up of 122 columns shaped like closed-bud papyrus plants (the shorter columns) that support the architraves on level one. These architraves in turn support the ceiling slabs on that level. There is also a central aisle of 12 taller, open-papyrus-shaped columns that hold up the architraves on level two (which is higher tan level one). These architraves support the ceiling slabs on level two.

Between levels 1 and 2 is the clerestory, which allows light into the Hall through slits in its walls. These openings or window grilles are supported by pillars with niches where the windows could be fitted and secured horizontally. These window grilles stand on the architraves of the lower level.

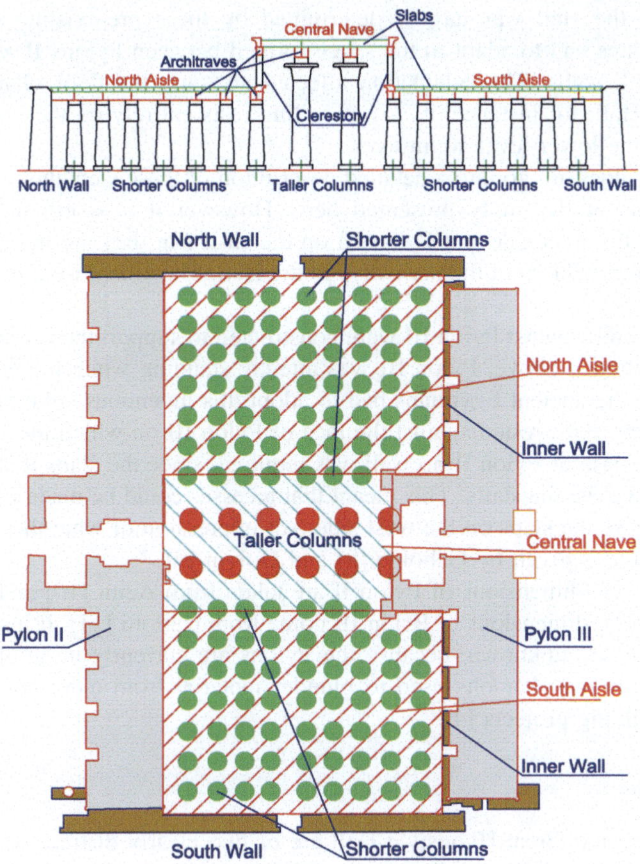

Fig. 1. Cross-section drawing (above) and plan view (below) of the parts of the Great Hypostyle Hall at Karnak.

5.1 Pylons and Walls

The Pylons are two large, trapezoid structures that form towers bridged together at their centres, allowing access from one to the other. Under this bridge was the doorway to the Hall. The Pylons are made of stone blocks laid one on top of the other, tapering as their height increases.

The thick, high walls are also made of stone blocks laid one on top of the other. They are held in place entirely by gravity and are structurally identical to modern stone walls except for the total absence of any kind of binding mortar.

The Pylons were not originally part of the Hypostyle Hall: they were already standing when the ancient Egyptians decided to construct the Hall [1]. However, they constitute an essential part of it, as they form its perimeter on the east and west sides.

The shape of the Hall was largely determined by these pre-existing elements: the ancient Egyptians had to adapt to the space formed between Pylons II and III and the north and south walls. This adaptation affected not only the floor plan but also the height of the Hall: for instance, its lower ceiling rests entirely on the north and south walls and on the lower-level architraves.

This paper does not go into a detailed description of these elements, as they are not the main focus of the study presented here. However it is worth mentioning that Pylon II is on the west side and Pylon III on the east, that they are trapezoid in shape and that they formed two of the doorways to the Hall. The others were in the north and south walls.

The walls built against Pylon III were constructed to support the lower-level ceiling slabs. As mentioned above, Pylon III was already standing when the Hypostyle Hall was built, and the ancient Egyptians had to adopt this ingenuous solution to solve the problem that the slabs rested against the face of Pylon III on which the flagstaffs were located. In the case of Pylon II no wall was required, since the slabs there rest against the side opposite the flagstaffs. This meant that recesses could be made in the stones of Pylon II to house the slabs on the west side. An impression of what these walls might have looked like is given by Gabolde [8] and Golvin [9].

The plan-view dimensions of Pylon II are taken from Azim [10] and Carlotti [11] and the plan-view dimensions of Pylon III come from Carlotti [11]. It must be stressed that the heights are unknown, because the Pylons are currently in an almost entirely dismantled state: the pharaohs used to plunder material from older constructions for their latest building projects [4].

5.2 Columns

The columns in the Great Hypostyle Hall are of two clearly distinct types: the lower ones are shaped like closed-bud papyrus plants, and the taller ones like open papyrus plants (see Figs. 2 and 3). Structurally, their job is to support the architraves and transfer stresses to the ground via their plinths, which act as above-ground foundations. The columns are made of semi-circular stone blocks joined together by double dovetail cramps.

Their dimensions and shapes are described by Carlotti [11]. Those dimensions that he did not measure are incorporated from data provided by Liarte [2] (see Fig. 3). With these measurements, it is now possible to draw up a 3D model of the columns (see Fig. 2).

The locations of the plinths in the overall floor-plan of the Hall are given by Carlotti [11]. Based on his drawing, we propose the layout of columns to scale shown in Fig. 3. Figure 3 show the dimensions of these elements according to Liarte [2], where the Open-papyrus area = 187 m^2 and closed-bud papyrus area = 71 m^2.

Fig. 2. Open-papyrus-shaped column (above left) and closed-bud papyruse-shaped column (above right) (photos belonging to Isidoro Ángel Liarte Fernando. June 2010), and 3D model of the columns in the Hypostyle Hall. Closed-bud papyrus (below left) and Open-papyrus (below right) shapes

5.3 Architraves

Nowadays these elements are known as roof beams. Structurally, their job is to transmit loads from roofs and upper floors to the columns below. They are therefore highly important, since they work mainly under bending stress, by contrast to the compression stress under which columns work.

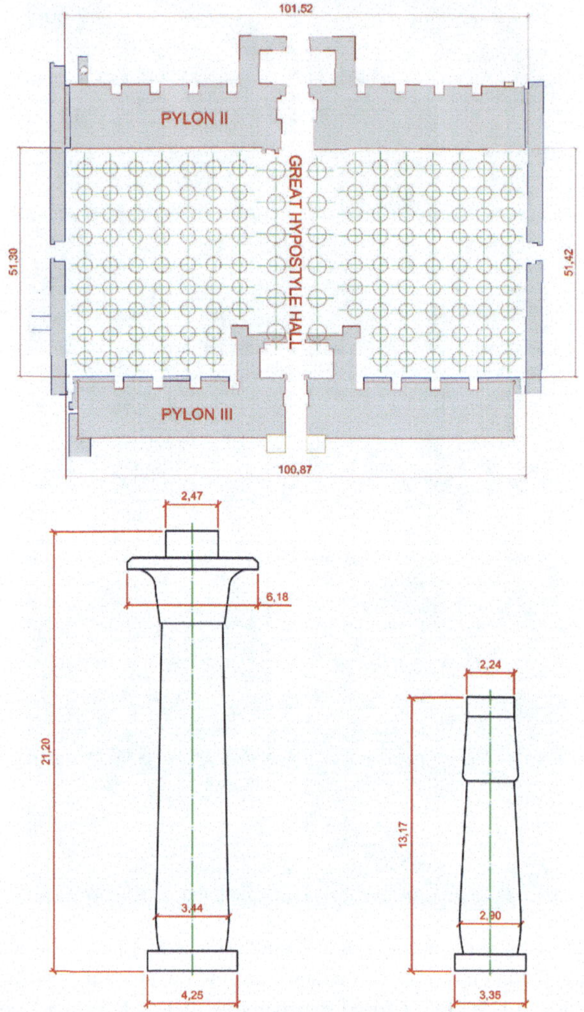

Fig. 3. Locations of the columns in the Great Hypostyle Hall (above) and Cross-sections of the columns (below), (dimensions in metres). Open-papyrus (left) and closed-bud papyrus (right) shapes. [2]

They work in two main ways, depending on the supports used: supported at both ends on the capitals of columns or on capitals and walls, or supported at one end and embedded at the other, as in the case of Pylon II here (see Fig. 4). An examination of the structural types and loads here shows that there are also architraves that are embedded at both ends: this is the case of those that support the clerestory area, the tops of which are held in place by the pillars (see Fig. 4).

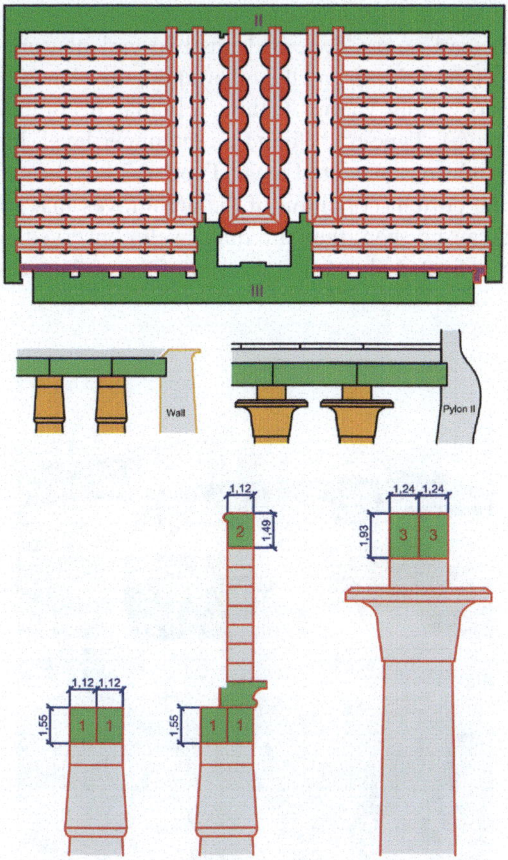

Fig. 4. Plan view of the architraves of the Hypostyle Hall (above), lengthwise section of the architraves (middle) and Cross-section of the architraves (below). (Dimensions in metres) [2]

Because they work under bending stress and because they are built of stone, the architraves here are highly susceptible to cracking: in general, the compression strength of stone is very high, but its tensile strength is very poor.

The architraves here are simple rectangular blocks. Their width can be easily determined, as two adjacent architraves match the total width of the capital of the columns on which they rest. Their length can also be determined, as they extend from centre to centre of the columns on which they rest.

The height of the architraves can be determined by measuring the scale drawings produced by Carlotti [11] and incorporating the data provided by Liarte [2]. Moreover, Carlotti [11] provides a plan view of the layout of the columns from which a plan view of the layout of the architraves can be produced (see Fig. 4).

Four different models of architrave can be distinguished, in terms of their size and locations in the Hall (see Fig. 4). Type nº 1 has two variants: one on the lower level, whose job is to support the ceiling, and the other under the clerestory area, whose job is to support the clerestory (this variant features circular decorative motifs facing the inside of the central nave); Type nº 2, found on the upper level, has the job of holding up the higher ceiling slabs of the central nave; Type nº 3 has the same job as Type nº 2.

Figure 5 shows the current condition of the architraves that support the clerestory area (Type nº 1). As can be seen, they are quite well conserved.

With all the data indicated above, 3D models of the architraves can now be drawn up (see Fig. 5).

Fig. 5. Current condition of the clerestory area (above). (Photo belonging to Isidoro Ángel Liarte Fernando. June 2010). 3D model of the architraves on the lower level (middle) and 3D model of the architraves on the upper level (below).

5.4 Ceiling Slabs

Based on the plan view in Fig. 6 and the dimensions estimated by Liarte [2], a proposed layout of the slabs that covered both ceilings of the Hypostyle Hall can now be given (see Fig. 6).

Like the architraves, the ceiling slabs are simply rectangular stone blocks. They rested on the architraves. Some also rested against the north and south walls, directly in recesses in Pylon II or on the inner walls built against Pylon III. These slabs are mostly supported at both ends, except for those fitted into the recesses in Pylon II, which are supported at one end and embedded at the other.

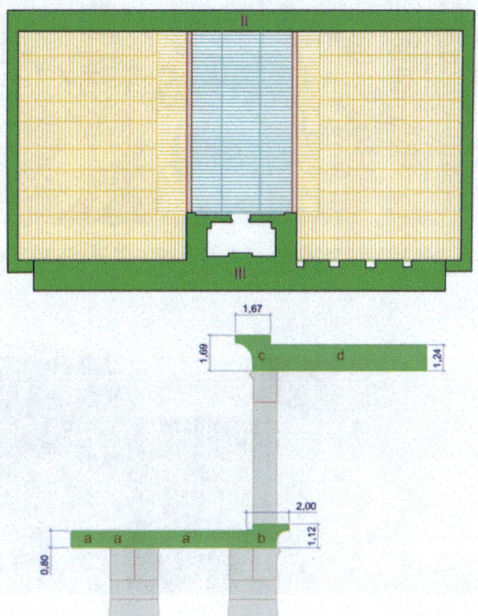

Fig. 6. Plan view of the ceiling slabs covering the Great Hypostyle Hall (above) and cross-section of the slabs (below) [2]. (Dimensions in metres)

5.5 Clerestory Area

The ancient Egyptians built the east-west line of the Hypostyle Hall higher than the rest. This was done so that gaps could be left in the side walls between the two ceiling levels (the clerestory area) through which daylight could enter the hall, supplementing the light that entered from the north and south between the outside walls and the slabs.

The second level is raised above the first by means of a system of pillars (uprights) and beams (architraves) so that the window grilles stand vertically on the architraves of the lower level and are held horizontally between the pillars. The upper architraves that bore the weight of the upper ceiling rested on these pillars (see Fig. 7).

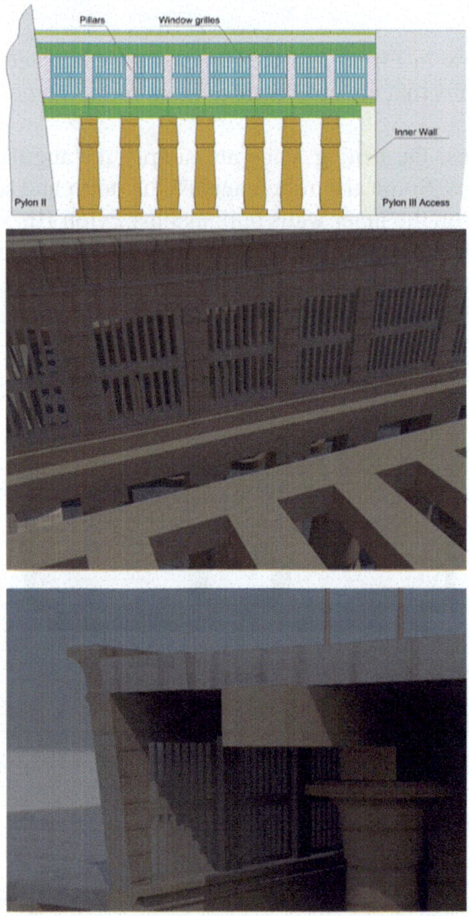

Fig. 7. Detail of the clerestory area (above) and 3D model of the architraves of the clerestory (outside view) (middle), (inside view) (below)

Based on the data provided by Liarte [2] we propose a 3D model of the clerestory area (Fig. 7). Below are shown the dimensions of the uprights and the window grilles. The weight of the window grilles is worked out per square metre (taking their thickness as 45 cm), since they are wall-panel type elements, while for the uprights the figures given are for the total weight.

The dimensions of uprights & window grilles are:

Upright → height 5,80 (m), width 1,15 × 1,12 (m) and volume < 7,47 m^3
Window grille → height 5,80 (m), width 1,00 × 0,45 (m) and area 1 m^2 (-openings).

6 Proposal for a 3D Structural Reconstruction of the Hypostyle Hall

Once all the dimensions of the structures that make up the Hypostyle Hall at Karnak are known, a full 3D model can be drawn up. The first step is to produce plan, elevation and cross-section views for use in working out the 3D model.

A knowledge of how the structural elements that make up the Hall were worked provides further information on its overall form with a view to modelling it in 3D.

6.1 Cross-Sections, Elevations and Plan Views

Elevations and cross sections are calculated taking into account the dimensions provided by Azin [10], Gabolde [8, 12] and by Golvin [9], and using the tables of dimensions produced by Carlotti [11] and the figures indicated by Liarte [2]. Some details were also worked out from the graphic documentation on the scale models on display in the halls at the entrance to the temple complex at Karnak itself, e.g. as regards the height of the pylons posited in the structural model of the Hypostyle Hall. These are mere descriptions, as their true heights are unknown. However, the fact that these pylons might have been taller or shorter than shown does not affect the structural analysis of the elements of the Hypostyle Hall because, as mentioned, they are independent of the pylons.

On this basis, the elevations and cross-sections of the Hypostyle Hall can be worked out as shown in Fig. 8.

Fig. 8. Lengthwise section seen from the north (above) and cross section, seen from the east (below)

6.2 The Whole Structure in 3D

Once all the plan views, elevations and cross-sections have been worked out, the next step is to draw up a model of the structure in three dimensions. This model will enable the volumes of all the structural elements of the Hall to be determined, and once that is done their weights can be worked out and a structural analysis can be conducted. This will be covered in another paper.

Figure 9 show the 3D reconstruction proposed from different views, and Fig. 10 shows what the inside of the Hall would look like without the ceiling slabs.

Fig. 9. 3D model of the Great Hypostyle Hall at Karnak. Seen from the south-west (above-left). Seen from the east (below-left). Seen from the north-west (above-right). Seen from the west (below-right).

Fig. 10. 3D model of the layout of the columns

7 Conclusions

With all the dimensional data and other details compiled here, it has proved possible to draw up an accurate 3D reconstruction of the Hypostyle Hall at Karnak. This model can be used in the future to aid restoration and conservation work in the Hall, especially on the columns, the architraves and the clerestory area. This latter area is clearly defined, as are the heights of the north and south walls.

With this geometrical reconstruction and proper knowledge of the properties of the material used, structural studies can be conducted as required to calculate estimations of all the elements that make up the Hall.

Moreover, the results of those calculations will provide important information on how the ancient Egyptians calculated and dimensioned these structures, which have survived for so long in spite of their size and complexity.

The study will also prove useful in future restoration work to return the monument to its former splendour and ensure the safety of the many visitors who come to see it.

References

1 Sullivan, E.: The Development of the Temple of Karnak. On Digital Karnak, Los Angeles (2010). https://dlib.etc.ucla.edu/projects/Karnak. Accessed Nov 2013
2. Liarte Fernando, I.Á.: Contraste y validez de modelos predictivos de rotura de materiales frágiles: Caso a estudio La Gran Sala Hipóstila de Karnak. Doctoral thesis, Universidad de La Rioja, Spain (2013)
3. Karnak 3D. https://www.karnak3d.net/. Accessed Nov 2013
4. Digital Karnak project. © 2008 All rights reserved. Regents of the University of California. https://dlib.etc.ucla.edu/projects/Karnak. Accessed Nov 2013
5. Recontitutions CAO et maquettes de monuments égyptiens, Gerard Homann. https://gerard.homann.free.fr/WEB_EGYPTE/egypte.html. Accessed Nov 2013
6. Autodesk AutoCAD 30-day demo version. https://www.autodesk.es. Accessed Mar 2019
7. Autodesk 3ds Max Design 30-day demo version. https://www.autodesk.es. Accessed Mar 2019
8. Gabolde, L.: À propos de deux obélisques de Thoutmosis II, dédiés à son père Thoutmosis I et érigés sous le règne d'Hatshepsout-pharaon à l'ouest du IVe pylône. Cahiers de Karnak **8**, 143–158 (1985)
9. Golvin, J.Cl.: La restauration antique du passage du IIIe pylône. Cahiers de Karnak **8**, 189–206 (1985)
10. Azim, M.: La structure des pylônes d'Horemheb à Karnak. Cahiers de Karnak **7**, 127–166 (1982)
11. Carlotti, J.Fr.: Contribution à l'étude métrologique de quelques monuments du temple d'Amon-Rê à Karnak. Cahiers de Karnak **10**, 65–125 (1995)
12. Gabolde, L.: La "cour de fêtes" de Thoutmosis II à Karnak. Cahiers de Karnak **9**, 1–100 (1993)

Automatic Detection of High-Voltage Power Lines in LiDAR Surveys Using Data Mining Techniques

Daniel Chasco-Hernández[1], José A. Sanz-Delgado[2], Víctor García-Morales[3], and Jesús Álvarez-Mozos[1](✉)

[1] Department of Engineering, Public University of Navarre, Los Tejos, Campus Arrosadia, 31006 Pamplona, Spain
jesus.alvarez@unavarra.es

[2] Department of Statistics, Computer Science and Mathematics, Public University of Navarre, Los Pinos, Campus Arrosadia, 31006 Pamplona, Spain

[3] Department of Engineering and Territorial Systems, TRACASA, Cabárceno 6, 31621 Sariguren, Spain

Abstract. The correct classification of power lines in LiDAR point clouds has attracted the interest of the mapping community in the last years. The objective of this research is the detection and automatic extraction of high-voltage transmission lines from LiDAR data using data mining techniques. With this aim, a Single Photon LiDAR (SPL) survey acquired over the region of Navarre (Spain) in 2017 was used, with a mean point density of 14 pt/m^2. Different data mining techniques were evaluated, including decision trees (C4.5 and CART) and ensemble learning algorithms (Random Forests, Bagging and AdaBoost). Fifteen test sites were studied corresponding to areas with high-voltage power lines over different conditions regarding the underlying vegetation and topography. For these sites 92,104 LiDAR points were identified as power lines and more than 4M points as not power lines using existing cartography. This dataset was randomly split in train and test sets and then balanced two obtain a similar amount of data for the two classes. The results obtained show the importance of balancing the training data with improvements in accuracy of $\sim$10% with respect to the imbalanced case. Accuracies higher than 87% were obtained in all balanced cases, with particularly successful results for ensemble learning techniques, being AdaBoost the technique with the highest accuracy 91%. These results suggest that the combination of SPL surveys and data mining tools can be successfully used for the operational mapping of high voltage power lines.

Keywords: LiDAR · Data mining · Single Photon LiDAR · Power lines · Supervised classification

1 Introduction

Power line mapping is a costly task for the organizations in charge of making cartography [1]. Therefore, methods to automatically identify and map power lines are required in order to update existing maps, create new ones for areas where such maps do not exit, and hence for enhancing the efficiency of cartographic organizations.

F. Cavas-Martínez et al. (Eds.): INGEGRAF 2019, LNME, pp. 568–575, 2020.
https://doi.org/10.1007/978-3-030-41200-5_62

Since its invention, Light Detection And Ranging (LiDAR) technology has had a dramatic impact in the cartographic methods and workflows. Indeed, airborne LiDAR is an efficient and cost-effective technique for the rapid and precise acquisition of massive 3D point clouds [2]. Furthermore, it enables the automation of the surveying process and the production of maps with a level of detail unattainable with previous cartographic techniques [3]. Before the onset of LiDAR, both topographic surveys and photogrammetric restitution processes required an enormous amount of expert hand labor, so producing detailed scale cartography was very costly [4]. With LiDAR the production of high resolution Digital Terrain Models (DTM) and cartography can be done much faster, cheaper and in a more objective (with less human intervention) and precise manner.

Data mining is a set of techniques and approaches designed to extract knowledge and patterns from large datasets [5], so they are ideal tools to efficiently process LiDAR datasets. The aim of this work is to apply state of the art data mining algorithms to automatically identify high voltage power lines in LiDAR point clouds, so as to obtain a classified point cloud that can produce a power line map. To attain this objective a Single Photon LiDAR (SPL) dataset was processed, which was acquired over the region of Navarre (Spain) in 2017. SPL is a very recent and innovative technology with significant improvements in terms of the number of points acquired. In the next sections, the materials and methods used are described and the obtained results are shown, finally some conclusions and recommendations are outlined.

2 Materials and Methods

An airborne LiDAR survey was carried out in Navarre (Spain) between September and November 2017 using the Leica SPL100 sensor. The SPL technology splits the laser beam into a 10×10 array achieving a much higher point density than its predecessors. Some other key differences with the more conventional LiDAR systems used so far are its operation wavelength at 532 nm, the possibility to record as many as 10 returns per laser pulse and its faster recovery time (1.6 ns). On the other hand its circular scanning pattern creates a very irregular point distribution with a higher point proportion at the borders of the swath, and the obtained point clouds are very noise with a large amount of noise points <50 m above the ground. For this project, a flight height of 3,900–6,300 m (asl) was set with an airspeed of 200 knots, leading to a swath width of 2.3 km and an average point density of 14 pt/m^2. The complete survey over Navarre comprised more than 500 billion points and $\sim$50 TB of data, and was freely distributed as 1×1 km tiles in LAS 1.4 format (ftp://ftp.cartografia.navarra.es). The LAS files contain information on UTM X, Y and Z in ellipsoidal elevation considering EPSG: 25830 (ETRS89 UTM 30 North) reference system, additionally, for each point the LiDAR intensity, return number and RGB reflectance are given (the latter obtained from a RCD30 half-format camera). Finally, each point was automatically assigned to a class using TerraScan software following the ASPRS standard class definition [6].

Using these SPL100 point clouds, two-class supervised classification algorithms were fit to predict whether a point corresponded to the class ‘high-voltage power line’ (class = 1) or not (class = 0). The point attributes used as descriptive features for the classification were the following (Table 1):

Table 1. Point attributes used as input for the classification.

Attribute	Description
Z	Point elevation
Intensity	Intensity of the laser return
Return number	Number of return of the point
Number of returns	Total number of returns for the pulse
R	Digital number of the red channel
G	Digital number of the green channel
B	Digital number of the blue channel

A ground truth dataset was built to train and test the classification algorithms. For this, fifteen 1 × 1 km tiles were selected on areas with high-voltage power lines present over different land uses (6 over forest, 4 over agriculture and 5 over urban areas) (Fig. 1). Then, existing power line cartography was overlaid (Fig. 1) and a 20 m buffer width around the lines was established to mask out the remaining areas. Using a proprietary algorithm (TerraScan), points belonging to power lines were classified. The algorithm is based on a recursive fitting of the catenary curve for all the points in the cloud [7], so it was very expensive in terms of computing time. Besides, it required a significant number of parameters that had to be set for each area, being the outcome very sensitive to some of them, so visual inspection of the results was mandatory and parameter values had to be fine-tuned for each area to avoid false positive and incorrect results. This process was very labor intensive but necessary to obtain a detailed training dataset to build the models upon (Fig. 2).

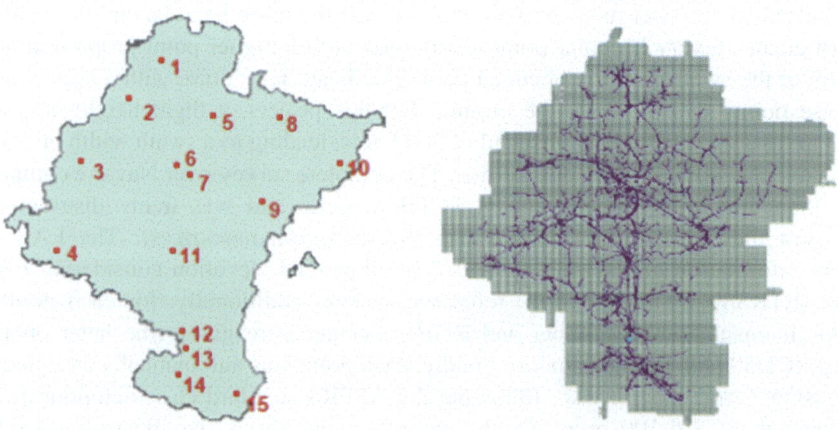

Fig. 1. Left: Location of the fifteen tiles used to obtain the train and test datasets. Right: High-voltage power line cartography overlaid over the SPL LiDAR tiles.

Fig. 2. Sample result of the training dataset used.

A dataset of 4,527,727 points was built out of which 92,104 corresponded to power lines. This dataset was saved as a csv file and randomly split in 75% for training and 25% for testing using the 'hold out' method, and thus ensuring that instances used for training the algorithms were not used for evaluating them and vice-versa. Different data mining approaches were used for classification:

1. Decision trees
   a. C4.5
   b. CART
2. Ensemble learning techniques
   a. Random Forests
   b. Bagging
   c. AdaBoost

A decision tree (DT) is a classification algorithm that represents a set of organized decisions following a hierarchical structure depending on a set of attributes or descriptive features [5]. DTs are automatically built through a learning process based on the training dataset that is successively split in nodes and branches until leaves are reached (a terminal node where all the elements belong to the same class). At the nodes the dataset is split using the attribute that maximizes an impurity criterion, this criterion was the Gini index in CART and the Information Gain Ratio in C4.5. The obtained DT were pruned to avoid overfitting [5].

Ensemble learning techniques follow the assumption that an ensemble of several classifiers (e.g., DTs) outperforms the results of each classifier alone [8]. Bagging is an ensemble method whereby a classifier is built using new sets of instances obtained applying the Bootstrapping technique, yielding a lower variance prediction if compared to each of the subsets alone and avoiding overfitting [9]. Adaboost updates a first simple DT (weak classifier), by training it iteratively so as to improve the previous version by giving a higher and lower weight to instances incorrectly and correctly classified, respectively [2].

Finally, Random Forest (RF) [10] is an ensemble learning technique that combines the Bagging concept [9] and the Random Subspace concept [11]. On a RF each tree is built upon a different subset of data and using a random subspace of attributes to select the best splitting attribute at each node. This way, the forest is composed of DT that are varied since they learn from a different subset of data and a different subspace of attributes. The final prediction is obtained by majority voting of the different DTs. The generalization error of RF generally depends on the proportion of the final vote assigned to the winning class and on the correlation between the DT that compose the forest.

Supervised data mining methods are severely affected by data imbalance [12], i.e., they tend to over-predict the most frequent classes in the training set. Therefore, in severely imbalance cases like this, training data needs to be balanced first, so as to obtain a training set with the same number of class 1 and class 0 instances. Two approaches can be followed: under-sampling the most frequent class or over-sampling the less frequent one. In this work an over-sampling approach called SMOTE was applied that creates synthetic instances of the minority class by interpolating two instances of the original dataset [13]. Then to reduce the eventual noisy instances created in the borders of the two classes the Tomek Links (TL) method was used [14].

Finally, the test dataset was used to evaluate the quality of the predictions obtained, using a set of performance metrics calculated from the confusion matrix (Eqs. 1–5).

$$Accuracy = \frac{TP + TN}{TP + TN + FP + FN} \tag{1}$$

$$Recall = TPR = \frac{TP}{TP + FN} \tag{2}$$

$$Precision = \frac{TP}{TP + FP} \tag{3}$$

$$Specificity = TNR = \frac{TN}{TN + FP} \tag{4}$$

$$Geometric\ mean = GM = \sqrt{TPR \times TNR} \tag{5}$$

where, TP corresponds to the true positive predictions (power lines correctly classified as power lines), TN true negative, FP false positive and FN false negative predictions. The Accuracy (1) represents the proportion of correctly classified points. The Recall or True Positive Ratio (2) corresponds to the proportion of power line points classified as such. The Precision (3) represents the proportion of points classified as power line that actually corresponded to this class. The specificity represented the True Negative Ratio (4), and finally, the Geometric Mean (5) represented the proportion of correctly classified points for both classes. The latter can be used as an overall performance metric to evaluate the behavior of each algorithm.

3 Results and Discussion

The application of the trained algorithms to the test data produced in general quite accurate results (Fig. 3). After evaluating the confusion matrices for each case the performance metrics were computed (Table 2). We must point out that SMOTE followed by Tomek Links (SMOTE+TL) was applied only to balance the set of training points, which was subsequently used to train the different models. The performance of the trained methods was always obtained using the original set of test points (without being balanced).

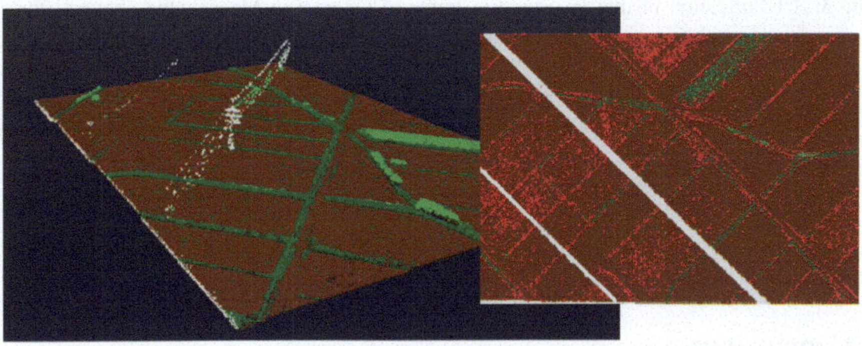

Fig. 3. Sample of a subset of a test area with points classified as power lines represented in white.

Table 2. Performance metrics in testing (all in %) obtained for each classification algorithm using both the imbalanced and balanced datasets.

Classification algorithm	Imbalanced problem				Balanced problem (SMOTE+TL)			
	Acc.	Rec.	Prec.	Spec.	Acc.	Rec.	Prec.	Spec.
C4.5	95.2	62.8	59.7	97.3	98.7	76.9	66.7	99.2
CART	95.6	65.0	56.1	97.2	98.7	76.9	65.0	99.1
Random Forests	99.3	65.6	65.9	98.0	99.3	80.4	85.2	99.7
Bagging	96.6	62.9	72.2	98.6	99.2	81.7	81.0	99.6
Adaboost	97.1	66.2	75.7	98.8	99.1	84.5	82.9	99.5

In general, it can be observed that results improved after balancing, with the largest improvements for Recall and Precision metrics, so both commission and omission errors for the class of interest diminished after balancing. Ensemble algorithms were more benefited by balancing than C4.5 or CART DT. So, it can be said that SMOTE followed by TOMEK LINK was a successful balancing method. Having said that, it can be observed (Table 2) that all algorithms made accurate predictions after balancing the points, with the ensemble learning techniques leading to slightly higher metrics, especially in Recall and Precision, whereas Accuracy and Specificity obtained high rates in all cases.

As already explained, the Geometric Mean is the metric that offers a compromise between the correct identification of class 1 and class 0 points, and can be used as a reference for selecting the best performing method (Table 3). Although differences were minor the Adaboost algorithm was the first in the ranking, followed by Bagging, Random Forests at an intermediate position and simple DT algorithms C4.5 and CART a step behind. The main difference was that these last methods had a higher rate of false positives.

Table 3. Classification results in testing in terms of Geometric Mean value obtained for each classification algorithm using both the imbalanced and balanced datasets.

Classification algorithm	Imbalanced	Balanced
C4.5	78.15%	87.36%
CART	79.49%	87.32%
Random Forests	80.14%	89.53%
Bagging	78.73%	90.22%
Adaboost	80.90%	91.73%

4 Conclusions

Several data mining techniques were implemented on a high-voltage power line detection application based on SPL LiDAR point cloud data. The main conclusion is that adequate results were obtained in most cases. Yet, class imbalanced appeared as a key issue and significant improvements were obtained in all cases after training data were balanced, with largest improvements in Recall and Precision metrics, especially in ensemble techniques. Ensemble techniques achieved higher precision metrics than simple decision trees like C4.5 and CART, in particular in the avoidance of false positives. Although differences were not large the highest accuracies, in terms of the Geometric Mean metric, were achieved by Adaboost followed by Bagging.

Using machine learning techniques for detecting high-voltage power lines can be interesting not only for providing accurate results but also for being less computationally expensive than other techniques relying on the computation and fitting of geometric constraints.

Future research efforts in this line should explore the generalization ability of the techniques investigated, validating them over a larger amount of test points. Also, new attributes obtained from LiDAR data, in particular contextual variables between each point and its neighbors should be investigated for their eventual added value.

Acknowledgments. The Government of Navarre and Tracasa are acknowledged for the provision of the SPL LiDAR data, the TerraScan software license and their expertise.

References

1. Zhu, L., Hyyppä, J.: Fully-automated power line extraction from airborne laser scanning point clouds in forest areas. Remote Sens. **6**(11), 11267–11282 (2014)
2. Axelsson, P.: Processing of laser scanner data—algorithms and applications. ISPRS J. Photogramm. Remote Sens. **54**, 138–147 (1999)
3. Jwa, Y., Sohn, G., Kim, H.B.: Automatic 3D powerline reconstruction using airborne LiDAR data. Int. Arch. Photogramm. Remote Sens. **38**, 105–110 (2009)
4. Liston, R.L.: Photogrammetric methods for mapping resource data from high altitude panoramic photography. Photogramm. Eng. Remote Sens. **48**(5), 725–732 (1982)
5. Rokach, L., Maimon, O.: Data Mining with Decision Trees. Theory and Applications, 2nd edn. World scientific Publishing, Singapore (2008)
6. American Society for Photogrammetry and Remote Sensing (ASPRS): LAS specification version 1.4 – R13. https://www.asprs.org. Accessed 05 Apr 2019
7. TerraSolid: TerraScan User's Guide. https://www.terrasolid.com. Accessed 05 Apr 2019
8. Scikit-Learn User Guide. https://scikit-learn.org/stable/_downloads/scikit-learn-docs.pdf. Accessed 05 Apr 2019
9. Breiman, L.: Bagging predictors. Mach. Learn. **24**(2), 123–140 (1996)
10. Breiman, L.: Random Forests. Mach. Learn. **45**(1), 5–32 (2001)
11. Ho, T.K.: The random subspace method for constructing decision forests. IEEE Trans. Pattern Anal. Mach. Intell. **20**(8), 832–844 (1998)
12. Chawla, N.V., Japkowicz, N., Kolcz, A.: Editorial: special issue on learning from imbalanced data sets. SIGKDD Explor. **6**(1), 1–6 (2004)
13. Chawla, N.V., Bowyer, K.W., Hall, L.O., Kegelmeyer, W.P.: SMOTE: Synthetic Minority Over-sampling TEchnique. J. Artif. Intell. Res. **16**, 321–357 (2002)
14. Visa, S., Ralescu, A.: Learning imbalanced and overlapping classes using fuzzy sets. In: Workshop on Learning from Imbalanced Datasets II (ICML 2003), pp. 91–104 (2003)

Comparison of Digital Terrain Models Obtained with LiDAR and Photogrammetry

Martínez-Agirre Alex[1], García-Morales Víctor[2], and Álvarez-Mozos Jesús[1(✉)]

[1] Department of Engineering, Public University of Navarre, Los Tejos, Campus Arrosadia, 31006 Pamplona, Spain
jesus.alvarez@unavarra.es

[2] Department of Engineering and Territorial Systems, TRACASA, Cabárceno 6, 31621 Sarriguren, Spain

Abstract. Airborne LiDAR sensors capture three-dimensional information of the Earth, useful for obtaining high accuracy Digital Terrain Models (DTM). The Spanish National Plan for Aerial Orthophotography (PNOA) is an initiative of the Spanish Geographical Institute whereby nationwide LiDAR datasets are periodically acquired and made available to the public as .las files and value added products (e.g., DTM). The objective of this study is to assess the added value of PNOA LiDAR DTMs by comparing them to DTMs obtained through classical photogrammetric techniques. With this aim, four areas of interest were selected in Navarre (north of Spain), in areas with challenging characteristics such as forests, karst landforms, agricultural terraces and ravines. A 5 × 5 m DTM obtained with classical photogrammetry in 2008 was compared with a LiDAR DTM of the same pixel size obtained in 2011, assuming no significant changes occurred in this time. Height differences were evaluated, as well as slope, aspect and curvature differences. Besides, a multiresolution analysis was carried out to quantify how DTM smoothing affected height variations between neighbor pixels, measured with the standard deviation on a 5 × 5 window. The results obtained showed that the LiDAR DTMs provided an enhanced description of topography, particularly under forests and in areas with complex topography.

Keywords: LiDAR · Photogrammetry · DTMs · Cartography · Geomorphology

1 Introduction

Digital Terrain Models (DTM) represent the topographic surface as a grid of regularly spaced pixels whose values correspond to the bare earth height (removing man-made objects and vegetation) [1]. Although some other types of representations might also be used (e.g., irregular point clouds or TINs), the above mentioned raster model is the most frequent and convenient format to work with [1]. DTMs are extremely valuable datasets for engineering related activities occurring at the earth surface [2], as well as for many other disciplines (e.g., geomorphology, ecology, hydrology and forest management) [3]. DTMs can be obtained through different techniques [3], but at

F. Cavas-Martínez et al. (Eds.): INGEGRAF 2019, LNME, pp. 576–585, 2020.
https://doi.org/10.1007/978-3-030-41200-5_63

detailed scales (1:5000 or higher) aerial photogrammetry was the standard for several decades, until Light Detection And Ranging (LiDAR) surveys came up [4]. LiDAR datasets consist of an enormous amount of points with known X, Y, Z coordinates, corresponding basically to the bare earth surface, vegetation elements and man-made objects. After a filtering process whereby vegetation and man-made objects are masked out, raster DTMs can be easily obtained at typical resolutions of 5 m and higher depending on the density of the LiDAR point cloud.

In many countries, LiDAR surveys are becoming the standard data source for DTMs, as they have a number of advantages compared to photogrammetry. For example, the laser pulse can penetrate through vegetation and reach the soil beneath forests. In addition, photogrammetry frequently requires the intervention of a skilled analyst to delineate contour lines or to define the lines obtained (semi-)automatically [5]. However, the cost of LiDAR surveys might still be too high for under developed countries, limiting the adoption of this technology.

The objective of this study is to quantitatively compare DTMs obtained with LiDAR and photogrammetry considering areas with particular topographic features where the added value of LiDAR might be particularly apparent.

2 Materials

A 1:5000 scale DTM obtained through photogrammetry was used. This DTM was created through manual restitution of contour lines, key points and other planimetric elements useful for representing the topography. Different photogrammetric flights were used to build the DTM. A first model was built in 2000. It was later updated and improved with successive flights in 2003, 2004, 2006 and 2008. All flights were carried out in summer time, typically flying at a height of $\sim$5000 m over ground level with a 60% overlap and obtaining ground sampling distances (GSD) of 25–50 cm. The final product was a DTM with a 5×5 m pixel size in UTM-30N ETRS-89 projection system and orthometric heights.

On the other hand, a LiDAR was flown over Navarra between February 2011 and May 2012 as part of the Spanish National Plan for Orthophotography (PNOA) using a Leica ALS60 sensor. The flight specifications (height, speed, FoV) were set up so as a required point density of >0.5 pt/m^2 and a vertical accuracy of <20 cm attending to the PNOA specifications. Obtained LAS point clouds were ground filtered using TerraScan software and a DTM of 5×5 m pixel was interpolated and published in UTM-30N ETRS-89 projection system and orthometric heights. This DTM is publicly available at the Spanish National Geographic Institute (www.cnig.es).

Four 4 × 4 km areas of interest were selected in Navarra to reflect challenging conditions for DTM extraction, regarding relief abruptness, particular landforms or dense vegetation cover (Table 1 and Fig. 1).

Table 1. Areas of interest.

Zone	Name	Xmin–Xmax	Ymin–Ymax	Description
1	Oskotz	600000–604000	4756000–4760000	Dense forest
2	Uxue	620000–624000	4706000–4710000	Agricultural terraces
3	Larra	680000–684000	4755000–4759000	Karst relief
4	Rincón del Bú	620000–624000	4665000–4669000	Arid ravines

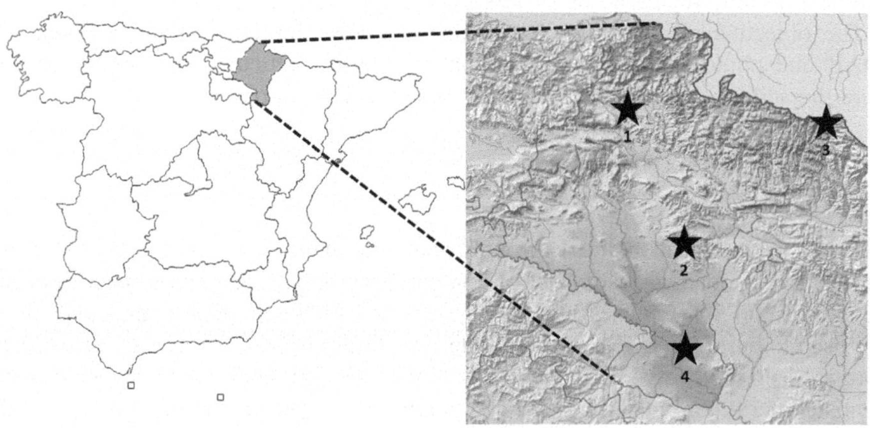

Fig. 1. Location of the four study sites in Navarre. See Table 1 for more details.

3 Methods

The applied methodology comprised several processes (Fig. 2). DTMs were clipped and terrain features (i.e. slope, aspect and curvature) were computed following [6] for both the LiDAR and photogrammetric DTMs (from now on referred to as DTM_L and DTM_P, respectively). Raster differencing was applied to quantitatively evaluate DTM differences. Height differences were computed (z_{diff}), as well as slope (slp_{diff}), aspect (asp_{diff}) and curvature (cur_{diff}) differences. The mean and standard deviation of the differences were evaluated for each area of interest.

Finally, a multiresolution analysis was carried out to evaluate the level of detail of elevation data included on each DTM (Fig. 2). For this, the standard deviation of heights (s_z) on a 5 × 5 window (25 × 25 m) was used as a descriptor of the abruptness of relief or topographic roughness. Then, the original DTMs were smoothed using box-car filters of increasing $k \times k$ window size (with $k = 3, 5, \ldots, 19$), and s_z was successively computed for the filtered DTMs. The trend of $s_z(k)$ was plotted, being the slope of this trend an indicator of the elevation information lost when neighbor pixels were averaged.

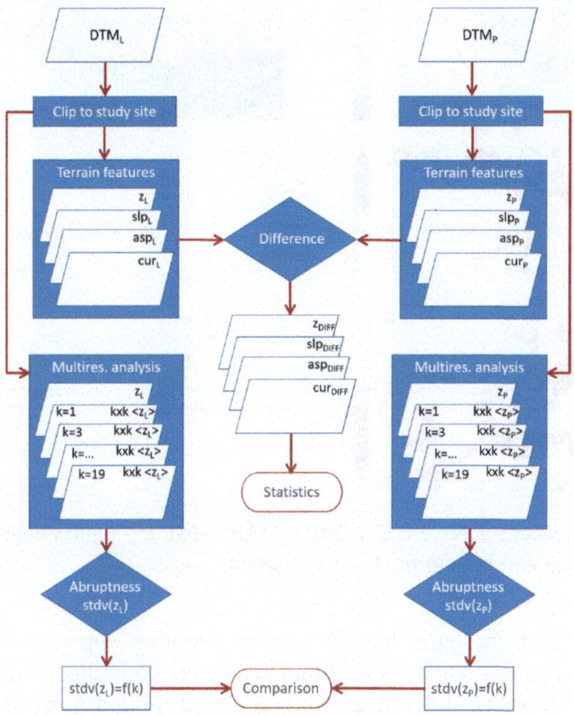

Fig. 2. Flowchart describing the methodology followed for the analysis.

4 Results

Figures 3, 4, 5 and 6 represent the results obtained for Zone 1, similar figures were obtained for the other three zones but only summary results were given (Table 2) for the sake of brevity. At a first sight DTMs obtained with LiDAR and photogrammetry looked quite similar (Fig. 3 top row), however a closer inspection revealed differences that were clearly depicted in z_{diff} (Fig. 3 bottom row). In zone 1, z_{diff} took mostly negative values, corresponding to areas where DTM$_L$ was below DTM$_P$, mostly due to the dense forest canopy in the area.

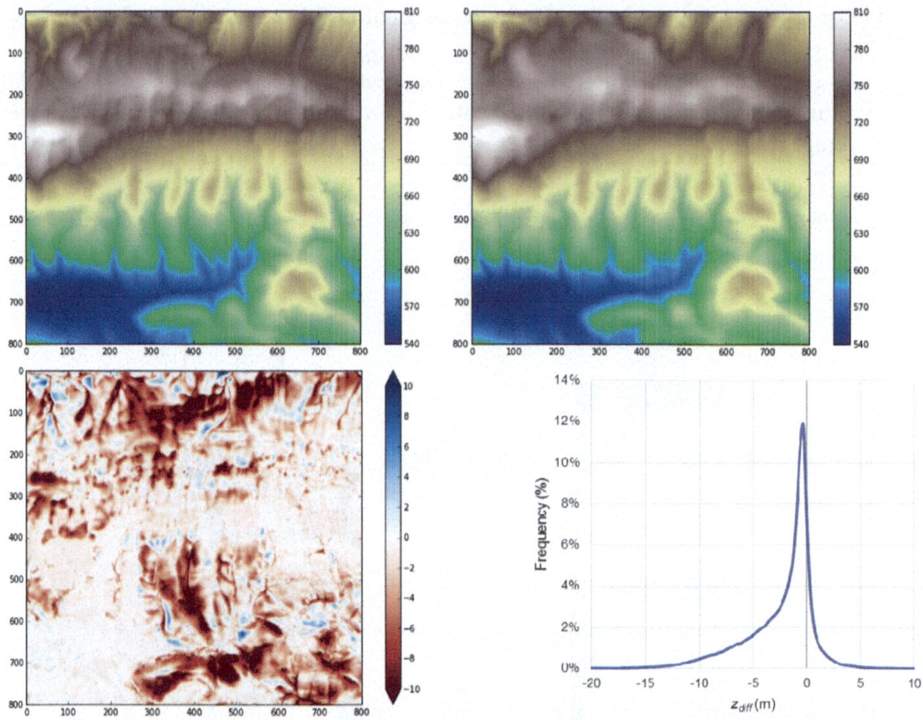

Fig. 3. DTM differences for Zone 1. Top-left: LiDAR DTM, top-right: Photo DTM, bottom-left DTM difference (z_{diff}) and bottom-right z_{diff} histogram.

Slope, aspect and curvature had also quite a similar appearance, but significant differences were apparent when raster differences were computed (Figs. 4, 5 and 6). In the case of slope (Fig. 4), differences were centered around a mean value of 1.68% (Table 2) and positive values were more frequent, indicating, in general, steeper slopes in DTM_L than in DTM_P.

For the aspect (Fig. 5) the histogram of differences was rather symmetric, revealing both positive and negative aspect variations. These differences mostly corresponded to sinkholes frequent in the highlands of the mountain range and to steep valleys in the northern part of the area. As explained, this site was densely forested area and thus photogrammetry was unable to describe correctly these terrain features. The central part of zone 1, mostly occupied by grasslands, had a better agreement between photogrammetry and LiDAR. Finally, curvature differences (Fig. 6) were mostly located along linear features such as streams, forest roads and cattle tracks, where LiDAR data revealed more subtle details of topography.

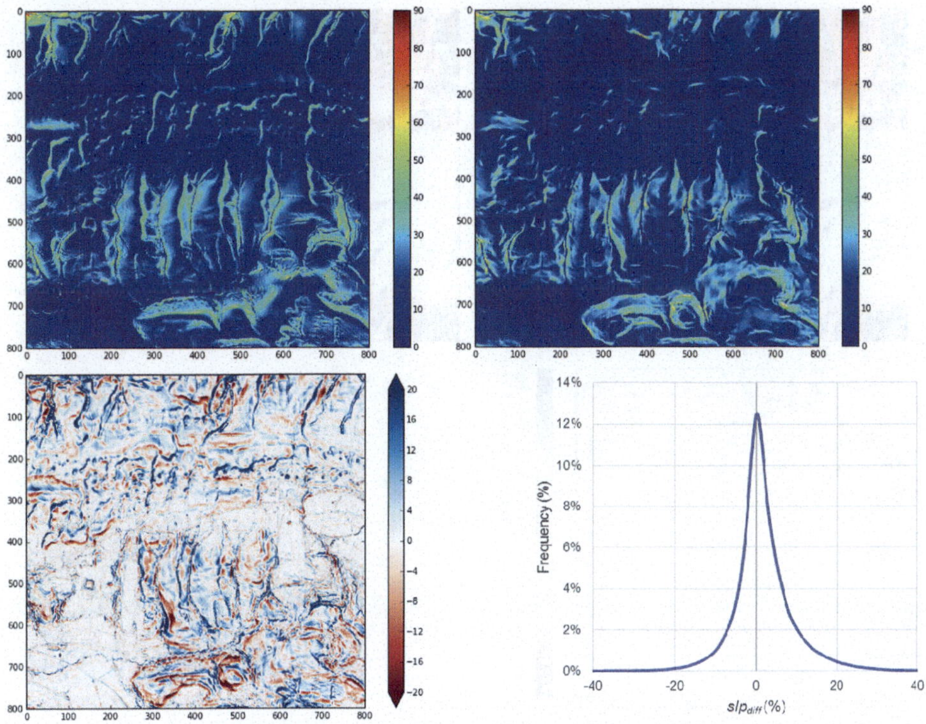

Fig. 4. Slope differences for Zone 1. Top-left: LiDAR slope, top-right: Photo slope, bottom-left slope difference (slp_{diff}) and bottom-right slp_{diff} histogram.

When analyzing the results obtained for the other study zones (Table 2), it can be observed that height differences were in general smaller than for Zone 1 and with positive sign, indicating a higher z value for DTM_L. Height differences and their variations (s value in brackets in Table 2) were particularly large for Zone 3, a very abrupt karstic mountain area. In this area, slope differences were also very relevant with much steeper slope values obtained for DTM_L (average value 11.30% higher). Aspect and curvature differences were also notable. Zone 2 corresponded to a terraced agricultural area, where mean differences were moderate but standard deviation values were relevant, especially for the slope, indicating that terrace limits were not adequately delineated in DTM_P. Finally, Zone 4 showed the least differences between LiDAR and photogrammetry, since it corresponded to a rather flat and bare area were arid ravines exist. In this case, aspect differences were quite relevant (high standard deviation values) probably due to the difficulty of correctly computing the aspect for mostly flat areas.

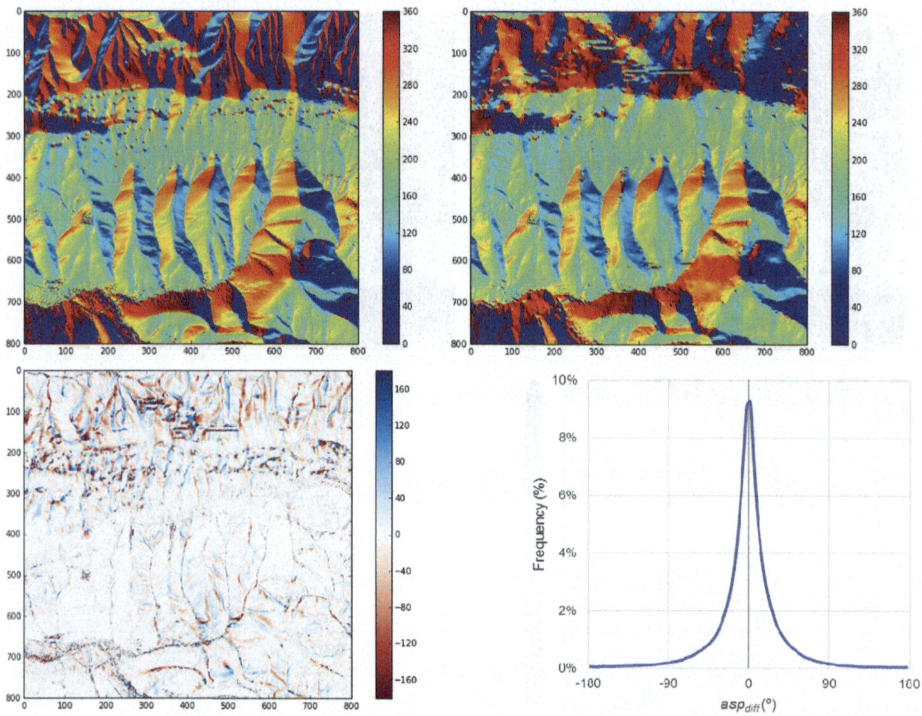

Fig. 5. Aspect differences for Zone 1. Top-left: LiDAR aspect, top-right: Photo aspect, bottom-left aspect difference (asp_{diff}) and bottom-right asp_{diff} histogram.

Table 2. Mean and standard deviation (s in brackets) of height, slope, aspect and curvature differences (z_{diff}, slp_{diff}, asp_{diff} and cur_{diff}, respectively) for the four study zones.

Variable	Units	Zone 1	Zone 2	Zone 3	Zone 4
z_{diff}	(m)	−2.37 (3.20)	0.92 (1.38)	1.63 (4.04)	0.45 (1.08)
slp_{diff}	(%)	1.68 (7.21)	0.62 (12.79)	11.30 (26.29)	0.56 (7.76)
asp_{diff}	(°)	−0.53 (37.94)	−0.06 (26.19)	0.46 (49.32)	−0.04 (61.99)
cur_{diff}	(m^{-1})	−2.5E−5 (0.05)	1.0E−5 (0.15)	3.8E−5 (0.26)	−6.5E−6 (0.10)

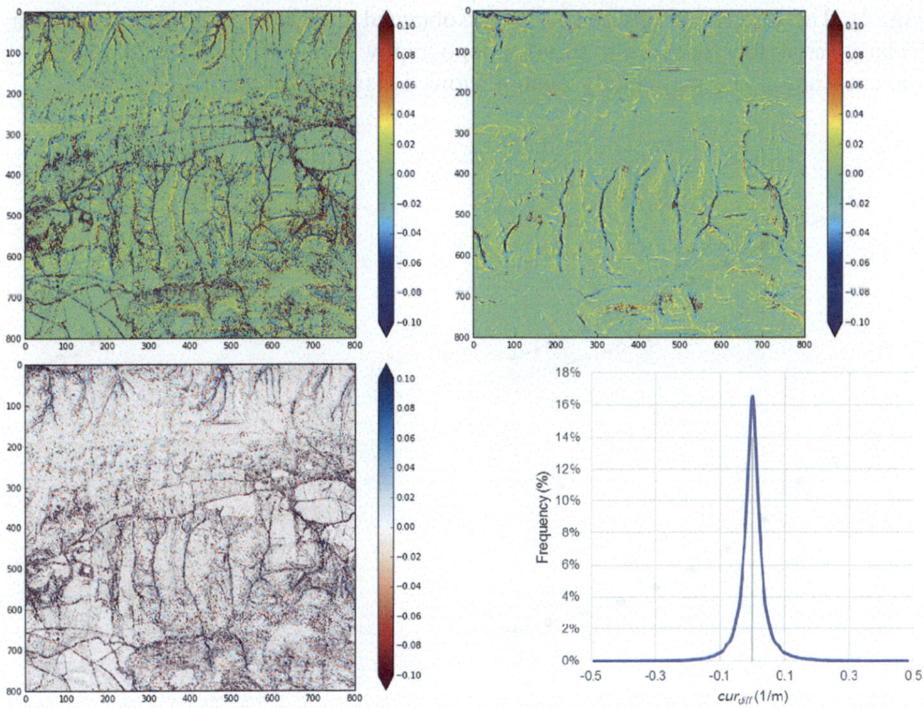

Fig. 6. Curvature differences for Zone 1. Top-left: LiDAR curvature, top-right: Photo curvature, bottom-left curvature difference (cur_{diff}) and bottom-right cur_{diff} histogram.

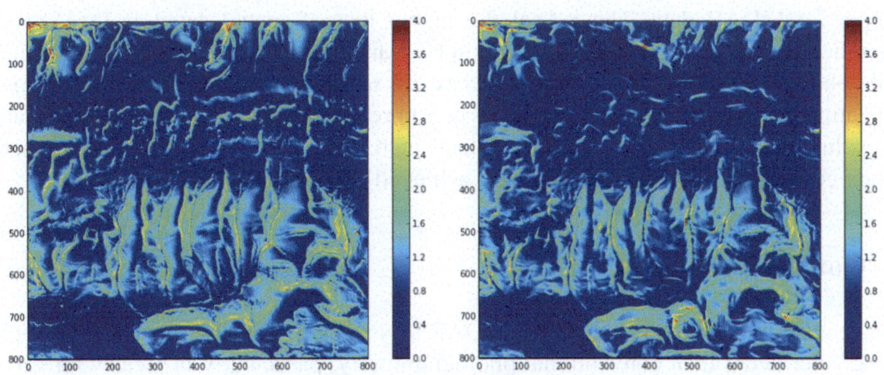

Fig. 7. Topographic roughness (s_z) of Zone 1, as obtained with LiDAR (left) and photogrammetry (right).

Finally, the topographic roughness was evaluated by means of s_z. It can be observed that s_z took higher values for DTM_L, illustrating a stronger variability of elevation values around neighbor pixels and thus a finer spatial detail of elevation data

(Fig. 7). The difference between s_z values obtained for DTM_L and DTM_P was particularly large for zones with stronger topography (Zone 3) and forests (Zone 1), whereas Zone 2 and particularly Zone 4 showed less differences (Fig. 8).

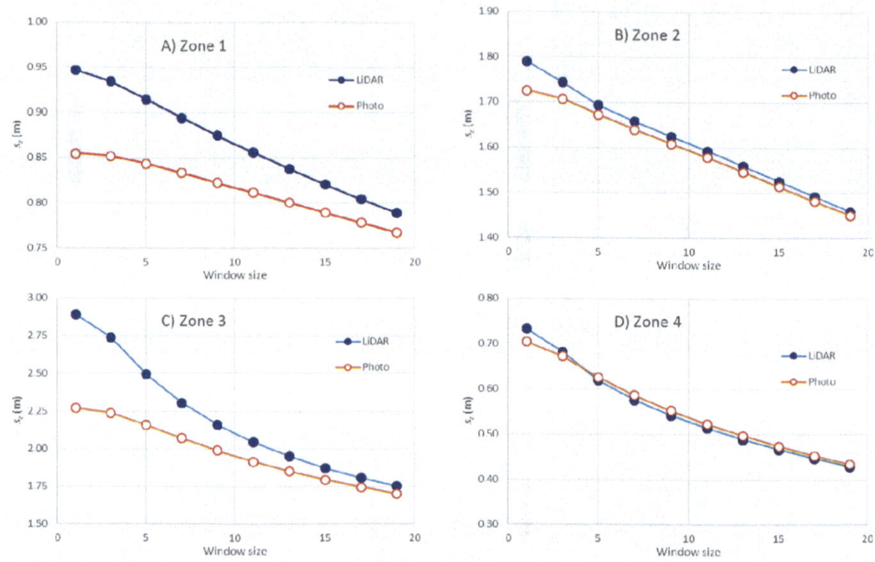

Fig. 8. Results of the multiresolution analysis for the four zones. s_z variations for increasing window size (k) are plotted for both the LiDAR and photogrammetric DTMs.

The multiresolution analysis (Fig. 8) showed a decreasing trend of $s_z(k)$, which was steeper for LiDAR data. This revealed a higher loss of spatial detail as DTMs were smoothed, particularly in areas with abrupt terrain (Zones 1 and 3). DTMs obtained with photogrammetry had a gentler decay of $s_z(k)$, especially when k was small, indicating that the topographic roughness at 5 m resolution was mostly similar to that at a resolution of 25 m ($k = 5$). As k values increased, LiDAR and photogrammetry trends converged, this convergence was achieved faster in Zones 2 and 4.

5 Conclusion

The results obtained demonstrate that LiDAR based DTMs can describe the topography at a greater detail than conventional photogrammetry, particularly in areas with strong topography and dense vegetation cover (forests). Elevation differences, as well as slope, aspect and curvature, revealed finer details in LiDAR based DTMs. These differences were confirmed when computing the topographic roughness and its decay as DTMs were smoothed with a low pass filter, showing a smoother appearance of photogrammetry based DTMs for most cases. These results recommend the use of LiDAR surveys for DTM extraction for any mapping and engineering related activity, particularly for areas with strong topography and forest cover.

Acknowledgments. The research work reported here was partly funded by project CGL2016-75217-R (MINECO/FEDER, EU).

References

1. Li, Z., Zhu, Q., Gold, C.: Digital Terrain Modeling Principles and Methodology, 1st edn. CRC Press, Boca Raton (2004)
2. Petrie, G., Kennie, T.J.M.: Terrain modelling in surveying and civil engineering. Comput. Aided Des. **19**(4), 171–187 (1987)
3. Maune, F.: Digital Elevation Model Technologies and Applications: The DEM Users Manual, 3rd edn. American Society for Photogrammetry and Remote Sensing, Bethesda (2001)
4. Wack, R., Wimmer, A.: Digital terrain models from airborne laser scanner data-a grid based approach. Int. Arch. Photogramm. Remote. Sens. Spat. Inf. Sci. **XXXIV**(Pt. 3B), 293–296 (2002)
5. Gil, A.L., Núñez-Casillas, L., Isenburg, M., Alonso Benito, A., Rodrigo Bello, J.J., Arbelo, M.: A comparison between LiDAR and photogrammetry digital terrain models in a forest area on Tenerife Island. Can. J. Remote Sens. **39**(5), 396–409 (2013)
6. Zevenbergen, L.W., Thorne, C.R.: Quantitative analysis of land surface topography. Earth Surf. Proc. Land. **12**, 47–56 (1987)

3D Digitization of the Archaeological and Palaeontological Heritage Through Non-contact Low-Cost Scanners. Comparative Analysis

N. Santamaría-Hoyos[1(✉)], J. Santamaría-Peña[1], J. M. Valle[2], and F. Sanz-Adán[1]

[1] Departamento de Ingeniería Mecánica, Universidad de La Rioja, C/San José de Calasanz 31, 26.004 Logroño, Spain
natalia.santamaria@unirioja.es

[2] Lab. de Documentación Geométrica Del Patrimonio,UPV/EHU, Vitoria-Gasteiz, Spain

Abstract. This study analyses different 3D scanning technologies, in order to determine the optimal solution to geometrically document and reconstruct replicas of the found objects from archaeological or palaeontological sites. This allows not to touch them, without changing their conservation status, using low cost models. The technologies studied for this purpose have been structured light, with two variants (LED and Infrared), and convergent photogrammetry. The model under analysis has been a dinosaur footprint reconstructed model.

To get the best quality possible with each technology, it is highlighted the importance of controlling previous aspects such as the illumination of the model or the correct setting of the scanning equipment. There should also be considered postprocessing aspects such as the data postprocessing software parameters, the need of correctly align point clouds or the meshes generation.

Based on the three methodologies comparison, it is concluded that it is possible to obtain precisions in a range of millimetre tenths. This validates the methods for the register "in situ" of other elements in the same range of precision, as well as the manufacturing of archaeological and palaeontological replicas using photogrammetric techniques and structured light low-cost scanners. The technology to be used will depend on the complexity of the object to be registered (e.g. with epi-relief) and the precision required in the dimensions and shapes of the model.

Keywords: Geometric Documentation · Archaeological heritage · 3D digitization · Reverse engineering · 3D laser scanner · Structured light · Photogrammetry

1 Introduction

The Geometric Documentation of the Heritage has evolved along with the best technologies available at each moment. To get a procedure that provides, in an efficient way, the optimal solution for each casuistry is essential to use the best infographic technology available. "Currently, the 3D laser scanning technology and photogrammetry improve

F. Cavas-Martínez et al. (Eds.): INGEGRAF 2019, LNME, pp. 586–596, 2020.
https://doi.org/10.1007/978-3-030-41200-5_64

enormously the three-dimensional documentation about objects from the Historic-Artistic Heritage [1].

The amount of information (resolution) that can be acquired with these tools, along with huge precision of the superficial geometry (coordinates of each point), allows to detect little changes in the scanned object shape, reduce project deadlines and improve its quality. "Features as non-contact measurement, and the null allowed variation in the analysed object make that, in some occasions these systems are the only choice viable to obtain a three-dimensional model which allows doing the necessary operations to its study in a reduced period of time, and with similar precision to the one obtained through traditional topographical methods" [2, 3].

The present study analyses the viability of using a low-cost scanner to obtain the point cloud of the object "in situ", "minimizing the intervention on an object or a work, following as premise what Cesare Brandi said in his Theory of Conservation", and the stance that has been supported since the first Athens Letter [4, 5].

"The introduction of 3D techniques in heritage context is a fully accepted reality as it proves the emergence of the denominated Virtual Archaeology. Following the same tendency, the studies centred in the Palaeolithic have experimented a great revolution in their investigation and documentation works about remains". Currently, one of the most interesting methods in Heritage conservation is making security digital copies that could be mechanized in case of total or partial loss of the original one [6–8].

Regarding the study case, 3D Geometric Documentation of a dinosaur footprint, there were just analysed non-contact systems: with LED structured light "Scan in a Box", with infrared structured light "Sense 3D", and with digital camera "Photogrammetry". In the three of them with the condition of being low-cost systems, as well as "Free software" for the scanned point cloud postprocessing.

1.1 Structured Light Technology

LED Light: The measurement principle of the LED structured light scanners is based on detecting the deformation suffered by a light pattern projected against the target surface, which is captured by a camera. The projected pattern consists in multiple stripes, allowing the simultaneous acquisition of several samples. An algorithm determines the distance to the object depending on the scanned data through triangulation [9].

Infrared Light: Infrared structured light scanners technology, works in a similar way to the LED one. In this case, it is an infrared projector the one in charge of projecting coded light patterns. An infrared camera captures the values, generating after, depth infrared video frames. At the same time, a colour camera with an integrated chromatic sensor captures the model colours and applies them into the mesh [10, 11].

1.2 Close Range Photogrammetry

Also known as convergent photogrammetry, whose algorithmics is described as "Structure for Motion (SfM)", consists in defining the shape, dimensions and position in the space of an arbitrary object through homologous beams from different digital photographs of that object [9, 12]. This means it reconstructs close models by photographing each point of them, in a redundant way, from different angles, in order to get a suitable intersection of those homologous beams (Fig. 1).

Process:

- Get multiple overlapped images of the target surface from different angles.
- Extract the points of interest in the object surface to easily do the matching process.
- Determining the XYZ coordinates of the point from the surface by setting the beam.
- Postprocessing software

Fig. 1. Successive photograph took

1.3 Specifications of Each Technology

A summary table with the technical specifications of the technologies compared in this study is included in Table 1. Note that *Photogrammetry* allows to scan objects of indefinite size and that *Sense 3D* allows to obtain the mesh instantaneously.

Table 1. Analysed 3D scanners specifications

Scanner	Scan in box [17]	Sense 3D [15, 16]	Photogrammetry [14]
Owner:	FARO Technologies	3D Systems	Reflex calibrated camera
Object size:	From 0,1 m to 2 m	From 0,2 m to 2 m	Undefined
Data acquisition:	*.txt; *.csv; *.ply	*.txt; *.csv; *.ply	32 photographs
Resolution:	From 0,08 mm to 0,4 m	±0,9 mm in x–y ± 1 mm in z	72 dpi in x–y–z
Precision:	±0,1% Object dimension	±1 mm	±0,3 mm.
Software:	IDEA	Sense	PhotoScan standard
Mesh format:	*.obj; *.stl; *. wrl	*.ply; *.obj; *.stl	*.obj: mesh; *.png: textura; *.mtl: materials
Price:	2.390€	498€	850€

2 Materials and Methods

The studied model was a dinosaur footprint replica. The plate that sheltered the footprint was 0,8 m × 0,6 m (Fig. 1).

There were two main goals: checking the precision of different methods and determining the optimal technology for the reconstruction of the objects located in archaeological sites, whose disassembly and transport to a laboratory is not recommended.

Methodology:

- Obtaining the point clouds with the three technologies.
- Mesh generation.
- Comparison of the deviations between the different point clouds and the reference mesh with the software "Cloud Compare". The mesh obtained with "Scan in a Box" was set as the reference one because it was the one considered more accurate of them (Table 1).

2.1 Scanning Process

Sense 3D: Using infrared structured light technology, it consists in a colour camera, an infrared camera, and an infrared laser projector for measuring distances. This is the quickest and easiest process of the three:

- Connect the scanner to a computer with the software "Sense" installed.
- Turn around the object to capture its entire geometry.
- Natural light gets better results than the artificial one.
- Mesh is automatically gene-rated by the software "Sense".

Scan in a Box: The process is more complex than the "Sense 3D" one (Fig. 2):

- Parameters calibration to get the optimal scanning: scene light, exposure, object position, scanning background.
- Object positioning opposite the scanner.
- Capturing object data from different angles (partial scans).
- Point Clouds alignment (by points, manually or automatically).

Photogrammetry: This is the process that requires more time because:

- The number of photographs is proportional to the complexity of the geo-metry, the size of the object and the precision required.
- It is necessary to have environmental conditions in which it is possible to perceive the object.
- There must be aligned all the photographs to generate the point cloud.
- The point cloud needs to be postprocessed.

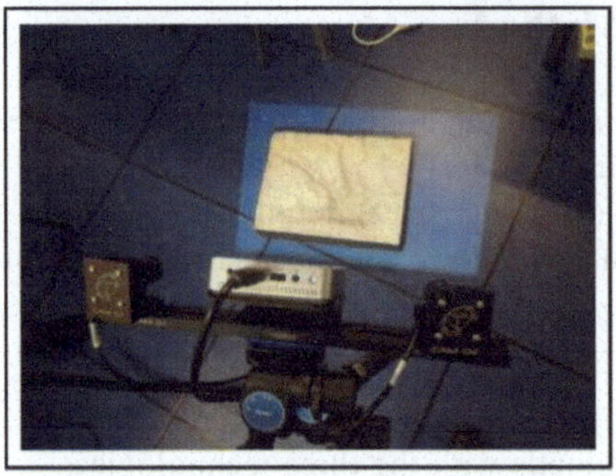

Fig. 2. Reference scanner "Scan in Box"

2.2 Acquisition and Processing of the Scanned Data

The acquisition, processing and comparison of the scanned data was the following:

- Acquisition of the point cloud of each scanner in txt format.
- Cleaning of the point cloud and noise removal.
- Mesh generation, in polygonal formats (*ply, obj, stl, wrl*).
- Mesh alignment and scaling.
- Calculation of the distances between the point clouds and the reference mesh.
- Deviations statistical analysis.
- Main deviations location.
- Results interpretation.

2.3 Meshes Comparison

After generating the mesh, the process begins by importing the polygonal meshes from the point clouds of the three scans; in *ply* format the one from "Scan in a Box", in *obj* the one from "Sense 3D" and in *obj* the one from photogrammetry. In photogrammetry it is possible to also import the materials (in *mtl* format) and the texture file (in *png* format), if in addition to the geometry there want the object polychromy.

After importing the files, the meshes must be scaled and aligned with each other to proceed with the comparison (Fig. 3).

Fig. 3. Mesh alignment process

The scaling process is based on taking the distance between two specific points for each mesh and determining the scaling factor respect to the reference model.

On the other hand, the alignment consists in two phases, using the Cloud Compare software. The first is a preliminary alignment by picking pair points, the second is an automatic alignment employing the ICP Algorithm (Iterative Closest Point) to get fully overlapped models.

3 Results and Discussion

To check the existing deviations between the point cloud and the selected reference mesh, the minimum distances to the mesh of all registered points were calculated.

This distance results from projecting each of the compared mesh vertex into the surface of the nearest triangle of the reference mesh orthogonally. The "Cloud to Mesh distance" option of Cloud Compare was used.

To analyse the precision of the compared technologies the process bellow was followed:

- Drawing a Pareto Diagram that shows the repeatability of the obtained deviations (y axis), depending on their values (these have been grouped in ±1 mm intervals in x axis).
- Generation of the normal distribution (Gaussian Curve) of the point set depending on the deviation (x axis) and its repeatability (y axis). The mean and the standard deviation are calculated (it is considered that the model is similar to the reference one in every point inside $\pm 2\sigma$ interval).
- Generation of a color map of the compared model to determine the deviations in each region (using a scalar field of 256 colours).

3.1 Infrared Light-LED Light Comparison

The Pareto Diagram of Fig. 4, shows the repeatibility of the deviation values. The accumulative line shows that the 81, 6% of the points had a deviation, in absolute value, lower than 2 mm.

On the other hand, Fig. 5 shows that the deviations adopt a Normal Distribution, almost symetric, the mean located in 0, 087 mm, and a standard deviation $\sigma = 1{,}7$ mm.

Combining the data from both Figs. (4 y 5), it is obtained that the 93, 67% of the points had a deviation lower than a $\pm\ 2\sigma$, and the 98, 21% are lower than $\pm 3\sigma$.

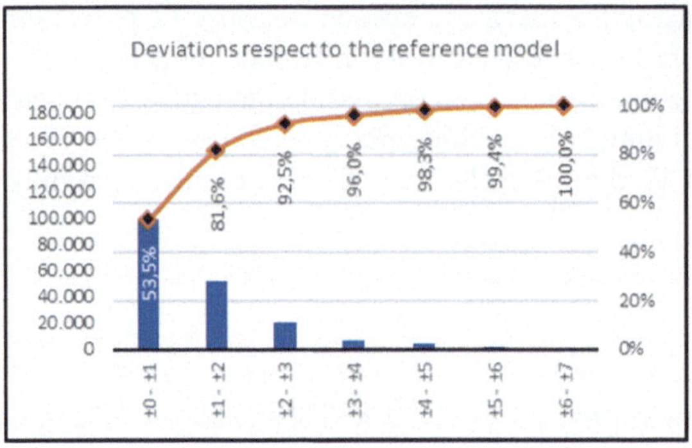

Fig. 4. Repeatability of the deviations (Inf-LED)

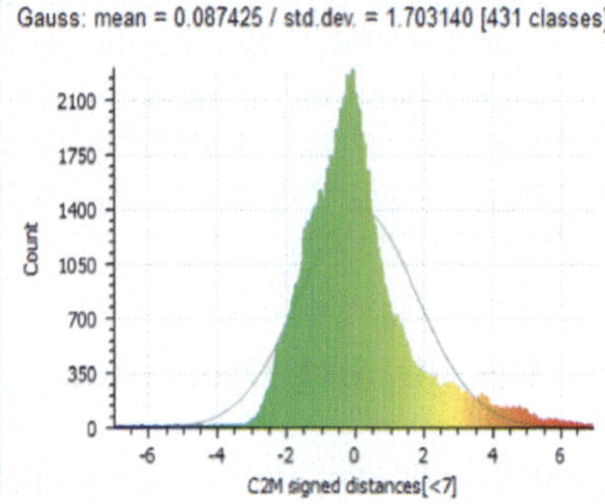

Fig. 5. Deviations distribution (Inf-LED)

As important as knowing the value and way of the deviations, is knowing where they are located to discern whether they affect to the result or to determine the corrective measures to be applied. Figures 5 and 6 show that:

- All the deviations above +4 mm (red scale) are in the boundaries, what is easily attributable to a careless data acquisition or to a worse point clouds alignment in that region. In addition, all those points (the 4% of the total) are located outside the measured object (the footprint), which means that all the points that reconstruct the footprint are inside the $\pm 2\sigma$ range (Fig. 5).
- In the footprint surface itself, the greatest deviations appear in the deepest region, with values between +2.5 mm and +4 mm.
- Figure 7 shows that the rest of the footprint points have similar deviations in absolute value, and inside the ± 1 mm range, which corresponds with the tolerance imposed by the technique specifications of Table 1.

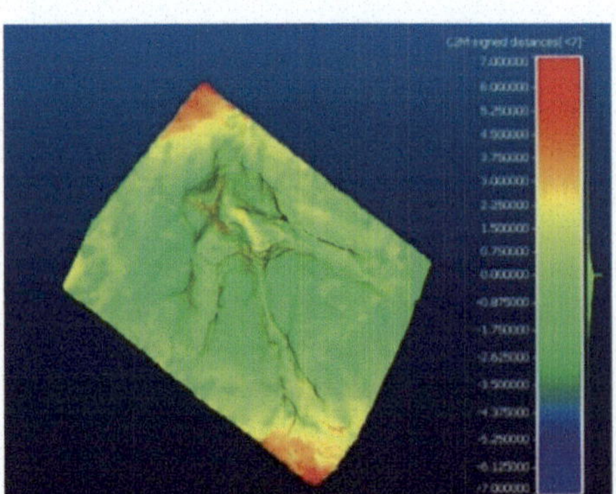

Fig. 6. Deviations map (Inf-LED)

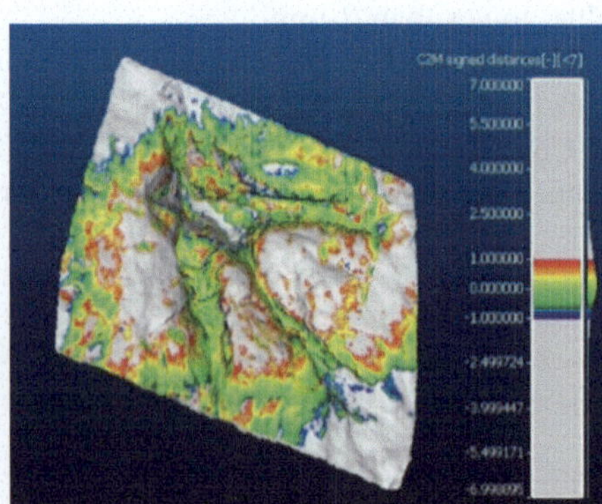

Fig. 7. Deviations location (Inf-LED)

3.2 Photogrammetry-LED Light Comparison

Same process was followed with Photogrammetry: calculating the deviations of the point cloud with respect to the reference mesh (Scan in a Box). Results are shown by its graphics (Figs. 8 y 9) and its comparative table with the obtained results (Table 2, Subsect. 3.3).

The process won't be repeated due to its similarity with the one applied in Subsect. 3.1.

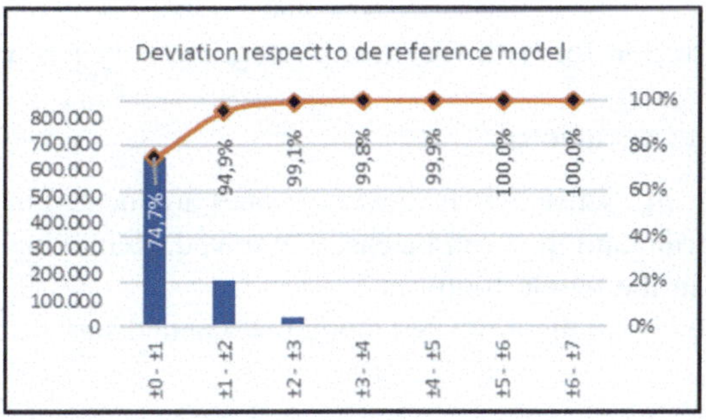

Fig. 8. Repeatability of the deviations (Phot-LED)

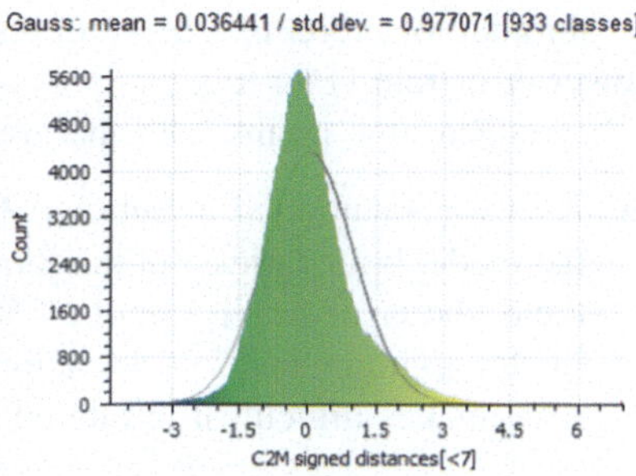

Fig. 9. Deviations distribution (Phot-LED)

In this case, the deviations are not so striking or so concentrated as with "Sense 3D". They are also lower, having almost all the points a deviation under ±2 mm.

Regarding the study in detail (Figs. 9 and 10), there is a relative uniform distribution in regions inside the stabilised limits. Figure 11 shows that the most part of the model is under the ±1 mm limit, being the greater deviations in the boundaries and in the deepest region of the model.

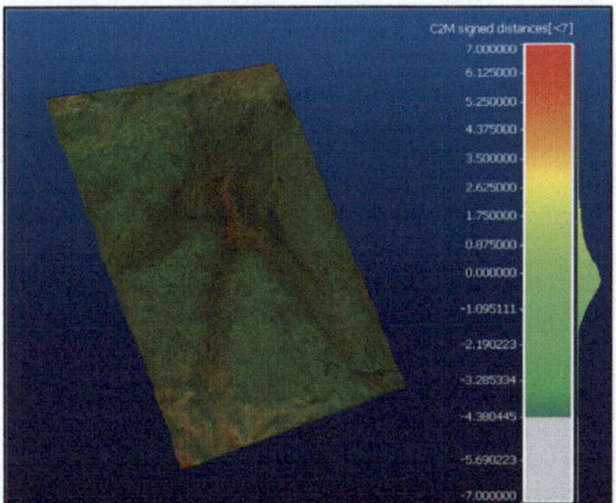

Fig. 10. Deviations map (Phot-LED)

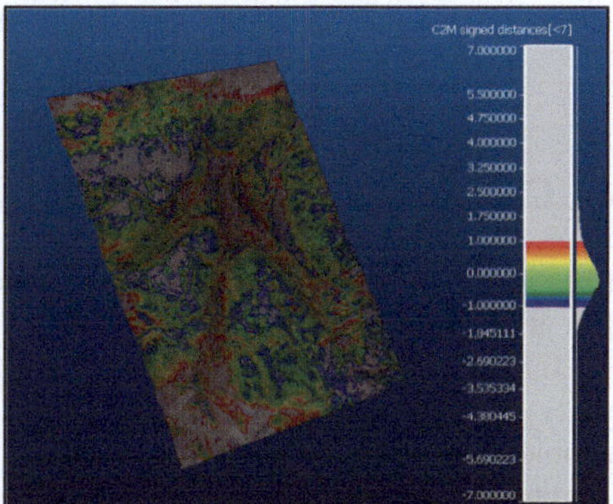

Fig. 11. Deviations location (Phot-LED)

3.3 Statistical Results Comparison

Comparing Figs. 4, 5, 6, 7, 8, 9, 10 and 11 and analysing the results from Table 2, it is observed that:

Regarding the standard deviation, photogrammetry has a lower standard deviation (σ = 0, 98) so, it could be considered more precise.

If the standard was the error rate outside the $\pm 2\sigma$ range (±1,95 for photogrammetry and ±3,4 for "Sense 3D") then, the first one is more precise than the first one due to the fact that just 6% of the points are outside the range, against the 6, 3% using Sense 3D (Table 2).

In addition, forcing a ±3 mm restriccion for both systems, photogrammetry has a lower error rate (1%) too.

Analising the points with the greater deviations:

- As it was mentioned in Subsect. 3.1, in "Sense 3D" its deviation ones are located in the model boundaries, out the footprint, and they can be easily removed. Removing them, the error rate is about a 2% in the whole footprint.
- In "Photogrammetry" its deviation ones are disseminated along the scanned surface, what makes difficult it is filtered.

Table 2. Comparative table of the obtained results

Technology	Infrared Light/LED light	Photogrammetry/LED
Deviations pareto	< ±2 mm: 81,6% points.	< ±2 mm: 94,9% points
Gaussian curve	Mean = 0,087; σ = 1,7 Deviations: For <$\pm 2\sigma$: 93,7% points For <$\pm 3\sigma$: 98,2% points	Mean = 0,036; σ = 0,98 Deviations: For <$\pm 2\sigma$: 94% points For <$\pm 3\sigma$: 98,9% points
Deviations map	Error rate: For $\pm 2\sigma$ = 6,3% Removing boundaries points ≈ 0%	Error rate: For $\pm 2\sigma$ = 6% Removing boundaries points ≈ 6%
Deviations location	Deepest region points: +1,5/+4 mm Rest of the points: ≤±1 mm	Disseminated points: ±3 mm Rest of the points: <±1 mm

4 Conclusions

LED structured light "Scan in a Box" requires artificial light, so it must be used in field just at night.

Infrared scanner "Sense 3D" is quicker and more versatile for the field labours in every Ambiental conditions than photogrammetry. This last one requires specific Ambiental conditions, apart from a much longer postprocessing process.

Photogrammetry is not limited by the object size unlike "Sense 3D". That is why is also used for building's reconstruction applications.

"Sense 3D" scanner as well as photogrammetric reconstruction are economic methods that generate enough precise models to geometrically document and reconstruct every single archaeological remain without disassembling or transporting it to a laboratory.

Photogrammetry be able to generate high resolution models in a relatively fast way. It can be done using even smartphones, but an expensive camera is required to get the best results possible. In addition, despite there are free software available, better results could be achieved by using payment software (for example PhotoScan Professional).

In addition, photogrammetry must respect colour gestion usual protocols related to camera profiles and ICC characterization profiles. Therefore, field work must be done by professionals.

After sending this communication, comparative analysis is still improving by:

- Scanning other objects whose precision is known.
- Quantification of errors associated with scaling and alignment of point cloud employing different methods.
- Reconstructing CAD models from meshes generated through low- cost technologies.
- Influence of lighting on data acquisition by non-contact scanning.

References

1. Rovira, D.: Nuevas tecnologías aplicadas al estudio y conservación de bienes culturales: Estado de la cuestión en museos españoles. Marq. Arqueología y museos 0, 39–46 (2005)
2. Luhmann, S.R.: Close Range Photogrammetry Principles, techniques and applications. Whittles Publishing, Dunbeath (2006)
3. Salvadori, F.: Three-dimensional scanning techniques applied to 3D modelling of pottery finds. In: Archäologie und Computer, Workshop, Vienna, vol. 7 (2002)
4. Brandi, C.: Teoría de la restauración. Ed. Alianza (2000)
5. Carta de Atenas. Sobre la universalidad del arte de los pueblos (1931). https://icomos.org
6. Duque, J., De Francisco, S.: Tridimensional Archaeology. 3D techniques applied to the paleolithic record. In: BSAA Arqueología, LXXXI, pp. 9–53 (2015). ISSN: 1888-976X
7. Tejado, J.M.: Escaneado en 3D y prototipado de piezas arqueológicas: Nuevas técnicas en el registro, conservación y difusión del Patrimonio Arqueológico. Iberia **8**, 135–158 (2005)
8. Valle Melón, J.M., et al.: El modelo 3D como base para la documentación y difusión de los elementos patrimoniales. Aplicación a mausoleo romano denominado "la Sinagoga" de Sádaba (Zaragoza). RA (1), 4–19 (2016). https://doi.org/10.13128/ra-19316
9. Robledano-Arillo, J., Moreno-Pelayo, V., Pereira-Uzal, J.M.: Aproximación experimental al uso de métricas objetivas para la estimación de calidad cromática en la digitalización de patrimonio documental gráfico. Revista Española de Documentación Científica, **39**(2) (2016). https://doi.org/10.3989/redc.2016.2.1249
10. Intel RealSense Camera SR300. https://software.intel.com/en-us/realsense/sr300. Accessed 15 May 2019
11. Bell, T., Beiwen, L., Zhang, S.: Structured light techniques and applications. Wiley Encyclopedia of Electrical and Electronics Engineering, pp. 1–24. Jhon Willey & Songs (2016)
12. Rüther, H., Smit, J., Kamamba, D.: A comparison of close-range photogrammetry to terrestrial laser scanning for heritage documentation. South Afr. J. Geomatics. **1**(2), 149–162 (2012)

13. Pereira-Uzal, J.M.: New perspectives in rock art recording. Bizkaiko Foru Aldundia-Diputación Foral de Bizkaia (2017). ISSN 0214–7971
14. Ortiz, P., Del Pino, B.: Heritage documentation, photogrammetry, videogrammetry, temple diana. Virtual Archaeol. Rev. **4**(8), 90–94 (2013). ISSN:1989–9947
15. Lievendag, N.: 3DScanExpert. https://3dscanexpert.com/3d-systems-cubify-sense-3d-scanner-review/. Accessed 18 May 2019
16. https://es.3dsystems.com/3d-scanners/sense-scanner. Accessed 30 Apr 2019
17. www.scanninbox.com. Accessed 30 Apr 2019

Optimization of the Management and Maintenance of Urban Elements Through Geolocation and Linking of Interactive Database to the Construction Information Model

M. J. García-Granja[1(✉)], Óscar D. de-Cózar-Macías[2], E. B. Blázquez-Parra[2], and A. B. Gutiérrez-Sánchez[3]

[1] Art and Architecture Department, Higher Technical School of Architecture, University of Malaga, Pza. El Ejido, 29013 Malaga, Spain
mjggranja@uma.es

[2] Graphic Expression, Design and Projects Department, School of Industrial Engineering, University of Malaga, C/Doctor Ortiz Ramos, 29071 Malaga, Spain

[3] Apogea, C/Steve Jobs 2, Promalaga Building, Office 15, PTA, 29071 Malaga, Spain

Abstract. *BIMoPOLIS* emerges as an idea, awarded with a Spin-Off prize, based on the development of digital content in the field of architecture and infrastructures, making use of the latest technologies in the fields of surveying, modeling in three dimensions (3D), geolocation and virtual and augmented reality (VR) (AR).

The objective of these digital contents is their inclusion in applications designed to transform the way in which management and maintenance operations in these areas are carried out, optimizing protocols in search of the concept of Smart City.

To this end, the usual procedure involves raising and geopositioning the infrastructure in question, and then incorporating the relevant information for routine maintenance operations such as measurements, materials, distributors' contact information, prices, detail views, or other indicators of interest to the 3D model.

The linking of these data to the virtual model entails numerous advantages since, through the use of a single application, it is possible to report incidents, locate faults, contact operators and suppliers, report finished jobs, etc., resulting in great savings in time and costs of management and communication in maintenance operations.

This article focuses on the first project developed by *BIMoPOLIS*, commissioned by the Vice-rectorate of Smart-Campus of the University of Málaga (UMA) and consisting of arboreal geo-positioning and linking of an interactive database to a building information model depicting the main section of the boulevard Louis Pasteur, located in the Teatinos Campus.

Keywords: Smart City · Geopositioning · Building information modeling

F. Cavas-Martínez et al. (Eds.): INGEGRAF 2019, LNME, pp. 597–604, 2020.
https://doi.org/10.1007/978-3-030-41200-5_65

1 Introduction

During the last 15 years, Malaga has received heavy investments directed towards transforming the city into a Smart City, by means of improved sustainability and energy efficiency, in order to present it to the world as an urban laboratory that attracts and accelerates the creation of R&D&i companies.

At the same time, the University of Malaga is making great efforts to create a Smart Campus, in which the main boulevard of Teatinos aspires to become the new green heart of the university. The intention is to address, in a comprehensive manner, the mobility, the environment, socio-cultural activity and technological development, transforming the boulevard into an articulating element of the university and the city.

The germ of the idea of *BIMoPOLIS* is born in this environment, specifically in the framework of innovation and entrepreneurship of the inauguration of the Digital Content Pole of Malaga, in which the University of Malaga has representation, with its Network of Strategic Chairs, as result of an agreement with the City Council.

From then on, we began to work together on several lines of research, focused on the development of 3D digital content in the fields of architecture and infrastructures. We employed the latest technologies in the fields of Geomatics and BIM methodology in order to develop thematic applications, dedicated to large urban centers with pretensions to become Smart Cities [1].

The previously mentioned BIM methodology consists of 3D parametric modeling, collaborative in real time, for buildings and infrastructural constructions. It is possible to incorporate all types of information to each element in these models, which allows the realization of calculations of costs, facilities, structures and energy, among others [2].

Although the utilities derived from the use of these technologies can be multiple, we focus on two of them:

On the one hand, 3D laser scanning [3] and photogrammetry for the BIM modeling of the city's real estate [4], together with the development of applications for tablets and smartphones, which provide written and audiovisual information in augmented reality [5]. These can be applied to achieve both an immersive tourism experience [6] and to optimize the work of conservation and restoration of monuments.

On the other hand, the BIM modeling [7] and 3D geo-positioning of urban infrastructures aimed at improving and optimizing the processes of pre-management and maintenance in order to reduce the time and personnel required [8].

Once the objectives of the project were defined, we presented it to the XXI edition of the Spin-Off competition, organized each year by the Vice-Rector for Social Innovation and Entrepreneurship of the University of Malaga, under the slogan "*BIMoPOLIS. 3D Smart solutions for the Smart City*", being awarded in the category of teaching and research staff.

The first opportunity for the implementation of the research carried out arises as a result of the II Technology Transfer Conference *TTAndalucía 2017*, where companies, research groups and public entities that are active in the field of research and technology have the opportunity to meet and exchange ideas.

Specifically, the interest in these technologies surfaced in a meeting with the representatives of the Vice-Rector of Smart-Campus of the UMA. This event leads to subsequent meetings, which resulted in the project for the arboreal geo-positioning and linking of an interactive database to the construction information model corresponding to the main section of the Louis Pasteur Boulevard, located in the Teatinos Campus of the UMA.

2 Materials, Methods and Results

Specifically in the case of the commission for the Louis Pasteur Boulevard project, its main objective is defined by the commissioner and is restricted to performing a 3D planimetry, in digital format, in which the existing tree species are to be geotagged. Every tree in the field of action defined, which covers 87.000 m^2, is assigned an icon that, when clicked on, displays its image, as well as updatable information on its coordinates, health status and main magnitudes (height, diameter of trunk and crown, etc.) [9].

In general, the biggest challenge posed by this needs program is the requirement to introduce the information in a format that, while not being hosted on a server, must be sufficiently standardized, so that it can be stored on external devices and managed by both specialized technicians and maintenance personnel, who lacks technical knowledge. This way, this documentation can be updated over time and, in turn, serve as support for future interventions on campus, related to the automation of its management and other planned projects. The aforementioned restrictions force us to discard the first alternative, namely using a hosting platform in the cloud for the BIM model [10].

At the technical level, the first decision we face is the choice of the most appropriate system for data collection. After analyzing the various methods available and weighing, both the scope of the project and its budget, we opted to perform the topographic survey and the geolocation of trees through Global Positioning System (GPS) [11].

Upon retrieval of the field information, we needed to discuss the most suitable procedure for the creation of the 3D model and the way of massively introducing the collected information [12]. After studying the different options [13], the most appropriate, given the conditions imposed, turns out to be the connection of the model with an external database that can be hosted on a local server and can be directly modified with a simple document manager and that can even be included in the university's document manager application.

However, at this point of the investigation, we are confronted by the difficulty of linking the new information with the one already existing in the registers of the UMA. For this, we use the published advances on methodologies for the automation of the generation process of 3D city models, in the City Geography Markup Language format (CityGML) [14], which we take as starting base for this project.

In our case, the technologies and working methods employed differ from the previous ones in the following aspects:

We performed the BIM modeling of the field of action at the first development level (LOD 1) [15]. For this, we used a georeferencing and automatic generation of 3D locations application, which was programmed in C# by BiMMate, employing Cadastre data. We decided to use images of the digital model of the field with a 5 m mesh (MDT05) appertaining to the National Aerial Orthophotography Plan (PNOA), in geodetic reference system ETRS89 and projection in Mercator Universal Transverse coordinate system (UTM), in spindle 30.

For the topographic survey and the geolocation of trees –in the ETRS89 reference system– the Global Navigation Satellite System (GNSS) Geotronics Global Positioning System (GPS) is used, in particular a GPS L1/L2 RTK7/GPRS with subscriber identification module card (SIM), of the Trimble range, which permits to conduct operations in real time, in connection with the Andalusian Positioning Network (RAP).

With the data obtained, we proceed to the BIM modeling of the scope of action, in level of development 1 (LOD 1), including in the planimetry of more than 525 trees, each of which is associated to its file by means of a hyperlink hosted in an Access table, like the one shown in Fig. 1. In this way, by clicking on any family of vegetation, it is possible to access a file with all its information.

Hoja10

Nombre vulgar	Nombre cientifico	Altura	Diametro copa	Diametro tronco	Estado sanit	Accion a	Coordenada X	Coordenada Y	Coordenada Z	
Almendro	Prunus dulcis	4,010	4,870	0,240	Bueno	Mantener	3678779,500	40.645.530,750	43,366	(1)
Almendro	Prunus dulcis	3,340	3,220	0,230	Bueno	Mantener	3678809,935	40.645.478,899	43,435	(1)
Almendro	Prunus dulcis	3,230	3,430	0,180	Bueno	Mantener	3678825,055	40.645.405,491	43,512	(1)
Almendro	Prunus dulcis	3,410	2,810	0,240	Bueno	Mantener	3678758,178	40.645.567,095	43,379	(1)
Olivo	Olea europaea	4,410	3,620	0,170	Bueno	Mantener	3678735,027	40.645.401,030	44,375	(1)
Palmera canaria	Phoenix canariensis	2,010	2,120	0,310	Bueno	Mantener	3678692,831	40.645.560,258	44,093	(1)
Palmera canaria	Phoenix canariensis	2,410	2,960	0,170	Bueno	Mantener	3678693,140	40.645.537,010	44,391	(1)
Palmera canaria	Phoenix canariensis	2,890	3,410	0,280	Bueno	Mantener	3678708,706	40.645.523,794	43,997	(1)
Palmera canaria	Phoenix canariensis	1,750	2,710	0,310	Bueno	Mantener	3678729,997	40.645.475,595	43,926	(1)
Palmera canaria	Phoenix canariensis	1,710	2,860	0,260	Bueno	Mantener	3678752,557	40.645.438,214	43,935	(1)
Palmera canaria	Phoenix canariensis	0,980	1,710	0,110	Bueno	Mantener	3678713,740	40.645.435,680	44,376	(1)
Palmera canaria	Phoenix canariensis	1,420	1,810	0,110	Bueno	Mantener	3678702,372	40.645.418,762	44,998	(1)
Olivo	Olea europaea	3,850	3,050	0,310	Bueno	Mantener	3678717,320	40.645.428,810	44,404	(1)
Palmera canaria	Phoenix canariensis	1,230	1,930	0,190	Bueno	Mantener	3678758,932	40.645.387,327	44,250	(1)

Fig. 1. Information entered in the access database of each geolocated tree (own elaboration)

For the elaboration of the mentioned arboreal records, we carried out an interdisciplinary work with biologists and agricultural technical engineers of the UMA, whose collaboration was essential for the identification of the more than 80 existing tree species in the section of the boulevard to be documented, as can be seen in Fig. 2.

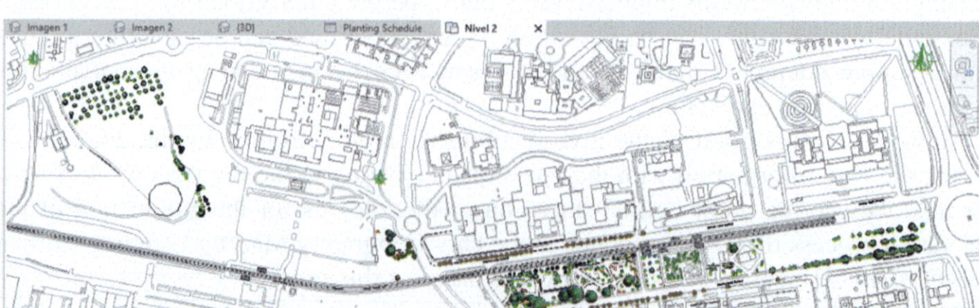

Fig. 2. Plan view of the BIM model generated with Autodesk Revit (own development)

Regarding the BIM modeling of the environment, we employed an application to facilitate the generation of locations, programmed in C# language and developed by the open collaborative professional platform BiMMate, which allows us to automatically generate a 3D model using the planimetric and altimetric information contained in the Cadastre [16]. We streamlined this information to facilitate access to it and, in this way, to highlight the purpose of the project, such as in the example of Fig. 3.

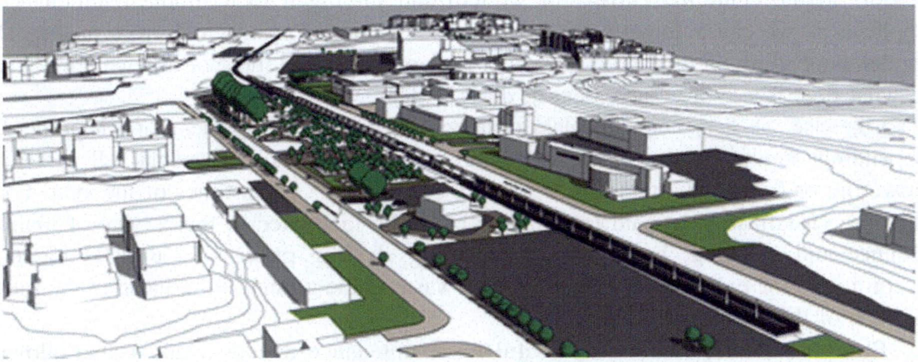

Fig. 3. BIM model in 3D view of Louis Pasteur blvd. and its surroundings (own development)

And, finally, we introduce each plant species, as a parametric element linked to its file, in the 3D model, for which we use the Autodesk Revit modeling software as can be seen in Fig. 4. In this way, this updatable information is associated to each parametric family, so that, if we modify its dimensional data in the table, its 3D visualization is modified. For example, if we change the height of a tree in the database, this information is updated in the 3D model, where its height will be changed according to the data introduced [17]. By virtue of this, our model constitutes a reflection of the reality, alive in time.

```
Parametros Compartidos.txt: Bloc de notas
Archivo  Edición  Formato  Ver  Ayuda
# This is a Revit shared parameter file.
# Do not edit manually.
*META   VERSION MINVERSION
META    2       1
*GROUP  ID      NAME
GROUP   1       Arboles
*PARAM  GUID    NAME    DATATYPE        DATACATEGORY    GROUP   VISIBLE DESCRIPTION     USERMODIFIABLE
PARAM   663f2f02-be10-496f-9876-07b412e60893    Nombre Vulgar   TEXT            1       1
PARAM   1a3c9c09-722b-4648-9ac0-7844f52bb273    EStado Sanitario        TEXT            1       1
PARAM   dfa4272c-3ce2-4ac5-a58a-5d69d744e113    Diametro Tronco LENGTH          1       1
PARAM   8c5e6d63-250a-4c8e-97e8-9d49465fbdea    Número de Punto NUMBER          1       1
PARAM   fce4b663-91c1-4d06-8bf5-64fe1d52ae30    Coordenada Y    NUMBER          1       1
PARAM   a1139883-82e7-4a7f-92cd-ac646056746f    Observaciones   TEXT            1       1
PARAM   9eec3499-0816-47e6-a12a-1959c1b0647e    Diametro Copa   LENGTH          1       1
PARAM   d5670b9b-7c54-4d76-972f-8e4712727e66    Altura Arbol    LENGTH          1       1
PARAM   9f62d9ab-7115-4d91-b1a2-e330f090c422    Nombre Cientifico       TEXT            1       1
PARAM   49e864ae-8181-4c42-8941-ace636980963    Código de Punto TEXT            1       1
PARAM   755745d8-3982-4486-af0b-e331aafe0614    Acción a realizar       TEXT            1       1
PARAM   ff0d44f2-4334-4478-9145-7cd98bf73895    Imagenes        IMAGE           1       1
PARAM   1d899af5-7b1f-4234-b54e-5dcce56c072a    Coordenada Z    NUMBER          1       1
PARAM   a6bb49fe-a29e-4ce2-89d0-6a90e387d5e1    Coordenada X    NUMBER          1       1
```

Fig. 4. List of parameters shared between the parametric family of each tree and the BIM model

As a result, we observe that any information associated with urban infrastructures can be introduced in a BIM model that, connected to sensors, allows us to manage an intelligent city through applications, both existing in the market and programmed by us [18].

Using the Louis Pasteur boulevard project as a practical example, we have a 3D model for it, with all the vegetation geolocated and linked to a database which provides a file with the technical information of each tree, shrub, etc., so, just by including sensors to the centralized irrigation system and linking it to a simple maintenance application, we can automatically control, from a control post, which trees in the 3D model, depending on their species, size or time of year, need more or less watering and at what time it is more convenient to do it [19].

Likewise, it can help to plan pruning, according to species, or cleanliness, depending on the existence, in certain areas, of trees whose seasonal fall of leaf, flowering or fruit, make them foreseeably more dirty at certain times of the year. For this, the visualization of said information could be facilitated to the service of maintenance and cleaning, by means of including shading patterns in the 3D model, according to the expected degree of vegetal waste in the ground, introduced in the files of each of the species [20].

But this advance is not only useful for maintenance management, it also allows static or dynamic solar studies to simulate the projection of shadows at any time of the year, of both building volumes and trees. The link of these to the characteristics of their species, allows to predict their growth rate, height and crown, of each of them, which has multiple applications [21].

On the other hand, from the point of view of urban design, it is useful to determine at the present time, and predict in the future, the areas where the provision of furniture or urban infrastructures such as benches, pergolas or awnings is more necessary. So as to help decide what type of tree species to plant, for example through simulations of shadow maps according to age, growth rate and type of deciduous or perennial leaf of each tree [22].

Moreover, it is possible to program an algorithm and associate it to the parameters of each species' file. This would allow to know the growth (fast or slow) of each one according to its typology and determine its eco-efficiency, depending on the ratio between the value addition of what has been produced (greater or lesser absorption of CO2, shadow projection, etc.) and the added environmental impact that it took to produce it (greater or lesser need for irrigation, maintenance by pruning, etc.). This ratio can be very useful to compare and decide between different options of trees to plant.

It also allows for more precise energy efficiency studies of buildings, by taking into account their own shadows and those cast by everything that surrounds them, throughout the year [23].

Currently, the project is serving as a support to continue with the development program of the Knowledge Forest, where it is possible to sponsor and name a geolocated tree on the UMA website, with the aim of contributing to reforestation.

3 Conclusions

In conclusion, this project gives us the opportunity to investigate and put into practice ideas on how the development of the latest technologies in positioning and geolocation, massive data capture, mobile mapping, together with database management, allow for real control of the city, both at the inventory level and at the level of maintenance and resource management.

The practical implications of this project show future research lines related to the optimization of the maintenance of cities where, thanks to the new available technologies, an inventory of elements of urban furniture, infrastructures, vegetation, etc. could be made, storing that information in a BIM model of the city linked to a database, creating what we could call the city's book, which could be reviewed and managed by municipal public agencies and services.

The updating of such information could combine two types of procedures: one, directly, with the administration itself introducing those changes and/or new pertinent information; and another indirectly, that is, allowing the citizens inform of any deteriorated urban element, uploading the photo and location thereof, to a web for repair requests. This information, once verified by the operators of the body responsible for maintenance, would be included in the database, which would result in a saving of resources, by having live information on the state of maintenance of urban elements in the city book. This second alternative would allow to improve the management of the city, involving the citizen in the solution of the problems of their environment and involving them in their care.

All these options could be part of the development of a roadmap for the construction of smart cities whose urban development was based on sustainability, through the use and integration of such technologies [24].

References

1. Bello-Perez, F.O., Pérez-Castillo, J.N.: Interoperatividad entre los dominios de la arquitectura, la ingeniería y la construcción y los sistemas de información geográfica. Ingeniería y Universidad (2012)
2. Bryde, D., et al.: The benefits of building information modelling (BIM). Int. J. Proj. Manage. **31**(7), 917–980 (2013)
3. Tang, P., Huber, D., Akinci, B., Lipman, R., Lytle, A.: Automatic reconstruction of as-built building information models from laser-scanned point clouds: a review of related techniques. Autom. Constr. **19**(7), 829–843 (2010)
4. Isikdag, U.: BIM an IoT: a synopsis from GIS perspective. In: The International Archives of the Photogrammetry, Remote Sensing and Spatial Information Sciences, vol. XL-2/W4, pp. 33–37 (2015)
5. Ramos, F., Trilles, S., Torres-Sospedra, J., Perales, F.J.: Nuevas tendencias en el uso de aplicaciones de realidad aumentada para contextos de ciudades inteligentes. ISPRS Int. J. Geo-Inf. **7**, 418 (2018). https://doi.org/10.3390/ijgi7120478
6. Liang, M., Min, H., Luo, R., Zhu, J.: Simultaneous recognition and modeling for learning 3-D object models from everyday scenes. IEEE Trans. Cybern. **45**(10), 2237–2248 (2015)

7. Bradley, A., Li, R., Dunn, S.: BIM for infrastructure: an overall review and constructor perspective. Autom. Constr. **71**(2), 139–152 (2016)
8. Chen, H.M., Chang, K.C., Lin, T.H.: A cloud-based system framework for performing online viewing, storage, and analysis of big data of massive BIMs. Autom. Constr. **71**, 34–48 (2016)
9. Vertockt, S., Gerke, M., Kerle, M.: Geolocalization of crowdsourced Images for 3-D modeling of city points of interest. IEEE Geosci. Remote Sens. Lett. **12**(8), 1670–1674 (2015)
10. Kolbe, T.H., Grogër, G.: Unified representation of 3D city models. Geoinf. Sci. J. **4**(1) (2004)
11. Wong, J., Wang, X., Li, H., Chan, G., Li, H.: A review of cloud-based BIM technology in the construction sector. J. Inf. Technol. Constr. **19**, 281–291 (2014)
12. Han, K.K., Golparvar-Fard, M.: Potential of big visual data and building information modeling for construction performance analytics: an exploratory study. Autom. Constr. **73**, 184–198 (2017)
13. Degbelo, A., Granell, C., Trilles, S., Bhattacharya, D., Casteleyn, S., Kray, C.: Apertura de ciudades inteligentes: desafíos centrados en los ciudadanos y oportunidades de GIScience. ISPRS Int. J. Geo-Inf. **5**, 16 (2016)
14. Kolbe, T.H., Grogër, G., Plümer, L.: CityGML – interoperable access to 3D city models. In: Oosterom, P., Zlatanova, S., Fendel, E.M. (eds.) Proceedings of the International Symposium on Geo-information for Disaster Management, Delf. Springer (2005)
15. Yalcinkaya, M., Singh, V.: Patterns and trends in building information modeling (BIM) research: a latent semantic analysis. Autom. Constr. **59**, 68–80 (2015)
16. BiMMate-Presentación de aplicaciones-Generación de emplazamiento. https://www.youtube.com/watch?v=5vAmQaBVmP4. Accessed 16 Apr 2019
17. Leite, F., Cho, Y., Behzadan, A., Lee, S., Choe, S., Fang, Y., et al.: Visualization, information modeling, and simulation: grand challenges in the construction industry. J. Comput. Civil Eng. **30**(6), 04016035 (2016)
18. Brenner, C.: Towards Fully Automatic Generation of City Models. In: IAPRS, vol. XXXIII, Part b3/1, Comm. III, ISPRS Congress, Amsterdam (2000)
19. Kolbe, T.H., Grogër, G.: Towards unified 3D city models. In: Schiewe, J., Hahn, M., Madden, M., Sester, M. (eds.) Challenges in Geospatial Analysis, Integration and Visualization, II. Proceeding of Joint ISPRS Workshop, Stuttgart (2003)
20. Rodríguez, R., Álvarez, M., Miranda, M., Díez, A., Papí, F., Rodríguez, P.: Generation of three-dimensional urban models. Informes de la Constr. **65**, 229–240 (2003). https://doi.org/10.3989/ic.12.041
21. Chen, K., Lu, W., Peng, Y., Rowlinson, S., Huang, G.Q.: Bringing BIM and building: From a literature review to an integrated conceptual framework. Int. J. Proj. Manage. **33**(6), 1405–1416 (2015)
22. Volk, R., Stengel, J., Schultmann, F.: Building information modeling (BIM) for existing buildings-literature review and future needs. Autom. Constr. **38**, 109–127 (2014)
23. Oti, A.H., Kurul, E., Cheung, F., Tah, J.H.M.: A framework for the utilization of Building Management System data in building information models for building design and operation. Autom. Constr. **72**(2), 195–210 (2016)
24. Musa, S.: Smart Cities - Una hoja de ruta para el desarrollo. Potenciales IEEE **37**, 19–23 (2018)

From Opportunity Identification to Concept Generation of Sustainable Modular Buildings

Guido Bruni, Luigi De Napoli(✉), Vincenzo Maria Mattanò, and Antonino Morabito

Department of Mechanical, Energy and Management Engineering (DIMEG), University of Calabria, Ponte Pietro Bucci 46/C, 87036 Rende, CS, Italy
luigi.denapoli@unical.it

Abstract. The paper discusses the process that leads to choosing a modular building system and how this has suggested a framework to make the right decision in clarifying the task and in generating and selecting new concepts. After illustrating the genesis of the need to build in a modular way, attention shifts to the criteria for evaluating the characteristics of modular buildings. The evaluation criteria were identified and sorted by category and, after an assessment of what already exists on the market and in literature, the guidelines for the design of a new modular building system were provided. The paper also illustrates some conceptual solutions that have arisen from the application of the identified criteria.

Keywords: Sustainable modular buildings · MCDA · Concept generation · Concept assessment · Product assessment

1 Introduction

Over the last years, prefabrication has been acquiring increasing importance in both the structural sector and the construction of houses. This trend, which is linked to many different factors, has led researchers to adopt new approaches to the accommodation standard. The development of prefabricated houses can represent a good alternative to the traditional housing, as it ensures a high level of customization of the project at reasonable costs, within the context of a specific real estate market characterized by a steady increase in the houses' price.

If compared to the building processes of traditional housing, prefabrication also allows improving environmental sustainability. Managing all the production stages regularly, construction in the plant allows to decrease the number of waste, besides reducing the human and material impact on the place where the house is to be installed, thus creating an ecological alternative.

From the above considerations, it can be understood that a prefabricated house should be characterized by a very low environmental impact and at the same time by the creation of an internal healthy and comfortable environment. Particularly important is also the building of an energy-efficient structure able to reduce consumptions by selecting safe and sustainable materials, and choosing sites and installation methodologies that minimize the climate and environmental impact (especially exposure) [1].

F. Cavas-Martínez et al. (Eds.): INGEGRAF 2019, LNME, pp. 605–614, 2020.
https://doi.org/10.1007/978-3-030-41200-5_66

These first aspects just mentioned suggesting that the criteria for evaluating a modular building project are many and not simple to manage and to visualize.

In this work, first of all, the aspects to be considered have been tried to categorize.

Once carried out an analysis of what the market offers according to the identified criteria, it was inspired by this to formulate some conceptual alternatives for a new system of modular buildings, of which only partial reference will be made here, showing, for example, some concepts of the structural part and giving some examples of architectural solutions.

2 Evolutionary Typological Analysis and Future Perspectives of Prefabricated Modular Structures

The development of prefabrication, in its first stages, has been based on the transition from merely pre-assembled elements to the creation of components produced employing an actual industrial process.

In temporal terms, such steps can be traced back to the Industrial Revolution. The first significant example in the building context is the construction of the Crystal House by J. Paxton.

Subsequently, early in the 20th century, the idea of industrial production of houses was developed. This topic, which is still relevant today, has attracted the interest of several leading architects, who carried out countless research activities having considerably different characteristics and peculiarities from each other [2].

In some cases, these experimentations have a strong design value (Maison Citrohan, 1920, by Le Corbusier). Whereas, many projects are focused on construction requirements (i.e. Manning Portable Cottage, H. Manning, 1837). Some other projects are based on the corporate brand instead (i.e. BoKlok, a joint venture between IKEA and Skanska, 1996-present).

The analysis of the experiences made and projects implemented within the framework of the production of prefabricated houses shows that it is difficult to detect all the strengths and weaknesses of this type of housing. In general, it is possible to identify a non-exhaustive list of strengths for this type of housing, namely: simplicity of construction with commonly used methodologies; project adaptability in terms of appearance and functionality; competitiveness of the product on the market.

Its weaknesses could be construction and project complexity; design shortcomings on the life cycle of the building; wrong estimations of its commercial value; top-down design.

Irrespective of the factors analyzed above, an essential aspect is the market demand for these type of products [3].

Based on these considerations, it is necessary to take into account urbanization and future demographic trends. According to the data published by the United Nations Department of Economic and Social Affairs, in 2050 the world population is estimated to amount to 9.55 billion people, compared to the current 7 billion, and the average urbanization rate should be equal to about 65%, compared to the current 50% (WUP 2014, [4]).

It is envisaged that by 2025 one billion new homes will be required worldwide (UN-Habitat 2016, [5]).

Therefore, the urbanization process will continue to be directly linked to economic growth. For land consumption, it should be taken into consideration that, according to the urbanization models implemented until today, the growth rate of urbanized land is more than double as the population urbanization rates [6]. On average, a city that increased its population by 20% needed to also increase urbanized land by 46%.

The production capacity of the building sector is strongly influenced by the rise in the cost of labor as well as by the progressive decrease in the workers' skills level. Such aspects can facilitate the development of the prefabricated construction sector, also considering that, if properly designed, prefabricated typologies could promote the reversibility process of land use [7].

3 Methodology for a Comparative Analysis of Prefabricated Houses to Identify the Project Requirements

To effectively address the conceptual design stage of a prefabricated house, it is necessary to carry out a careful analysis of the context in which the project will be implemented. To this end, an analysis of the state of the art has been accomplished and the results have been shown employing a matrix relating the most relevant factors that should be taken into account in the project to meet the market requirements (see Fig. 1).

The proposed matrix summarizes, along a dimension, the criteria identified to evaluate the solutions proposed by the market and the design alternatives that are developed by a design team; along the other dimension, there are the product/project alternatives. One hundred requirements have been identified that have been clustered in the following subsections: architecture, functional quality, industrial production, sustainability, land use, emergency management, environmental and social impact, plant engineering, energy, and economy (see Fig. 2). By way of example, Fig. 2 shows the criteria identified for each cluster.

Each requirement has been evaluated positively or negatively to obtain an overall percentage of the area in which that requirement falls. The project expectation is that all requirements are met and that each area is 100% valued.

In Fig. 1 are collected the evaluations of each criterion for 47 items (either product already on the market or just projects). As can be seen, no product/project meets all the criteria.

At the moment, each criterion has the same weight. Soon, this research will continue with the implementation of possibly varying the weights associated with each criterion and the overall evaluation will be then made according to the most appropriate MCDM methods.

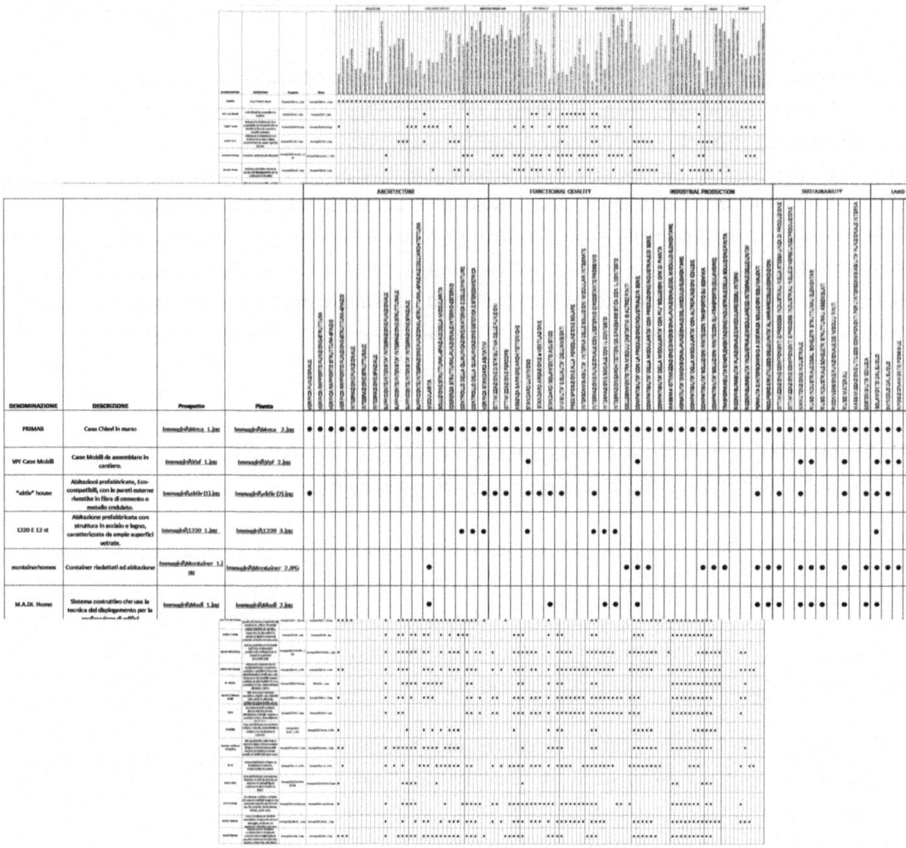

Fig. 1. Evaluation matrix for the comparative analysis of prefabricated houses.

To test the effectiveness of this procedure, several projects regarded as significant and in line with the current needs linked to prefabricated houses and project/customer requirements have been selected.

Evaluation can be considered exclusively as an analytical tool aimed at guiding the choices in the concept being elaborated.

As it is understood, the table of comparison is subject to continuous updating and in-depth analysis, intending to provide the most up-to-date picture of the state of the art.

4 Results and Discussion

The comparative analysis of the examined cases provides a significant overview of the current trends and allows, based on the prepared evaluation criteria, to make some considerations (see Fig. 3).

Fig. 2. All the criteria, sorted by typology.

The radar chart in Fig. 3 shows the percentage of satisfied criteria in each cluster by each alternative.

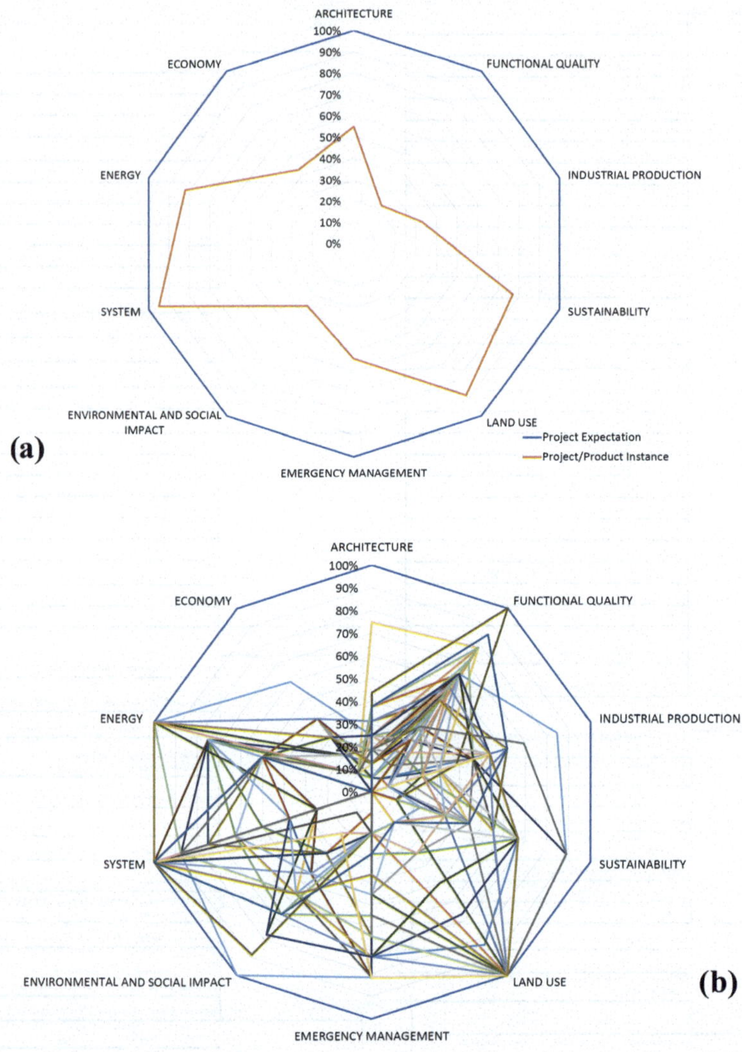

Fig. 3. Radar chart of state-of-art modular buildings vs clustered criteria. (a) Radar chart of a single instance. (b) Radar chart of all the examined instances.

Figure 3a shows the behavior of a single examined case compared to the "ideal" one. Figure 3b shows the behavior of all the examined instances.

The seeming confusing representation, however, allows highlighting which areas are less covered by existing solutions. This evidence has addressed consequently the generation of new concepts.

Furthermore, what emerges is an increasing interest in the residential industrialization, which contrasts with a lacking development of modular solutions able to respect housing standards. An inadequate optimization of the functionality/transportability ratio, due to transportation needs, is also highlighted.

In terms of sustainability, the reluctance to elaborate standardized elementary modules is reflected in little or no possibility of re-using and maintaining the structures. Their universality, namely their applicability also in other construction sectors, would increase the sustainability of this solution and the recovery not only of the material but also of entire assembled structural parts, i.e. whole structural skeletons, in the same or other applications.

Besides, there is insufficient development of solutions designed for prompt use, re-use and transportability, and reversibility of the land use, regardless of the soil where they are installed.

5 New Concepts of Modular Building Systems

The survey on the state of the art of prefabricated modular buildings conducted with the support of the aforementioned evaluation matrix suggested guiding the design of new systems to meet the requirements not yet covered by most systems currently on the market.

Special attention has been paid to the aspects of environmental sustainability (for example, the use of easily reusable materials and equipment), to the transportability of a building in semi-definitive configuration, to the ease of installation in situ, to the low impact on the territory (in particular, in view of the possible subsequent dismantling). In particular, at the moment, environmental sustainability issues have been addressed by putting into practice the most up-to-date eco-design practices.

The generated concepts have involved both the structural part and the architectural components. The following figures present some concepts generated in several stages.

The concepts presented here as examples have the metal structure in common. Mostly it is steel, in various configurations (profiles, welded sheets, corrugated panels, etc.).

In Fig. 4 a structure made by only welded sheets is shown.

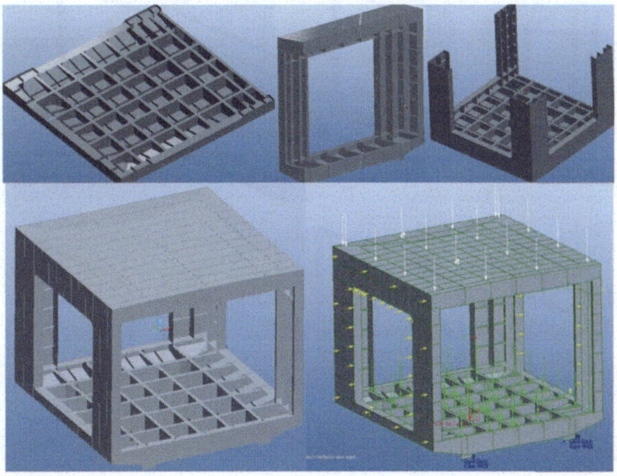

Fig. 4. Structure of a single module composed only by welded sheets.

In Fig. 5 there is an example of the metal structure of a single module based on standard profiles, corrugated panels and angular elements made by welded sheets. In this case, the corrugated panels also perform a structural function, either in the roof and on the floor.

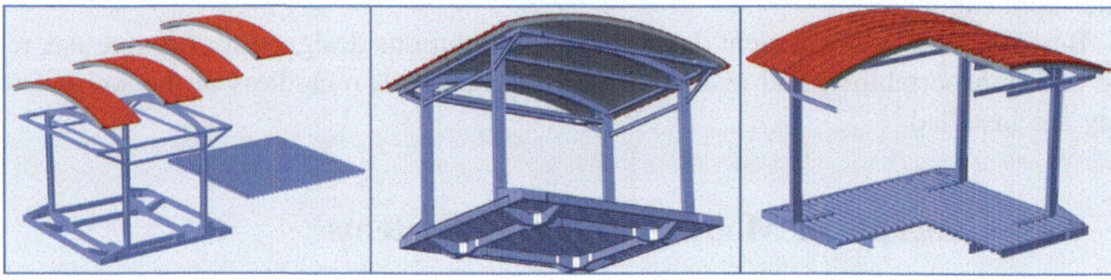

Fig. 5. Structure of a single module: standard profiles, corrugated panels, and welded sheets.

In Fig. 6 there are many examples of metal structures of a single module based on standard profiles and angular elements made by welded sheets. In all these nine concepts (and also in the concept shown in Fig. 5) the single module rests, at four points, on ground screws, easily removable; this would avoid the use of concrete slabs that, in the phase of disposal, represent one of the major environmental problems.

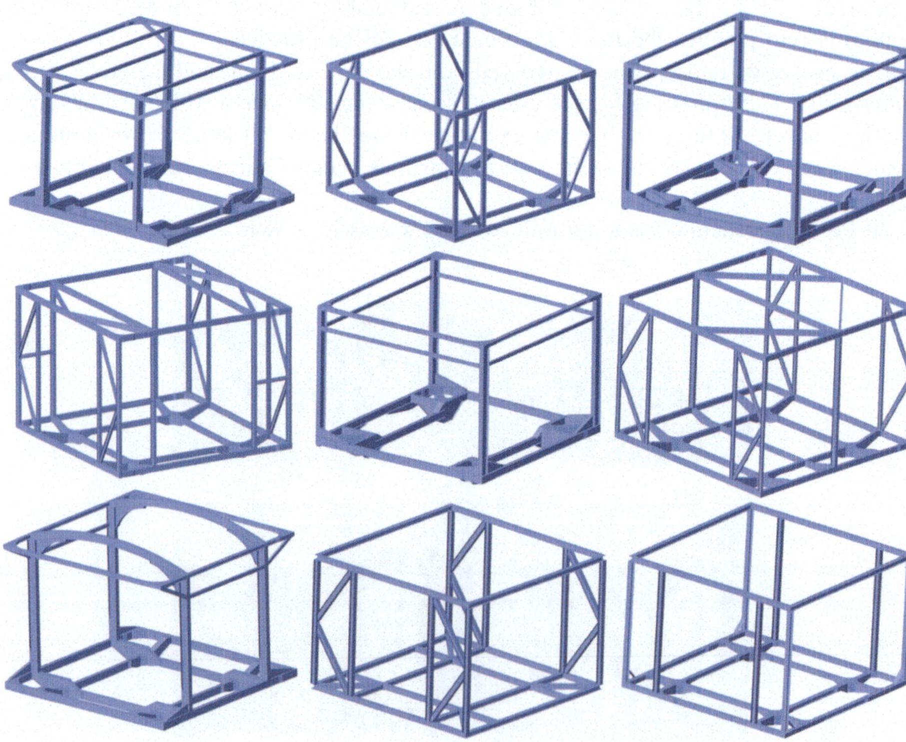

Fig. 6. Structures of a single module: standard profiles and angular joints with welded sheets.

In Fig. 7 there is an example of a survey on architectural elements, in which four individual modules are assembled to obtain a complete residential unit.

Fig. 7. Example of architectural components.

Some preliminary FEM analyses have been carried out to eliminate some too bold solutions. In Fig. 8 an example is shown in which a metal structure is verified according to the current standard regulations.

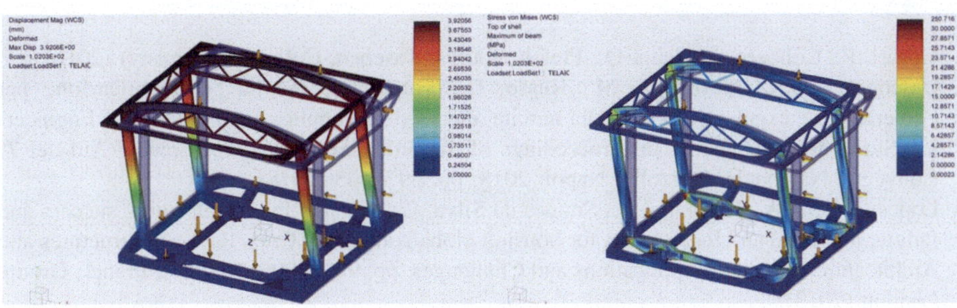

Fig. 8. Example of FEM analyses of the structure.

The final choice of the concept, which will continue in the embodiment design phase up to the realization of the physical prototype, will then be performed on the basis of an ad hoc selection made on the basis of criteria not yet highlighted and appropriately weighted, but in any case belonging to the group of those already identified for the evaluation matrix mentioned above.

Since the buildings are modular, particular attention has been paid to the joints between single modules, depending on the case, welded or bolted. In some alternatives, the use of ad hoc corner joints has been conceived.

6 Conclusions and Future Works

A methodology to investigate prefabricated modular buildings has been presented in this work. It has been employed either to assess the existing solutions and to guide the creation of conceptual design alternatives.

In the next future, the methodology will be improved by implementing a tool based on the up-to-date MCDA (Multi-Criteria Decision Analysis) procedures.

Particular attention will be paid to the evaluation of environmental aspects. In addition to the use of the most up-to-date eco-design guidelines, as far as possible the environmental impact will be quantified with LCA-based techniques, if not on the whole module, at least on the subsystems. Since in the early stages of design, precisely the conceptual one, sufficiently complete data on the product life cycle are not yet available, in the case of the evaluation of conceptual projects of a single module or of a modular building, a simplified LCA procedure is being implemented, that should take into account all the lifecycle gates.

Acknowledgments. The research work reported here has been possible by the financial support of Regione Calabria, through the POR Project CALABRIA FERS FSE 2014-2020 (project acronym: PRIMAB).

References

1. Gössel, P., Cobbers, A., Jahn, O.: Prefab Houses. Taschen, Cologne (Germany) (2010)
2. Cascone, S.M., Caporlingua, M., Russo, G., Tomasello, N.: La "prefabbricazione per l'emergenza": excursus storico dalla nascita alle moderne applicazioni. History of Engineering Storia dell'Ingegneria. In: Proceedings of the 3rd International Conference- Atti del 7° Convegno Nazionale, Cuzzolin, Napoli 2018, pp. 597–605 (2018)
3. Lopes Correia, A., Murtinho, V., Simões da Silva, L.: Housing industrialization, success and failure, universal and local: limits for housing globalization. In: Cruz, P.J. (ed.) Structures and Architecture: Concepts Applications and Challenges, pp. 999–1006. Taylor & Francis Group, London (2013)
4. https://esa.un.org/unpd/wup/publications/files/wup2014-report.pdf. Accessed 15 Apr 2019
5. http://wcr.unhabitat.org/main-report/. Accessed 15 Apr 2019
6. Seto, K.C., Guneralp, B., Hutyra, L.R.: Global forecasts of urban expansion to 2030 and direct impacts on biodiversity and carbon pools. In: Proceedings of the National Academy of Sciences (USA), vol. 109, pp. 16083–16088 (2012)
7. Mussinelli, E., Tartaglia, A., Di Pasquale, J.: Typological and technological innovation for the application of hybrid systems to housing construction: between technological culture and application testing. TECHNE – J. Technol. Archit. Environ. (13), 287–294 (2017)

Correction to: TDT-L0 a Test-Based Method for Assessing Students' Prior Knowledge in Engineering Graphic Courses

Barbara Motyl, Gabriele Baronio, Domenico Speranza, and Stefano Filippi

Correction to:
Chapter "TDT-L0 a Test-Based Method for Assessing Students' Prior Knowledge in Engineering Graphic Courses" in: F. Cavas-Martínez et al. (Eds.):
***Advances in Design Engineering*, LNME,**
https://doi.org/10.1007/978-3-030-41200-5_50

The book was inadvertently published with an error in Chapter 50. The author's name has been changed from 'Barbara Moty' to 'Barbara Motyl'.

The updated version of this chapter can be found at
https://doi.org/10.1007/978-3-030-41200-5_50

F. Cavas-Martínez et al. (Eds.): INGEGRAF 2019, LNME, p. C1, 2020.
https://doi.org/10.1007/978-3-030-41200-5_67

Correction to: Digitization of the Mechanical Articulator: Virtual Articulator

José Antonio Oriozabala-Brit, Xabier Amezua-Lasuen,
Xabier Garikano-Osinaga, Mikel Iturrate-Mendieta,
Iñaki Martin-Amundarain, and Eneko Solaberrieta-Mendez

Correction to:
Chapter 23 in: F. Cavas-Martínez et al. (Eds.):
***Advances in Design Engineering*, LNME,**
https://doi.org/10.1007/978-3-030-41200-5_23

In the published version of the book, the names and surnames of the authors were reversed, and it has been corrected now.

The updated version of this chapter can be found at
https://doi.org/10.1007/978-3-030-41200-5_23

F. Cavas-Martínez et al. (Eds.): INGEGRAF 2019, LNME, p. C2, 2024.
https://doi.org/10.1007/978-3-030-41200-5_68

Author Index

F. Cavas-Martínez et al. (Eds.): INGEGRAF 2019, LNME, pp. 615–617, 2020.
https://doi.org/10.1007/978-3-030-41200-5

The manufacturer's authorised representative in the EU is Springer Nature Customer Service Centre GmbH, Europaplatz 3, 69115 Heidelberg, Germany. If you have any concerns regarding our products, please contact ProductSafety@springernature.com

Printed and bound by CPI Group (UK) Ltd, Croydon, CR0 4YY

25/09/2026

02222853-0017